Neuronal Calcium Sensor Proteins

Neuronal Calcium Sensor Proteins

Pavel P. Philippov
and
Karl-Wilhelm Koch
Editors

Nova Science Publishers, Inc.
New York

For permission to use material from this book please contact us:
Telephone 631-231-7269; Fax 631-231-8175
Web Site: http://www.novapublishers.com

Library of Congress Cataloging-in-Publication Data

Neuronal calcium sensor proteins / Pavel P. Philippov and Karl-Wilhelm Koch (editors).
p. cm.
Includes index.
ISBN 1-59454-978-8
1. Neurochemistry. 2. Calcium--Physiological effect. 3. Cellular signal transduction. 4. Nerve tissue proteins. I. Philippov, Pavel P. II. Koch, Karl-Wilhelm.
QP356.3.N476 2006
612.8'042--dc22 2005038050

Published by Nova Science Publishers, Inc. ✦ New York

CONTENTS

PREFACE

Calcium signalling is an astonishing example how a simple cation can trigger and regulate an enormous variety of cellular and physiological responses. Ca^{2+}-signalling routes very often involve Ca^{2+}-binding proteins that sense changes in intracellular [Ca^{2+}] and trigger cellular responses by regulating specific targets. One fascinating group among these Ca^{2+}-sensors are the neuronal calcium sensor (NCS) proteins, named for their localization in neuronal tissue (although there are reports of their expression in nonneuronal tissues as well). While recent excellent reviews have covered key aspects of this protein group, the field expanded in recent years making it more and more difficult to represent every facette of this ongoing research endeavor. Therefore, we felt the time was right to represent properties of NCS proteins in a book.

Pioneering work related to NCS proteins was initiated by the identification of recoverin and its chicken and frog orthologues, visinin and S-modulin respectively, in visual cells. These papers for the first time showed the presence of a new group of Ca^{2+}-binding proteins in a neuronal system. However, classification of NCS proteins as a new family of proteins only followed after more subfamilies were discovered and after their predominant neuronal localization was demonstrated.

Recently NCS proteins have become the subject of scientific interest of a number of researchers around the world. For this reason we thought it would be important to present an overview of the various points of view of the researchers in the field. That is why in the book, several chapters might be dedicated to each of the NCS protein family members despite the possible risk of partial overlapping of information. We were pleased to learn that many of our colleagues working in the field shared our enthusiasm and provided us with excellent manuscripts. In addition, we also included contributions that deal with the application of biophysical and chemical methods to understand NCS protein function.

With new genome sequences published every year, we can be sure that the present phylogenetic dendrograms that represent the evolutionary relationship between NCS proteins will be modified in the coming years. However, some fundamental principles of NCS protein functions as outlined in this book will hopefully persist in a world of changing paradigms.

PART I: NCS-1 (FREQUENIN)

In: Neuronal Calcium Sensor Proteins
Editors: Pavel Philippov and Karl Koch, pp. 3-17

ISBN 1-59454-978-8

Chapter I

FREQUENIN: A HIGHLY CONSERVED PROTEIN WITH A CONSERVED FUNCTION?

Malte C. Stockebrand and Olaf Pongs*
Center for Molecular Neurobiology Hamburg, Institute for Neural Signal Transduction, Falkenried 94, 20251 Hamburg, Germany.

Abstract

Originally, frequenin or neuronal calcium sensor-1 (NCS-1) has been cloned from *Drosophila melanogaster T(X:Y) V7* mutants. The phenotype of these mutants is associated with a frequency-dependent facilitation of neurotransmitter release at neuromuscular junctions of third instar *Drosophila* larvae. Subsequently, comparable phenotypes were observed in transgenic *C. elegans* worms, in embryonic nerve-muscle cultures of *Xenopus* overexpressing NCS-1 and in cultured rodent neurons transiently transfected with NCS-1 cDNA. Based on genetic and biochemical evidence, it has been shown that the yeast orthologue of NCS-1 (Frq1) functions as a modulatory subunit of phosphatidylinositol 4-kinase (Pik1). Pik1 activity in yeast is required for Golgi function and vesicle trafficking. Thus, frequenin/NCS-1 may play an essential role in phosphatidylinositol 4-phosphate synthesis and phosphatidylinositol 4-phosphate related cell signalling. In recent years other proteins have been proposed as possible frequenin/NCS-1 targets. Their relevance to *in vivo* functions of frequenin/NCS-1 is, however, unclear. Possibly, the very high sequence conservation which is observed between yeast and mammalian frequenin/NCS-1 proteins reflects also a highly conserved functional role of frequenin *in vivo*.

Keywords: Frequenin, Frq1, Neuronal Calcium Sensor-1, Pik1

---

* Correspondence concerning this article should be addressed to Olaf Pongs, Center for Molecular Neurobiology Hamburg, Institute for Neural Signal Transduction, Falkenried 94, 20251 Hamburg, Germany. e-mail address pointuri@uke.uni-hamburg.de.

Abbreviations

ARF1	ADP-ribosylation factor 1
DAG	diacylglycerol
D2 receptor	D2 dopamin receptor
GAP	GTPase activating protein
GC	guanylyl gyclase
GCAP	guanylyl cyclase activating protein
GDI	guanidine nucleotide-dissociation inhibitor
GRAM	Glucosyltransferases, Rab-like GTPase activators, and Myotubularian
I_{Ca}	Calcium current
$InsP_3$	phosphatidylinositol 1,4,5-trisphosphate
KChIP	K^+-Channel Interacting Protein
NCS-1	neuronal calcium sensor-1
NCX	Na^+/Ca^{2+} exchanger
NMR	Nuclear Magnetic Resonance
NOS	nitric oxide synthase
PDE	3':5'-cyclic nucleotide phosphodieasterase
PI4K	Phosphatidylinositol 4-kinase
PI4KIIIβ	phosphatidylinositol 4-kinase type IIIβ
PtdIns(3)P	phosphatydilinositol 3-phosphate
PtdIns(4)P	phosphatidylinositol 4-phosphate
$PtdIns(4,5)P_2$	phosphatidylinositol 4,5-bisphosphate
PtdIns(5)P	phosphatidylinositol 5-phosphate
SNARE	soluble N-ethyl-maleimide-sensitive fusion factor attachment protein receptor

1. Introduction

Molecular biological characterization of the *T(X;Y)V7* rearrangement in Drosophila has originally lead to the identification of frequenin as a novel calcium-binding protein that modulates synaptic efficacy in the *Drosophila* nervous system [1]. Frequenin derived its name from the observation that its overexpression in *T(X;Y)V7* mutants as well as in transgenic flies elicited an enhanced, frequency-dependent facilitation of neurotransmitter release at neuromuscular junctions of third instar larvae. The *T(X;Y)V7* rearrangement in *Drosophila* was recognized because it is associated with a hyperexcitability phenotype comparable to the one of *Shaker* flies. The behavioural and electrophysiological phenotype of *T(X;Y)V7* flies implied that frequenin was involved in Ca^{2+}-dependent modulation of synaptic efficacy.

Subsequent cloning of frequenin-related genes (then referred to as neuronal calcium sensor-1, NCS-1) from organisms as divergent as yeast and man showed that frequenin/NCS-1 protein sequences have been highly conserved throughout evolution including an N-terminal myristoylation signal. The sequences of known vertebrate and amphibian frequenin/NCS-1 proteins have a sequence identity of nearly 100%, the sequences of vertebrate and invertebrate frequenin/NCS-1 proteins still have an identity that is close to 75% and that between human NCS-1 and the yeast *Saccharomyces cerevisiae* orthologue Frq1 is around 60%. We hypothesize that the high sequence conservation among frequenin/NCS-1 proteins may reflect similar functions of NCS-1 proteins in various cellular settings.

Consistent with the original observation, that frequenin modulates synaptic efficacy in *Drosophila*, effects of NCS-1 have been reported on neurotransmission and secretion in other systems (Fig. 1). A role of frequenin/NCS-1 was described for some type of learning and memory behaviour in *C.elegans* [2], for short term synaptic plasticity in cultured rat hippocampal neurons [3], for synaptic efficacy in embryonic *Xenopus laevis* motor neurons [4], for regulation of secretion in PC12 cells [5-8], bovine chromaffine cells [9,10], cultured mast cells [11], and 3t3L1 adipocytes [12].

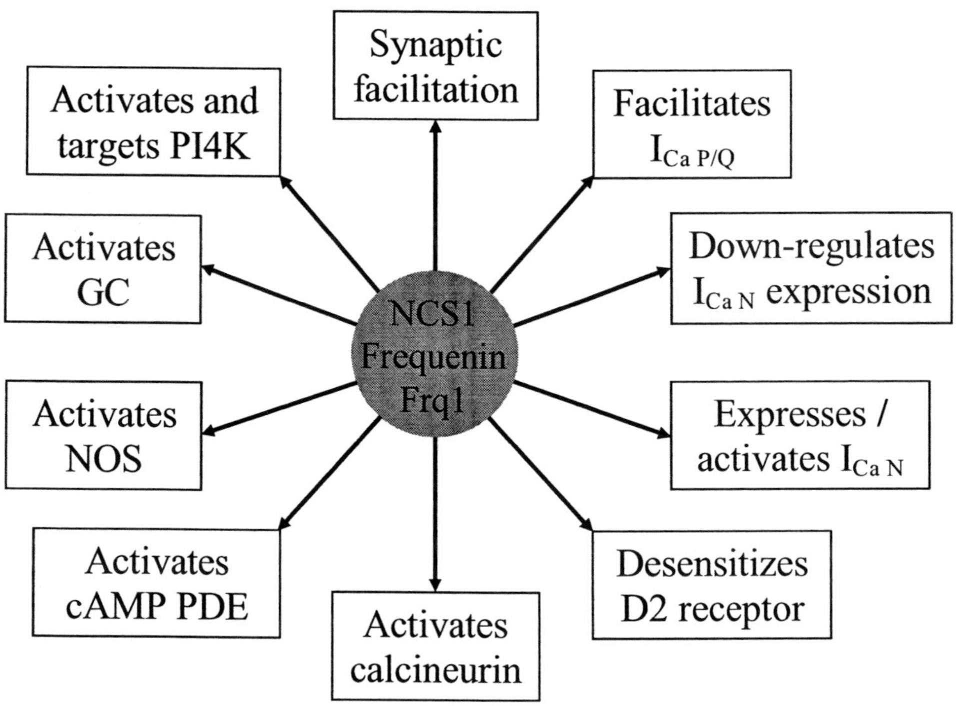

Figure 1. Overview on reported/proposed targets and functions of frequenin homoloques. I_{Ca} – Calcium current (P/Q- or N-type); D2 receptor – D2 dopamin receptor; PDE – cAMP phosphodiesterase; NOS – nitric oxide synthase; GC – guanylyl cyclase; PI4K – phosphatidylinositol 4-kinase.

However, identification of molecular targets and unequivocal characterization of frequenin/NCS-1 target interactions at the molecular and cellular level has remained enigmatic. The bewildering list of proposed targets includes, e.g., voltage-activated Ca^{2+}-channels, A-type K^{+}-channels [13-21], Na^{+}/Ca^{2+} exchanger (NCX) [22], D2 dopamine receptor [23,24], the small GTPase ARF1 [25], calcineurin, 3':5'-cyclic nucleotide phosphodiesterase (PDE), nitric oxide synthase (NOS) [26], interleukin 1 receptor accessory protein like (IL1RAPL) [27], phosphatidylinositol 4-kinase Pik1 [28], phosphatidylinositol 4-kinase type IIIβ (PI4KIIIβ) [29-32]. In the light of the very high sequence conservation of frequenin/NCS-1 proteins, this promiscuous list of possible functional frequenin/NCS-1 protein targets is clearly surprising.

2. *Saccharomyces Cerevislae* Frq1 Non-Catalytic Subunit of Phosphatidylinositol 4-Kinase Pik1

Frq1 represents the orthologue of frequenin/NCS-1 in yeast *S. cerevisiae*. The temperature sensitive *FRQ1* mutant *frq1*^{ts} is lethal at restrictive temperature. The lethal phenotype is rescued by Pik1, phosphatidylinositol 4-kinase, the only essential Frq1 target. Pik1 is a type III phosphatidylinositol 4-kinase, resistant to adenosine inhibition and homologous to mammalian type III phosphatidylinositol 4-kinaseβ. Pik1 is a soluble protein of 1066 amino acids. It catalyses the synthesis of phosphatidylinositol 4-phosphate, PtdIns(4)P. PtdIns(4)P is the precursor of phosphatidylinositol 4,5-bisphosphate, PtdIns(4,5)P_2, which, in response to extracellular stimuli, can be hydrolysed to produce the important second messengers inositol 1,4,5-trisphosphate ($InsP_3$) and diacylglycerol (DAG). Furthermore, PtdIns(4,5)P_2 by itself can have key regulatory functions such as binding to regulatory proteins playing important roles in cell signalling, vesicle trafficking, cell proliferation, and cytoskeleton organization as well as modulating the activity of various ion channels and NCX [for review see refs. 33, 34]. Pik1 is essential for viability of yeast *S. cerevisiae* and is involved in the regulation of late stages of vesicular protein trafficking in secretion and endocytosis [35]. Frq1 directly interacts with Pik1 and thereby regulates its activity and membrane targeting [28,36]. Since *FRQ1* overexpression does not rescue lethality of a *pik1Δ* mutation, Pik1 acts downstream of Frq1.

Yeast *S. cerevisiae* possesses two genes encoding type III phosphatidylinositol 4-kinases, *STT4* and *PIK1* [37,38], and furthermore the non-essential type II phosphatidylinositol 4-kinase *LSB6* [39,40]. Both type III phosphatidylinositol 4-kinases are essential for yeast viability, deletion of either type III phosphatidylinositol 4-kinase is lethal and overexpression of one of the two cannot compensate for the loss of the other [41]. Thus, it may not be surprising that overexpression of *STT4* is not able to rescue the *frq1*^{ts} phenotype. Stt4 and Pik1 represent the major phosphatidylinositol 4-kinases in yeast having very different functions and producing distinct pools of PtdIns(4)P [35]. Stt4 activity is required for the maintenance of vacuole morphology, cell wall integrity, and actin cytoskeleton organization [35], and it is an essential component in phosphatidylserine metabolism [42]. Pik1 is localized to the nucleus and the cytoplasm, where it is concentrated at the Golgi. Nuclear Pik1 probably functions in cytokinesis [43]. Pik1 is essential for normal secretion, Golgi and vacuole membrane dynamics, and endocytosis. Some *pik1*^{ts} mutants exhibit defects in secretion of Golgi-modified secretory pathway cargos. The *pik1*^{ts} cells accumulate unusual membrane structures like “Berkely bodies” probably representing hypertrophy of Golgi cisternae due to block of exit of transport vesicles [35,44-46]. This phenotype is independent of the defective actin cytoskeleton that has been observed in certain other *pik1*^{ts} mutant strains [45].

The defect of endocytosis in *pik1^{ts}* mutants is related to a defect in endosome-to-vacuole transport [45]. Thus, a *pik1^{ts}ypt31Δ* double mutant displays at permissive temperatures defects in the trafficking of proteins requiring continual recycling via early endosomes to maintain their proper subcellular localization, like the SNARE Snc1 protein or the cell wall enzyme Chs3 [46]. Interestingly, deletion of *MSS4*, the only PtdIns(4,5)P_2 synthesizing enzyme in yeast, does not affect morphology and secretory function of the Golgi [47,48]. It follows that the protein trafficking phenotype of *pik1^{ts}* mutants is independent of changes in PtdIns(4,5)P_2 concentration, but is apparently associated with a lack of PtdIns(4)P. Furthermore, other phosphatidylinositol 4-kinaseinases, but not Pik1, produce PtdIns(4)P as an essential intermediate of PtdIns(4,5)P_2. This observation implies that PtdIns(4)P can function as a signalling molecule in its own right and that a decrease in PtdIns(4)P is likely the underlying cause of the phenotype of *pik1^{ts}* and *frq1^{ts}* mutants. In fact, recent reports have ascribed to PtdIns(4)P a specific signpost role for protein recruitment to Golgi and to plasma membrane compartments, e.g. PtdIns(4)P is necessary to localize the clathrin adapter molecule AP1 to the trans-Golgi [49].

A more than 25000-fold purification of Pik1 by ammonium sulphate fractionation and subsequent column chromatography showed that Frq1 firmly associates with Pik1, even in nominally Ca^{2+}-free buffer conditions. Frq1 stimulates the activity of Frq1-free Pik1 (prepared from *frq1Δ* cells, highly overexpressing *PIK1*), depending on the experimental setup three- to ten-fold. Apparently, Frq1 functions as a non-catalytic subunit of Pik1. The interaction of Frq1 with Pik1 is not stimulated by increasing Ca^{2+}. Frq1 and N-myristoylated Frq1 stimulate Pik1 activity to a similar degree. *In vivo* the Pik1 substrate phosphatidylinositol is inserted into the membrane. It is not clear whether this affects Frq1 stimulation of Pik1 [28]. Co-immunoprecipitation experiments of recombinant myc-tagged Pik1 and Frq1 demonstrated a direct physical interaction of Frq1 with Pik1. In agreement with the other biochemical experiments, the interaction was independent of Ca^{2+}. Likewise the G2A mutation, that prevents N-myristoylation of Frq1, did not affect Frq1-Pik1 interaction. Thus, neither Ca^{2+} nor post-translational modification of Frq1 was required for the interaction of Frq1 and Pik1 *in vitro*. However, *in vivo* N-myristoylation does contribute to Frq1 function. Non-myristoylated Frq1(G2A) expressed from the *FRQ1* promoter was much less efficient than native Frq1 in suppressing the temperature-sensitive *pik1^{ts}* allele [28]. Pull-down assays show that NCS-1 *in vitro* binds N-terminal Pik1 fragments much more efficiently when it is N-myristoylated [50].

3. Components of Pik1/PtdIns(4)P Signalling Pathway

A few other proteins have been proposed in addition to Frq1 to functionally interact with Pik1. The data have arisen from synthetic genetic interactions, identified when a *PIK1* mutant was screened for second-site mutations that had altered the original phenotype [51,52]. Although synthetic genetic effects between two genes do not necessarily indicate a direct interaction of their products, it is still a useful approach to explore potential genetic interactions. The *pik1^{ts}* cells exhibit synthetic lethality or growth and transport defects with deletion mutants of the small GTPase *ARF1* [45,35] and of the Rab GTPase guanidine nucleotide-dissociation inhibitor *GDI1* (*SEC19*) [45]. In addition, the temperature sensitive phenotype of phosphatidylinositol transport protein *sec14*^{ts} mutations could be suppressed by *PIK1* overexpression [44]. The PtdIns(4)P 5-kinase mutant *mss4*^{ts} is synthetically lethal with *pik1*^{ts} [35]. Yet *mss4Δ* mutant strains do not show secretory defects at Golgi to plasma membrane stage of secretion, where Pik1 function plays a pivotal role. Apparently Pik1 synthesizes PtdIns(4)P which functions in cellular signalling independent of subsequent conversion to PtdIns(4,5)P_2.

In an attempt to identify candidate components of the Pik1/PtdIns(4)P signalling pathway at the Golgi, Sciorra et al. [46] performed a synthetic genetic array analysis of Pik1. The query mutation (*pik1*^{ts}) was crossed to an array of ~4700 deletion mutants representing the complete set of viable haploid deletion mutants in *S. cerevisiae*. Four genes with previously ascribed roles in Golgi anterograde trafficking to the plasma membrane were identified: *TRS33*, *KRE11* (2 nonessential subunits of the transport protein particle complex II), *DRS2*, an amino-phospholipid translocase required for generating clathrin coated exocytic vesicles, and the Rab GTPase Ypt31, the yeast homologue of mammalian rab11, required for secretory transport. PtdIns(4)P level and Pik1-GFP localization are unaffected by *YPT31/32* inactivation or by *YPT31* overexpression. Overexpression of neither *YPT31* nor the GTPase deficient mutant *ypt31Q72L* rescued *pik1*^{ts} temperature-sensitivity [46]. It is unlikely for Ypt31 to act directly downstream of Pik1. Consistent with this observation is that transport protein particle complex II functions upstream of Ypt31. However, both Pik1 and Ypt31 function is required for targeting of proteins like SNARE Snc1 to sites of polarized growth and membrane fusion [46]. This indicates an involvement of Pik1 and Ypt31 in common signalling pathways, facilitating Golgi transport function. Interestingly, the GAP Gyp2, a Ypt31 activating protein, possesses a GRAM (Glucosyltransferases, Rab-like GTPase activators, and Myotubularian) domain which binds PtdIns(3)P, PtdIns(4)P, and PtdIns(5)P. In this context, it is noteworthy that the mammalian small GTPase ARF1 was co-localized with NCS-1 in immunohistochemical experiments with transiently transfected tissue culture cells [53]. Also, ARF1 can be co-immunoprecipitated with NCS-1 from bovine brain cytosol [25] and mediates recruitment of mammalian phosphatidylinositol 4-kinase IIIβ to the Golgi complex [54]. Since PtdIns(4)P is the product of enzymatic Pik1 activity, the synthetic genetic array analysis results are consistent with the idea that proteins acting downstream of Pik1 respond to changes in PtdIns(4)P concentration.

4. Functional and Structural Conservation of Frequenin/NCS-1

Frq1 and human NCS-1 have a high degree of sequence and structural conservation. Apparently, the structural conservation is correlated with a functional one. Genetic experiments showed that expressing *Xenopus* frequenin or human NCS-1 in yeast *S. cerevisiae* cells rescued the inviability of *frq1Δ* mutant *S. cerevisiae* spores [28,50]. The results demonstrated that vertebrate NCS-1 can functionally replace Frq1 in yeast cells. In contrast, closely related NCS proteins like human KChIP2 or bovine recoverin did not rescue *frq1Δ* mutant spore inviability [50,28]. As for Frq1, neither N-myristoylation nor Ca^{2+} binding ability was essential for a successful *frq1Δ* complementation by NCS-1. However, the experiments also showed that NCS-1 is not completely equivalent to native yeast Frq1, because under certain stringent conditions NCS-1 expression was not able to compensate loss of Frq1 function.

Frq1 and human NCS-1 bind in a comparable manner to Pik1 *in vivo* and *in vitro* [36,50]. The *in vitro* interaction of Frq1 with Pik1 has been extensively characterized. The stoichiometry of Frq1 binding to Pik1 is 1:1 as revealed by dynamic light scattering and fluorescence quenching experiments. Pull-down experiments with glutathione S-transferase-Pik1 fusion proteins as well as competition binding experiments with N-terminal Pik1 peptides indicated that a Frq1 interaction domain was localized to the Pik1 amino-terminus within a region of 73 amino acid residues. Further *in vitro* binding studies combined with mutational analyses showed that the amino-terminal Pik1 binding domain could be subdivided into two smaller subdomains, localized to Pik1 amino acid residues 125-136 (-FQVARRVLNNLQ-) and 157-169 (-APALVLSSMIMSA-), respectively. The notion that two neighbouring, presumably α-helical domains form a complex with NCS-1 is reminiscent of the structure described for some complexes between calmodulin and certain calmodulin binding domains [55].

NMR analysis of Pik1-Frq1 complexes indicates conformational changes in Frq1 upon binding Pik1 N-terminus, which seems to make intimate contact with Frq1. The Frq1-Pik1 interaction is endothermic and thus largely entropically driven consistent with the proposal that interactions between Frq1 and Pik1 are mainly hydrophobic [36,50]. The *in vitro* binding experiments were largely done with recombinant unmyristoylated frequenin. It is not clear if myristoylation affects NCS-1 binding properties. Preliminary pull-down experiments showed that unmyristoylated NCS-1 does not bind to PI4KIIIβ N-terminus, consistent with published *in vitro* myristoylation-dependent NCS-1 PI4KIIIβ interaction [32]. The effect of myristoylation on Pik1 binding of NCS-1, mentioned above, is in agreement with the assumption of a function for amino terminal myristoic acid in target binding. Differences in binding capacity for Pik1 fragments in *in vitro* pull-down experiments were most profound with Pik1 fragments which included also the more N-terminal subdomain, previously identified as being important for Frq1 binding (Pik1 amino acid residues 125-136) [50].

The combined genetic and biochemical experiments in yeast have strong implications with respect to possible functions and targets of frequenin/NCS-1. Yeast Frq1 is a subunit of the Pik1 enzyme and stimulates its catalysis of PtdIns(4)P synthesis. Pik1 activity is required for secretory function of the Golgi and for trafficking of proteins like the SNARE Snc1 [46]. The protein trafficking phenotype of $pik1^{ts}$ mutants is independent of changes in PtdIns(4,5)P_2 concentration (observed in some but not all $pik1^{ts}$ mutants), but is apparently associated with a lack of PtdIns(4)P [35]. It follows that downstream targets of the Frq1/Pik1 enzyme complex are proteins involved in PtdIns(4)P sensitive cellular pathways like certain GTPases and GAPs. Given the high degree of structural and functional conservation, it is likely that most aspects of the yeast phenotype may have been conserved during evolution. In fact, the secretory phenotypes that have been described for frequenin overexpression in *Drosophila* and for overexpression of NCS-1 protein in *Xenopus laevis* and in transiently transfected rat hippocampal neurons are consistent with a similar role of NCS-1 in

Drosophila and in vertebrates as compared to the one of Frq1 in yeast. However, there is an apparent discrepancy between the many different and seemingly unrelated molecular NCS-1 targets that have been described in the literature and the genetic yeast data. It is not clear what the reasons are for this obvious discrepancy. Some of the reasons may originate from the structural features of frequenin/NCS-1, which make it a difficult molecule to work with.

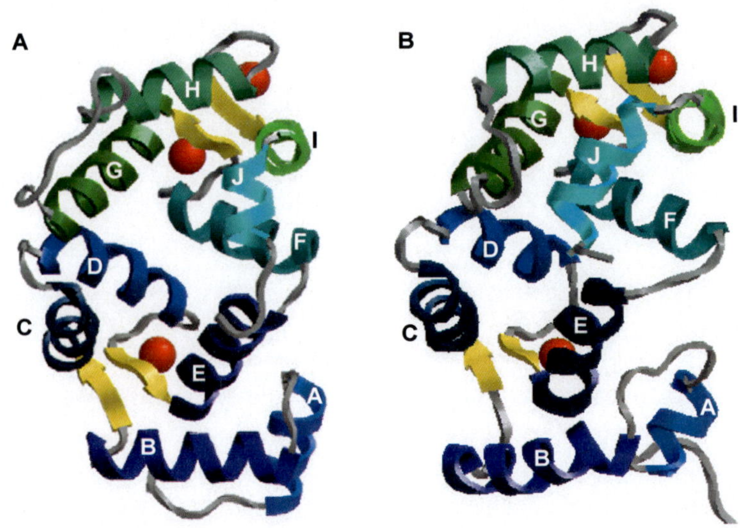

Figure 2. Ribbon diagrams of mammalian NCS-1 (PDB: 1G8I) [53] and yeast frequenin Frq1 (PDB: 1FPW) [56]. A: NCS-1, The large hydrophobic crevice is made up of 46 residues, provided by helices D, E, and F on the floor of the crevice, C and G on one side and J on the other. B: Frq1, as in NCS-1 two tightly folded domains (^{1}M-^{92}T and ^{100}K-^{190}I) are packed against each other forming an accessible crevice lined primarily with hydrophobic side chains. A-J, Helices A-J, Yellow arrows: beta sheets forming the EF-hand motifs, red balls: Ca^{2+}-ions.

5. Ca^{2+}- Binding Properties

NCS-1 or frequenins display four EF-hand Ca^{2+}-binding motifs which match very well the canonical helix-loop-helix structure of EF-hand motifs. Only three of the four motifs – EF2, EF3 and EF4 – are functional Ca^{2+}-binding sites as determined by *in vitro* binding studies [56]. The data are supported by structural analysis which revealed three bound Ca^{2+} ions in the NCS-1 crystal structure (Fig. 2) [53]. NCS-1 binds Ca^{2+} in a cooperative manner with apparent K_D values between 0.4 and 10 μM. Importantly, N-myristoylated Frq1 binds Ca^{2+} with an apparent K_D of 3 μM and a positive Hill coefficient of 1.25 shows that the myristoyl group in Frq1 might serve as a built-in allosteric effector on Ca^{2+}-binding. Fluorescence spectroscopy monitoring the intrinsic tryptophan fluorescence during Ca^{2+}-induced conformational changes suggests two conformational states of Frq1 [56]. The threshold for Ca^{2+}-binding of NCS-1 may be only slightly above the cellular resting Ca^{2+} level. At physiological concentrations of about 0.5 mM free Mg^{2+}, an apparent K_D for Ca^{2+} binding of roughly 0.3 μM was observed for two of the three functional EF-hands [57]. NCS-

1 may bind Ca^{2+} with much higher affinity than for example calmodulin which binds Ca^{2+} with a K_D in the micromolar range.

6. NCS-1 Structure

Ames et al.[56] presented a model for the Frq1 structure, based on NMR data and homology to recoverin. This structure is in good agreement with the crystal structure of NCS-1, which has been determined and refined to a resolution of 1.9 Å by the molecular replacement method, using neurocalcin as a search model [53]. The structure consists of residues Asn^5 – Val^{190}, the overall dimensions are 35 x 60 x 40 Å. The four EF hands present the typical architectural short β-strand. They localize pairwise and the two pairs are connected via a hinge loop (helices E-F, see Fig. 2). The structural arrangement resembles the ones seen in other members of the NCS family and Ca^{2+}-myristoyl switch proteins like neurocalcin [58], recoverin [59,60] and GCAP [61]. NCS-1 contains 10 α-helices labelled A-J. Eight of the ten helices provide residues making up a hydrophobic crevice. The EF-hands are located on one side, the hydrophobic crevice on the opposite side of the molecule. This hydrophobic crevice is the dominant striking feature of the NCS-1 structure, caused by a large positional shift of the C-terminal helix J, compared with its position in related structures (e.g. recoverin or neurocalcin). The hydrophobic crevice is completely solvent-exposed. It very likely participates either in binding of protein targets by hydrophobic interactions, or in targeting the protein and possibly binding partners to membranous structures within the cell. About one fourth of the molecule lines the hydrophobic crevice with C-terminal as well as N-terminal regions being part of it.

In NMR-experiments, a myristoyl switch was not observed for yeast Frq1 [56]. Whether different experimental conditions or other NCS-1 proteins will show the same result, is not known. According to the NMR data, the N-myristoyl group in Frq1 is solvent-exposed in the absence of Ca^{2+} and does not undergo a calcium-induced conformational change. The presence of the N-myristoyl group is, however, necessary for high-affinity interaction of Frq1 with biological membranes [56]. Ca^{2+}-binding appears to induce conformational changes that additionally contribute to the ability of Frq1 to associate with membrane. In the absence of Ca^{2+}, association of Frq1 with membrane depends on the presence of the N-myristoyl moiety as revealed by subcellular fractionation experiments. Non-myristoylated Frq1 only weakly associated with membrane in the presence of Ca^{2+}. The data indicate that Frq1 lacking the N-terminal myristoyl group retains some capacity to associate with membranes in a Ca^{2+}-dependent manner. Apparently, Frq1 does not behave like a classical Ca^{2+}-myristoyl switch protein, which only associates with membrane when myristoylated and Ca^{2+} is present.

Since human NCS-1 can functionally substitute yeast Frq1 [50], we mapped the non-conservative amino acid exchanges between the primary sequences of yeast Frq1 and human NCS-1 (Fig. 3) onto the crystal structure of NCS-1 (Fig. 4). The hydrophobic crevice located on one side of the frequenin surface (Fig. 4 A, D) is a hallmark of frequenin/NCS-1 [53,56]. Most likely, it corresponds to the interacting site for cellular targets of frequenin/NCS-1 [36, 50]. The non-conservative amino acid exchanges between yeast Frq1 and human NCS-1 are found outside of the hydrophobic crevice and are mostly confined to the hydrophilic surface on the opposite side of the molecule where the calcium binding sites line up (Fig. 4 B, E). In the hydrophobic crevice, only two conservative amino acid exchanges appear between mammalian NCS-1 and yeast Frq1: V68L and I128V. Thus, the hydrophobic crevice of frequenin/NCS-1 has remained virtually unchanged during evolution. It underlines the potential importance of the hydrophobic crevice for binding to other proteins. For comparison, we mapped also the non-conservative amino acid exchanges between the close NCS relative of frequenin/NCS-1, the potassium channel interacting protein KChIP2, and NCS-1 onto the NCS-1 structure (Fig. 4E). In the yeast complementation assays, KChIP2 was unable to functionally substitute Frq1 [50]. The structural comparison clearly shows that the hydrophobic crevice of NCS-1 has not been conserved in KChIP2. Most likely, the non-conservative amino acid replacements in the hydrophobic crevice are correlated with a differing target specificity of KChIPs and NCS-1. The KChIP1 crystal structure shows, that albeit a hydrophobic region is present on one side of the molecule this region is not shaped as a crevice [62].

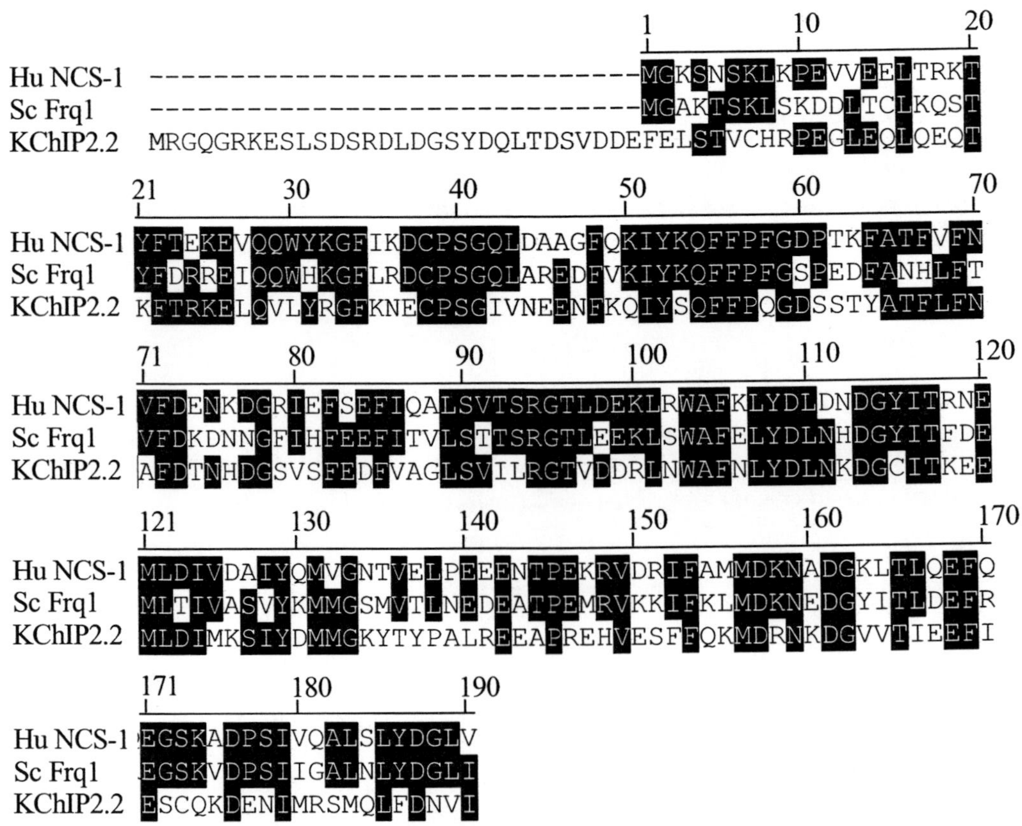

Figure 3. Amino acid sequence alignment of mammalian NCS-1 (Hu NCS-1) with yeast Frq1 (Sc Frq1) and KChIP2.2. Invariant residues are highlighted in white on a black background. Numbers refer to amino acid positions in both the NCS-1 and Frq1 sequences.

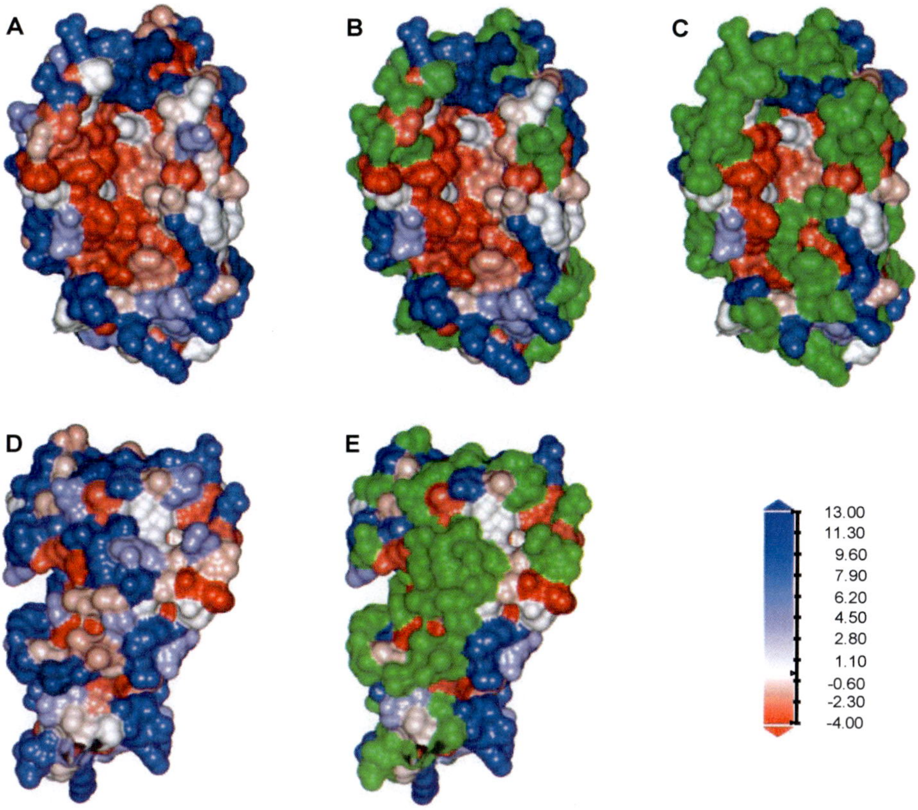

Figure 4. Conservation of amino acids in the three-dimensional NCS-1 Structure. A) The NCS-1 protein structure is coloured according to the hydrophobicity of the amino-acid residues with blue being more hydrophilic and red more hydrophobic. The large hydrophobic crevice is made up of 46 residues, provided by helices D, E, and F on the floor of the crevice, C and G on one side and J on the other (see also Fig. 2). Non-conservative amino acid exchanges between mammalian NCS-1 and yeast frequenin (B) or between mammalian NCS-1 and human KChIP2 (C) are coloured in green. Conservative amino acid exchanges were classified as follows: A,V,I,L; M,C; F,W,Y; D,E; N,Q; H,K,R; S,T. D and E are as A and B, turned 90 degree around the vertical axis with the hydrophobic crevice facing right. The colour code for hydrophobicity shows the free energies (kcal/mol) for the transfer of an amino acid residue in α-helix from water to membrane interior with an assumed dielectric constant of 2.

The hydrophobic properties of NCS-1 were exploited to purify NCS-1 on phenyl-Sepharose columns by hydrophobic interaction chromatography [53,63]. The property of NCS-1 to readily interact with hydrophobic surfaces can, however, become quite an obstacle in biochemical and cellular assays. Therefore, careful and unbiased controls are to be recommended, when one tries to characterize functional and biochemical binding properties of NCS-1. The hydrophobic nature of NCS-1 implies that only high-affinity interactions like the one determined for Frq1/NCS-1 binding to the Pik1 N-terminus (K_D ~100 nM) may be considered as true NCS-1 targets with some confidence. In summary, NCS-1 is a very hydrophobic and high-affinity Ca^{2+}-binding protein. Unfortunately, few reports have addressed these properties in overexpression experiments which may give rise to unspecific effects due to Ca^{2+}-buffering or NCS-1 stickiness. Furthermore, NCS-1 may be mislocalized to cellular compartments, if overexpressed in transiently transfected cells. For instance, observations that myristoylation of NCS-1 does not influence cellular targeting of NCS-1 depend entirely on the transfection and cellular system used. In stably transfected cell lines with a moderate expression of NCS-1, NCS-1 is targeted predominantly to the plasma membrane and this targeting completely depends on myristoylation of NCS-1. Another obstacle may stem from the fact that members of the NCS protein family exhibit a high degree of sequence conservation. It is likely, that the high degree of sequence conservation may give rise to interactions with protein binding partners, which normally do not interact with NCS-1. This situation may especially be relevant in studies which heterologously over express NCS-1 *in vitro* or in transiently transfected cells.

7. Conclusion

A combination of genetic and biochemical experiments have characterized in great detail the function of frequenin/NCS-1 in yeast *S. cerevisiae* and the structure of frequenin/NCS-1 by NMR spectroscopy and X-ray analysis. Yeast Frq1 is a non-catalytic subunit of the phosphatidylinositol 4-kinase Pik1 and thereby is involved in the phosphorylation of a membrane lipid. Pik1 is an essential kinase in yeast which synthesizes PtdIns(4)P. PtdIns(4)P appears to serve as a signalling molecule for recruitment of specific proteins to distinct membrane compartments. The phenotype of Pik1 mutants indicates that this activity is related to trafficking of proteins with a rapid turnover. Human NCS-1 can functionally substitute Frq1 in yeast. It is, however, still uncertain if mammalian NCS-1 proteins are similarly involved in PtdIns(4)P synthesis and the respective cellular processes like yeast Frq1.

Acknowledgement

We thank the Deutsche Forschungsgemeinschaft for generous support.

References

[1] Pongs, O., Lindemeier, J., Zhu, X.R., Theil, T., Engelkamp, D., Krah-Jentgens, I., Lambrecht, H.-G., Koch, K.W., Schwemer, J., Rivosecchi, R., Mallart, A., Galceran, J., Canal, I., Barbas, J.A., and Ferrús A. 1993, *Neuron, 11*, 15.

[2] Gomez, M., De Castro, E., Guarin, E., Sasakura, H., Kuhara, A., Mori, I., Bartfai, T., Bargmann, C., Nef, P. 2001, *Neuron, 30*, 241.

[3] Sippy, T., Cruz-Martin, A., Jeromin, A., Schweizer, F.E. 2003, Nat. Neurosci., 6, 1031.

[4] Olafsson, P., Wang, T., and Lu, B. 1995, Proc. *Natl. Acad. Sci. U.S.A., 92*, 8001.

[5] Koizumi, S., Rosa, P., Willars, G.B., Challiss, R.A.J., Taverna, E., Francolini, M., Bootman, M.D., Lipp, P., Inoue, K., Roder, J., and Jeromin, A. 2002, *J. Biol. Chem., 277*, 30315.

[6] McFerran, B.W., Weiss, J.L., Burgoyne, R.D., 1999, *J. Biol. Chem., 274*, 30258.

[7] McFerran, B.W., Graham, M.E., and Burgoyne, R.D. 1998, *J. Biol. Chem., 273*, 22768.
[8] Scalettar, B.A., Rosa, P., Taverna, E., Francolini, M., Tsuboi, T., Terakawa, S., Koizumi, S., Roder,J., and Jeromin, A. 2002, *J. Cell. Sci., 115*, 2399.
[9] Pan, C.-Y., Jeromin, A., Lundstrom, K., Yoo, S.H., Roder, J., and Fox, A.P., 2002, *J. Neurosci., 22*, 2427.
[10] Weiss, J.L., Archer, D.A., and Burgoyne, R.D. 2000, *J. Biol. Chem., 275*, 40082.
[11] Kapp-Barnea, Y., Melnikov, S., Shefler, I., Jeromin, A., and Sagi-Eisenberg, R. 2003, *J. Immunol. 171*, 5320.
[12] Mora, S., Durham, P.L., Smith, J.R., Russo, A.F., Jeromin, A., and Pessin, J.E. 2002, *J. Biol. Chem., 277*, 27494.
[13] Guo, W., Malin, S.A., Johns, D.C., Jeromin, A., and Merbonne, J.M. 2002, *J. Biol. Chem., 277*, 26436.
[14] Nakamura, T.Y., Sturm, E., Pountney, D.J., Orenzoff, B., Artman, M., and Coetzee, W.A. 2003, *Pediatr. Res., 53*, 554.
[15] Ren, X., Shand, S.H., and Takimoto, K. 2003, *J. Biol. Chem., 278*, 43564.
[16] Zhang, Y., MacLean, J.N., An, W.F., Lanning, C.C., and Harris-Warrick, R.M. 2003, *J. Neurophysiol., 89*, 1902.
[17] Rousset, M., Cens, T., Gavarini, S., Jeromin, A., and Charnet, P., 2003, *J. Biol. Chem., 278,* 7019.
[18] Tsujimoto, T., Jeromin, A., Saitoh, N., Roder, J.C., and Takahashi, T. 2002, *Science, 295*, 2276.
[19] Wang, C.-Y., Yang, F., He, X., Chow, A., Du, J., Russell, J.T., and Lu, B.2001, *Neuron, 32*, 99.
[20] Weiss, J.L., and Burgoyne, R.D. 2001, *J. Biol. Chem, 276*, 44804.
[21] Poulain, C., Ferrús, A., and Mallart, A. 1994, *Pflugers Arch., 427*, 71.
[22] Rivosecchi, R., Pongs, O., Theil, T., and Mallart, A. 1994, *J. Physiol., 474*, 223-32.
[23] Kabbani, N., Negyessy, L., Lin, R., Goldman-Rakic, P., and Levenson, R. 2002, J. *Neurosci., 22*, 8476.
[24] Bergson, C., Levenson, R., Goldman-Rakic, P.S., and Lidow, M.S. 2003, *Trends Pharmacol. Sci., 24*, 486.
[25] Haynes, L.P., Thomas, G.M., and Burgoyne, R.D. 2005, *J Biol Chem 7*, 6047.
[26] Schaad, N.C., De Castro, E., Nef, S., Hegi, S., Hinrichsen, R., Martone, M.E., Ellisman, M.H., Sikkink, R., Rusnak, F., Sygush, J., and Nef, P. 1996, *Proc. Natl. Acad. Sci. U.S.A., 93*, 9253.
[27] Bahi, N., Friocourt, G., Carrie, A., Graham, M.E., Weiss, J.L., Chafey, P., Fauchereau, F., Burgoyne, R.D., Chelly, J. 2003, *J. Mol. Genet., 12*, 1415.
[28] Hendricks, K.B., Wang, B.Q., Schnieders, E.A., and Thorner, J. 1999, *Nat. Cell Biol. 1*, 234-41.
[29] Taverna, E., Francolini, M., Jeromin, A., Hilfiker, S., Roder, J., and Rosa, P. 2002, *J. Cell. Sci., 115*, 3909.
[30] Zheng, Q., Bobich, J.A., Vidugiriene, J., McFadden, S.C., Thomas, F., Roder, J., and Jeromin A. 2005, *J. Neurochem., 92*, 442.
[31] Rajebhosale, M., Greenwood, S., Vidugirienes, J., Jeromin, A., and Hilfiker, S. 2003, *J. Biol. Chem., 278*, 6075.

[32] Zhao, X., Várnai, P., Tuymetova, G., Balla, A., Tóth, Z.E., Oker-Blom, C., Roder, J., Jeromin, A., and Balla, T. 2001, *J. Biol. Chem., 276*, 40183.
[33] Odorizzi, G., Babst, M., and Emr, S.D. 2000, *Trends Biochem Sci, 25*, 229.
[34] Hilgemann, D.W., Feng, S., and Nasuhoglu, C. 2001, *Sci. STKE 111*, re19.
[35] Audhya A., Foti M., and Emr, S.D. 2000 *Mol. Biol. Cell, 11*, 2673.
[36] Huttner, I.G., Strahl, T., Osawa, M., King, D.S., Ames, J.B., and Thorner, J. 2003, *J. Biol. Chem., 278*, 4862.
[37] Yoshida, S., Ohya, Y., Goebl, M., Nakano, A., and Anraku, Y. 1994, *J. Biol. Chem., 269*, 1166.
[38] Flanagan, C.A., and Thorner, J. 1992 *J. Biol. Chem. 267*, 24117-25.
[39] Han, G.-S., Audhya, A., Markley, D.J., Emr, S.D., and Carman, G.M. 2002, *J. Biol. Chem., 277*, 47709.
[40] Shelton S.N., Barylko, B., Binns, D.D., Horazdovsky, B.F., and Albanesi, J.P., 2003, *Biochem. J., 371*, 533.
[41] Cutler, N.S., Heitman, J., and Cardenas, M.E. 1997, *J. Biol. Chem., 272*, 27671.
[42] Trotter, P.J., Wu, W.-I., Pedretti, J., Yates, R., and Voelker, D.R. 1998, *J. Biol. Chem., 273*, 13189.
[43] Garcia-Bustos, J.F., Marini, F., Stevenson, I., Frei, C. and Hall, M.N. 1994, *EMBO J 13*, 2352.
[44] Hama H., Schnieders, E.A., Thorner, J., Takemoto, J.Y., and DeWald, D.B. 1999 *J. Biol. Chem. 274*, 34294.
[45] Walch-Solinema, C., and Novick, P. 1999 *Nat. Cell Biol. 1*: 523.
[46] Sciorra, V.A., Audhya, A., Parsons, A.B., Segev, N., Boone, C., and Emr, S.D. 2005, *Mol. Biol. Cell 16*: 776.
[47] Audhya, A., and Emr, S.D. 2002, *Dev. Cell 2*, 593.
[48] Stefan, J.C., Audhya, A., and Emr, S.D. 2002, *Mol. Biol. Cell, 13*, 542.
[49] Wang, JY.J., Wang, J., Sun, H.Q., Martinez, M., Sun, Y.X., Macia, E., Kirchhausen, T., Albanesi, J.P., Roth, M.G., and Yin, H.L. 2003, *Cell, 114*, 299.
[50] Strahl, T., Grafelmann, B., Dannenberg, J., Thorner, J., Pongs, O., 2003, *J. Biol. Chem. 278*, 49589.
[51] Tong, A.H.Y., Evangelista, M., Parsons, A.B., Xu, H., Bader, G.D., Pagé, N., Robinson, M., Raghibizadeh, S., Hogue, C.W.V., Bussey, H., Andrews, B., Tyers, M., and Boone, C., 2001, *Science, 294*, 2364.
[52] Hartman IV, J.L., Garvik, B., and Hartwell, L. 2001, *Science, 291*, 1001.
[53] Bourne, Y., Dannenberg, J., Pollmann, V., Marchot, P., and Pongs, O. 2001, *J. Biol. Chem. 276*, 11949.
[54] Godi, A., Pertile, P., Meyers, R., Marra, P., Di Tullio, G., Iurisci, C., Luini, A., Corda, D., and De Matteis, M.A. 1999, *Nat. Cell Biol., 1*, 280.
[55] Hoeflich, K.P. and Ikura, M. 2002, *Cell, 108*, 739.
[56] Ames, J.B., Hendricks, K.B., Strahl, T., Huttgner, I.G., Hamasaki, N., and Thorner, H. 2000, *Biochemistry, 39*, 12149.
[57] Cox, J.A., Durussel, I., Comte, M., Nef, S., Nef, P., Lenz, S.E., and Gundelfinger, E.D. 1994, *J. Biol. Chem., 269*, 32807.
[58] Vijay-Kumar S. and Kumar, V.D. 1999, *Nat. Struct. Biol., 6*, 80.

[59] Flaherty, K.M., Zozulya, S., Stryer, L., McKay, D.B. 1993, *Cell, 75*, 709.

[60] Ames, J.B., Ishima, R., Tanaka, T., Gordon, J.I., Stryer, L., and Ikura, M. 1997, *Nature, 398*, 198.

[61] Ames, J.B., Dizhoor, A.M., Ikura, M., Palczewski, K., and Stryer, L. 1999, *J. Biol. Chem., 274*, 19329.

[62] Scannevin, R.H., Wang, KW., Jow, F., Megules, J., Kopsco, D.C., Edris, W., Carroll, K.C., Lü, Q., Xu, W., Zhangbao, X., Katz, A.H., Olland, S., Lin, L., Taylor, M., Stahl, M., Malakian, K., Somers, W., Mosyak, L., Bowlby, M.R., Chanda, P., and Rhodes, K.J. 2004, *Neuron, 41*, 587.

[63] Fisher, J.R., Sharma, Y., Kuliano, S., Piccioti, R.A., Krylov, D., Hurley, J., Roder, J., and Jeromin, A. 2000, *Protein Expr. Purif., 20*, 66.

In: Neuronal Calcium Sensor Proteins
Editors: Pavel Philippov and Karl Koch, pp. 19-43
ISBN 1-59454-978-8

Chapter II

SENSING THE DIFFERENCE: NEURONAL CALCIUM SENSOR-1 AND ITS INVOLVEMENT IN VESICLE BIOGENESIS, TRAFFICKING AND FUSION

Elena Fdez and Sabine Hilfiker*
Institute of Parasitology and Biomedicine "Lopez-Neyra", Consejo Superior de Investigaciones Cientificas (CSIC), Granada, Spain.

Abstract

Many aspects of brain function are regulated by Ca^{2+} signals. Temporally and spatially distinct Ca^{2+} signals are detected by a variety of proteins, including EF-hand-containing Ca^{2+} binding proteins. The neuronal Ca^{2+} sensor (NCS) protein family has emerged as a key player in modulating neuronal function and synaptic plasticity, even though the precise mechanism of action of any member of this family remains largely unknown. Neuronal calcium sensor-1 (NCS-1), also named frequenin, was originally identified in *Drosophila* in a screen for neuronal hyperexcitability mutants. Overexpression of NCS-1 has been shown to enhance evoked neurotransmitter release, paired-pulse facilitation and exocytosis in several neuronal and neuroendocrine cell types, indicating an important role for this protein in modulating synaptic efficacy. NCS-1 is an N-terminally myristoylated protein that contains four EF-hand motifs, three of which are able to bind Ca^{2+} in the submicromolar range. NCS-1 is constitutively associated with membranes even under resting Ca^{2+} conditions, suggesting that a Ca^{2+}/myristoyl switch does not operate to regulate the subcellular localization of NCS-1. Instead, binding of Ca^{2+} to NCS-1 may alternatively affect its interaction with, or modulation of, a variety of downstream binding partners. Several NCS-1-interacting proteins have been described,

* Correspondence concerning this article should be addressed to Sabine Hilfiker Institute of Parasitology and Biomedicine "Lopez-Neyra", Consejo Superior de Investigaciones Cientificas (CSIC), Granada, Spain. sabine.hilfiker@ipb.csic.es.

including ion channels, receptors and type III phosphatidylinositol 4-kinase β (PI4Kβ). The latter interaction suggests that NCS-1 may enhance secretion by modulating vesicle biogenesis, vesicle trafficking and/or vesicle fusion events in a phosphoinositide-dependent manner. The present review describes our current knowledge about NCS-1 and the molecular mechanism(s) by which NCS-1 may enhance neurotransmission and synaptic plasticity.

Keywords: neuronal calcium sensor proteins, NCS, calcium, synaptic transmission, synaptic plasticity, PI4K, phosphoinositides, vesicle trafficking, dopamine

Abbreviations

ARF	ADP-ribosylation factor
CNS	central nervous system
DA	dopamine
DLPFC	dorsolateral prefrontal cortex
DRIP	dopamine receptor-interacting protein
FAPP	four-phosphate-adaptor protein
GCAP	guanylate cyclase-activating protein
GPCR	G protein-coupled receptor
GRK	G protein-coupled receptor kinase
IL1RAPL	interleukin-1 receptor accessory protein-like protein
KChIP	K^+-channel-interacting protein
NCS	neuronal Ca^{2+}sensor
PI4K	phosphatidylinositol 4-kinase
PI4Kβ	type III phosphatidylinositol 4-kinase β
PtdIns(4)P	phosphatidylinositol 4-phosphate
PtdIns(4,5)P_2	phosphatidylinositol (4,5)-bisphosphate
TGN	trans-Golgi network
VILIP	visinin-like protein.

1. Introduction

The cytosolic Ca^{2+} concentration in cells is tightly regulated, since changes in intracellular Ca^{2+} can regulate many different aspects of cell function. In neurons, Ca^{2+} ions regulate processes such as neurotransmission, cytoskeletal dynamics, signal transduction and gene expression. Neuronal Ca^{2+} signals can vary drastically both in their temporal as well as spatial characteristics [1]. For example, neurotransmitter release is triggered by submillisecond increases of Ca^{2+} levels within nanometers of open Ca^{2+} channels. Other Ca^{2+} signals include localized rises lasting for tens of milliseconds, spatially more extensive Ca^{2+} elevations for example in dendrites, Ca^{2+} waves propagating over a couple of seconds, or global and prolonged Ca^{2+} elevations throughout the cell. Such distinct Ca^{2+} signals are transduced into changes of neuronal function mediated by Ca^{2+}-binding proteins, which may act either as Ca^{2+} buffers or as Ca^{2+} sensors. In either case, these proteins must be able to respond rapidly and selectively to defined Ca^{2+} signals. Such selectivity is thought to be achieved by the distinct localization of various Ca^{2+} sensing proteins as well as by their different kinetics (on-rate) and affinity towards Ca^{2+}, allowing them to transduce distinct Ca^{2+} signals into varied changes in neuronal function.

2. The NCS Family

The neuronal calcium sensor (NCS) proteins are Ca^{2+} sensors with micromolar or submicromolar affinities and change their conformation upon Ca^{2+} binding, which allows them to modulate the interaction with their targets and thus act as effector molecules to transduce Ca^{2+} signals into appropriate downstream events. Whilst it is becoming increasingly clear that the NCS proteins regulate a variety of neuronal processes [2-5], the exact functions of most NCS family members are still poorly understood.

Table 1. The NCS family of Ca^{2+}-sensor proteins.

Protein	Class	Expression	Splice variants
NCS-1	A	Brain, retina, non-neuronal	None
VILIP-1	B	Brain, retina	None
VILIP-2	B	Brain	None
VILIP-3	B	Brain	None
Hippocalcin	B	Brain	None
Neurocalcin δ	B	Brain, retina	None
Recoverin	C	Retina	None
GCAP1	D	Retina	None
GCAP2	D	Retina	None
GCAP3	D	Retina	None
KChIP1	E	Brain	1a, 1b
KChIP2	E	Brain, cardiac myocytes	2, 2a, 2b
KChIP3	E	Brain	3a, 3b
KChIP4	E	Brain	4a, 4b

Abbreviations: GCAP, guanylate cyclase-activating protein; KChIP, K^+ channel-interacting protein; NCS, neuronal calcium sensor; VILIP, visinin-like protein.

The human genome encodes at least 14 genes for NCS proteins, and such diversity is further increased by the existence of splice variants (Table 1). Based on evolutionary appearance and sequence similarity, the NCS protein family has been subdivided into five subfamilies (class A-E) [2-5] (Table 1). Class A consists of NCS-1, which was originally identified as frequenin in *Drosophila* [6], and which constitutes the primordial member of the NCS family, being present in yeast. Class B includes the visinin-like proteins (VILIPs), including VILIP-1, VILIP-2, VILIP-3, neurocalcin δ and hippocalcin, and a member of this subfamily is represented in the genome of *Caenorhabditis elegans* and evolutionarily later species. The members of classes C and D, the recoverin and the guanylate cyclase-activating proteins (GCAPs), respectively, are present in amphibia and evolutionarily subsequent species, whilst the class E proteins, the K^+-channel-interacting proteins (KChIPs), are represented in genomes from fish onwards [5].

The different NCS class proteins also tend to differ in their expression patterns [2] (Table 1). For example, NCS-1 is widely expressed in neuronal as well as non-neuronal cells, whilst most other NCS proteins are neuron-specific. On the other hand, some NCS proteins are expressed in only a subset of neuronal cell types; for example, hippocalcin is predominantly expressed in hippocampal pyramidal neurons, whilst VILIP-3 is mostly expressed in cerebellar Purkinje cells [5]. Studies of such types have established that certain neuronal cell types express several or all of the NCS proteins, but that the overall expression pattern is unique for each protein. Thus, the NCS proteins are likely to perform distinct functions specific for each particular neuronal cell type.

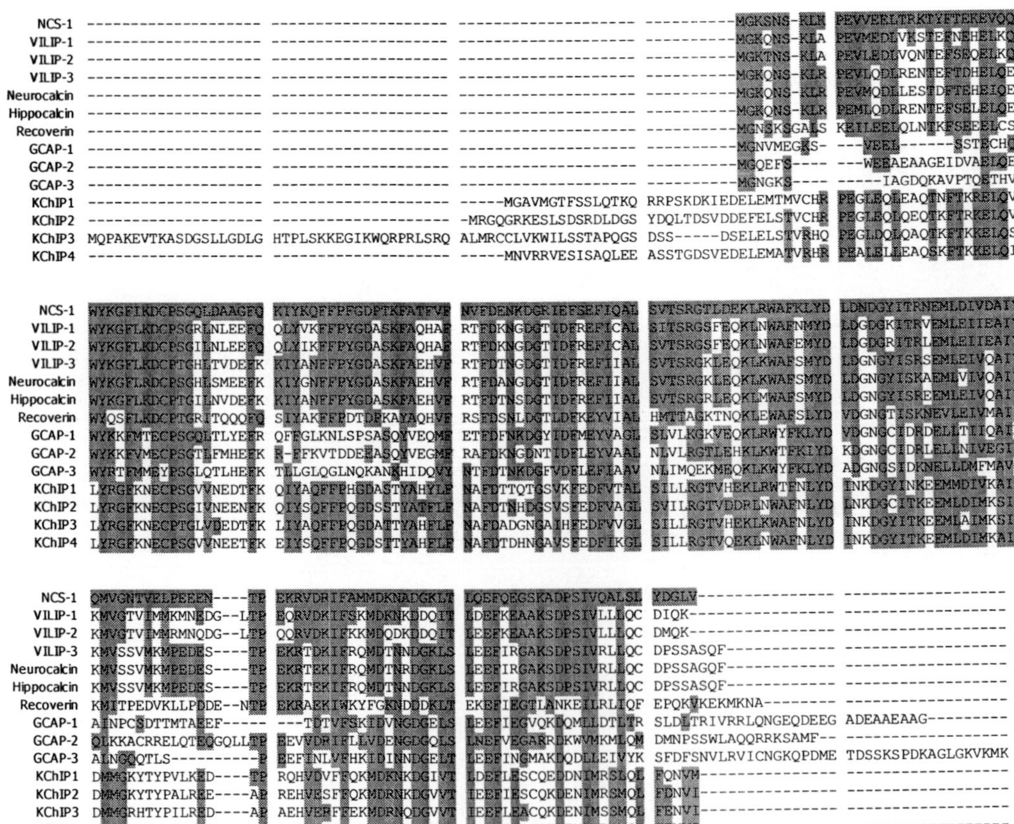

Figure 1. Alignment of human NCS protein sequences. Amino acid sequence alignment of human NCS proteins illustrates high overall sequence homology. Amino acids highlighted in blue are identical, whilst amino acids highlighted in grey are similar to those of human NCS-1, respectively. The database accession numbers for the sequences are as follows: NCS-1 (P62166), VILIP-1 (NP_003376), VILIP-2 (Q5TG97), VILIP-3 (P62748), Neurocalcin (P61601), Hippocalcin (NP_002134), Recoverin (P35243), GCAP-1 (P43080), GCAP-2 (Q9UMX6), GCAP-3 (O95843), KChIP1 (NP_055407), KChIP2 (Q9HD11), KChIP3 (Q9Y2W7) and KChIP4 (Q9H2A4).

A major impediment in trying to study the individual roles of the different NCS proteins has been their high sequence similarity (Figure 1). NCS proteins are all of similar size, possess around 50% or more of sequence identity with each other, have N-terminal myristoylation consensus sequences (except KChIPs 2-4), and show high-affinity Ca^{2+} binding *in vitro* not far above basal Ca^{2+} concentrations (100 nM). In addition, all members of the family possess four EF-hand motifs (Figure 1). Generally, EF hands are comprised of 29-residue helix-loop-helix motifs, with 12 residues forming a Ca^{2+}-binding loop [7]. However, only three (or two in the case of recoverin and KChIP1) are able to bind Ca^{2+}. In all of the proteins, the first EF-hand is predicted to be unable to bind Ca^{2+} due to the presence of a conserved cysteine-proline substitution in the Ca^{2+}-binding loop (Figure 1). Instead, EF-1 seems to be involved in interactions with the myristoyl group in the Ca^{2+}-free forms, or possibly interactions with binding partners in the Ca^{2+}-bound forms of the proteins, respectively.

The high sequence similarity amongst NCS proteins raises important issues of apparent functional redundancy. For example, whilst multiple NCS proteins may interact with the same protein partner(s) when assayed *in vitro*, the affinity of such interactions may vary significantly, and moreover, specificity *in vivo* may be achieved by the distinct cellular and subcellular localizations of the different NCS proteins and/or their protein targets. Therefore, *in vitro* interaction studies, or cellular overexpression studies to elucidate the function(s) and molecular mode(s) of action of individual NCS proteins, have to be interpreted with great care. In the present chapter, we will describe current knowledge about the cellular functions and possible molecular mechanism(s) of action of NCS-1.

3. Cellular and Subcellular Localization of NCS-1

Localization studies of NCS-1 have been generally hampered by the compromised specificity of the antibodies used. For example, only few of the affinity-purified anti-NCS-1 antibodies currently available have been tested for their cross-reactivity with other members of the NCS family by Western blotting. However, where anti-NCS-1 antibodies have been immunodepleted against other NCS protein species, the localization data obtained have been reliable [8]. In addition, in situ hybridization studies to detect NCS-1 mRNA have clearly established that NCS-1 gene expression is pan-neuronal [9]. Whilst initially thought to be neuron-specific [10], NCS-1 has subsequently been shown to be expressed in most non-neuronal tissues as well, even though at reduced levels [11]. Thus, its expression was found in glands, kidney, skin, heart, breast, uterus, pancreas, colon and prostate [11]. Similarly, NCS-1 expression has been found in neuroendocrine cells [12] as well as in several mammalian tissue culture cell lines, such as COS-7 [13], MDCK [14], HEK-293 [15], AtT-20 [16], 3T3L1 [17] and HL-60 cells [18], respectively. Together, these findings suggest that NCS-1 may perform more general, rather than solely neuron-specific functions.

The subcellular distribution of NCS-1 has been investigated by various techniques. Subcellular fractionation assays using rat brain material suggest that NCS-1 is broadly distributed, being partially cytosolic and partially associated with the membranes of the trans-Golgi network (TGN) and the endoplasmic reticulum [19]. Immunocytochemical and electron-microscopic histochemistry techniques also indicate that endogenous NCS-1 displays perinuclear staining and is associated with the TGN [8,13,19-21]. Various studies indicate the presence of NCS-1 in both pre- and postsynaptic compartments [8,21-24], and small amounts have also been detected on vesicular structures and at the plasma membrane [19,25]. Whilst initially suggested to be localized to small synaptic-like microvesicles in neuroendocrine PC12 cells [12,26,27], subsequent studies clearly established that only minute amounts of NCS-1 colocalize with small synaptic-like microvesicles [19]. In addition, NCS-1 does not co-localize with large dense-core vesicles in PC12 cells (Figure 2) or in chromaffin cells [20]. Thus, the co-localization of NCS-1 with synaptophysin in differentiated PC12 cells [26], but its virtual absence from highly-purified small synaptic vesicle preparations of rat brain [19] may indicate the presence of NCS-1 on immature transport organelles (containing synaptophysin), rather than on mature vesicles (see also below).

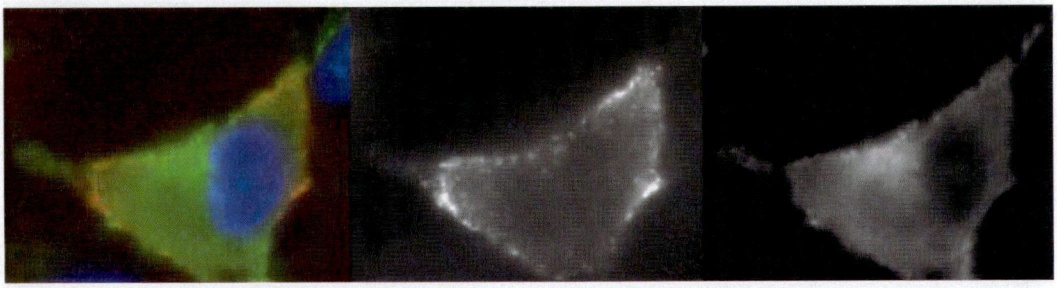

Figure 2. Localization of NCS-1 in transfected PC12 cells. Non-differentiated PC12 cells were co-transfected with constructs encoding for NCS-1 and human growth hormone (hGH), which is packaged into large dense-core vesicles. The localization of NCS-1 and hGH was assessed by double-immunocytochemistry using anti-NCS-1 chicken polyclonal antibodies at a dilution to preferentially visualize NCS-1-overexpressing cells, and anti-hGH rabbit polyclonal antibodies to visualize hGH [31]. Cells were fixed and analysed as described [31]. HGH (red) is mainly localized to large dense-core vesicles situated in the cell periphery, whilst overexpressed NCS-1 (green) is partially cytosolic and partially localized to the nuclear periphery, typical of TGN staining.

Finally, localization studies of NCS-1 performed upon overexpression of the protein have to be interpreted with care, as dependent on the expression levels, perinuclear staining is also increasingly accompanied with staining at and/or close to the plasma membrane [28-31]. This may indicate that, dependent on the level of overexpression, intracellular membrane binding sites become saturated, and that the protein starts to associate with other membranes in a non-specific manner. In sum, studies of endogenous NCS-1 localization suggest a broad distribution, with NCS-1 being present in cell bodies as well as pre- and postsynaptic compartments, and being partially cytosolic and partially associated with membranes of the TGN, and possibly with TGN-derived transport vesicles.

4. Myristoylation and Subcellular Targeting of NCS-1

Controlling the intracellular localization of NCS-1 is important for its ability to appropriately respond to differing Ca^{2+} signals, and thus for its function in regulating neuronal events. How is the specific membrane association of NCS-1 achieved? With the exception of KChIPs 2-4 [32,33], all NCS proteins are N-terminally myristoylated. Myristoyl tails are 14 carbon acyl chains added to the N-termini of proteins, which then act as membrane anchors [34]. This modification is determined by a vaguely-defined 17-residue motif at the extreme N-terminus of proteins [35]. Myristoylation can be required, but must not necessarily be sufficient for membrane anchoring [36], and other protein features, such as regions rich in basic residues, can mediate further membrane attraction [37]. Also, myristoylation does not always serve a constitutive membrane anchoring function. The fatty acid can switch between folding back to a domain of the acylated protein, and extending to the outside again, both controlled by the binding of Ca^{2+}. Such a Ca^{2+}/myristoyl switch mechanism for reversible membrane association has been described in detail for some NCS proteins, such as recoverin [34]. If Ca^{2+} is low, recoverin cradles its myristoyl group in a hydrophobic pocket created by the N-terminus of the protein. The binding of two Ca^{2+} ions by cytosolic recoverin leads to a large conformational change, which is followed by the extrusion of the myristoyl group previously buried within the hydrophobic pocket [38,39]. This results in translocation and myristoyl-dependent association of Ca^{2+}-bound recoverin to intracellular membranes [34,40]. Further, basic residues in the N-terminal domain electrostatically interact with the negatively charged phospholipid headgroups at the membrane surface, and such electrostatic interaction may contribute to the overall energetics of membrane binding [41]. Importantly, the Ca^{2+}-induced rotation of two domains of the protein exposes many hydrophobic residues that otherwise contact the myristoyl group in the Ca^{2+}-free state. The generation of such a hydrophobic crevice may generate a potential binding site in recoverin for its target protein, rhodopsin kinase.

Whilst hippocalcin [25,30], neurocalcin δ [42,43], VILIP-1 and VILIP-3 [44,45] all seem to possess a Ca^{2+}/myristoyl switch mechanism for reversible membrane association, such mechanism does not apply for NCS-1. In contrast, biochemical studies suggest that the myristoyl group of NCS-1 is exposed even in the absence of Ca^{2+} [46,47]. Further, analysis of full-length NCS-1 fused to GFP in live cells [25], or endogenous NCS-1 in fixed cells [8,21] indicates that NCS-1 is associated with TGN membranes even under basal or lowered Ca^{2+} concentrations, or if all Ca^{2+}-binding EF-hands are mutated. Intriguingly, intracellular increases in Ca^{2+} can evoke the translocation of cytosolic NCS-1 to membranes [19], suggesting that a (small) portion of cytosolic NCS-1 may be recruited to membranes via a mechanism different from the Ca^{2+}/myristoyl switch. However, in all cases, membrane association requires N-terminal myristoylation.

What are the determinants for the absence of a Ca^{2+}/myristoyl switch in certain NCS proteins, such as NCS-1, KChIP1, GCAP1 and GCAP2? Structural comparisons between NCS proteins have not provided immediate clues as to the basis of this difference (see also below), and for example the features within the sequence of recoverin that determine the Ca^{2+}/myristoyl switch mechanism, such as the inactivated first EF-hand and the hydrophobic residues that cradle the myristoyl group in the Ca^{2+}-free form, are highly conserved across all NCS proteins [48]. An elegant recent study has used sequence and structural comparison between NCS proteins in an attempt to identify candidate residues that determine the absence of a Ca^{2+}/myristoyl switch mechanism for NCS-1 [48]. These studies indicate that the N-terminus and first EF-hand domain determine the inability of NCS-1 to sequester the myristoyl group in the Ca^{2+}-free state. Specifically, two acid (E14 and E15) and two basic residues (R18 and K19) within a key motif (EELTRK) are part of an α-helix in the crystal structure of NCS-1 which joins the myristoyl tail to EF-hand 1. These residues undergo hydrogen-bond interactions which could stabilize the helix in a rigid conformation to keep the N-terminus of NCS-1 in an open conformation, with the myristoyl group exposed [48] (Figure 3). Indeed, mutagenesis studies indicate that these residues are sufficient to lock the myristoyl group in an open conformation. Interestingly, these residues are not conserved in NCS-1 from lower species, including *S. cerevisiae* (Figure 4), and the fission yeast homologue of NCS-1 does seem to undergo a Ca^{2+}/myristoyl switch mechanism [49], suggesting that the ancestral form of the protein may have displayed a switch mechanism, which may have been lost in NCS-1 from higher organisms.

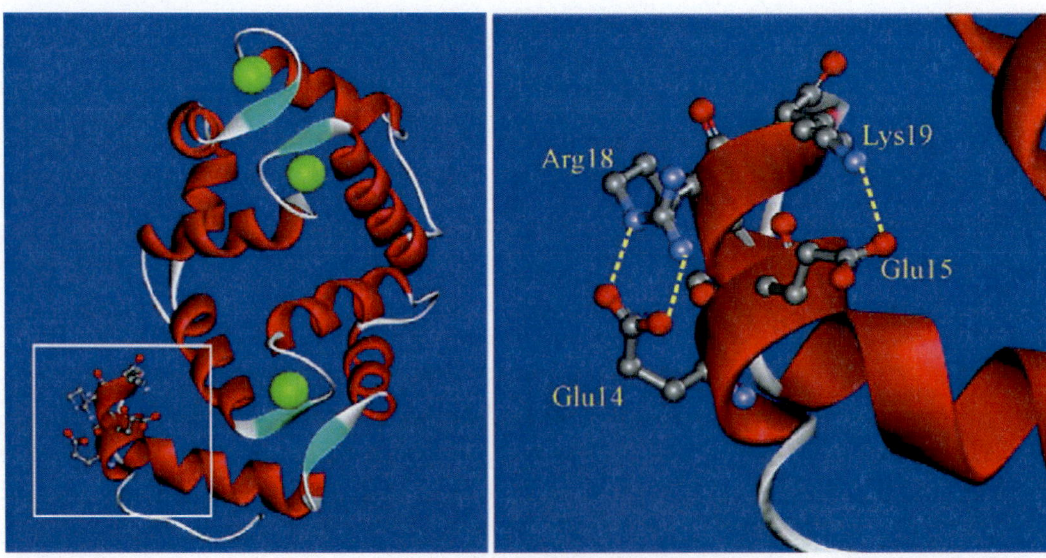

Figure 3. Structure of NCS-1 indicating residues that may determine the absence of a Ca^{2+}/myristoyl switch mechanism. *Left*, ribbon representation of the crystal structure of human NCS-1, with alpha helices in red, loop regions in light blue, and the three bound Ca^{2+} ions represented by green balls. Residues Glu14, Glu15, Arg18 and Lys19 are shown as ball-and-stick representation in the boxed region. *Right*, expansion of the boxed region showing the hydrogen-bonding interactions between the indicated residues. Reproduced from *J. Biol. Chem.* (2004) 279, 14347-14354 by copyright permission of the Journal of Biological Chemistry.

Since the myristoyl anchor is not sufficient to confer specificity of the association of myristoylated proteins with distinct intracellular membranes, additional targeting features have to exist. To determine such features, a minimal myristoylation motif derived from hippocalcin (amino acid residues 1-14) was fused to GFP. This protein construct was targeted

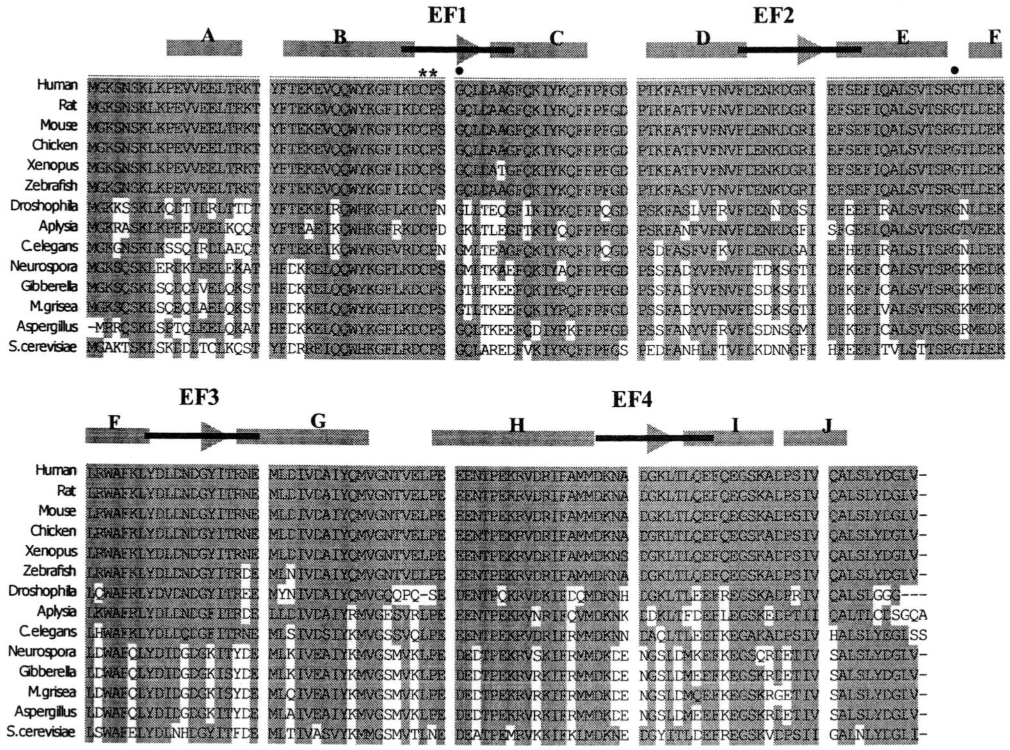

Figure 4. Alignment of NCS-1 proteins from different species. Amino acid sequence alignment of NCS-1 proteins illustrates high overall sequence homology across species. Amino acids highlighted in blue are identical, whilst amino acids highlighted in grey are similar to those of human NCS-1, respectively. The secondary structural elements (alpha helices A to J, with helix J corresponding to the C-terminal helix mentioned in the text, and β-sheets indicated by arrows) are shown in gray, with the four EF hands labeled EF1 to EF4. The consensus sequence for the 29-residue EF-hand motif is indicated by black lines. The conserved cysteine-proline substitution in EF-1 is indicated by asterisks, and the two rotational glycine residues by black dots, respectively. The database accession numbers for the sequences are as follows: human (P62166), rat (P62168), mouse (Q8BNY6), chicken (P62167), Xenopus (Q6GQJ0), Zebrafish (Q53A16), Drosophila (P37236), Aplysia (Q16981), C. elegans (P36608), Neurospora (EAA28220), Gibberella (EAA77148), M. grisea (Q52FT6), Aspergillus (Q4WLA4) and S. cerevisiae (Q06389).

to the TGN and to the plasma membrane when overexpressed in cells, suggesting that the N-terminal residues contain all the intrinsic targeting information necessary for proper intracellular localization [30]. Further, basic residues at positions 3, 7 and 9 were found to be essential for the efficient targeting of this motif to the TGN and the plasma membrane, and mutation of all three amino acids to non-charged residues led to the association of the motif with distinct intracellular organelles [30]. Unfortunately, this approach was not performed with the N-terminal residues of NCS-1, and indeed basic residues at positions 3, 7 and 9 are not fully conserved across NCS-1 from different species (Figure 4). In addition, other NCS proteins that are targeted to the TGN and the plasma membrane, such as VILIP-1, do not contain basic residues at all three positions (see Figure 1). Finally, VILIP-3, which does possess the three basic residues predicted to be important for proper targeting, displays calcium-independent Golgi localization, and shows a slight membrane association even in its non-myristoylated form [45]. Together, these data suggest that whilst residues at the extreme N-terminus clearly are important for proper intracellular localization of NCS proteins, additional protein-dependent localization mechanisms, or lipid preferences together with differences in the lipid composition of individual membranes [50], may aid in the distinct localization of various NCS proteins, and thus their ability to sense different Ca^{2+} signals.

In general, all members of the NCS family display high-affinity Ca^{2+} binding and cooperativity of Ca^{2+} binding *in vitro*. For example, Ca^{2+} binding to NCS-1 is half-maximal below 1 μM free Ca^{2+} and co-operative, with a Hill coefficient of approximately 2 [51]. Thus, NCS-1, as well as other NCS proteins, are able to transduce very small changes in intracellular Ca^{2+} concentrations in a range that would activate only a small proportion of calmodulin, and a fraction of the NCS proteins are likely active under basal conditions. In such a manner, the presence or absence of a Ca^{2+}/myristoyl switch mechanism, together with the targeting of different NCS proteins to distinct intracellular domains, allows for spatial and temporal control of Ca^{2+} sensing. For example, NCS proteins such as hippocalcin, neurocalcin δ and VILIP-1, which possess a Ca^{2+}/myristoyl switch mechanism, will require slow, prolonged and global Ca^{2+} elevation to translocate from the cytosol to their respective intracellular membrane compartments. In contrast, membrane-associated NCS-1 will be able to respond to brief, rapid and local Ca^{2+} transients mainly close to the TGN.

5. Structural Features of NCS-1: The Hydrophobic Crevice

The structures of several NCS proteins, including unmyristoylated, Ca^{2+}-bound human and yeast NCS-1 [13,46], unmyristoylated Ca^{2+}-bound GCAP-2 [52], unmyristoylated Ca^{2+}-bound neurocalcin δ [53] and unmyristoylated Ca^{2+}-bound KChIP1 complexed with the N-terminus of distinct K^+ channels [54,55], have been solved by different techniques such as X-ray crystallography or NMR. Recoverin has been most extensively characterized by such means, with structures of the protein available in its myristoylated, Ca^{2+}-free form [56,57], its myristoylated, Ca^{2+}-bound form [39] and in intermediate forms [58,59] thought to reflect a transient structure following the binding of the first Ca^{2+} ion. Such studies have shown that Ca^{2+}-free recoverin exists in a compact structure, with the myristoyl group buried within a hydrophobic pocket formed by residues from EF-1 and additional hydrophobic residues contributed by other helices [56]. In addition, recoverin was found to possess two distinct domains, with EF-1 and EF-2 intimately interacting to form the N-terminal domain, EF-3 and EF-4 forming the C-terminal domain, and a U-shaped linker between these domains positioning the four EF-hands in a compact tandem array. Importantly, comparison of the Ca^{2+}-free with the Ca^{2+}-bound forms of myristoylated recoverin demonstrated extensive conformational changes upon Ca^{2+} binding as a result of rotation of the backbone at two glycine residues [39]. On the one hand, rotation at Gly 42, located in the loop between the helices of EF-1, was found to expose the myristoyl group towards the aequeous solution and to expose a hydrophobic surface potentially able to interact with target proteins. On the other hand, rotation at Gly 96, in the U-shaped linker between the two domains, further leads to differences in helix interactions between EF-2 and EF-3 which account for the cooperativity of Ca^{2+} binding, whilst minor structural changes occur in the very C-terminal domain [39].

These structural studies of recoverin have yielded great insights into the molecular mechanics of the Ca^{2+}/myristoyl switch mechanism, and superposition of the main chain structures have revealed large overall similarities in the structural fold of the different NCS proteins, including NCS-1 [13,52,53]. However, many of the hydrophobic residues that clamp the myristoyl group in the Ca^{2+}-free state of recoverin are conserved between all NCS family members, and thus cannot be used to predict the presence or absence of a Ca^{2+}/myristoyl switch mechanism. Similarly, upon Ca^{2+} binding, the structures of NCS proteins all reveal a hydrophobic surface, formed by highly conserved residues from the N-terminal as well as C-terminal domains (Figure 5). This solvent-exposed hydrophobic crevice has been suggested to serve as a possible binding site for downstream protein targets. Indeed, mutagenesis studies for some NCS proteins have revealed that distinct residues within the crevice are important for downstream target regulation [60-62]. However, the high sequence conservation of these residues amongst members of the NCS family indicates that whilst they may contribute to target interactions, they are not determinants of the specificity of such interactions, and alternative roles for the crevice, for example in mediating dimerization of NCS proteins, have been proposed as well [52,53,57,63].

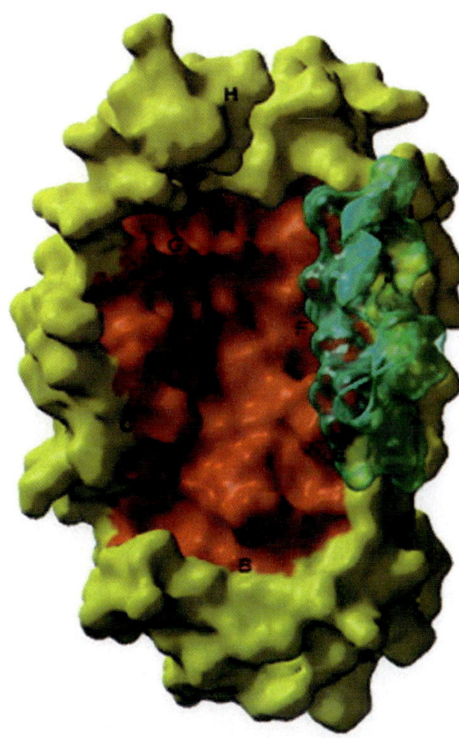

Figure 5. Molecular surface structure of human NCS-1. Molecular surface structure of human NCS-1, viewed down the large hydrophobic crevice (red). The secondary structural elements are labeled (A-J), and the C-terminal helix J is displayed in cyan. Reproduced from *J. Biol. Chem.* (2001) 276, 11949-11955 by copyright permission of the Journal of Biological Chemistry.

There also exist profound differences in the structural details of NCS proteins which may underlie their distinct functional roles. For example, the very N-terminus shows virtually no sequence similarity between NCS-1 and recoverin, neurocalcin δ, the GCAPs or the KChIPs (see Figure 1), and structural studies reveal the existence of an N-terminal helix of variable length and orientation in the different NCS proteins [52,56]. Importantly, the extreme C-terminus, also relatively variable in primary sequence between different NCS proteins, is comprised of an additional C-terminal helix. This helix interacts with helices of EF-3 and EF-4 in recoverin and GCAP-2 in a manner similar to the interaction of calmodulin with its helical target peptides [64], and thus has been proposed to enhance the specificity of NCS proteins by blocking their adventitious binding to calmodulin targets. Interestingly, the C-terminal helix in KChIP1 has been proposed to interact with the N-terminus of the K^+ channel Kv4.2, suggesting that this helix may directly be involved in target interactions [55]. Similarly, the extreme C-terminus of NCS-1 has been shown to be essential for its interaction with one specific downstream target [65], and introduction of a peptide derived from the very C-terminus of NCS-1 into presynaptic terminals can abolish a certain type of facilitation [66]. Therefore, the relative structural positioning ot the C-terminal helix may affect the shape of the hydrophobic crevice and thus aid in determining the specificity by which different NCS proteins, including NCS-1, associate with their target proteins (Figure 5).

Structural studies on NCS-1 have determined additional features unique to this protein as well. For example, the C-terminal helix in human, non-myristoylated, Ca^{2+}-bound NCS-1 is positioned in such a manner that it exposes a much larger hydrophobic patch than the one for example seen in recoverin or neurocalcin δ [13]. However, the hydrophobic crevice of the yeast counterpart of NCS-1 is less than half the size of that of human NCS-1 [46], mainly because of the relative position of the C-terminal helix with respect to the rest of the protein structure. The functional significance of this (especially with respect to differences in the function of yeast versus mammalian NCS-1) remains unknown. In addition, the myristoyl group of NCS-1 is solvent-exposed in the absence of Ca^{2+}, in agreement with the observation that myristoylated NCS-1 is membrane-associated even in the absence of Ca^{2+}. Instead, the myristoyl group of NCS-1 seems to be involved in conferring some degree of cooperativity in Ca^{2+} binding [46,67]. Thus, Ca^{2+}-induced structural changes in NCS-1 are predicted to increase its propensity to interact with membranes by means independent of the N-myristoyl group.

In conclusion, the most prominent structural effect produced by Ca^{2+} binding to different NCS proteins seems to involve 'flipping out' of the myristoyl group (for NCS proteins which contain a Ca^{2+}/myristoyl switch mechanism), and in all cases exposure of a hydrophobic crevice in the protein. This crevice, in conjunction with the relative orientation of the C-terminal helix, may confer specificity to accommodate distinct downstream protein targets. However, additional structural studies with myristoylated NCS proteins will be necessary to more rigorously determine the structural impact of the myristoyl group on the proteins in the presence and absence of Ca^{2+}. Ultimately, the structures of individual NCS proteins bound to their respective target proteins will need to be determined to fully understand the selectivity and mechanism of action of distinct NCS proteins.

6. Physiological Importance of NCS-1: A Modulator of Synaptic Efficacy

While structure-function, localization and protein interaction studies may give a confusing picture as to the exact role(s) of the distinct NCS proteins in the nervous system, genetic manipulations of individual NCS proteins have clearly demonstrated their physiological importance in regulating distinct neuronal functions. NCS-1 was originally discovered in *Drosophila*, and its overexpression was shown to cause a frequency-dependent facilitation of neurotransmitter release at the neuromuscular junction [6]. Subsequently, overexpression of NCS-1 in *Xenopus* was found to enhance both spontaneous and evoked transmission at the neuromuscular junction [68]. In addition, direct loading of NCS-1 into the presynaptic nerve terminal at the calyx of Held synapse mimicked activity-dependent facilitation of P/Q-type Ca^{2+} currents during repetitive stimulation, whilst a C-terminal peptide derived from NCS-1 abolished such facilitation [66]. The latter result suggests that the C-terminal fragment of NCS-1 may interfere with a direct or indirect interaction of NCS-1 with P/Q-type Ca^{2+} channels, even though such an interaction has yet to be described.

In general, the above-mentioned facilitating effects of NCS-1 may have been indirect, caused for example by effects on channels or Ca^{2+} regulatory processes. However, a recent study at hippocampal neurons in culture has clearly shown that overexpression of NCS-1 in glutamatergic synapses more directly and specifically facilitates evoked transmission to paired pulses, thereby increasing the reliability of postsynaptic activation [69]. Calcium currents were unaffected in these cells, and there was no effect of NCS-1 on basal transmission. In addition, down-regulation of native NCS-1 (but presumably a series of additional proteins as well) was accompanied by a reduction in facilitation, suggesting that normal facilitation may also be produced by the action of native, endogenous NCS-1 [69]. These data suggest that NCS-1 may act as a Ca^{2+}-binding target modulating secretion to mediate facilitation at some synapses, and in addition basal release at other synapses, by regulating the underlying molecular process(es) in different ways [70]. Finally, a role for NCS-1 in associative learning and memory has been suggested by disruption of its expression in *C. elegans* [71], and the description of NCS-1 loss-of-function genetics in mice is greatly awaited. Together, the currently available studies clearly establish an important role for NCS-1 in enhancing synaptic transmission in different model systems. Thus, elucidating the mechanism(s) by which NCS-1 performs its potentiating role(s) has since become a major focus of investigation.

7. Mechanism of NCS-1 Action

How does NCS-1 work to facilitate transmitter release? Conceptually, NCS-1 may enhance transmitter release in a variety of ways [70,72], such as by increasing the number of functional synaptic contacts. Alternatively, NCS-1 may increase the number of vesicles per synaptic contact. This may mobilize additional vesicles to docking sites to increase the size of the readily releasable vesicle pool. NCS-1 may act to 'prime' docked vesicles, making them more available for release by increasing the probability of release. Such an effect may be mediated by an NCS-1-dependent change in the lipid or protein composition of the vesicle and/or plasma membrane, thereby affecting the efficiency with which a vesicle fuses upon Ca^{2+} influx. Finally, NCS-1 may directly affect the fusion machinery to increase the sensitivity of secretion to a local increase in Ca^{2+} elevation as intracellular residual Ca^{2+} builds up, or may enhance secretion indirectly, for example by increasing the amount and/or activity of presynaptic channels and/or postsynaptic receptors.

Evidence supporting several of these possible mechanistic explanations for the NCS-1-mediated enhancement of transmission have been accumulating. For example, overexpression of NCS-1 in *Xenopus* neuromuscular synapses in culture has been found to increase the number and size of synaptic contacts [73]. Similarly, overexpression of NCS-1 in the NG108-myocyte co-culture model led to enhanced synapse formation, probably due to an increase in the number of cholinergic synaptic vesicles [28]. In an anterior pituitary cell line, NCS-1 overexpression was found to increase the storage pool of adrenocorticotrophin, indicating that NCS-1 may act to increase the number of regulated secretory vesicles in this cell line [16]. Overexpression of NCS-1 in PC12 or chromaffin cells was found to enhance release [12,20,27,31], either by increasing the number of vesicles available for release or by increasing the fusion-competence of individual vesicles. Indeed, a recent study of insulin secretion by pancreatic β cells suggested that NCS-1 increases exocytosis by promoting the priming of secretory granules for release and increasing the number of granules in the readily releasable pool [74]. Finally, a series of studies suggested that NCS-1 can increase the surface expression of presynaptic Ca^{2+} channels and regulate postsynaptic receptor trafficking (see below). So how does NCS-1 work? Whilst only careful *in vivo* experiments (together with quantitative electron microscopy to evaluate changes in the number of vesicles) will be able to tease apart some of these possibilities, additional clues as to the mechanism of NCS-1 action have come from studying NCS-1-protein interactions.

8. NCS-1 and its Interacting Partners: Multifunctional or Promiscous?

The high sequence similarity amongst different NCS proteins has not only hampered the determination of their exact cellular and subcellular localizations, but also the determination of their interactions with specific downstream targets. For example, even though only 21% homologous to calmodulin, NCS-1 can substitute for calmodulin in activating several Ca^{2+}/calmodulin-dependent enzymes, such as the protein phosphatase calcineurin or cyclic nucleotide phosphodiesterase, both *in vitro* and *in vivo* [75]. Similarly, recoverin, but also several other NCS family members including NCS-1, can inhibit the phosphorylation of rhodopsin by G protein-coupled receptor kinase 1 (GRK1; also known as rhodopsin kinase) [76-79]. In addition, several NCS proteins including NCS-1 can attenuate the GRK2-mediated phosphorylation of distinct receptors [80,81] (see also below). These data support the hypothesis that NCS proteins may be direct regulators of G protein-coupled receptor (GPCR) signaling [82], by interacting with different subtypes of GRKs to regulate the desensitization of different receptors. In this manner, NCS proteins may exert calcium sensitivity on the signaling properties of GPCRs. However, this would not ensure specificity, and indeed the ubiquitous calcium sensor calmodulin has been shown to regulate the activity of GRKs as well [83,84]. It seems possible that the apparent promiscuity of some of these described interactions *in vitro* is due to the high sequence similarity amongst these EF-hand-containing proteins, and is not necessarily reflecting a multifunctional nature of individual NCS proteins in regulating multiple downstream events. Specificity *in vivo* may then be achieved by the detailed subcellular localization of individual NCS proteins and individual GRKs. A detailed analysis of identified NCS-target protein interactions, using a more complete array of different NCS proteins, together with detailed subcellular localization studies are necessary to clearly establish which NCS-target protein interactions are likely to be relevant *in vivo*.

Finally, other described NCS-1-protein interactions, such as with distinct K^+ channels [15,85], could similarly be due to the high sequence similarity between NCS-1 and KChIPs, rather than representing a specific, high-affinity interaction of mechanistic importance to explain the facilitating effects of NCS-1 in synaptic transmission. In addition to those currently described, additional specific binding partners for NCS-1 likely exist [47], although their exact nature remains to be determined. Below we discuss the relevance of two well-established NCS-1 interactions in more detail.

9. The NCS-1-Disease Connection

Abnormal activity of the dopamine (DA) system has been implicated in several psychiatric and neurological illnesses, such as schizophrenia, bipolar disorder, or attention deficit hyperactivity disorder [86], even though the precise sites of dopamine dysfunction in these disorders are currently unknown. In the CNS, DA modulates neuronal excitability by regulating ligand- and voltage-gated ion channels. The actions of DA are mediated by a family of seven-transmembrane G-protein-coupled receptors (GPCRs), D1-D5 receptors, which display unique properties with respect to their effects on downstream signaling cascades. Receptor desensitization, characterized by a decline in receptor responsiveness to agonist, represents a critical adaptation mechanism that protects against receptor overstimulation. The desensitization of activated GPCRs is mediated by the phosphorylation of residues within the intracellular domains of receptors and is mediated both by second messenger-dependent kinases and by GRKs, followed by receptor internalization. In addition, the intracellular activity of the individual DA receptor subtypes are regulated by the actions of a cohort of cytoskeletal, adaptor and signaling proteins called dopamine receptor-interacting proteins (DRIPs). DRIPs seem to regulate most stages of the life-cycle of a DA receptor, including biosynthesis, trafficking to the plasma membrane, and internalization [86]. The recent discovery that the levels of NCS-1 are upregulated in patients with schizophrenia or bipolar disorders [87], and that NCS-1 expression shows cell-specific changes in brains of schizophrenics [88], together with the identification of NCS-1 as a DRIP for the D2 receptor [80] suggests the possibility that defects in Ca^{2+} homeostasis (mediated by NCS-1) might contribute to abnormalities in the brain DA system in several neurological diseases [89].

NCS-1 was identified as a DRIP for the D2 receptor in a screen of a brain cDNA library using the very C-terminal tail of the D2 receptor. The D2 receptor-interacting region in NCS-1 was mapped to the N-terminal 71 residues, and the interaction was shown to occur in a Ca^{2+}-independent manner [80]. When expressed in mammalian cells, NCS-1 attenuated the DA-induced D2 receptor internalization by a mechanism which involved a reduction in D2 receptor phosphorylation. NCS-1 exerted its effect on D2 receptor signaling through an interaction with GRK2, a kinase also associated with the D2 receptor. The effect of NCS-1 in inhibiting GRK2-dependent desensitization of D2 receptors was Ca^{2+}-dependent [80]. The difference in the Ca^{2+} dependency between the NCS-1/D2 receptor interaction and the NCS-1/GRK2 interaction suggests the presence of two distinct complexes, and the possibility that NCS-1 and GRK2 may compete for binding to the D2 receptor. In this scenario, elevated levels of NCS-1, combined with abnormalities in Ca^{2+} homeostasis, may potentiate signaling through the D2 receptor by preventing GRK2-mediated D2 receptor internalization (Figure 6). However, the reported (albeit weak) inhibition of GRK2-mediated M2 muscarinic receptor phosphorylation by VILIP-1 and neurocalcin δ [81] leaves open the possibility that the observed effects on D2 receptor internalization are mediated by an NCS protein distinct from NCS-1, and further experiments will be necessary to determine the exact nature of the NCS protein involved.

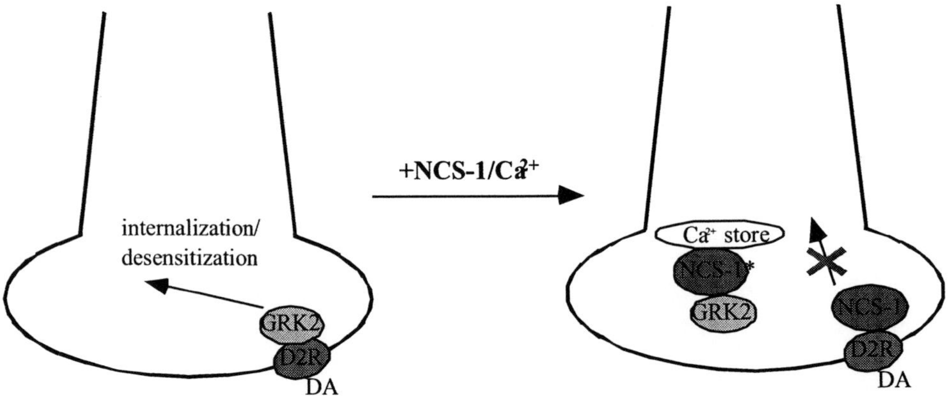

Figure 6. Model for the role of NCS-1 in D2 receptor signaling. The interaction between the dopamine (DA) D2 receptor (D2R) and GRK2 results in desensitization of the receptor via internalization, either pre- and/or postsynaptically (left). NCS-1 interacts with the D2 receptor in a Ca^{2+}-independent manner, presumably at the plasma membrane. In contrast, the interaction between NCS-1 and GRK2 seems to be Ca^{2+}-dependent (indicated by NCS-1*) [80]. Thus, Ca^{2+}-bound NCS-1 may recruit GRK2, thereby inhibiting GRK2 to interact with the D2 receptor (right). Such NCS-1/GRK2 interaction may occur on intracellular organelles and may be promoted by Ca^{2+} release from intracellular stores (e.g. Golgi) or by Ca^{2+} entry across the plasma membrane. In either case, abnormal Ca^{2+} homeostasis and/or an increase in the levels of NCS-1 would result in the potentiation of signaling through the D2 receptor.

In either case, how may an increase in NCS-1 levels in the dorsolateral prefrontal cortex (DLPFC) relate to schizophrenia? Dysfunction in the DLPFC is largely responsible for the debilitating deficits in working memory-dependent executive functions observed in schizophrenic patients [86]. NCS-1 inhibits the desensitization of the D2 receptor, and thus enhances the influence of D2 receptors on cell activity [80]. This is expected to result in a depressive effect on excitatory transmission and thus a decreased activation of this cortical area, an outcome which would explain the efficacy of D2 receptor antagonists in treating symptoms of schizophrenia [86]. Additional experiments will be needed to establish a conclusive link between NCS-1, D2 receptors and neurological illnesses such as schizophrenia or bipolar disorder.

NCS-1 has also recently been described as an interacting protein of interleukin-1 receptor accessory protein-like protein (IL1RAPL) [65]. Mutations in IL1RAPL are responsible for a form of X-linked mental retardation [90]. Yeast two-hybrid analysis suggested that the 16 C-terminal residues of NCS-1 (but not of hippocalcin or calmodulin) interact with ILRAPL, and *in vivo* experiments further established that this interaction occurs in a Ca^{2+}-independent manner [65]. Finally, overexpression of IL1RAPL was found to inhibit secretion from neuroendocrine cells, suggesting a role for this protein in regulating neurotransmitter release. The implication of a gene product involved in mental retardation, in addition to its role in the regulation of synaptic transmission is similar to that found for αGDI (GDP dissociation inhibitor) [91,92]. This protein is a regulator of Rab3 activities, proteins that participate in synaptic vesicle fusion [93]. Further studies will be necessary to determine whether genes such as IL1RAPL that are involved in mental retardation act by regulating novel aspects of synaptic activity (e.g. mediated by NCS-1), the dysfunction of which may result in an impairment of the development of cognitive functions.

10. The NCS-1-Trafficking Connection

Important insight into the possible mechanism of action of NCS-1 has come from studies performed in yeast. Frequenin, the orthologue of NCS-1 and the only NCS family member identified in *S. cerevisiae,* was found to interact with PIK1, one of the two non-redundant phosphatidylinositol 4-kinases (PI4Ks) in yeast [94]. Binding was found to be of high-affinity, and purification of the frequenin-PIK1 complex revealed a 1:1 stoichiometry [95]. Mapping experiments suggested that frequenin binds to a hydrophobic domain within PIK1, but unfortunately the corresponding interacting domain within frequenin was not determined in these studies [95,96]. The binding of frequenin to PIK1, while Ca^{2+}-independent, enhanced the activity of PIK1 by 3-fold [94], indicating that fequenin may act as a regulator of PIK1 activity. Importantly, overexpression of PIK1 was found to suppress the lethality of frequenin-mutant cells, indicating that PIK1 is the sole essential target for frequenin in yeast cells [94]. PIK1 is localized to the cytoplasm and to *trans*-Golgi membranes, and a mutation in PIK1 causes an accumulation of Golgi structures and a concomitant defect in Golgi-to-cell-surface and Golgi-to-vacuole trafficking [97,98]. Thus, at least in yeast, frequenin acts to modulate the activity of PIK1 and thereby regulates intracellular membrane trafficking events.

Evidence for a similar role of NCS-1 in mammalian cells has been accumulating as well. For example, expression of human NCS-1 in *S. cerevisiae* (but not of recoverin or KChIP2) can rescue the inviability of frequenin-mutant cells, suggesting that the function of NCS-1 in mammals may closely resemble that of frequenin in yeast [96]. Furthermore, endogenous mammalian NCS-1 seems to interact with PI4Kβ (the PIK1 homologue) in neuronal and non-neuronal cells [14,19,20,29], stimulate the activity of PI4Kβ *in vitro* [29,99], and colocalize with PI4Kβ on the TGN [13].

PI4Ks are responsible for the synthesis of phosphatidylinositol 4-phosphate (PtdIns(4)P), the precursor of phosphatidylinositol (4,5)-bisphosphate (PtdIns(4,5)P_2). On the one hand, the ability of NCS-1 to enhance agonist-evoked secretion could be attributed to its activation of PI4Kβ, leading to the upregulation of Ins(1,4,5)P_3 receptor signalling and Ca^{2+} mobilization [31,100]. On the other hand, both PtdIns(4)P and PtdIns(4,5)P_2 are known to play important roles in several membrane trafficking steps, including vesicle budding from the TGN, exocytosis and endocytosis. In sum, the data suggest that PI4Kβ constitutes an *in vivo* downstream target of NCS-1, and indicate that the NCS-1-mediated modulation of phosphoinositide levels may regulate intracellular membrane trafficking events. If indeed the case, alterations in the number and/or fusion ability of vesicles, mediated by changes in membrane phosphoinositide composition, may underlie the observed facilitating effects of NCS-1 on synaptic transmission *in vivo*.

11. Phosphoinositides and Membrane Trafficking

Phosphoinositides have emerged as important regulators of a variety of membrane trafficking processes. For example, *de novo* synthesis of PtdIns(4,5)P_2 has been shown to be important for an ATP-dependent priming step preceding the Ca^{+}-dependent fusion of regulated secretory vesicles. PtdIns(4,5)P_2 also plays a positive role in regulating clathrin-mediated endocytosis from the plasma membrane and in regulating an actin-based vesicle motility mechanism [101]. In addition, PtdIns(4)P and PtdIns(4,5)P_2 are necessary for the exit of transport carriers from the TGN.

The TGN is a major sorting station within the cell, in which distinct vesicle populations, destined for various locations, are formed. A key aspect of this process is the formation of protein complexes that recognize sorting signals on the cytosolic portion of cargo proteins and make a coat that is necessary to collect cargo and form transport intermediates [102]. Sorting signals are recognized only on certain membrane surfaces, and the mechanisms that determine where cytosolic coats form on membranes are poorly understood. However, recent data strenghten the hypothesis that phosphoinositides [103] determine the membrane locations where coat proteins may assemble [104]. In this scenario, dual interactions of distinct coat proteins with specific phosphoinositides as well as with additional protein binding sites (such as small GTPases) control their efficient and specific recruitment to membranes [105].

In mammals, the Golgi complex contains three PI4Ks, with PI4Kβ being recruited to the Golgi complex in response to ADP-ribosylation factor (ARF; a small GTPase) activation [106]. Recent studies have identified two effectors of PtdIns(4)P, FAPP1 and FAPP2 (four-phosphate-adaptor proteins 1 and 2), which interact with both PtdIns(4)P and ARF and which control the generation of constitutive post-Golgi vesicle carriers [107]. Thus, an ARF-dependent recruitment of PI4Kβ, followed by the production of PtdIns(4)P and the consecutive recruitment of FAPPs may generate distinct TGN membrane domains from which constitutive TGN-derived carriers emerge [104,105,108,109].

How does NCS-1 fit into this picture? A recent study suggests that NCS-1 is able to recruit ARF1 to the TGN in a Ca^{2+}-dependent manner [110]. Whilst both NCS-1 and ARF1 can activate PI4Kβ in isolation, an antagonistic mode of action was uncovered when both proteins were present in combination with PI4Kβ. These data suggest that activation of PI4Kβ can proceed in the presence of ARF1 or NCS-1 alone, but not if both effectors are present simultaneously, and further indicate that ARF1-NCS-1 may form a mutually exclusive complex in preference to binding to PI4Kβ. Thus, the inhibitory interaction between the ARF1 and NCS-1 pathways may function to prevent the simultaneous activation of PI4Kβ by both effectors, and thereby may assist in the generation of segregated trafficking domains at the TGN [110]. However, clearly additional work is necessary to corroborate these findings.

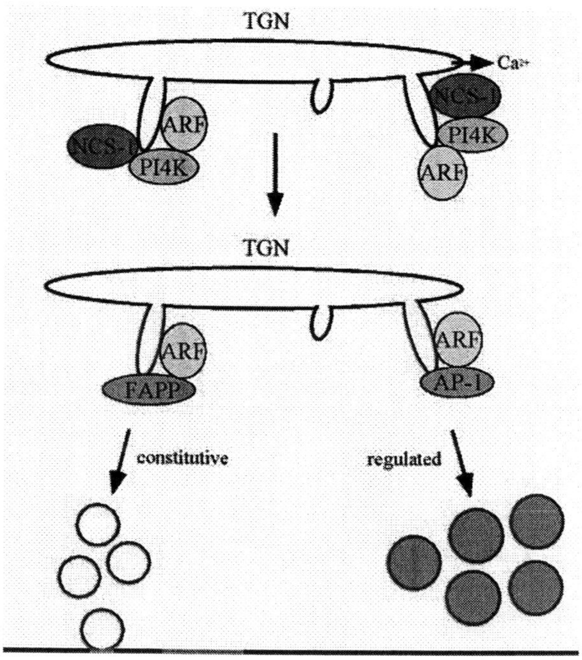

Figure 7. Model for the role of NCS-1 in vesicle trafficking out of the TGN. NCS-1 (with or without ARF), possibly upon Ca^{2+} release from the TGN, can recruit PI4Kβ (PI4K) to the TGN. This results in the production of PtdIns(4)P (red lines) at defined domains of the TGN, followed by the ARF-mediated and PtdIns(4)P-dependent recruitment of distinct coat proteins (FAPPs for the generation of constitutive post-Golgi vesicle carriers, and AP-1 for the generation of regulated vesicle carriers, respectively). An increase in constitutive vesicle carriers is followed by a concomitant increase in the levels of PtdIns(4,5)P_2 at the plasma membrane (blue lines), which seems to affect the size of a readily releasable vesicle pool. On the other hand, an increase in regulated secretory vesicle carriers (e.g. large dense-core vesicles) may be reflected in a concomitant increase in the actual number of vesicles available for release.

An alternative mechanism of NCS-1 function may involve the possibility that NCS-1 (with or without ARF1) helps in the recruitment and/or maintenance of PI4Kβ at the Golgi complex (Figure 7). This will further enhance the production of PtdIns(4)P and thus the recruitment of FAPP coat proteins, leading to an increase in the generation of constitutive post-Golgi transport vesicles. The fusion of these constitutive vesicles containing elevated levels of PtdIns(4)P with the plasma membrane, followed by conversion to PtdIns(4,5)P_2, would lead to an increase in the levels of PtdIns(4,5)P_2 at the plasma membrane. Indeed, plasmalemmal PtdIns(4,5)P_2 levels have been shown to increase the size of the readily releasable vesicle pool [111], an effect which is similar to that observed when NCS-1 is overexpressed in certain cell systems. On the other hand, the NCS-1-mediated enhanced production of PtdIns(4)P at the Golgi may result in the increased generation of transport carriers destined for the regulated secretory pathway [72], which could be more directly reflected in the number of secretory vesicles available for release and/or their ability to fuse with the plasma membrane [112], again an effect observed or proposed for the action of NCS-1 in enhancing neurotransmission.

Whatever the exact mechanism, an NCS-1-mediated effect on trafficking out of the Golgi complex adds an additional layer of Ca^{2+}-dependent regulation. Indeed, the Golgi can act as a Ca^{2+} store [113], and local Ca^{2+} release has been shown to be important for traffic from a pre-Golgi compartment to the Golgi complex [114] and within the Golgi complex [115]. Thus, NCS-1 may act as a Ca^{2+} sensor with a specific role in post-Golgi membrane traffic.

Conclusion

While far from being understood at a molecular level, current data suggest that NCS-1 may potentiate secretion from a variety of cell types by enhancing the trafficking of TGN-derived transport carriers in a PI4Kβ-specific manner. In addition, the established role of NCS-1 in D2 receptor desensitization, together with the finding that it can interact with various receptors and/or channels either directly and/or indirectly, suggest additional roles for NCS-1 in regulating a variety of neuronal functions. Whether the latter effects are indirect and due to alterations in membrane trafficking, or whether NCS-1 stays bound to carrier vesicles and performs additional roles by stabilizing and/or regulating the activity of a defined subset of proteins on these vesicles remains to be determined. In either case, NCS-1 is an important molecule which may constitute a major regulatory site for modification of synaptic properties and comprise an important target for understanding and treating neurological illnesses such as schizophrenia or bipolar disorder.

Acknowledgements

Work in our laboratory is supported by a grant from the Spanish Ministry of Education and Science. To whom correspondence should be addressed: sabine.hilfiker@ipb.csic.es.

References

[1] Augustine, G.J., Santamaria, F. and Tanaka, K. 2003, *Neuron, 40*, 331.

[2] Braunewell, K.-H. and Gundelfinger, E.D. 1999, *Cell Tissue Res., 295*, 1.
[3] Burgoyne, R.D. and Weiss, J.L. 2001, *Biochem. J., 353*, 1.
[4] Burgoyne, R.D. 2004, *Biochim. Biophys. Acta, 1742*, 59.
[5] Burgoyne, R.D., O'Callaghan, D.W., Hasdemir, B., Haynes, L.P. and Tepikin, A.V. 2004, *Trends Neurosci., 27*, 203.
[6] Pongs, O., Lindemeier, J., Zhu, X.R., Theil, T., Engelkamp, D., Krah-Jentgens, I., Lambrecht, H.-G., Koch, K.W., Schwemer, J., Rivosecchi, R., Mallart, A., Galceran, J., Canal, I., Barbas, J.A. and Ferrús, A. 1993, *Neuron, 11*, 15.
[7] Moncreif, N.D., Kretsinger, R. H. and Goodman, M. 1990, *J. Mol. Evol., 30*, 522.
[8] Martone, M.E., Edelmann, V.M., Ellisman, M.H. and Nef, P. 1999, *Cell Tissue Res., 295*, 395.
[9] Paterlini, M., Revilla, V., Grant, A.L. and Wisden, W. 2000, *Neuroscience, 99*, 205.
[10] Olafsson, P., Soares, H.D., Herzog, K.-H., Wang, T., Morgan, J.I. and Lu, B. 1997, *Mol. Brain Res., 44*, 73.
[11] Gierke, P., Zhao, C., Brackmann, M., Linke, B., Heinemann, U. and Braunewell, K.-H. 2004, *Biochem. Biophys. Res. Commun., 323*, 38.
[12] McFerran, B.W., Graham, M.E. and Burgoyne, R.D. 1998, *J. Biol. Chem., 273*, 22768.
[13] Bourne, Y., Dannenberg, J., Pollmann, V., Marchot, P. and Pongs, O. 2001, *J. Biol. Chem., 276*, 11949.
[14] Weisz, O.A., Gibson, G.A., Leung, S-M., Roder, J. and Jeromin, A. 2000, *J. Biol. Chem., 275*, 24341.
[15] Guo, W., Malin, S.A., Johns, D.C., Jeromin, A. and Nerbonne, J.M. 2002, *J. Biol. Chem., 277*, 26436.
[16] Guild, S.B., Murray, A.T., Wilson, M.L., Wiegand, U.K., Apps, D.K., Jin, Y., Rindler, M., Roder, J. and Jeromin, A. 2001, *Mol. Cell. Endocrinol., 184*, 51.
[17] Mora, S., Durham, P.L., Smith, J.R., Russo, A.F., Jeromin, A. and Pessin, J.E. 2002, *J. Biol. Chem., 277*, 27494.
[18] Brochetta, C., Perrotta, M.G., Jeromin, A., Romano, M., Vita, F., Soranzo, M.R., Borelli, V., Roder, J. and Zabucchi, G. 2003, *Inflammation, 27*, 361.
[19] Taverna, E., Francolini, M., Jeromin, A., Hilfiker, S., Roder, J. and Rosa, P. 2002, *J. Cell Sci., 115*, 3909.
[20] Pan, C.-Y., Jeromin, A., Lundstrom, K., Yoo, S.H., Roder, J. and Fox, A.P. 2002, *J. Neurosci., 22*, 2427.
[21] Negyessy, L. and Goldman-Rakic, P.S. 2005, *J. Comp. Neurol., 488*, 464.
[22] Jinno, S., Jeromin, A., Roder, J. and Kosaka, T. 2002, *Neuroscience, 113*, 449.
[23] Wilkinson, B.L., Jeromin, A., Roder, J. and Hyson, R.L. 2003, *Neuroscience, 117*, 957.
[24] Bergmann, M., Grabs, D., Roder, J., Rager, G. and Jeromin, A. 2002, *J. Comp. Neurol., 449*, 231.
[25] O'Callaghan, D.W., Ivings, L., Weiss, J.L., Ashby, M.C., Tepikin, A.V. and Burgoyne, R.D. 2002, *J. Biol. Chem., 277*, 14227.
[26] Scalettar, B.A., Rosa, P., Taverna, E., Francolini, M., Tsuboi, T., Terakawa, S., Koizumi, S., Roder, J. and Jeromin, A. 2002, *J. Cell Sci., 115*, 2399.

[27] Koizumi, S., Rosa, P., Willars, G.B., Challiss, A.J., Taverna, E., Francolini, M., Bootman, M.D., Lipp, P., Inoue, K., Roder, J. and Jeromin A. 2002, *J. Biol. Chem., 277*, 30315.
[28] Chen, X.-L., Zhong, Z.-G., Yokoyama, S., Bark, C., Meister, B., Berggren, P.-O., Roder, J., Higashida, H. and Jeromin, A. 2001, *J. Physiol., 532*, 649.
[29] Zhao, W., Varnai, P., Tuymetova, G., Balla, A., Toth, Z.E., Oker-Blom, C., Roder, J., Jeromin, A. and Balla, T. 2001, *J. Biol. Chem., 276*, 40183.
[30] O'Callaghan, D.W. and Burgoyne, R.D. 2003, *J. Cell Sci., 116*, 4833.
[31] Rajebhosale, M., Greenwood, S., Vidugiriene, J., Jeromin, A. and Hilfiker, S. 2003, *J. Biol. Chem., 278*, 6075.
[32] An, W.F., Bowlby, M.R., Betty, M., Cao, J., Ling, H.P., Mendoza, G., Hinson, J.W., Mattsson, K.I., Strassle, B.W., Trimmer, J.S. and Rhodes, K.J. 2000, *Nature, 403*, 553.
[33] Holmqvist, M.H., Cao, J., Hernandez-Pineda, R., Jacobson, M.D., Carroll, K.I., Sung, M. A., Betty, M., Ge, P., Gilbride, K.J., Brown, M.E., Jurman, M.E., Lawson, D., Silos-Santiago, I., Xie, Y., Covarrubias, M., Rhodes, K.J., Distefano, P.S. and An, W.F. 2002, *Proc. Natl. Acad. Sci. USA, 99*, 1035.
[34] Zozulya, S. and Stryer, L. 1992, *Proc. Natl. Acad. Sci. USA, 89*, 11569.
[35] Maurer-Stroh, S., Eisenhaber, B. and Eisenhaber, F. 2002, *J. Mol. Biol., 317*, 523.
[36] Resh, M.D. 1994, *Cell, 76*, 411.
[37] Resh, M.D. 1999, *Biochim. Biophys. Acta, 1451*, 1.
[38] Tanaka, T., Ames, J.B., Harvey, T.S., Stryer, L. and Ikura, M. 1995, *Nature, 376*, 444.
[39] Ames, J.B., Ishima, R., Tanaka, T., Gordon, J.I., Stryer, L and Ikura, M. 1997, *Nature, 389*, 198.
[40] Dizhoor, A.M., Chen, C.K., Olshevskaya, E., Sinelnikova, V.V., Phillipov, P. and Hurley, J.B. 1993, *Science, 259*, 829.
[41] Valentine, K.G., Mesleh, M.F., Opella, S.J., Ikura, M. and Ames, J.B. 2003, *Biochemistry, 42*, 6333.
[42] Ladant, D. 1995, *J. Biol. Chem., 270*, 3179.
[43] Ivings, L., Pennington, S.R., Jenkins, R., Weiss, J.L. and Burgoyne, R.D. 2002, *Biochem. J., 363*, 599.
[44] Spilker, C., Dresbach, T. and Braunewell, K.-H. 2002, *J. Neurosci., 22*, 7331.
[45] Spilker, C. and Braunewell, K.-H. 2003, *Mol. Cell. Neurosci., 24*, 766.
[46] Ames, J.B., Hendricks, K.B., Strahl, T., Huttner, I.G., Hamasaki, N. and Thorner, J. 2000, *Biochemistry, 39*, 12149.
[47] McFerran, B.W., Weiss, J.L. and Burgoyne, R.D. 1999, *J. Biol. Chem., 274*, 30258.
[48] O'Callaghan, D.W. and Burgoyne, R.D. 2004, *J. Biol. Chem., 279*, 14347.
[49] Hamasaki-Katagiri, N., Molchanova, T., Takeda, K. and Ames, J.B. 2004, *J. Biol. Chem., 279*, 12744.
[50] Senin, I.I., Hoeppner-Heitmann, D., Polkovnikova, O.O., Churumova, V.A., Tikhomirova, N.K., Philippov, P.P. and Koch, K.-W. 2004, *J. Biol. Chem., 279*, 48647.
[51] Cox, J.A., Drussel, I., Comte, M., Nef, S., Nef, P., Lenz, S.E. and Gundelfinger, E.D. 1994, *J. Biol. Chem., 269*, 32807.
[52] Ames, J.B., Dizhoor, A.M., Ikura, M., Palczewski, K. and Stryer, L. 1999, *J. Biol. Chem., 274*, 19329.

[53] Vijay-Kumar, S. and Kumar, V.D. 1999, *Nat. Struct. Biol., 6*, 80.

[54] Scannevin, R.H., Wang, K.-W., Jow, F., Megules, J., Kospco, D.C., Edris, W., Carroll, K.C., Lu, Q., Xu, W., Xu, Z., Katz, A.H., Olland, S., Lin, L., Taylor, M., Stahl M., Malakian, K., Somers, W., Mosyak, L., Bowlby, M., Chanda, P. and Rhodes, K.J. 2004, *Neuron, 41*, 587.

[55] Zhou, W., Qian, Y., Kunjilwar, K., Pfaffinger, P.J. and Choe, S. 2004, *Neuron, 41,* 573.

[56] Tanaka, T., Ames, J.B., Harvey, T.S., Stryer, L. and Ikura, M. 1995, *Nature, 376*, 444.

[57] Flaherty, K.M., Zoulya, S., Stryer, L. and McKay, D.B. 1993, *Cell, 75*, 709.

[58] Ames, J.B., Hamashima, N. and Molchanova, T. 2002, *Biochemistry, 41*, 5776.

[59] Weiergraber, O.H., Senin, I.I., Philippov, P.P., Granzin, J. and Koch, K.-W. 2003, *J. Biol. Chem., 278*, 22972.

[60] Olshevskaya, E.V., Boikov, S., Ermilov, A., Krylov, D., Hurlay, J.B. and Dizhoor, A.M. 1999, *J. Biol. Chem., 274*, 10823.

[61] Tachibanaki, S., Nanda, K., Sasaki, K., Ozaki, K. and Kawamura, S. 2000, *J. Biol. Chem., 275*, 3313.

[62] Ermilov, A.N., Olshevskaya, E.V., Dizhoor, A.M. 2001, *J. Biol. Chem., 276*, 48143.

[63] Hwang, J.-Y., Schlesinger, R. and Koch, K.-W. 2004, *Eur. J. Biochem., 271*, 3785.

[64] Ikura, M., Clore, G.M., Gronenborn, A.M., Zhu, G., Klee, C.B. and Bax, A. 1992, *Science, 256*, 632.

[65] Bahi, N., Friocourt, G., Carrie, A., Graham, M.E., Weiss, J.L., Chafey, P., Fauchereau, F., Burgoyne, R.D. and Chelly, J. 2003, *Hum. Mol. Genet., 12*, 1415.

[66] Tsujimoto, T., Jeromin, A., Saitoh, N., Roder, J.C. and Takahashi, T. 2002, *Science, 295*, 2276.

[67] Jeromin, A., Muralidhar, D., Parameswaran, M.N., Roder, J., Fairwell, T., Scarlata, S., Dowal, L., Mustafi, S.M., Chary, K.V.R. and Sharma, Y. 2004, *J. Biol. Chem., 279*, 27158.

[68] Olafsson, P., Wang, T. and Lu, B. 1995, *Proc. Natl. Acad. Sci. USA, 92*, 8001.

[69] Sippy, T., Cruz-Martin, A., Jeromin, A. and Schweizer, F.E. 2003, *Nat. Neurosci., 6*, 1031.

[70] Zucker, R.S. 2003, *Nat. Neurosci., 6,* 1006.

[71] Gomez, M., De Castro, E., Guarin, E., Sasakura, H., Kuhara, A., Mori, I., Bartfai, T., Bargmann, C.I. and Nef, P. 2001, *Neuron, 30*, 241.

[72] Hilfiker, S. 2003, *Biochem. Soc. Trans., 31*, 828.

[73] Wang, C.-Y., Yang, F., He, X., Chow, A., Du, J., Russell, J.T. and Lu, B. 2001, *Neuron, 32*, 99.

[74] Gromada, J., Bark, C., Smidt, K., Efanov, A.M., Janson, J., Mandic, S.A., Webb, D.-L., Zhang, W., Meister, B., Jeromin, A. and Berggren, P.-O. 2005, *Proc. Natl. Acad. Sci. USA, 102*, 10303.

[75] Schaad, N.C., De Castro, E., Nef, S., Hegi, S., Hinrichsen, R., Martone, M.E., Ellisman, M.H., Sikkink, R., Rusnak, F., Sygush, J. and Nef, P. 1996, *Proc. Natl. Acad. Sci. USA, 93*, 9253.

[76] Calvert, P.D., Klenchin, V.A. and Bownds, M.D. 1995, *J. Biol. Chem., 270*, 24127.

[77] Chen, C.K., Ingelese, J., Lefkowitz, R.J. and Hurley, J.B. 1995, *J. Biol. Chem., 270*, 18060.

[78] De Castro, E., Nef, S., Fiumelli, H., Lenz, S.E., Kawamura, S. and Nef, P. 1995, *Biochem. Biophys. Res. Commun., 216*, 133.

[79] Faurobert, E., Chen, C.-K., Hurley, J.B. and Teng, D. H.-F. 1996, *J. Biol. Chem., 271*, 10256.

[80] Kabbani, N., Negyessy, L., Lin, R., Goldman-Rakic, P. and Levenson, R. 2002, *J. Neurosci., 22*, 8476.

[81] Kato, M., Watanabe, Y., Iino, S., Takaoka, Y., Kobayashi, S., Haga, T. and Hidaka, H. 1998, *Biochem. J., 331*, 871.

[82] Iacovelli, L., Sallese, M., Mariggio, S., and De Blasi, A. 1999, *FASEB J., 13*, 1.

[83] Chuang, T..T., Paolucci, L. and De Blasi, A. 1996, *J. Biol. Chem., 271*, 28691.

[84] Pronin, A.N., Satpaev, D.K., Slepak, V.Z. and Benovic, J.L. 1997, *J. Biol. Chem., 272*, 18273.

[85] Nakamura, T.Y., Pountney, D.J., Ozaita, A., Nandi, S., Ueda, S., Rudy, B. and Coetzee, W.A. 2001, *Proc. Natl. Acad. Sci. USA, 98*, 12808.

[86] Bergson, C., Levenson, R., Goldman-Rakic, P.S. and Lidow, M.S. 2003, *Trends Pharmacol. Sci., 24*, 486.

[87] Koh, P.O., Undie, A.S., Kabbani, N., Levenson, R., Goldman-Rakic, P.S. and Lidow, M.S. 2003, *Proc. Natl. Acad. Sci. USA, 100*, 313.

[88] Bernstein, A.M., 2002, *NeuroReport, 13*, 393.

[89] Carlsson, A., Waters, N., Holm-Waters, S., Tedroff, J., Nilsson, M. and Carlsson, M.L. 2001, *Annu. Rev. Pharmacol. Toxicol., 41*, 237.

[90] Carrie, A., Jun, L., Bienvenu, T., Vinet, M.C., McDonell, N., Couvert, P. Zemni, R., Cardona, A., Van Buggenhout, G., Frints, S., Hamel, B., Moraine, C., Ropers, H.H., Strom, T., Howell, G.R., Whittaker, A., Ross, M.T., Kahn, A., Fryns, J.P., Beldjord, C., Marynen, P. and Chelly, J. 1999, *Nat. Genet., 23*, 25.

[91] D'Adamo, P., Menegon, A., Lo Nigro, C., Grasso, M., Gulisano, M., Tamanini, F., Bienvenu, T., Gedeon, A.K., Oostra, B., Wu, S.K., Tandon, A., Valtorta, F., Balch, W.E., Chelly, J. and Toniolo, D. 1998, *Nat. Genet., 19*, 134.

[92] Bienvenu, T., des Portes, V., Saint Martin, A., McDonell, N., Billuart, P., Carrie, A., Vinet, M.C., Couvert, B., Toniolo, D., Ropers, H.H., Moraine, C., van Bokhoven, H., Fryns, J.P., Kahn, A., Beldjord, C. and Chelly, J. 1998, *Hum. Mol. Genet., 7*, 1311.

[93] Geppert, M. and Sudhof, T.C. 1998, *Ann. Rev. Neurosci., 21*, 75.

[94] Hendricks, K.B., Wang, B.Q., Schieders, E.A. and Thorner, J. 1999, *Nat. Cell Biol., 1*, 234.

[95] Huttner, I.G., Strahl, T., Osawa, M., King, D.S., Ames, J.B. and Thorner, J. 2003, *J. Biol. Chem., 278*, 4862.

[96] Strahl, T., Grafelmann, B., Dannenberg, J., Thorner, J. and Pongs, O. 2003, *J. Biol. Chem., 278*, 49589.

[97] Walch-Solimena, C. and Novick, P. 1999, *Nat. Cell Biol., 1*, 523.

[98] Audhya, A., Foti, M. and Emr, S.D. 2000, *Mol. Biol. Cell, 11*, 2673.

[99] Zheng, Q., Bobich, J.A., Vidugiriene, J., McFadden, S.C., Thomas, F., Roder, J. and Jeromin, A. 2005, *J. Neurochem., 92*, 442.

[100] Kapp-Barnea, Y., Melnikov, S., Shefler, I., Jeromin, A. and Sagi-Eisenberg, R. 2003, *J. Immunol., 171*, 5320.

[101] Martin, T.F.J. 2001, *Curr. Opin. Cell Biol., 13*, 493.
[102] Robinson, M.S. 2004, *Trends Cell Biol., 14*, 167.
[103] Downes, C.P., Gray, A. and Lucocq, J.M. 2005, *Trends Cell Biol., 15*, 259.
[104] De Matteis, M.A. and Godi, A. 2004, *Nat. Cell Biol., 6*, 487.
[105] Itoh, T. and De Camilli, P. 2004, *Nature, 429*, 141.
[106] Godi, A., Pertile, P., Meyers, R., Marra, P., Di Tullio, G., Iurisci, C., Luini, A., Corda, D. and De Matteis, M.A. 1999, *Nat. Cell Biol., 1*, 280.
[107] Godi, A., Di Campli, A., Konstantakopoulos, A., Di Tullio, G., Alessi, D.R., Kular, G.S., Daniele, T., Marra, P., Lucocq, J.M. and De Matteis, M.A. 2004, *Nat. Cell Biol., 6*, 393.
[108] Shin, H.-W. and Nakayama, K. 2004, *Trends Biochem. Sci., 29*, 513.
[109] Roth, M. 2004, *Nat. Cell Biol., 6*, 384.
[110] Haynes, L.P., Thomas, G.M.H. and Burgoyne, R.D. 2005, *J. Biol. Chem., 280*, 6047.
[111] Milosevic, I., Sorensen, J.B., Lang, T., Krauss, M., Nagy, G., Haucke, V., Jahn, R. and Neher, E. 2005, *J. Neurosci., 25*, 2557.
[112] Gong, L.-W., Di Paolo, G., Diaz, E., Cestra, G., Diaz, M.-E., Lindau, M., De Camilli, P. and Toomre, D. 2005, *Proc. Natl. Acad. Sci. USA, 102*, 5204.
[113] Pinton, P., Pozzan, T. and Rizzuto, R. 1998, *EMBO J., 17*, 5298.
[114] Chen, J.-L., Ahluwalia, J.P. and Stamnes, M. 2002, *J. Biol. Chem., 277*, 35682.
[115] Porat, A. and Elazar, Z. 2000, *J. Biol. Chem., 275*, 29233.

In: Neuronal Calcium Sensor Proteins
Editors: Pavel Philippov and Karl Koch, pp. 45-64

ISBN 1-59454-978-8

Chapter III

NCS-1 as a Regulator of Kv4 K^+ Channels

***T.Y. Namamura*[1], *L. Parachuru*[4], *S. Nandi*[4], *L. Porter*[2], *D.J. Pountney*[3] and *W.A. Coetzee*[4,*]**

[1]National Cardiovascular Center Research Institute, Suita, Osaka, Japan;
[2]Genious Targeting Labs. Inc, USA;
[3]GlaxoSmithKline, Harlow, United Kingdom;
[4]New York University School of Medicine, USA.

Abstract

The role of K^+ channels in genetically-related diseases, such as the long-QT syndrome, is increasingly being realized [1]. As an important repolarizing current, the A-type or transient outward K^+ current (I_{to}) plays a key role in several cell types. In neurons, A-type currents determine somatodendritic signal integration and in the heart, I_{to} determines the action potential shape and duration (and as such also controls the contractile force). The I_{to} also contributes to refractoriness, which prevents re-excitability and the occurrence of life-threatening arrhythmias. The cloning of K^+ channel subunits [2-5] provided a molecular framework for the understanding of K^+ channel function and diversity. Several subfamilies of voltage-activated K^+ (Kv) channels have been described. Channels responsible for the A-type currents and the transient outward K^+ current are now generally believed to be tetrameric proteins comprised of Kv4 subunits [6-9].

The regulation of K^+ channel expression and function by an increasing number of auxiliary (β) subunits is a powerful way in which the already large K^+ channel diversity is further magnified [10]. A report that a group of Ca^{2+}-binding proteins KChIPs interact with Kv4 channels led us to seek novel Kv4 auxiliary subunits expressed in the heart. In

* Correspondence concerning this article should be addressed to William A. Coetzee, DSc, Pediatric Cardiology, NYU School of Medicine, 560 First Avenue, TCH-407 New York, NY 10514. Tel: 212-263-8518; Fax: 212-263-5100; e-mail: william.coetzee@med.nyu.edu.

our search of publicly available sequence databases (as part of the human and mouse genome projects), we came across several potential candidates. We focused on one of these proteins, NCS-1 (mammarian ortholog of frequenin), and found it to be highly expressed in the heart. We performed a study examining the developmental time-course of NCS-1 expression in the heart. Immunoblot analysis demonstrates NCS-1 protein to be expressed in adult mouse ventricle at levels comparable to that found in some brain regions. Interestingly, cardiac NCS-1 protein expression levels are significantly higher in fetal and neonatal mouse hearts than in the adult. Immunocytochemical analysis of isolated neonatal mouse ventricular myocytes demonstrates co-localization of NCS-1 and Kv4.2 proteins at the sarcolemma. NCS-1 exhibits each of the characteristics that identify it as a *bone fide* Kv4 auxiliary subunit. NCS-1 (a) specifically increases Kv4 current amplitudes and modifies its inactivation time course, (b) increases surface trafficking of Kv4.2 proteins, (c) co-immunoprecipitates with native Kv4.2 proteins and (d) co-localizes with Kv4.2 proteins in mouse brain and cardiac myocytes. Thus, NCS-1 should be considered as an important potential Kv4 regulatory subunit, in the heart and in neurons.

Keywords: NCS-1, Potassium channels, Kv4 channels.

Abbreviations

NCS-1	Neuronal Ca^{2+} sensor -1 (also referred to as frequenin)
Kv channels	Voltage-gated K^+ channels
Kvβ subunit	Accessory (or β) subunits of Kv channels
KChIPs	K^+ channel interacting proteins
KChAP	K^+ channel-associated protein
VILIP	Visinin-like proteins
GCAPs	Guanylyl cyclase-activating proteins
EST	expressed sequence tag

1. Molecular Architecture of Potassium Channels

Electrophysiological recordings from cells reveal a large diversity of K^+ channels (of the ion channels, K^+ channels are by far the most diverse [10,11]. Functionally, K^+ channels can be grouped into several main classes [12]. Those that are activated by voltage are termed voltage-gated K^+ (Kv) channels. After their initial opening, many K^+ channels spontaneously close (a property referred to as inactivation). Kv channels can loosely be segregated into those that inactivate rapidly (sometimes referred to as A-type K^+ channels) and those that show relatively little inactivation with sustained depolarization (these channels are also termed delayed rectifiers).

The large molecular diversity of K^+ channel principal subunits is further magnified by the presence of regulatory auxiliary (β) subunits, which alter the function and/or expression levels of a variety of K^+ channel principal subunits. The best-described auxiliary subunit of voltage-gated K^+-channels is the Kvβ subunit, which is a cytosolic protein that associates specifically with Kv1 α-subunits. Kvβ-subunits were reported to enhance the trafficking to the surface membrane of the Kv α-subunits [13,14] and also to participate in the inactivation process (N-type inactivation) [15-17]. Several other auxiliary subunits for other K^+ channels have also been described, including KChAP (K^+ channel associated protein; which interacts with Kv2.1 and Kv1.3 subunits), KCR and minK that respectively modulate *eag* and KvLQT1 channels. This chapter focuses on a novel role of an intracellular Ca^{2+}-sensor protein NCS-1, regulating the expression levels and activity of Kv4 channels.

2. The Transient Outward Current and Kv4 Channels in the Heart

Following the initiation of an action potential, several types of voltage-dependent K^+ channels are activated, including the rapidly inactivating transient outward K^+ current (I_{to}). We will focus on this current, which plays an important role in determining the action potential duration and hence contractile force. I_{to} also affects the refractoriness of cardiac tissue and may therefore be involved in the onset and prevention of arrhythmias.

2.1. The Molecular Composition of the Transient Outward K^+ Current Channels in the Heart

The large diversity of subunits and subunit interaction at the molecular level [11] is difficult to reconcile with the more limited (although diverse in its own right) complement of native K^+ channels present in heart cells. There is mounting evidence that Kv4 subfamily members are primary subunits of cardiac I_{to} in rat, dog and human ventricular myocytes [18-21]. Evidence for a relationship between Kv4 channels and I_{to} includes use of antisense oligonucleotides and dominant negative constructs. Using antisense oligonucleotides that specifically eliminate Kv4.2 and Kv4.3 transcripts, we and others demonstrated a large and significant reduction of rat ventricular I_{to} [6,7]. Other molecular approaches, including the use of transgenic mice that overexpress dominant negative Kv4.2 constructs in the heart [8,9] or targeted deletion of the Kv1.4 allele [22] further support the view that Kv4 proteins are important contributors to the cardiac I_{to}.

2.2. Auxiliary Subunits Regulating the Expression Levels and Kinetics of Kv4 Channels

There is increasing evidence suggesting that auxiliary subunits regulate the expression levels and functional properties of many K^+ channel proteins [12,23]. From mRNA fractionation experiments, the existence of Kv4-specific auxiliary subunits had been postulated several years ago [24,25]. However, the molecular identification of these regulatory proteins had remained elusive until the recent description of a group of three Ca^{2+}-binding proteins, KChIP1 to KChIP3 (a 4th KChIP has subsequently been identified), that specifically alter the expression levels and kinetics of Kv4.2 and Kv4.3 currents [26]. These KChIPs were identified in a yeast-screening assay as proteins that bind to the N-terminus of the Kv4.2 protein. Functionally, KChIPs modulate the density and inactivation kinetics of Kv4 channels when heterologously co-expressed in mammalian cells or in Xenopus oocytes. They also co-localize and co-immunoprecipitate with brain Kv4 principal subunits, and thus are integral components of native neuronal Kv4 channel complexes.

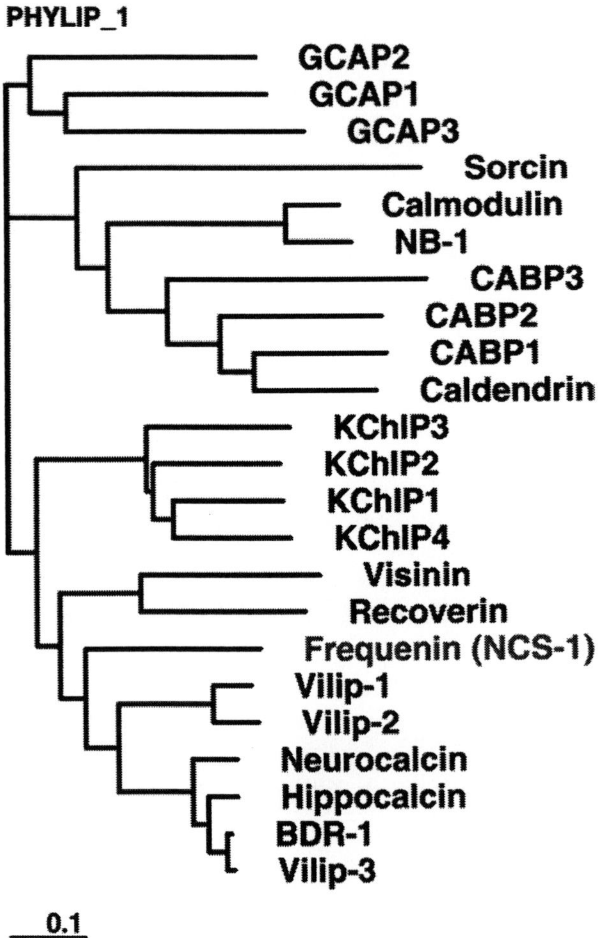

Figure 1. Phylogenetic tree of Ca^{2+} binding protein family members.

3. NCS-1 – Prototype of the Neuronal Ca^{2+}-Sensor Proteins

Calcium is the most versatile and important intracellular messenger in cells, regulating a variety of cellular processes such as neurotransmission, contraction and signal transductions. The various actions of Ca^{2+} are mediated by a large family of EF-hand Ca^{2+}-binding proteins, which may act as Ca^{2+}-sensors. The neuron-specific EF-hand Ca^{2+} binding proteins (neuronal

Ca^{2+} sensor proteins; NCS, Fig. 1) [27,28] are all small (~ 200 amino acids) globular intracellular proteins with four identifiable EF-hand motifs, although in general only two or three of these EF-hands can bind Ca^{2+} [28]. This subfamily includes as subgroups the recoverins and GCAPs (guanylyl cyclase-activating proteins), which are primarily expressed in retinal photoreceptor cells, and the NCS-1s and VILIPs (visinin-like proteins), which are widely but differentially expressed in the nervous system.

3.1. Evolutionary Conservation of NCS-1

Through molecular genetics studies, frequenin has been identified in *Drosophila* as a regulator of synaptic neurotransmission [29]. The V7 *Drosophila*, which overexpresses frequenin, was found to exhibit an abnormal motor phenotype (*Shaker*-like behavior). Electrophysiological analysis of these mutants has shown altered action potential duration, spiking frequency and abnormal neurotransmitter release at neuromuscular junctions [29]. Frequenin was postulated to be the evolutionary ancestor of the neuronal Ca^{2+}-binding protein family from which the other members evolved [28]. It is a highly conserved protein with orthologs found in yeast, plants, *C. elegans*, *Xenopus*, and mammals (with 75 % amino acid identity between species as evolutionary distantly related as yeast and man). The mammalian ortholog of frequenin is also referred to as NCS-1 [28,30]. However, throughout this chapter we will interchangingly use the terms "NCS-1" and "frequenin", given the fact that the expression of NCS-1 is not neural specific and the implications that this protein may have in excitability and frequency regulation of the heart beat.

3.2. The Relationship Between KChIPs and NCS-1

Although KChIPs and NCS-1 belong to the same subfamily of Ca^{2+}-binding proteins, they are only distantly related to each other (see Fig. 1). The sequence identity between NCS-1 and KChIPs is 31-37 %. Other subfamily members (including the VILIPs, neurocalcin and hippocalcin) are more closely related to NCS-1 (around 60% amino acid identity). Thus, KChIPs and NCS-1 are very different proteins and can reasonably be expected to have divergent biological functions. Surprisingly, VILIPs, neurocalcin and hippocalcin have no effect on Kv4 currents [31].

3.3. Ca^{2+} Binding by NCS-1

Of the 4 EF hand motifs of NCS-1, the first EF-hand does not contribute to Ca^{2+}-binding, since key amino acids in EF1 are not conserved. Two of the remaining EF-hands (EF-3 and EF-4) bind Ca^{2+} with an affinity well within the biological range of Ca^{2+} concentrations ($K_{EF3} = K_{EF4} = 200\text{-}400$ nM) while EF-2 binds Ca^{2+} rather weakly ($K_{EF2} = 10$ μM) [32,33]. Since normal resting Ca^{2+} levels are about 100 nM in many mammalian cells, a small proportion of NCS-1 proteins are expected to be in the Ca^{2+}-bound state at rest. The Ca^{2+} binding of NCS-1 is strongly cooperative (i.e. the binding of one Ca^{2+} vastly increases the Ca^{2+} affinity of the remaining Ca^{2+} binding sites [32], resulting in a steep dependence of NCS-1 on Ca^{2+} levels). Thus, the activity of NCS-1 is expected to be steeply dependent on increases in intracellular Ca^{2+} throughout the biological range of Ca^{2+} concentrations (100 nM to 1 μM). NCS-1 is therefore exquisitely fine-tuned to rapidly respond to intracellular Ca^{2+} transients, as occurs during the heart beat.

3.4. Membrane Targeting of NCS-1

In some members of the neuronal Ca^{2+}-binding proteins, membrane targeting is promoted by a Ca^{2+}-myristoyl switch mechanism, in which an N-terminal myristoyl acid moiety is uncovered from the hydrophobic protein core by a conformational change resulting from Ca^{2+} binding. The increased N-terminal hydrophobicity conferred by myristoylation results in targeting of the protein to membranes (or membrane-associated proteins) [34-36]. In the case of NCS-1, the myristoylated N-terminus is perpetually solvent-exposed [32,37], suggesting that this switch mechanism does not occur. Membrane targeting can therefore occur in the absence of Ca^{2+} [30]. Upon Ca^{2+} binding, however, NCS-1 undergoes a conformational change [30,32] and its binding to membranes is greatly increased by Ca^{2+} binding [32]. A consensus site for N-terminal myristoylation is present in KChIP1, but not in KChIP2-KChIP4, suggesting that their membrane targeting mechanisms (if any) may differ.

3.5. The Functional Roles of NCS-1

A survey of the literature shows that the functional effects have been assigned to NCS-1, most (if not all) relate to specific forms of subcellular and Golgi trafficking. Early reports showed facilitation of neurotransmission in the frequenin-overexpressing V7 *Drosophila* mutant, which resulted mainly from a pre-synaptic defect [29]. It is now appreciated that in vertebrates, NCS-1 enhances neuronal synaptic transmission [38], transmitter release at the vertebrate neuromuscular synapse [39] and exocytosis in neuroendocrine cells [29]. NCS-1 also stimulates neurosecretion in mammalian cells, as demonstrated by increase in stimulated exocytosis of dense-core granules in adrenal chromaffin and PC12 cells [40]. Exocytosis of neurotransmitters during neurosecretion is triggered by a rise in the cytosolic Ca^{2+} concentration, which through interaction with Ca^{2+}-sensor proteins regulates both secretory vesicle recruitment and membrane fusion. The mechanism behind some of these events is becoming increasingly clear. It is now known that NCS-1 binds to, and stimulates the activity of phosphatidylinositol 4-kinase β [41] (a key enzyme in the phosphoinositol synthesis pathway) [42], which in turn is an essential factor in secretion [43] and Golgi trafficking of specific surface proteins [44,45]. Biochemical assays have shown that NCS-1 binds multiple proteins, both in a Ca^{2+}-dependent and Ca^{2+}-indpependent manner [30] and NCS-1 is likely to have a diverse (nevertheless specific) array of target proteins. It is possible that NCS-1 (and possibly also KChIPs) may affect Kv4 channels in part through this emerging mechanism.

3.6. The 3-Dimentional Crystal Structure of NCS-1

The crystal structure of unmyristoylated Ca^{2+}-bound human NCS-1 was refined to 1.9 Å resolution [32,37]. The protein shares the typical α-helical fold found in homologous proteins (recoverin and neurocalcin), with overall dimensions of 35×60×40 Å. NCS-1 contains 10 α-helical folds that typifies the recoverin proteins (Fig. 2). Four EF-hand helix-loop-helix folds are present in two pairs; an N-terminal pair (EF1, EF2) and a C-terminal pair (EF3, EF4), which are connected by the hinge loop 93-97 and related by an approximate 2-fold axis. The four EF-hands are arranged on one side of the molecule in a tandem linear array. This structural arrangement is similar to that of other members of the myristoyl-switch Ca^{2+}-binding proteins (e.g. neurocalcin, recoverin, and GCAP) and is different from the dumbbell arrangement found in calmodulin and troponin C. This structural analysis has shown that of the four EF-hands, calcium ions bind only to EF2, EF3 and EF4 (see Fig. 2).

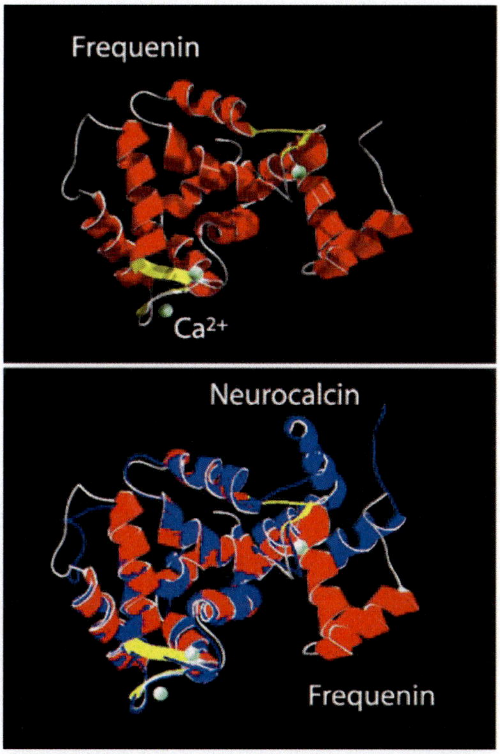

Figure 2. Structures of NCS-1 (shown in red) and neurocalcin (shown in blue).

4. Importance of Finding the Regulatory Subunits of I_{to} Channel Proteins in the Heart

Because of its contribution to action potential repolarization [46], modulation of the transient outward K^+ current is of fundamental importance in the regulation of excitability. The current density and biophysical properties are altered by a variety of physiological conditions, including temperature [47], stimulation rate [48,49], cytosolic calcium [50], α_1-adrenergic stimulation [51-55], protein kinase C [6,56] and G-proteins [57]. There are important transmural differences in I_{to} density across the ventricular wall [58-60] as well as regional differences between left/right ventricles [61] and in the atria [62]. Large changes (e.g .current density) also occur during development [63-68]. Pathophysiological conditions that have been described to regulate I_{to} properties include anoxia [69], metabolic inhibition and oxidant stress [69-72], acidosis [73], hypothyroidism [74], diabetes [75-77], hypertrophy [78] and failure [79,80]. Genetic approaches identified a potential role of I_{to} in arrhythmogenesis [9] and I_{to} is also a target for a number of antiarrhythmic agents [81-86]. Identifying the auxiliary subunits of I_{to} channel proteins and how they contribute to the regulation of I_{to} is of paramount importance in understanding the role of this K^+ current in health and disease.

5. Computer Search for Ca^{2+}-Binding Proteins Related to KChIPs

With the description of KChIPs as Kv4 channel β-subunits [26], we were interested in identifying other Ca^{2+}-binding protein family members expressed in the heart, with the aim of determining whether (and how) they may modulate Kv4 channel function. The strategy that we employed was to search public sequence databases for sequences cloned from heart cDNA libraries. KChIP2 was used as the query to perform a tblastn[1] searches against EST (expressed sequence tagged) databases. In this manner, we identified several interesting sequences, including EST W81153, which originated from a human fetal heart library. Further sequence comparisons identified this EST as being human NCS-1 (Genbank accession number: NM_014286). We obtained this EST from the IMAGE Consortium. Sequencing on both strands confirmed the identity of the clone and revealed that it contained the entire open reading frame of human NCS-1. Subsequent experiments identified NCS-1 as a *bone fide* regulatory subunit of Kv4 currents. NCS-1 (a) specifically increases Kv4 current amplitudes and modifies their inactivation time course, (b) increases surface trafficking of Kv4.2 proteins, (c) co-immunoprecipitates with native Kv4.2 proteins and (d) co-localizes with Kv4.2 proteins in mouse brain [31]. Some of the key findings are recapitulated below.

6. NCS-1 is Expressed in Cardiac Tissue

First described in chicken [87] and later in rodents [88], NCS-1 was described to be specifically expressed in the central and peripheral nervous system. In mouse, NCS-1 mRNA and proteins were described to be localized in the hippocampus and cerebellum [88-90] and subcellularly in neurites (dendrites and axons) [88-90].We also found expression of both Kv4.2 and NCS-1 in the dendritic processes of pyramidal and granule cells of the

[1] This algorithm compares a protein query sequence against the nucleotide EST sequence database dynamically translated in all reading frames.

hippocampal formation as well as the dendrites of cerebellar granule cells [31]. Recent data demonstrate that NCS-1 expression is not restricted to neurons in the central nervous system; for example, NCS-1 expression also occurs at neuromuscular junctions [91], glial cells [88] and neuroendocrine cells [40].

We examined the expression of NCS-1 in mouse heart by performing Western analysis, using a commercial chicken anti-NCS-1 antibody. Initially, we ascertained that this antibody specifically detects NCS-1 in lysates from transfected COS cells, and not other Ca^{2+} binding proteins, including VILIP-1, neurocalcin or KChIPs (data not shown). We examined NCS-1 protein expression in the mouse heart at three developmental stages (14d fetal, 1d neonatal and adult; Fig. 3A), and found NCS-1 to be expressed at high levels in fetal heart with declining protein levels after birth. Similar data were obtained using rabbit anti-NCS-1 antibodies (kind gifts from Dr. Bai Lu, NIH and Dr. Jeromin, Baylor College of Medicine). This declining expression after birth is in contrast to what we observe in brain tissue, where NCS-1 levels are low at fetal stages, increase after birth, and then declines again with further maturation (data not shown). When comparing NCS-1 expression in different mouse tissues and different regions of the brain (Fig. 3B), our data are consistent with available literature: NCS-1 levels are undetectable in liver, skeletal muscle, spleen and thymus. High expression is seen in the brain (with the highest levels detected in the olfactory bulb and forebrain; defined here as the proximal third of the brain mass). Expression in adult heart is comparatively weaker, but on par with expression of NCS-1 in the midbrain and cerebellum. Interestingly, we did not detect NCS-1 expression in mouse atria.

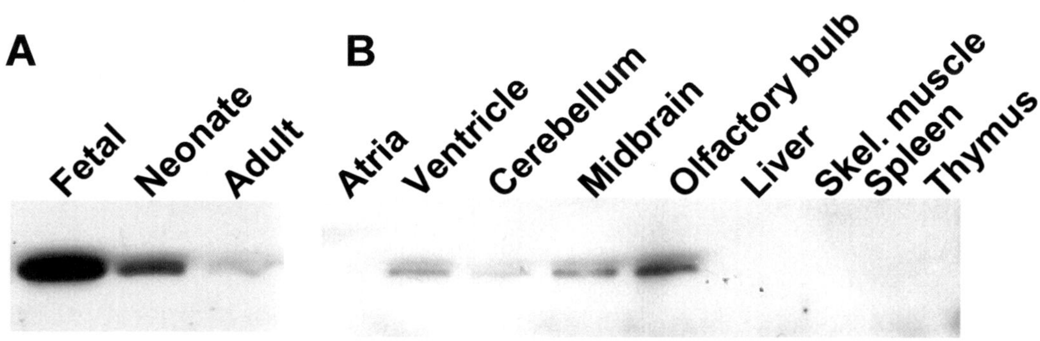

Figure 3. A. Immunoblot analysis of NCS-1 expression at different developmental stages in mouse heart. B. Expression of NCS-1 in different tissues of adult mouse.

7. NCS-1 Expression in Neonatal Mouse Ventricular Myocytes: CO-Localization with Kv4.2 Proteins

The immunoblot data shown above do not distinguish between NCS-1 expression in different cell types present in the heart (e.g. myocytes, endothelial cells, or neuronal cells in neuromuscular junctions). In answer to the question whether NCS-1 expresses in cardiac myocytes, we performed immunocytochemistry experiments using isolated ventricular myocytes. We chose neonatal mice for this purpose, since NCS-1 expresses at relatively high levels at this age. Isolated myocytes were fixed, permeabilized and co-stained with chicken anti-NCS-1 antibodies and rabbit anti-Kv4.2 antibodies. Detection occurred respectively with CY3 conjugated goat-anti chicken (red) and FITC-conjugated goat anti-rabbit (green) antibodies. Shown in an overlay image (Fig. 4), NCS-1 and Kv4.2 proteins are both expressed and importantly, appear to co-localize at the cell surface (yellow punctate staining). These data extend our previous finding of co-localization of Kv4.2 and NCS-1 protein [31] and further demonstrate that, at a subcellular level, NCS-1 is localized in heart cells in a manner that interaction with Kv4 proteins is likely to occur. Thus, we anticipate NCS-1 to be a modulator of cardiac Kv4 currents (I_{to}) in a manner analogous to what was observed in *Xenopus* oocytes (see below).

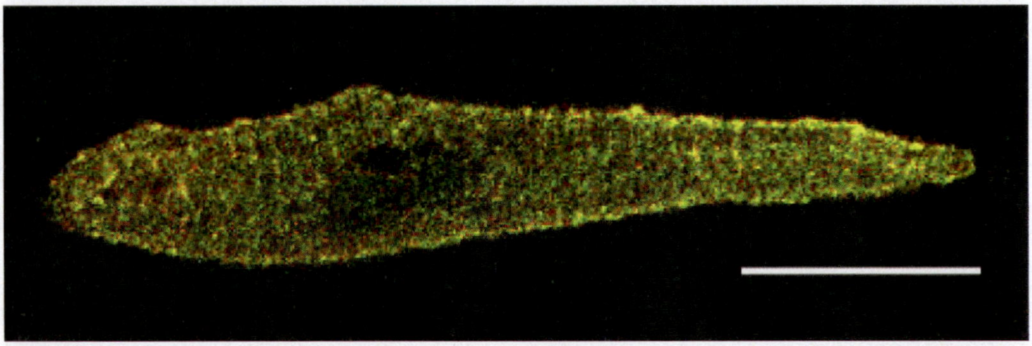

Figure 4. Expression of NCS-1 (red) and Kv4.2 (green) proteins in an isolated neonatal mouse ventricular myocyte. Scale bar = 10μm.

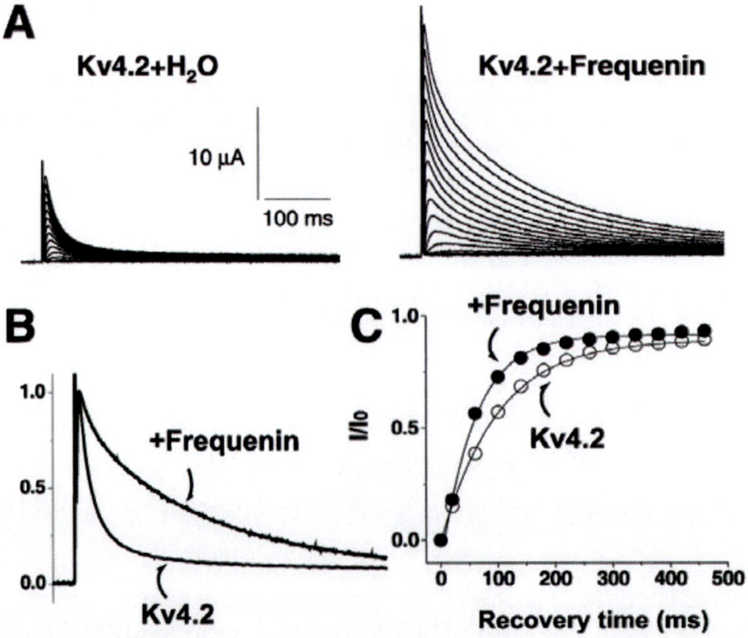

Figure 5. Effect of NCS-1 on Kv4.2 currents.

8. NCS-1 Modulates Kv4.2 Currents in *Xenopus* Oocytes

When *Xenopus* oocytes were co-injected with Kv4.2 and NCS-1 cRNAs, the expressed currents were substantially larger than that of oocytes injected only with Kv4.2 cRNA (plus H_2O to assure a constant injection volume; Fig. 5A). NCS-1 also markedly slowed the time course of Kv4.2 current inactivation, as illustrated by normalizing Kv4.2 current traces in the presence and absence of NCS-1 (Fig. 5B).

To determine how NCS-1 alters inactivation kinetics, we examined the decay of individual outward currents assuming a sum of two exponential terms [6]. NCS-1 altered neither the fast nor the slow time constants of inactivation, but instead, reversed their relative contribution [31]. Thus, the slower inactivation of Kv4.2 currents was due to an enhanced relative contribution of the slow inactivation process. NCS-1 also accelerated the recovery from inactivation of Kv4.2 currents (Fig. 5C) without causing a shift in steady-state inactivation voltage-dependence [31]. Similar data have been reported by another group [92].

9. Effects of other Ca^{2+} Binding Proteins on Kv4.2 Currents

We examined the effects on Kv4 currents of other Ca^{2+} binding proteins with sequence similarity to NCS-1 (see Fig. 1). VILIP1, hippocalcin, neurocalcin or calmodulin were without effect on Kv4 currents in *Xenopus* oocytes [31]. Thus, interaction with Kv4 proteins appears not to be a general feature of neuronal Ca^{2+} binding proteins; this feature seems to be bestowed only on KChIPs and NCS-1. It is interesting that neurocalcin, which has a very similar three-dimensional structure to that of NCS-1 [32,37,93] lacks effects on Kv4 currents; this may open the door to future comparative structure-function studies.

10. NCS-1 Enhances Membrane Trafficking of Kv4 Subunits

To investigate the mechanism of the action of NCS-1 on the increase in Kv4.2 current amplitude, we examined the effect of NCS-1 on the membrane trafficking of Kv4.2 channel. COS cells were transiently transfected with Kv4.2 or GFP-tagged NCS-1 cDNA, or co-transfected with both cDNAs and subjected to immunocytochemistry (Fig. 6). When COS cells were transfected with Kv4.2 cDNA alone, Kv4.2 proteins (as detected with anti-Kv4.2 antibody from Sigma) were predominantly expressed in perinuclear regions (left). Likewise, NCS-1 by itself also predominantly expressed in perinuclear compartments (middle). In co-transfected cells, expression of both Kv4.2 and NCS-1 proteins was shifted to the outer margins of the cells (right panel). These data show that NCS-1 enhances membrane trafficking of Kv4 subunits in COS7 cells. Since NCS-1 trafficking is also affected by the presence of Kv4 proteins, the possibility is raised that NCS-1 and Kv4 proteins associate and that the efficiency of membrane trafficking of the resulting protein complex is enhanced, as is the case for some subunits of inwardly-rectifying potassium channels (Kir6) and their regulatory subunits (sulfonylurea receptors or SUR) [94]. The increase in Kv4.2 current amplitude may therefore be explained in part by the ability of NCS-1 to enhance the trafficking of Kv4.2 proteins to the plasma membrane. It should be noted that a subsequent study failed to observe Kv4.2 trafficking by NCS-1 in transfected mammalian cells [95]. The reasons for the different result are unclear and may include the use of a different cell type (HeLa cells were used in the latter study).

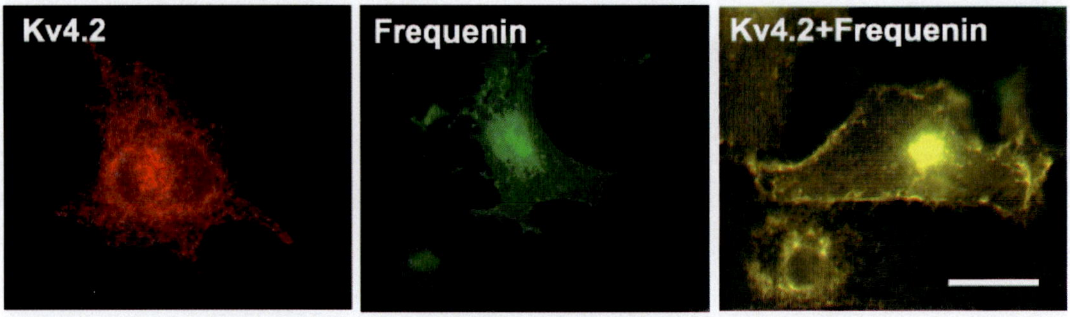

Figure 6. Trafficking of Kv4.2 and NCS-1 proteins in COS-7 cells. The titles indicate transfection cDNAs used.

11. The Effect of NCS-1 is Restricted to Kv4 Channels

We examined whether NCS-1 also affects other Kv subfamily members. We chose Kv1.4 and Kv3.4, both of which also express rapidly inactivating voltage-activated K^+ currents. NCS-1 had no effect on either these two currents [31]. Currents caused by the structurally unrelated K^+ channel subunit, Kir2.1, were similarly unaffected by NCS-1 (not shown). Thus, amongst those tested, the effect of NCS-1 was restricted to subunits of the Kv4 subfamily.

12. Effects of NCS-1 on other Members of the Kv4 Subfamily (Kv4.1 and Kv4.3)

When co-injected with NCS-1, Kv4.3 current amplitudes were also significantly larger and the inactivation time course was also considerably slower [31]. In contrast, NCS-1 had relatively modest effects on Kv4.1 currents. Although an increased Kv4.1 current amplitude was observed in some batches of oocytes, the overall effect was non-significant. Importantly, the Kv4.1 inactivation time course was unaffected by NCS-1.

13. The Kv4.2 N-Terminus is Responsible for the Differential Effects of NCS-1 on Kv4.2 and Kv4.1 Channels

We took advantage of the fact that NCS-1 had different effects on the Kv4.1 and Kv4.2 current inactivation time course to determine the structural elements of Kv4 proteins that are responsible for these differences. We constructed Kv4 chimeras by swapping the divergent N-termini and C-termini of Kv4.2 and Kv4.1 subunits. We refer to the construct containing the Kv4.1 N-terminus, the Kv4.2 core (S1-S6) and Kv4.2 C-terminus as 1-2-2 and conversely; chimera 2-1-1 is the Kv4.1 subunit having a Kv4.2 N-terminus. Data for several of these chimeras are shown in Fig 7. In each panel, data are shown for the chimeric current in the absence and presence of NCS-1. To facilitate comparison and to accentuate the effects of NCS-1 on the current time course, traces are overlayed after normalizing to the current amplitude. The time course of the 1-2-2 chimera current was slowed relative to wild-type Kv4.2 currents (Fig 7A). Interestingly, NCS-1 had no effect on the 1-2-2 chimera current. In contrast, the presence of the Kv4.2 N-terminus bestows NCS-1-sensitivity to the otherwise insensitive Kv4.1 current (NCS-1 slows 2-1-1 current inactivation; Fig. 7B). These effects of the Kv4.2 N-terminus were independent of which C-terminal sequence was present, since C-terminal chimeras 2-2-1 and 1-1-2 were modulated in a similar manner as wild type currents (Fig. 7C and D). Collectively, these results provide strong support for the notion that the Kv4.2 N-terminus (and likely also the Kv4.3 N-terminus) contains structural elements that mediate the differential effects of NCS-1 on these currents.

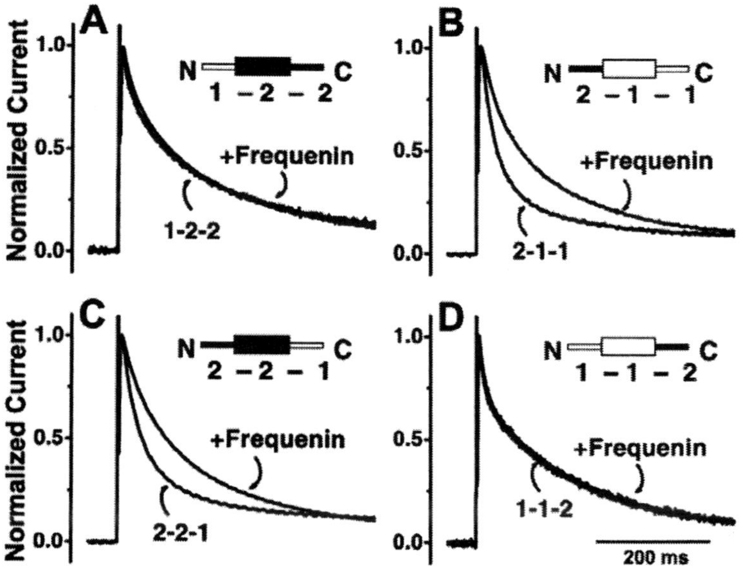

Figure 7. Effects of NCS-1 on Kv4 chimeras.

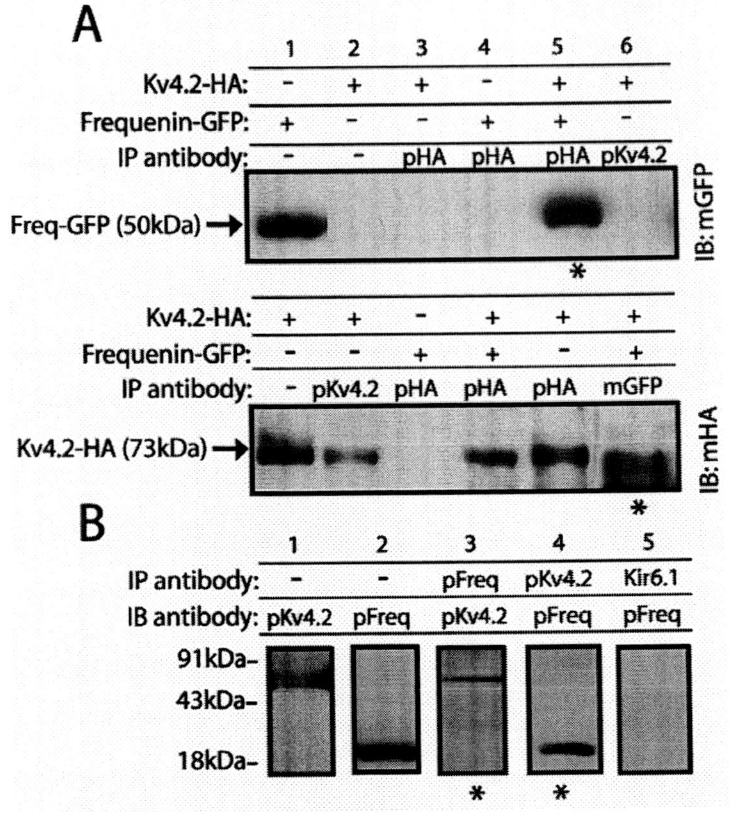

Figure 8. Co-immunoprecipitation of NCS-1 and Kv4.2 subunits.

14. NCS-1 Physically Interacts with Kv4.2 Channel Proteins

We used biochemical experiments to demonstrate that physical interaction occurs between NCS-1 and Kv4.2 channels. HEK-293 cells were co-transfected with HA-tagged Kv4.2 and GFP-tagged NCS-1 cDNAs. Kv4.2 proteins were immunoprecipitated from cell lysates using anti-HA antibodies (Covance) or anti-Kv4.2 antibodies (a gift from Dr G. Cheng). The immunoprecipitated proteins were separated by 12% SDS-PAGE and Western analysis was performed using an anti-GFP or anti-HA (clone 12CA5) monoclonal antibody. This resulted in detection of NCS-1 in immunoblots (Fig. 8A, lane 5). Conversely, Kv4.2 subunits were detected (using anti-HA antibodies) when NCS-1 was immunoprecipitated with anti-GFP antibodies (Fig. 8A, bottom panel). To determine whether this association also occurs in native tissue, we immunoprecipitated protein complexes with anti-Kv4.2 antibodies from mouse brain membrane fractions. NCS-1 could be detected in Kv4.2 immunoprecipitates, and reciprocally, Kv4.2 proteins were present when immunoprecipitating with anti-NCS-1 antibodies (Fig. 8B). These results demonstrate that association occurs between native Kv4.2 and NCS-1 proteins in the mouse brain. In the mouse heart, interaction between NCS-1 and Kv4 proteins has also been demonstrated [92].

These results are in apparent contrast with a subsequent report, which failed to observe interaction between NCS-1 and Kv4.3 proteins in heterologous expression systems [96]. A possible explanation for this contradictory result is that an N-terminal tagged GFP fusion protein of NCS-1 was used in the latter study. We find Kv4 currents in *Xenopus* oocytes not to be affected by N-terminal tagged versions of NCS-1, whereas untagged or C-terminal tagged NCS-1 consistently increases Kv4 currents and slow their inactivation time course (data not shown). For this reason, we have not performed our co-immunoprecipitation assays using N-terminal tagged NCS-1 constructs.

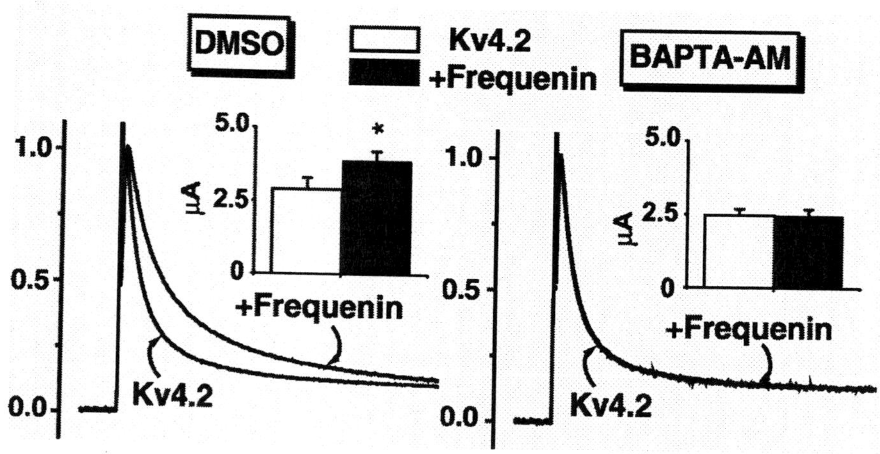

Figure 9. Effects of intracellular Ca^{2+} buffering.

15. Modulation of Kv4 Currents by NCS-1 Requires Intracellular Ca^{2+}

Since NCS-1 is a Ca^{2+} binding protein, we examined whether intracellular Ca^{2+} is required for its effects on Kv4 currents. After injection, *Xenopus* oocytes were incubated with membrane permeable Ca^{2+}-chelator, BAPTA-AM (10μM; or only with the solvent DMSO as a control; Fig. 9). In control DMSO-incubated oocytes, NCS-1 had the expected effects (cf. Figs. 5 & 9). The effects of NCS-1 on both current amplitude and Kv4.2 inactivation time course were totally prevented in BAPTA-AM incubated oocytes (Fig. 9). Thus, cytosolic Ca^{2+} is important in mediating the effects of NCS-1 on Kv4.2 currents. The fact that NCS-1 affects Kv4.2 currents under resting conditions in *Xenopus* oocytes suggests that NCS-1 was at least partially activated at the normal resting Ca^{2+} levels [~90 nM in *Xenopus* oocytes [97]]. It suggests that NCS-1 may also be partially activated at rest in cardiac myocytes, which have a intracellular diastolic Ca^{2+} activity in the same range [98,99]. Because of its high cooperativity of Ca^{2+} binding, it is likely that the activity of NCS-1 will be even higher when cytosolic Ca^{2+} levels are increased (e.g. during increases in heart rate or under certain pathophysiological conditions). One of the objectives of future studies should be to systematically examine the dependence of native A-type and the transient outward K^+ currents on intracellular Ca^{2+} levels.

Summary

We have shown that NCS-1 has all the characteristics that identify it as a *bone fide* regulatory subunit of Kv4 proteins. NCS-1 (a) specifically increases Kv4 current amplitudes and modifies its inactivation time course, (b) increases surface trafficking of Kv4.2 proteins, (c) co-immunoprecipitates with native Kv4.2 proteins and (d) co-localizes with Kv4.2 proteins in mouse brain and cardiac myocytes. The expression of NCS-1 in cardiac myocytes is of particular interest since it identifies NCS-1 as an additional potential accessory subunit of Kv4 channel in the heart that may modulate expression levels and function of cardiac I_{to} and neuronal A-currents.

Acknowledgements

These studies were supported in part by the National Institutes of Health (R01-HL71058), American Heart Association (Scientist Development Award, 013019N) and in part by the Seventh Masonic District Association, Inc.

References

[1] Sanguinetti, M. C., Jiang, C., Curran, M. E., and Keating, M. T. 1995, *Cell 81*, 299-307

[2] Tempel, B. L., Papazian, D. M., Schwarz, T. L., Jan, Y. N., and Jan, L. Y. 1987, *Science 237*, 770-775

[3] Kamb, A., Iverson, L. E., and Tanouye, M. A. 1987, *Cell 50*, 405-413

[4] Stuhmer, W., Stocker, M., Sakmann, B., Seeburg, P., Baumann, A., Grupe, A., and Pongs, O. 1988, *FEBS lett. 242*, 199-206

[5] Takumi, T., Ohkubo, H., and Nakanishi, S. 1988, *Science 242*, 1042-1045

[6] Nakamura, T. Y., Coetzee, W. A., Vega-Saenz de Miera, E., Artman, M., and Rudy, B. 1997, *Am. J. Physiol. 273*, H1775-H1786

[7] Fiset, C., Clark, R. B., Shimoni, Y., and Giles, W. 1997, *J. Physiol. (Lond.) 500*, 51-63

[8] Guo, W., Xu, H., London, B., and Nerbonne, J. M. 1999, *J Physiol 521 Pt 3*, 587-599

[9] Barry, D. M., Xu, H., Schuessler, R. B., and Nerbonne, J. M. 1998, *Circ. Res. 83*, 560-567

[10] Coetzee, W. A. and Rudy, B. (2003) K^+ Channels. In Cooper, D., editor. *Nature Encyclopedia of the Human Genome*, Nature Publishing Group, Macmillan Publishers Ltd, London

[11] Rudy, B. 1988, *Neuroscience 25*, 729-749

[12] Coetzee, W. A., Amarillo, Y., Chiu, J., Chow, A., McCormack, T., Moreno, H., Nadal, M., Ozaita, A., Pountney, D. J., Vega-Saenz de Miera, E., and Rudy, B. 1999, *Ann. N. Y. Acad. Sci. 868*, 233-285

[13] Baumgarten, C. M., Singer, D. H., and Fozzard, H. A. 1984, *Circ. Res. 54*, 65-73

[14] Belke, D. D., Wang, L. C., and Lopaschuk, G. D. 1998, *Biochim Biophys Acta 1391*, 25-36
[15] Leicher, T., Roeper, J., Weber, K., Wang, X., and Pongs, O. 1996, *Neuropharmacology 35*, 787-795
[16] Morales, M. J., Wee, J. O., Wang, S. M., Strauss, H. C., and Rasmusson, R. L. 1996, *Proc. Natl. Acad. Sci. USA 93*, 15119-15123
[17] Rasmusson, R. L., Morales, M. J., Wang, S. M., Liu, S. G., Campbell, D. L., Brahmajothi, M. V., and Strauss, H. C. 1998, *Circ. Res. 82,* 739-750
[18] Barry, D. M., Trimmer, J. S., Merlie, J. P., and Nerbonne, J. M. 1995, *Circ. Res. 77*, 361-369
[19] Dixon, J. E. and McKinnon, D. 1994, *Circ. Res. 75*, 252-260
[20] Dixon, J. E., Shi, W. M., Wang, H. S., McDonald, C., Yu, H., Wymore, R. S., Cohen, I. S., and McKinnon, D. 1996, *Circ. Res. 79*, 659-668
[21] Roberds, S. L. and Tamkun, M. M. 1991, *Proc. Natl. Acad. Sci. USA 88*, 1798-1802
[22] London, B., Wang, D. W., Hill, J. A., and Bennett, P. B. 1998, *J Physiol 509*, 171-182
[23] Martens, J. R., Kwak, Y. G., and Tamkun, M. M. 1999, *Trends Cardiovasc. Med. 9*, 253-258
[24] Rudy, B., Hoger, J. H., Lester, H. A., and Davidson, N. 1988, *Neuron 1*, 649-658
[25] Serodio, P., Kentros, C., and Rudy, B. 1994, *J. Neurophysiol. 72*, 1516-1529
[26] An, W. F., Bowlby, M. R., Betty, M., Cao, J., Ling, H. P., Mendoza, G., Hinson, J. W., Mattsson, K. I., Strassle, B. W., Trimmer, J. S., and Rhodes, K. J. 2000, *Nature 403*, 553-556
[27] Braunewell, K. H. and Gundelfinger, E. D. 1999, *Cell Tissue Res 295*, 1-12
[28] Burgoyne, R. D. and Weiss, J. L. 2001, *Biochem. J 353*, 1-12
[29] Pongs, O., Lindemeier, J., Zhu, X. R., Theil, T., Engelkamp, D., Krah-Jentgens, I., Lambrecht, H. G., Koch, K. W., Schwemer, J., Rivosecchi, R., and . 1993, *Neuron 11*, 15-28
[30] McFerran, B. W., Weiss, J. L., and Burgoyne, R. D. 1999, *J Biol. Chem 274*, 30258-30265
[31] Nakamura, T. Y., Pountney, D. J., Ozaita, A., Nandi, S., Ueda, S., Rudy, B., and Coetzee, W. A. 2001, *Proc Natl Acad Sci U S A 98*, 12808-12813
[32] Ames, J. B., Hendricks, K. B., Strahl, T., Huttner, I. G., Hamasaki, N., and Thorner, J. 2000, *Biochemistry 39*, 12149-12161
[33] Cox, J. A., Durussel, I., Comte, M., Nef, S., Nef, P., Lenz, S. E., and Gundelfinger, E. D. 1994, *J Biol. Chem 269*, 32807-32813
[34] Zozulya, S. and Stryer, L. 1992, *Proc. Natl. Acad. Sci. U. S. A 89*, 11569-11573
[35] Ames, J. B., Tanaka, T., Stryer, L., and Ikura, M. 1996, *Curr. Opin. Struct. Biol. 6*, 432-438
[36] Ames, J. B., Ishima, R., Tanaka, T., Gordon, J. I., Stryer, L., and Ikura, M. 1997, *Nature 389*, 198-202
[37] Bourne, Y., Dannenberg, J., Pollmann, V., Marchot, P., and Pongs, O. 2001, *J Biol. Chem 276*, 11949-11955
[38] Chen, X. L., Zhong, Z. G., Yokoyama, S., Bark, C., Meister, B., Berggren, P. O., Roder, J., Higashida, H., and Jeromin, A. 2001, *J Physiol 532*, 649-659

[39] Olafsson, P., Wang, T., and Lu, B. 1995, *Proc Natl Acad Sci U S A 92*, 8001-8005
[40] McFerran, B. W., Graham, M. E., and Burgoyne, R. D. 1998, *J Biol. Chem 273*, 22768-22772
[41] Zhao, X., Varnai, P., Tuymetova, G., Balla, A., Toth, Z. E., Oker-Blom, C., Roder, J., Jeromin, A., and Balla, T. 2001, *J Biol. Chem* (in press),
[42] Hendricks, K. B., Wang, B. Q., Schnieders, E. A., and Thorner, J. 1999, *Nat. Cell Biol. 1*, 234-241
[43] Hama, H., Schnieders, E. A., Thorner, J., Takemoto, J. Y., and DeWald, D. B. 1999, *J Biol. Chem 274*, 34294-34300
[44] Martin, T. F. 2001, *Curr. Opin. Cell Biol. 13*, 493-499
[45] Weisz, O. A., Gibson, G. A., Leung, S. M., Roder, J., and Jeromin, A. 2000, *J. Biol. Chem. 275*, 24341-24347
[46] Shibata, E. F., Drury, T., Refsum, H., Aldrete, V., and Giles, W. 1989, *Am. J. Physiol. 257*, H1773-H1781
[47] Hiraoka, M. and Hiraoko, M. 1973, *J. Physiol-London. 234*, 587-595
[48] Boyett, M. R. 1981, *J. Physiol-London. 319*, 1-22
[49] Kukushkin, N. I., Gainullin, R. Z., and Sosunov, E. A. 1983, *Pflugers Arch-Eur. J. Physiol. 399*, 87-92
[50] Dukes, I. D. and Morad, M. 1991, *J Physiol 435*, 395-420
[51] Tohse, N. 1988, *Hokkaido Igaku Zasshi - Hokkaido Journal of Medical Science 63*, 754-771
[52] Braun, A. P., Fedida, D., Clark, R. B., and Giles, W. R. 1990, *Journal of Physiology 431*, 689-712
[53] Fedida, D., Shimoni, Y., and Giles, W. R. 1990, *J. Physiol-London. 423*, 257-277
[54] Van Wagoner, D. R., Kirian, M., and Lamorgese, M. 1996, *Am. J. Physiol. Heart Circ. Physiol. 271*, H937-H946
[55] Pacioretty, L. M. and Gilmour, R. F., Jr. 1998, *Am. J. Physiol. 275*, H1599-H1605
[56] Tseng, G. N., Yao, J. A., and Tseng-Crank, J. 1997, *Am. J. Physiol. Heart Circ. Physiol. 272*, H1275-H1286
[57] Le Grand, B. 1993, *Biochemical & Biophysical Research Communications 196*, 1143-1147
[58] Furukawa, T., Myerburg, R. J., Furukawa, N., Bassett, A. L., and Kimura, S. 1990, *Circ. Res. 67*, 1287-1291
[59] Liu, D.-W., Gintant, G. A., and Antzelevitch, C. 1993, *Circ. Res. 72*, 671-687
[60] Wang, Z. G., Fermini, B., and Nattel, S. 1991, *Am. J. Physiol. 260*, H1501-6
[61] Di Diego, J. M., Sun, Z. Q., and Antzelevitch, C. 1996, *Am. J. Physiol. Heart Circ. Physiol. 271*, H548-H561
[62] Qi, A., Yeung-Lai-Wah, J. A., Xiao, J., and Kerr, C. R. 1994, *Am. J. Physiol. 266*, H643-9
[63] Kilborn, M. J. and Fedida, D. 1990, *J. Physiol-London. 430*, 37-60
[64] Wu, M. H., Su, M. J., and Lue, H. C. 1994, *Journal of Molecular & Cellular Cardiology 26*, 1167-1177
[65] Mansourati, J. and Le Grand, B. 1993, *Am. J. Physiol. 265*, H1466-70

[66] Sanchezchapula, J., Elizalde, A., Navarropolanco, R., and Barajas, H. 1994, *Am. J. Physiol. 266*, H1184-H1194

[67] Crumb, W. J., Jr., Pigott, J. D., and Clarkson, C. W. 1995, *Am. J. Physiol.* 268, H1335-42

[68] Wang, L. and Duff, H. J. 1997, *Circ. Res. 81*, 120-127

[69] Thierfelder, S., Hirche, H., and Benndorf, K. 1994, *Pflugers Archiv - European Journal of Physiology 427*, 547-549

[70] Pike, G. K., Bretag, A. H., and Roberts, M. L. 1993, *Journal of Physiology* 470, 365-382

[71] Ogbaghebriel, A. and Shrier, A. 1994, *Am. J. Physiol. 266*, H182-H190

[72] Nakamura, T. Y., Shattock, M. J., and Coetzee, W. A. 1995, *Biophys J* 68, A107.

[73] Xu, Z., Patel, K. P., and Rozanski, G. J. 1996, *Am. J. Physiol. Heart Circ. Physiol. 271*, H2154-H2161

[74] Sun, Z. Q., Ojamaa, K., Coetzee, W. A., Artman, M., and Klein, I. 2000, *Am J Physiol Endocrinol. Metab 278*, E302-E307

[75] Jourdon, P. and Feuvray, D. 1993, *J. Physiol-London.* 470411-429, -429

[76] Tsuchida, K. and Watajima, H. 1997, *Am. J. Physiol. Endocrinol. Metab. 273*, E695-E700

[77] Nishiyama, A., Ishii, D. N., Backx, P. H., Pulford, B. E., Birks, B. R., and Tamkun, M. M. 2001, *Am. J. Physiol Heart Circ. Physiol 281,* H1800-H1807

[78] Tomita, F., Bassett, A. L., Myerburg, R. J., and Kimura, S. 1994, *Circ. Res. 75,* 296-303

[79] Le Grand, B. L., Hatem, S., Deroubaix, E., Couetil, J. P., and Coraboeuf, E. 1994, *Cardiovasc. Res. 28*, 548-556

[80] Kaab, S., Dixon, J., Duc, J., Ashen, D., Nabauer, M., Beuckelmann, D. J., Steinbeck, G., McKinnon, D., and Tomaselli, G. F. 1998, *Circulation 98,* 1383-1393

[81] Beatch, G. N., Abraham, S., MacLeod, B. A., Yoshida, N. R., and Walker, M. J. A. 1991, *Br. J. Pharmacol. 102*, 13-18

[82] Castle, N. A. 1991, *Journal of Pharmacology & Experimental Therapeutics 257*, 342-350

[83] Delpon, E., Tamargo, J., and Sanchez-Chapula, J. 1992, *Br. J. Pharmacol. 106*, 464-469

[84] Gotoh, Y., Imaizumi, Y., Watanabe, M., Shibata, E. F., Clark, R. B., and Giles, W. R. 1991, *Am. J. Physiol. 260*, H1737-H1742

[85] Tsuchihashi, K. and Curtis, M. J. 1991, *J. Cardiovasc. Pharmacol. 18*, 445-456

[86] Wang, Z., Fermini, B., and Nattel, S. 1995, *Journal of Pharmacology & Experimental Therapeutics 272*, 184-196

[87] Nef, S., Fiumelli, H., De Castro, E., Raes, M. B., and Nef, P. 1995, *J Recept. Signal. Transduct. Res 15*, 365-378

[88] Olafsson, P., Soares, H. D., Herzog, K. H., Wang, T., Morgan, J. I., and Lu, B. 1997, *Brain Res Mol. Brain Res 44*, 73-82

[89] Martone, M. E., Edelmann, V. M., Ellisman, M. H., and Nef, P. 1999, *Cell Tissue Res 295*, 395-407

[90] Paterlini, M., Revilla, V., Grant, A. L., and Wisden, W. 2000, *Neuroscience* 99, 205-216

[91] Werle, M. J., Roder, J., and Jeromin, A. 2000, *Neurosci. Lett. 284*, 33-36

[92] Guo, W., Malin, S. A., Johns, D. C., Jeromin, A., and Nerbonne, J. M. 2002, *J. Biol. Chem. 277*, 26436-26443

[93] Flaherty, K. M., Zozulya, S., Stryer, L., and McKay, D. B. 1993, *Cell 75*, 709-716

[94] Zerangue, N., Schwappach, B., Jan, Y. N., and Jan, L. Y. 1999, *Neuron 22*, 537-548

[95] O'Callaghan, D. W., Hasdemir, B., Leighton, M., and Burgoyne, R. D. 2003, *J. Cell Sci. 2003. Dec. 1;116. (Pt. 23):4833. -45.* 116, 4833-4845

[96] Ren, X., Shand, S. H., and Takimoto, K. 2003, *J. Biol. Chem. 278*, 43564-43570

[97] Cork, R. J., Cicirelli, M. F., and Robinson, K. R. 1987, *Dev. Biol. 121*, 41-47

[98] Bers, D. M. and Ellis, D. 1982, *Pflugers Arch-Eur. J. Physiol. 393,* 171-178

[99] Haddock, P. S., Coetzee, W. A., Cho, E., Bers, D. M., and Artman, M. 1999, *Circ. Res. 85*, 415-427

PART II: HIPPOCALCIN, NEUROCALCIN AND VILIPs

In: Neuronal Calcium Sensor Proteins
Editors: Pavel Philippov and Karl Koch, pp. 67-81

ISBN 1-59454-978-8

Chapter IV

HIPPOCALCIN – A MULTIFUNCTIONAL MODULATOR IN THE HIPPOCAMPUS

Masaaki Kobayashi and Ken Takamatsu *
Department of Physiology, Toho University School of Medicine, 5-21-16 Ohmori-nishi, Ohta-ku, Tokyo, 143-8540, Japan.

Abstract

Hippocalcin (HPCA) is predominantly expressed in the hippocampus after the final stages of synaptic formation. HPCA undergoes Ca^{2+}- and myristoyl-dependent translocation from the cytosol to intracellular membranes of the trans-Golgi network and the plasma membrane in response to an increase in $[Ca^{2+}]i$. HPCA-null mutant (-/-) mice display no obvious abnormalities in the hippocampal structure and in the synaptic excitability of CA1 pyramidal cells; however, NMDA- and depolarization-induced phosphorylation of CREB and ERK are significantly attenuated in -/- hippocampal neurons. Spatial and associative learning abilities of -/- mice are impaired in the probe test on Morris water maze and in the visual discrimination learning task. HPCA is also expressed in the olfactory receptor neurons, and found to regulate the activity of ciliary adenylyl cyclases and particulate guaynylyl cyclases in a Ca^{2+}-dependent manner. HPCA inhibits GRK1 under high Ca^{2+} conditions, and regulates PKC-independent Ca^{2+}-induced phospholipase D activation in conjunction with Cdc42. HPCA binds MLK2 and -3, whereas no effect on kinase activity is apparent. HPCA also binds cytochrome b5 in a Ca^{2+}-dependent manner. The most convincing function of HPCA is a protecting effect against Ca^{2+}-induced cell death through its interaction with neuronal apoptosis inhibitory protein. Thus, HPCA may act as a multifunctional modulator in the Ca^{2+}-signaling pathways, such as those underlie long lasting neural plasticity, olfaction and neuronal cell death.

* Correspondence concerning this article should be addressed to Ken Takamatsu, Department of Physiology, Toho University School of Medicine, 5-21-16 Ohmori-nishi, Ohta-ku, Tokyo 143-8540, Japan. Tel: +81 3 3762 4151; Fax: +81 3 3762 8225; E-mail: physiken@med.toho-u.ac.jp.

Keywords: hippocalcin, calcium-binding protein, knockout mouse, hippocampus, visual discrimination learning, spatial learning, cAMP-responsive element-binding protein (CREB), extracellular signal-regulated protein kinase (ERK), phosphorylation.

Abbreviations

ACs	adenylyl cyclases
$[Ca^{2+}]i$	intracellular free calcium concentration
CREB	cAMP-response element-binding protein
ERK	extracellular signal-regulated kinase
fEPSP	field excitatory postsynaptic potential
GCs	guanylyl cyclases
GRK	G-protein-coupled receptor kinase
HPCA	hippocalcin
LED	light-emitting diode
LTP	long-term potentiation
MLK	mixed lineage kinase
MAP	mitogen activating protein
NAIP	neuronal apoptosis inhibitory protein
NCS	neuronal calcium sensor
NMDA	*N*-methyl-D-aspartate
ORNs	olfactory receptor neurons
PKC	protein kinase C
PLD	phospholipase D
PPF	paired pulse facilitation
PSD	postsynaptic density
TGN	trans-Golgi network
VGCC	voltage-gated calcium channel
VILIP	visinin-like protein.

1. Introduction

Hippocalcin (HPCA), a member of visinin-like protein (VILIP) subfamily, is a conserved protein having an identical amino acid sequence among rat, mouse and human [1-4]. HPCA has a Ca^{2+}-myristoyl switch mechanism within cells, allowing it to be cytosolic at resting $[Ca^{2+}]i$ and translocate to intracellular membranes of the trans-Golgi network (TGN) and the plasma membrane in response to an increase in $[Ca^{2+}]i$ [5-7]. Among neuronal calcium sensor (NCS) proteins, HPCA shows a very restricted tissue distribution pattern, exclusively in the brain and retina, as revealed by theoretical mRNA expression profiles derived from expression sequence tag and serial analysis of gene expression data from human tissues [8]. In the brain, HPCA is expressed predominantly in the hippocampus [1], as shown in Figure 1. In the hippocampus, HPCA is present at high levels in the pyramidal cells of Ammon's horn and moderately in the granule cells of the dentate gyrus. In the hippocampal pyramidal cells, HPCA is located in the cytoplasm and plasma membrane of cell bodies, dendrites and axons [9]. The expression level of HPCA in the hippocampal pyramidal cells is estimated to be more than 30μM [10]. During development, an abrupt increase in HPCA expression is consistent with the phase that synaptic formation occurs [11], indicating that HPCA may be involved in final steps of synaptogenesis and/or in synaptic functions in the matured hippocampus. Besides the hippocampus, HPCA is also found at considerable levels in other brain regions, including the neocortex, caudate-putamen, taenia tecti, claustrum, olfactory tubercle, anterior olfactory nucleus and olfactory bulb [9,12,13], in the ganglion cell layer and amacrine cell layer in the retina [14], and in the apical layer in the olfactory epithelium [15], indicating the additional roles for HPCA in the Ca^{2+}-signaling pathways *in situ*.

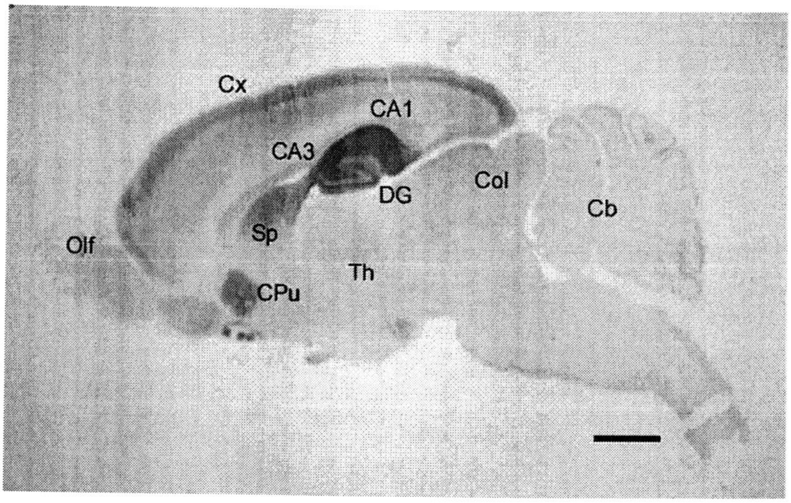

Figure 1. Distribution of hippocalcin in the brain. A Sagittal section of adult mouse (BalbC57BL/6J) brain was immunostained using an anti-hippocalcin antibody. CA1, CA1 region of Ammon's horn; CA3, CA3 region of Ammon's horn; Cb, cerebellum; Col, Colliculus; CPu, caudate-putamen; Cx, cerebral cortex; DG, dentate gyrus; Olf, olfactory bulb; Sp, septum; Th, thalamus. Scale bars= 1 mm.

2. Ca^{2+}-Myristoyl Switch in Hippocalcin

The Ca^{2+}-myristoyl switch, which enables the reversible localization of signaling proteins to membrane compartments including the cell surface and intracellular membrane specializations or to associated cytoskeletal components, has been postulated to be a signal transduction mechanism for the selective activation of downstream signaling cascades, such as receptors, receptor signaling complexes, and signal effector molecules [16]. The Ca^{2+}-myristoyl switch is a major biochemical feature among NCS proteins [4]. It has been revealed biochemically that HPCA undergoes Ca^{2+}- and N-myristoyl-dependent translocation [5]. The translocation of HPCA from the cytosol to membrane compartments has been examined in detail using HPCA-EYFP (enhanced yellow fluorescent protein) in living cells [6,7]. HPCA was present diffusely throughout the cell at resting $[Ca^{2+}]i$ and translocated to intracellular membrane of the TGN and plasma membrane in response to an increase in $[Ca^{2+}]i$. The heterologously expressed HPCA in HeLa cells showed a maximal rate of translocation with a time constant of around 1 sec. Half maximal translocation occurred at around 300 nM free Ca^{2+} with a dynamic range of Ca^{2+}-sensitivity of 200-800 nM $[Ca^{2+}]i$. This suggests that HPCA would be able to affect its target proteins on membranes, when $[Ca^{2+}]i$ is elevated barely above resting levels within the cell for sufficient time to allow translocation. HPCA translocation is more sensitive to increases in $[Ca^{2+}]i$ than calmodulin [17], but the full responses are on the time scale from one second to ten seconds. This time course is slower than Ca^{2+} transients seen in neurons due to action potentials, but is comparable to long-lasting Ca^{2+} elevations lasting up to hundreds of milliseconds in neurons for the induction of long-term potentiation or long-term depression. The observed reversible Ca^{2+}-myristoyl switch of HPCA is similar to that of VILIP-3 and neurocalcin δ in Ca^{2+}-dependent translocation to the TGN membrane and plasma membrane, but is different from that of VILIP-1 and NCS-1 in Ca^{2+}-independent accumulation to the membranous structures [18,19]. Since several NCS proteins are co-expressed in the same neurons, NCS proteins may allow neurons a highly selective response to various Ca^{2+} stimuli using their myristoyl groups in distinct ways in living cells.

3. Hippocalcin in the Hippocampus

3.1. Morphogenesis

Ontogenic analysis of the rat hippocampus revealed that HPCA appears in the hippocampal pyramidal cells at late developmental stages of dendritic arborization, and indicated that HPCA may play roles in final steps of synaptogenesis and/or in matured synapses [11]. However, the general structure of the adult brain of HPCA-null mutant (-/-) mice was normal at light microscopic levels [20]. There was no obvious morphological abnormality or cell loss in the hippocampal neurons or glial cells as compared with wild type (+/+) littermates. Amounts and distributions of synaptic markers, such as synaptophysin, microtubule-associated protein 2 and drebrin, were not different between -/- and +/+ mice. Moreover, the lack of HPCA was not compensated by up-regulation of other members of NCS family, such as VILIP-1, -2 and -3 in the -/- hippocampus and cerebral cortex. These results suggest that HPCA may not be essential for morphological development of the brain including the hippocampus.

3.2. Synaptic Efficacy

In the case of hippocampal pyramidal cells, HPCA is present throughout the cells in particular in the dendrites and axons, indicating the involvement of pre- and post-synaptic functions [9]. Synaptic neurotransmission efficacy was examined electrophysiologically using the hippocampal slices of -/- and +/+ mice [20]. Fast fEPSP recorded from CA1 region of -/- mice was indistinguishable from that of +/+ mice (Figure 2A). Application of single test stimuli of graded strength to the Schaffer collateral-commissural fibers showed that the baseline synaptic excitability was normal in the hippocampal CA1 of -/- mice. Concerning presynaptic functions, if HPCA simply acts as a Ca^{2+} buffer, paired pulse stimulus may augment the increase in $[Ca^{2+}]i$ and facilitate the transmitter release from presynaptic terminal and 2nd fEPSP in the -/- slices. If HPCA has a frequenin-like function, Ca^{2+}-dependent complex facilitation in transmitter release [21], attenuated paired pulse facilitation (PPF) may be detected in the -/- slices. The degree of PPF in the hippocampal CA1 as well as overall size and shape of the EPSPs was indistinguishable between -/- and +/+ mice at any interpulse intervals. Therefore, HPCA may not be solely a Ca^{2+} buffer and may not have a frequenin-like function in the presynaptic terminal.

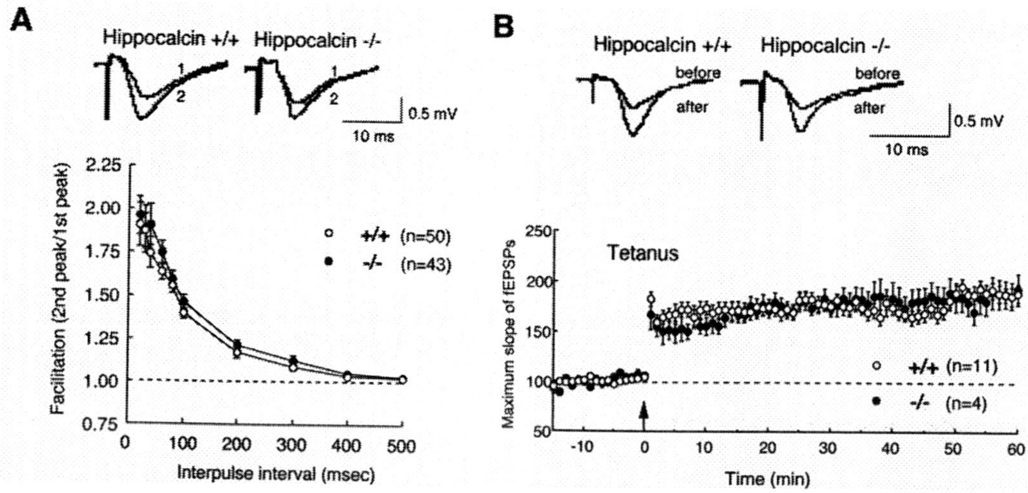

Figure 2. Hippocampal electrophysiology. (A) Paired pulse facilitation (PPF) in the hippocampal CA1 subfield following stimulation to the Schaffer collateral/commissural fiber pathway. (Upper panel) Representative fEPSP traces averaged from five traces of PPF are shown. PPF was elicited by paired-pulse stimulation at an interpulse interval of 100 ms. Field EPSPs evoked by the first and second stimuli are indicated as 1 and 2, respectively. (Lower panel) Averaged maximum slope of fEPSPs by paired-pulse stimulations at various interpulse intervals (10–500 ms). No significant alteration was observed in -/- mice. (B) LTP in the hippocampal CA1 subfield. (Upper panel) Typical fEPSP traces before (upper lines) and 1 min after (lower lines) tetanization. The averages of five traces are represented. (Lower panel) Averaged maximum slope of fEPSPs. This tetanization protocol induced a form of LTP that was indistinguishable in -/- mice and +/+ littermates. (Reproduced from [20]).

In the CA1 pyramidal cells, Ca^{2+} influx through NMDA receptor is an essential event for induction of LTP, followed by the phosphorylation of AMPA receptor channels, cytoskeletal proteins, and their regulators by resident protein kinases at postsynaptic densities (PSD) [22]. HPCA exists in postsynapse in the hippocampal neurons, especially in the CA1 pyramidal cell dendrites at a concentration of more than 30 μM [10]. The high affinity of HPCA for Ca^{2+} postulates that HPCA is linked to LTP [7]. However, no significant differences could be detected in both -/- and +/+ mice, neither in LTP induction as evaluated by the degree of potentiation 1 min after the second tetanus, nor in maintenance of LTP as evaluated by the potentiation 1 hr after the repeated and strong tetanization (Figure 2B), indicating that NMDA/AMPA receptor channels and phosphorylation events in the PSD are intact in -/- mice [20].

3.3. Activity-Dependent Gene Expression

Excitatory neurotransmission elevates the Ca^{2+} concentration in neuronal cells and activates transcription of immediate early genes, such as *c-fos*, *junB*, *c-jun* and *zif/268* [23,24]. Expression of these genes is controlled by upstream transcription factors including serum response factor, ternary complex factors such as Elk-1 or SAP-1, and cAMP-response element-binding protein (CREB). Among these transcription factors, CREB is known to be a key molecule underlying long-term memory [25,26]; however, CREB activation is not essential for hippocampal LTP produced by high frequency stimulation at 100 Hz [27-29]. In hippocampal CA1 neurons, activity-dependent phosphorylation of CREB on its Ser133 residue is absolutely dependent on Ca^{2+} influx via NMDA receptors and L-type VGCCs [30]. When these channels are blocked, other type of channels which cause similar robust increases in somatic and nuclear Ca^{2+} consistently fail to activate CREB [31]. Then, NMDA- and depolarization-induced CREB phosphorylation in the -/- hippocampal slices was assessed by immunohistochemical and immunoblot methods [20].

In the +/+ hippocampal slices, NMDA treatment caused intense nuclear phospho-CREB staining in a large number of CA1 neurons, moderate staining in CA3 neurons, and weak staining in granule cells of the dentate gyrus. However, only faint staining was observed in a limited number of CA1 neurons of the -/- slices. In the +/+ CA1 regions, NMDA stimulation led to an averaged 4.9-fold increase in phospho-CREB intensity over the basal level by immunoblot analysis (Figure 3A, B). In contrast, a significantly lower increase in CREB phosphorylation in the -/- CA1 regions was observed (approximately 2.6-fold over baseline). In the +/+ hippocampal slices, KCl stimulation led to intense phospho-CREB staining throughout the hippocampal pyramidal cells, but weak staining in the dentate granule cells. Similar to the finding with NMDA treatment, KCl treatment led to defective phospho-CREB formation in the -/- CA1 neurons. Immunoblot analysis confirmed a significantly lower amount of depolarization-induced phospho-CREB in the -/- CA1 regions (Figure 3C, D).

Phosphorylation of CREB is known to involve several protein kinases including CaM kinases, mitogen-activated protein (MAP) kinase and cAMP-dependent protein kinase [32,33]. It has been shown that HPCA facilitates the Ca^{2+}-dependent activation of phospholipase D (PLD) in conjunction with Cdc42, a small G protein of the Rho family, while HPCA alone does not directly affect PLD activity [34]. HPCA has also been shown to interact with mixed lineage kinase (MLK) 2, a MAP kinase kinase kinase under the control of small G proteins Rac1/Cdc42, but no effect on kinase activity was observed in HPCA-transfected cells [35]. The involvement of MAP kinase cascade, a pathway upstream of CREB, in the impaired CREB phosphorylation was investigated [20]. Both NMDA receptor- and VGCC-mediated phospho-CREB formations were partially reduced by the addition of PD98059, an inhibitor of the MAP kinase kinase MEK (MAPK/ERK kinase), and the intensity of phospho-CREB staining was indistinguishable between the two genotypes in this condition (see Figure 3). The NMDA- and KCl-induced phosphorylation of MAP kinase ERK (extracellular signal-regulated kinase) in the -/- slices was lower than that in the +/+ slices (unpublished data). The effect of calcium ionophore mimicked that of NMDA and KCl. On the contrary, treatments with 12-O-tetradecanoylphorbol-13-acetate, a protein kinase C-activator, and forskolin, a cAMP dependent protein kinase-activator, increased phosphorylation of ERK by similar levels in both genotypes. These findings raise the possibility that HPCA may be involved in the Ca^{2+}-dependent MAP kinase cascade in conjunction with a small G protein complex, leading to CREB-activation in the hippocampal neurons, and may contribute to long-lasting synaptic plasticity.

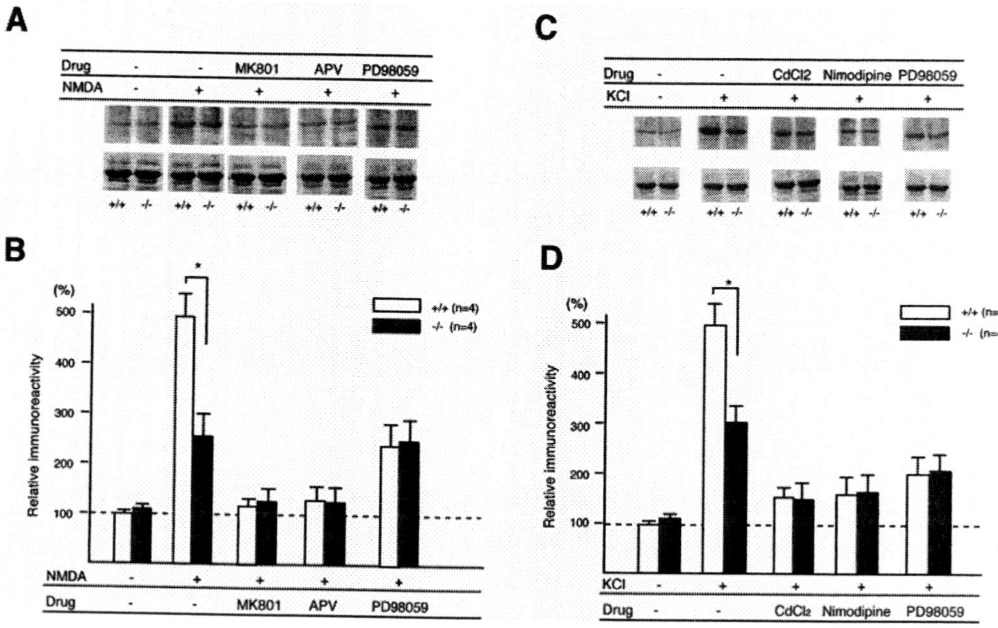

Figure 3. Quantification of phospho-CREB immunoreactivity in the CA1 region of NMDA- and KCl-perfused hippocampal slices. (A, B) NMDA-induced CREB phosphorylation. The hippocampal slices were incubated with both 200 μM MCPG and 5 μM nimodipine for 30 min, then stimulated with 100 μM NMDA for 10 min, or left unstimulated. For pharmacological studies, MK801 (10 μM), APV (100 μM), or PD98059 (20 μM) was added 30 min before stimulation. The CA1 area was dissected out and the extracts were used for immunoblot analysis with an anti-phospho-CREB antibody. Representative blot pattern (A) and relative phospho-CREB intensities obtained by densitometry in a computerized image analyzer system (B) are shown. (C, D) KCl-induced CREB phosphorylation. The slices were incubated in 200 μM MCPG and 100 μM MK801 for 30 min, then stimulated with 50 mM KCl for 10 min, or left unstimulated. Cadmium chloride (100 μM), nimodipine (10 μM), or PD98059 was added 30 min before stimulation for some studies. Representative blot pattern (C) and relative phospho-CREB intensities (D) are shown. * $P<0.05$ (unpaired Student's *t*-test), when comparing the intensity of -/- with +/+. (Reproduced from [20]).

3.4. Spatial and Associative Memory

Activity dependent gene expression followed by protein synthesis is fundamental for long lasting synaptic plasticity underlying the acquisition of memory. The hippocampus is thought to be an important cortical region for associative learning and memory, especially in the performance of spatial-positioning discrimination [36]. Indeed, damage to the hippocampus in animals results in impairments in a particular type of spatial learning and memory [37]. Selective deletion of the gene encoding the NMDA receptor in area CA1 of the hippocampus is sufficient to cause severe impairments in both LTP formation and spatial learning that relies on distal cues [38]. Two test paradigms, the Morris water maze test [39] and the visual discrimination learning task [40], were used to examine the abilities of HPCA -/- mice in spatial memory and working (trial-dependent) associative memory [20].

HPCA -/- mice are not different from wild type littermates with respect to behaviors in standard cage environment, spontaneous motor activities, rota-rod test, elevated plus-maze test, light/dark choice test and passive avoidance learning, indicating that HPCA may not contribute to the motor activity, motor coordination, anxiety, and aversive memory [20]. In the cued trials on Morris water maze test, both groups of mice swam well and could escape onto the visible platform (Figure 4). In the place trials, both groups showed substantial improvement in learning, as measured by escape latency and swimming distance, and did not differ significantly. In the probe trials, the -/- mice crossed over the previous platform location significantly less often that the +/+ mice, indicating impairments in spatial memory in HPCA -/- mice.

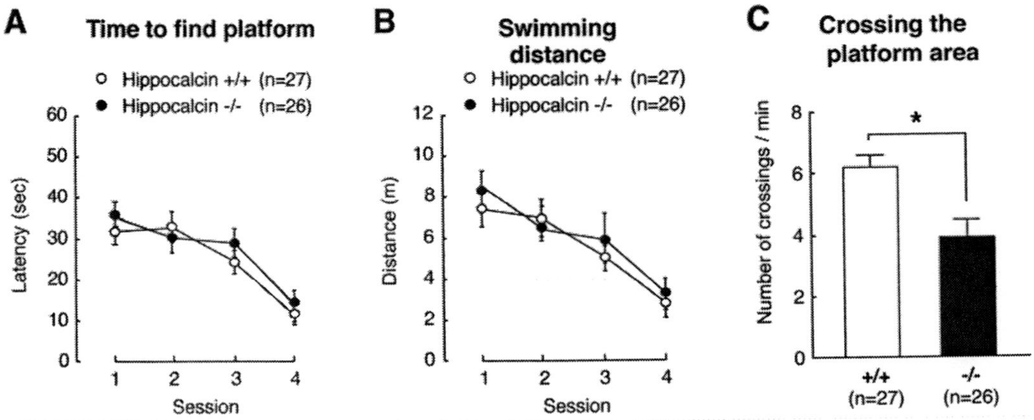

Figure 4. Learning ability on the Morris water maze test in wild-type (+/+) and hippocalcin-null (-/-) mice. The Morris water maze test was conducted in 3-month-old mice of both genotypes. Swimming behavior was continuously monitored by video and analyzed by a computerized tracking system. The time and swimming distance to find the platform was measured. (A, B) In the place trials, each mouse underwent one daily session of four 60 s trials for 4 consecutive days. Each value shows the mean latency (A) and averaged swimming distance (B). (C) Number of times the platform area was crossed in the probe trial. In this trial, the platform was removed, and the mouse was allowed to swim for 60 s. The average numbers of crossings over the former location of the platform are shown. All results are represented as means±S.E.M. (*n*, number of mice). In the probe test (C), significant difference was observed between the two groups showing fewer crossings of the former location of the platform in -/- mice. * $P<0.05$ (unpaired Student's *t*-test). (Reproduced from [20]).

In the operant test of visual discrimination learning, mice were required to learn on the basis of their ability to discriminate between right and left visual cues. In the original training, either right or left light-emitting diode (LED) was lit at random. A touch reaction ipsilateral to a lit LED was considered as 'correct' and the mouse could gain a food pellet under multiple variable extinction schedules. The +/+ littermate controls gradually improved in learning from the 3rd block (Figure 5A). In contrast, -/- mice performed this task at chance levels (50%) throughout the training. In the reversal training, a touch reaction contra-lateral to a lit LED was considered as 'correct'. The +/+ mice learned that the correct response had changed from the ipsilateral panel-touch reaction to the lit-light into the contralateral panel-touch reaction, and the correct response ratio increased from 6th block (Figure 5B). In contrast, and similar to the result on the original training, -/- mice performed at chance levels throughout the reversal training. Thus, the learning performance of -/- mice was significantly lower than that of +/+ mice. Together with the observations that disruption of the CREB gene causes severe impairments in spatial learning in mice [41,42], HPCA may be essential for hippocampal functions reflecting long-lasting plasticity including spatial and associative memory, as a part of the CREB activation pathway.

The visual discrimination learning is known to depend on the roles of many brain regions, such as the superior colliculus in association between visual motion direction and saccade direction, and the basal forebrain in a wide spectrum of attentional functions ranging from the detection, selection and discrimination of stimuli to the allocation of processing resources to competing tasks [43,44]. HPCA is found at considerable levels in many brain regions besides the hippocampus [9,12,13]. Lack of HPCA expression in the basal forebrain including the striatum may be partly responsible for the changes in visual discrimination learning in -/- mice.

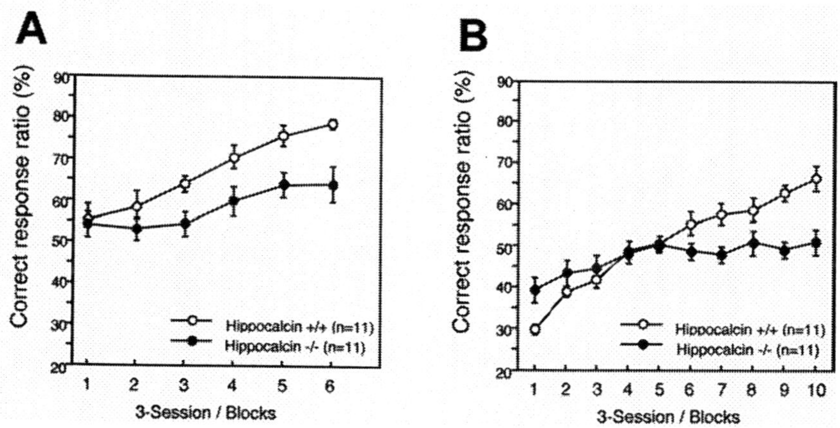

Figure 5. Learning ability on a visual discrimination task in wild-type (+/+) and hippocalcin-null (-/-) mice. (A) Correct response ratio in original training. A touch reaction ipsilateral to the randomly illuminated LED was judged to be "correct," and mice underwent one daily session of 40 trials for 18 consecutive days. The correct response ratio in each block was calculated as the ratio of correct responses to the total responses using data summed up from three sessions. Results are expressed as mean±S.E.M. A significant difference was observed between the two groups by ANOVA ($F_{(5,113)}$=2.82, P=0.021). (B) Correct response ratio in reversal training. The "correct" response was changed to be a touch on the side contralateral to the illuminated LED. Training was conducted for 30 consecutive days ($F_{(9,189)}$=7.85, P<0.0001). (Reproduced from [20]).

4. Hippocalcin and G-Protein Coupled Receptor Kinases in the Retina

Immunohistochemical analysis shows the presence of HPCA in the amacrine and ganglion cell layers of retina [14]. HPCA exists in the membranes of the inner plexiform layer but not detected in photoreceptor outer segments. HPCA, like recoverin, is able to inhibit GRK1 (rhodopsin kinase) phosphorylating activity in a Ca^{2+}-dependent manner, whereas in parallel experiments it does not interact with GRK2 (βARK) and GRK5 and is ineffective on rod outer segment phosphorylation by these kinase subtypes [45]. GRK1, like rhodopsin, is located in rod outer segments where it is closely linked with phototransduction and adaptation [46], and where HPCA is not located. Thus, HPCA appears to be unrelated to the physiology of phototransduction, indicating additional roles for HPCA other than GRK1 inhibition in the retina.

5. Hippocalcin and Adenylyl and Guanylyl Cyclases in the Olfactory Epithelium

HPCA immunoreactivity has been found to be predominantly localized to the apical layer of the olfactory epithelium where the olfactory knob and cilia of mature olfactory receptor neurons (ORNs) reside [15]. Immunoreactive signal was not detectable in soma and axons of ORNs. These results suggest that HPCA may be involved in initial olfactory signal transduction or adaptation. Immunoblot analysis showed that HPCA is also expressed in the olfactory bulb though at lower levels than in the olfactory epithelium, indicating additional roles for HPCA in olfactory processing. When odorants interact with specific receptor present on the cilia of ORNs, these G-protein coupled receptors activate adenylyl cyclases (ACs) to produce cAMP [47]. As intracellular cAMP levels increase, olfactory cyclic nucleotide-gated channels open to allow an influx of Na^{+} and Ca^{2+}, leading to the generation of an action potential. Olfactory signal transduction pathways are modulated by changes in $[Ca^{2+}]i$ and specific Ca^{2+}-binding proteins mediate these changes [47].

Effect of HPCA on cAMP signaling (ACs activity) in olfactory cilia was investigated [15]. HPCA increased ACs activity at low free Ca^{2+} concentration of 10 nM and this HPCA-mediated ACs activity decreased as free Ca^{2+} concentration increased above 300 nM. Since $[Ca^{2+}]i$ in the cilium is 40 nM at rest and increases up to 300 nM by odorant stimulation [48], HPCA may modulate ACs activity under physiological Ca^{2+} concentrations. An alternative effect has been reported for VILIP-1 [49]. *In vitro* recombinant VILIP-1 attenuated odorant-induced ACs activation in a Ca^{2+}-dependent manner. Thus, HPCA and VILIP-1 may fine tune the ACs activity in the adaptation of ORNs to odorant stimulation. Effect of the phosphorylation level of ciliary proteins on the modulation of ACs by HPCA has also been examined [15]. At low free Ca^{2+} levels, protein kinase treatment of cilia inhibited the effect of HPCA on ACs activity, whereas either native cilia or protein phosphatase-treated cilia showed increases in HPCA-mediated ACs activity. At high free Ca^{2+} conditions, only protein phosphatase-treated cilia showed significant increases in HPCA-mediated ACs activity. Thus, dynamic status of phosphorylation of ciliary proteins may play a critical roles in HPCA dependency of ciliary ACs. Calcium influx by odorant stimulation induces phosphorylation of ACs [50], that in turn may keep ACs activity at low levels and can be involved in the mechanisms for odor adaptation. Another second messenger, such as cGMP, is also produced by odorant stimulation [51]. In contrast to the effect on ACs, HPCA significantly inhibited the activity of particulate guanylyl cyclases (GCs) only within a limited range of free Ca^{2+}

concentrations of 1-10 nM [15]. The inhibitory effect of HPCA on GCs activity decreased as [Ca^{2+}]i concentrations increased. Thus, HPCA may play a role in keeping the level of cGMP low in ORNs at rest, lowering the activity of cGMP-dependent kinase. These data indicate that HPCA is present in ORNs, and is likely to regulate second messenger cascades in a Ca^{2+}-dependent manner. Several isoforms of AC and GC are present in the brain [52, 53], and thus represent other potential targets for regulation by HPCA.

6. Interacting Proteins with Hippocalcin

6.1. NAIP

HPCA was found to interact in a yeast two-hybrid screen and in direct biochemical assays with the neuronal apoptosis inhibitory protein (NAIP) [54,55]. In neuronal cell lines, overexpression of the baculovirus inhibitory repeat (BIR) domains of NAIP (NAIP-BIR1-3) counteracted the Ca^{2+}-induced cell death induced by ionomycin and thapsigargin. This protective effect is significantly enhanced when NAIP-BIR1-3 was co-expressed with HPCA. Over-expression of HPCA alone did not substantially enhance cell survival, but co-expression greatly increased their protective effects. Indeed, hippocampal neurons of -/- mice are frail against excitotoxicity caused by kainic acid [56]. In addition, an age-dependent degeneration in specific classes of neurons is seen in the -/- hippocampus and cerebral cortex [56]. The most convincing functional data on HPCA, as an inhibitor of apoptosis through its interaction with NAIP are reviewed in another chapter of this issue.

6.2. CDC42/PLD

During a search for PKC-independent Ca^{2+}-mediated PLD activation pathway, HPCA was isolated as a Ca^{2+}-induced PLD activator from rat brain [57]. While purified HPCA alone did not activate PLD activity, it increased PLD activity by 1.8-fold in conjunction with the small rho-like GTPase Cdc42 in the presence of 300 nM free Ca^{2+} [34]. In the absence of Ca^{2+} in the reaction mixture, the effect of HPCA to facilitate Cdc42-activated PLD activity was abolished. These results suggest that HPCA might be one of the regulatory proteins in the PKC-independent Ca^{2+}-induced PLD activation pathway in conjunction with the Cdc42 protein.

6.3. Cytochrome b5

During a search for a target protein of VILIP-3, microsomal cytochrome b5 was found to be a Ca^{2+}-dependent binding partner of VILIP-3 by a pull-down experiment using glutathione S-transferase-tagged VILIP-3 [58]. In those experiments, HPCA, as well as VILIP-3, could interact with microsomal cytochrome b5. The binding affinity of HPCA to cytochrome b5 is 1.3-fold stronger than that of VILIP3. Microsomal cytochrome b5 is one of the known heme-containing proteins and functions as an electron carrier for several membrane-bound oxygenases such as cytochrome p450 [59]. Three major isozymes of cytochrome b5 are known to exist: microsomal, mitochondrial and soluble forms. One of these is microsomal cytochrome b5, located in the endoplasmic reticulum. Calcium-dependent translocation to the TGN membrane of the perinuclear region has been observed in the case of HPCA and VILIP3 [6,7,18,19]. Therefore, microsomal cytochrome b5, which is located in the endoplasmic reticulum-perinuclear region, is well suited as a Ca^{2+}-dependent binding partner of HPCA and VILIP3. However, it remains difficult to assign specific physiological functions to HPCA and VILIP3 carried out as a result of the interaction with cytochrome b5, because the terminal enzyme involved in this context is still unknown.

6.4. MLK2 and -3

In an attempt to identify proteins that affect the activity and localization of MLK2 and -3 (mixed lineage Ser/Thr kinases), a yeast two-hybrid cDNA library was screened and HPCA was identified as a possible interaction partner [35]. MLKs are closely related to the mitogen-activated protein kinase kinase kinase (MAPKKK) family [60]. MLK2 and -3 interact with Rac and Cdc42, with motor proteins of the kinesin superfamily, and co-localize with the tublin cytoskeleton [35]. This suggests that they are involved in regulation of cytoskeleton dynamics in cells. After activation, MLK2 and -3 activate the JNK (c-Jun N-terminal kinase)/MAP kinase cascade, and activation of ERK and p38 by MLKs has also been reported [61]. Although the interaction of HPCA with MLK was not calcium-dependent and no influence on kinase activity or microtubule localization was observed [35], an additional involvement of HPCA in MAP kinase signaling has been suggested by studies in HPCA-null mutant mice [20]. The NMDA- and KCl-induced ERK phosphorylation was reduced in the -/- hippocampal slices compared to the +/+ slices, suggesting that HPCA seems to have a modulatory influence on MAP-kinase signaling. It is possible that HPCA may serve as an anchor of the MAPKKK family to enable complex functioning in the specialized intracellular component in a Ca^{2+}-dependent manner.

References

[1] Kobayashi, M., Takamatsu, K., Saitoh, S., Miura, M., and Noguchi, T. 1992, *Biochem. Biophys. Res. Commun., 189*, 511.

[2] Takamatsu, K., Kobayashi, M., Saitoh, S., Fujishiro, M., and Noguchi, T. 1994, *Biochem. Biophys. Res. Commun., 200*, 606.

[3] Masaki, T., Sakai, E., Furuta, Y., Kobayashi, M., and Takamatsu, K. 1998, *Gene, 225*, 117.

[4] Burgoyne, R.D., O'Callaghan, D.W., Hasdemir, B., Haynes, L.P., and Tepikin, A.V. 2004, *Trends. Neurosci., 27*, 203.

[5] Kobayashi, M., Takamatsu, K., Saitoh, S., and Noguchi, T. 1993, *J. Biol. Chem., 268*, 18898.

[6] O'Callaghan, D.W., Ivings, L., Weiss, J.L., Ashby, M.C., Tepikin, A.V., and Burgoyne, R.D. 2002, *J. Biol. Chem., 277*, 14227.

[7] O'Callaghan, D.W., Tepikin, A.V., and Burgoyne, R.D. 2003, *J. Cell Biol., 163*, 715.

[8] Gierke, P., Zhao, C., Brackman, M., Linke, B., Heinemann, U., and Braunewell, K.H. 2004, *Biochem. Biophys. Res. Commun., 323*, 38.

[9] Saitoh, S., Takamatsu, K., Kobayashi, M., and Noguchi, T. 1993, *Neurosci. Lett., 157*, 107.

[10] Furuta, Y., Kobayashi, M., Masaki, T., and Takamatsu, K. 1999, *Neurochem. Res., 24*, 651.

[11] Saitoh, S., Takamatsu, K., Kobayashi, M., and Noguchi, T. 1994, *Dev. Brain Res., 80*, 199.

[12] Grant, A.L., Jones, A., Thomas, K.L., and Wisden, W. 1996, *Neuroscience, 75*, 1099.

[13] Paterlini, M., Revilla, V., Grant, A.L., and Wisden, W. 2000, *Neuroscience, 99*, 205.

[14] Bastianelli, E., Takamatsu, K., Okazaki, K., Hidaka, H., and Pochet, R. 1995, *Exp. Eye Res., 60*, 257.

[15] Mamman, A., Simpson, P.J., Nighorn, A., Imanishi, Y., Palczewski, K., Ronnett, G.V., and Moon, C. 2004, *Biochem. Biophys. Res. Commun., 322*, 1131.

[16] Teruel, M.N., and Meyer, T. 2000, *Cell, 103*, 181.

[17] Persechini, A., and Cronk, B. 1999, *J. Biol. Chem., 274*, 6827.

[18] Ivings, L., Pennington, S.R., Jenkins, R., Weiss, J.L., and Burgoyne, R.D. 2002, *Biochem. J., 363*, 599.

[19] Spilker, C., and Braunewell, K.H. 2003, *Mol. Cell Neurosci., 24*, 766.

[20] Kobayashi, M., Masaki, T., Hori, K., Masuo, Y., Miyamoto, M., Tsubokawa, H., Noguchi, H., Nomura, M., and Takamatsu, K. 2005, *Neuroscience, 133*, 471.

[21] Rivosecchi, R., Pongs, O., Theil, T., and Mallart, A. 1994, *J. Physiol. (Lond), 474*, 223.

[22] Lynch, M.A. 2004, *Physiol. Rev., 84*, 87.

[23] Platenik, J., Kuramoto, N., and Yoneda, Y. 2000, *Life Sci., 67*, 335.

[24] Guzowski, J.F., McNaughton, B.L., Barnes, C.A., and Worley, P.F. 2001, *Curr. Opin. Neurobiol., 11*, 579.

[25] Mayford, M., and Kandel, E.R. 1999, *Trends Genet., 15*, 463.

[26] Silva, A.J. 2003, *J. Neurobiol., 54*, 224.

[27] Bach, M.E., Hawkins, R.D., Osman, M., Kandel, E.R., and Mayford, M. 1995, *Cell, 81*, 905.

[28] Pittenger, C., Huang, Y.Y., Paletzki, R.F., Bourtchouladze, R., Scanlin, H., Vronskaya, S., and Kandel, E.R. 2002, *Neuron, 34*, 447.

[29] Balschun, D., Wolfer, D.P., Gass, P., Mantamadiotis, T., Welzl, H., Schutz, G., Frey, J.U., and Lipp, H.P. 2003, *J. Neurosci., 23*, 6304.

[30] Macias, W., Carlson, R., Rajadhyaksha, A., Barczak, A., and Konradi, C. 2001, *Brain Res., 890*, 222.

[31] Deisseroth, K., Heist, E.K., and Tsien, R.W. 1998, *Nature, 392*, 198.

[32] Impey, S., Obrietan, K., Wong, S.T., Poser, S., Yano, S., Wayman, G., Deloulme, J.C., Chan, G., and Storm, D.R. 1998, *Neuron, 21*, 869.

[33] Waltereit, R., and Weller, M. 2003, *Mol. Neurobiol., 27*, 99.

[34] Hyun, J.K., Yon, C., Kim, Y.S., Noh, D.Y., Lee, K.H., and Han, J.S. 2000, *Mol. Cells, 10*, 669.
[35] Nagata, K., Puls, A., Futter, C., Aspenstrom, P., Schaefer, E., Nakata, T., Hirokawa, N., and Hall, A. 1998, *EMBO J., 17*, 149.
[36] Wood, E.R., and Dudchenko, P.A. 2003, *Nature neurosci., 6*, 546.
[37] Morris, R.G., Garrud, P., Rawlins, J.N., and O'Keefe, J. 1982, *Nature, 297*, 681.
[38] Tsien, J.Z., Huerta, P.T., and Tonegawa, S. 1996, *Cell, 87*, 1327.
[39] Morris, R. 1984, *J. Neurosci. Methods, 11*, 47.
[40] Tanaka, K., Wada, N., Hori, K., Asanuma, M., Nomura, M., and Ogawa, N. 1998, *J. Neurosci. Methods, 84*, 63.
[41] Kogan, J.H., Frankland, P.W., Blendy, J.A., Coblentz, J., Marowitz, Z., Schutz, G., and Silva, A.J. 1997, *Curr. Biol., 7*, 1.
[42] Gass, P., Wolfer, D.P., Balschun, D., Rudolph, D., Frey, U., Lipp, H.P., and Schutz, G. 1998, *Learn. Mem., 5*, 274.
[43] Turchi, J., and Sarter, M. 2001, *Eur. J. Neurosci., 14*, 103.
[44] Horwitz, G.D., Batista, A.P., and Newsome, W.T. 2004, *Neurosci. Lett., 366*, 315.
[45] Sallese, M., Iacovelli, L., Cumashi, A., Capobianco, L., Cuomo, L., and De Blasi, A. 2000, *Biochim. Biophys. Acta, 1498*, 112.
[46] Maeda, T., Imanishi, Y., and Palczewski, K. 2003, *Prog. Retin. Eye Res., 22*, 417.
[47] Ronnett, G.V., and Moon, C. 2002, *Annu. Rev. Physiol., 64*, 189.
[48] Leinders-Zufall, T., Rand, M.N., Shepherd, G.M., Greer, C.A., and Zufall, F. 1997, *J. Neurosci., 17*, 4136.
[49] Boekhoff, I., Braunewell, K.H., Andreini, I., Breer, H., and Gundelfinger, E. 1997, *Eur. J. Cell Biol., 72*, 151.
[50] Wei, J., Zhao, A.Z., Chan, G.C., Baker, L.P., Impey, S., Beavo, J.A., and Storm, D.R. 1998, *Neuron, 21*, 495.
[51] Duda, T., Fik-Rymarkiewicz, E., Venkataraman, V., Krishnan, A., and Sharma, R.K. 2004, Mol. Cell Biochem., 267, 107. *Erratum in: Mol. Cell Biochem., 273*, 225.
[52] Hanoune, J., and Defer, N. 2001, *Annu. Rev. Pharmacol. Toxicol., 41*, 145.
[53] Sharma, R.K. 2002, *Mol. Cell Biochem., 230*, 3.
[54] Mercer, E.A., Korhonen, L., Skoglosa, Y., Olsson, P.A., Kukkonen, J.P., and Lindholm, D. 2000, *EMBO J., 19*, 3597.
[55] Lindholm, D., Mercer, E.A., Yu, L.Y., Chen, Y., Kukkonen, J., Korhonen, L., and Arumae, U. 2002, Biochim. Biophys. *Acta, 1600*, 138.
[56] Korhonen, L., Hansson, I., Kukkonen, J.P., Brannvall, K., Kobayashi, M., Takamatsu, K., and Lindholm, D. 2005, *Mol. Cell Neurosci., 28*, 85.
[57] Han, J.S., Kim, H.C., Chung, J.K., Kang, H.S., Donaldson, J., and Koh, J.K. 1998, *Biochem. Mol. Biol. Int., 45*, 1089.
[58] Oikawa, K., Kimura, S., Aoki, N., Atsuta, Y., Takiyama, Y., Nagato, T., Yanai, M., Kobayashi, H., Sato, K., Sasajima, T., and Tateno, M. 2004, J. Biol. Chem., 279, 15142.
[59] Schenkman, J.B., and Jansson, I. 2003, *Pharmacol. Ther., 97*, 139.
[60] Katoh, M., Hirai, M., Sugimura, T., and Terada, M. 1995, *Oncogene, 10*, 1447.

[61] Tibbles, L.A., Ing, Y.L., Kiefer, F., Chan, J., Iscove, N., Woodgett, J.R., and Lassam, N.J. 1996, *EMBO J., 15*, 7026.

In: Neuronal Calcium Sensor Proteins
Editors: Pavel Philippov and Karl Koch, pp. 83-89
ISBN 1-59454-978-8

Chapter V

HIPPOCALCIN IN THE REGULATION OF CALCIUM INDUCED NEURONAL DEGENERATION

Dan Lindholm[1,3,*], Inga Hansson[1], Jyrki P. Kukkonen[2] and Laura Korhonen[1,3]

1Department of Neuroscience, Unit of Neurobiology, Uppsala University, Biomedical Centre, Box 587, S-751 23 Uppsala, Sweden;
2Department of Neuroscience, Unit of Physiology, Uppsala University, BMC, S-751 23 Uppsala, Sweden;
3Minerva Medical Research Institute, Biomedicum Helsinki, Haartmaninkatu 8, FIN-00290 Helsinki, Finland

Abstract

Hippocalcin belongs to the neuronal calcium sensor family and is expressed selectively by neurons in different parts of the nervous system, including the hippocampus. The exact physiological functions of hippocalcin in brain and in nerve cell signaling are not fully understood. Recent studies indicate that hippocalcin interacts with the neuronal apoptosis inhibitor protein (NAIP) that is an anti-apoptotic protein expressed by neurons. Hippocalcin in conjunction with NAIP was shown to render neurons more resistant towards cell death induced by high calcium levels. To learn more about the function of hippocalcin in cell death, we have analyzed the brain of mice deleted for the *hippocalcin* gene by means of gene targeting. In this chapter, we will present data showing that the lack of hippocalcin can influence the activation of other proteins such as caspase-12 in the hippocampal neurons, and increase their sensitivity towards excitotoxic injury induced by kainic acid.

* Correspondence concerning this article should be addressed to Dan Lindholm Department of Neuroscience, Unit of Neurobiology, Uppsala University, Biomedical Centre, Box 587, S-751 23 Uppsala, Sweden. E-mail Dan.Lindholm@neuro.uu.se.

Keywords: Cell death, calcium, hippocampus, neurons, caspase-12, endoplasmic reticulum

Abbreviations

BIR	baculovirus inhibitory repeat
ER	endoplasmic reticulum
IAP	inhibitor of apoptosis protein
NAIP	neuronal apoptosis inhibitor protein
XIAP	X-chromosome linked IAP

1. Introduction

Hippocalcin was cloned more than 10 years ago in the mouse and rodent brain and shown to have the ability to bind calcium [1,2]. This together with its expression in the nervous system suggested that hippocalcin have important roles in nerve cell physiology. Studies on the expression of hippocalcin have shown that mRNA encoding hippocalcin is present in neurons o different parts of the nervous system including the hippocampus [3-5]. This expression pattern is partly different from other calcium sensor (NCS) proteins suggesting that the NSC family proteins exhibit unique physiological functions in the nervous system. Hippocalcin has four EF-hand motifs of which three can bind Ca^{2+} that induces a conformational switch in the protein [1,2]. This can alter the function and intracellular distribution of hippocalcin within cells [6,7]. Recent studies on the subcellular localization and the function of the protein suggest that hippocalcin may be involved in the function of the synapse and in cell signaling [8].

Calcium ions are known to be important in the regulation of cell excitability and exocrine processes, especially in neuronal and endocrine cells. Tight control of the intracellular levels of calcium is also crucial for maintaining cell viability and for counteracting cell death. Thus, elevated intracellular calcium can lead to dysregulated metabolism that induce cell degeneration observed in the nervous system after stroke/ischemia and in excitotoxicity. This is often the result of a prolonged stimulation of neurons by excessive glutamate or other excitotoxins leading to calcium entry via glutamate receptors or other membrane channels [9]. Calcium levels are also regulated by uptake or the release from stores within the neurons and by other mechanisms [9].

2. Hippocalcin Interacts with the Anti-Apoptotic Protein NAIP

Hippocalcin in view of its ability to bind calcium can play a role in calcium-mediated neuronal excitability and synaptic transmission [8]. Apart from this, hippocalcin may be involved in the control of neuronal survival and influence cell death. In our laboratory, we have been interested in this issue regarding the function of hippocalcin in the nervous system. A starting point for this was our previous observation on hippocalcin in the yeast two-hybrid system to study putative protein interactions [10]. The results obtained showed that hippocalcin interacts with neuronal apoptosis inhibitor protein (NAIP) that belongs to the inhibitor of apoptosis protein (IAP) family [11]. The interaction of NAIP with hippocalcin was specific and found to be regulated by calcium levels [10].

NAIP like the other IAPs contains in its aminoterminal region common structural protein motifs so called baculovirus inhibitory repeat (BIR) [11]. The BIR regions have been crucial for the ability of the IAPs to inhibit cell death and activation of killer proteins, called caspases that are activated during apoptosis [11]. In X-chromosome linked IAP (XIAP) the BIR 2 and 3 domains are involved in inhibiting caspase3/7 and caspase-9 [12,13]. NAIP contains three BIR regions, and it has been reported that NAIP can bind caspase-3 and 7, but with a lower affinity than XIAP [14]. In our study, we observed that particularly the BIR3 in NAIP could interact with and bind hippocalcin [15]. Likewise, mutagenesis of the BIR3 region showed that specific amino acids and the zinc finger domain are required for the efficient interaction of NAIP with hippocalcin [15]. This data has helped to define some structural requirements by which the two proteins can interact with each other under different conditions and in a possible multiprotein assembly with still other cellular interactors [15]. In addition, using transfection of cDNAs encoding fusion proteins for hippocalcin and NAIP linked to green and red fluorescent protein showed that there is a functional interaction between NAIP-BIR3 and hippocalcin in neuroblastoma cells with an effect on the subcellular localization of the proteins [15].

3. Functional Roles of Hippocalcin in Neuroprotection Caused by NAIP

The exact role of NAIP in the nervous system is not clear, but mutations in the NAIP gene have been linked to the human genetic disease, spinal muscular atrophy leading to degeneration of the ventral motorneurons [16]. It has been shown that NAIP is a modifier of the severity of the phenotype of spinal muscular atrophy that is determined primarily by mutations in another protein, the survival motor neuron protein on human chromosome 5 [16,17]. It has also been shown that viral expression of NAIP in motorneurons can delay death caused by axotomy [18]. This suggests that NAIP is indeed a protective factor for motorneurons after injury and in neurodegeneration.

NAIP is, however, also expressed by other neuronal populations in the brain such as neurons of the hippocampus that also contain hippocalcin [19]. Overexpression of NAIP was shown to reduce damage in the hippocampus following ischemia, whilst NAIP gene deleted mice lacking the protein were more vulnerable to excitotoxic injury caused by kainic acid [19]. The mechanisms by which NAIP acts under these conditions have not been clearly defined. It is possible that the shown interaction of NAIP with hippocalcin play as both proteins protect against excitotoxic injury in the hippocampus.

In our study, using neuroblastoma and motorneuron cell lines, we observed that the interaction of NAIP with hippocalcin renders the cells more resistant towards death caused by elevated Ca^{2+} concentrations [10,15]. This effect was again associated with the BIR3 region in NAIP that binds hippcocalcin [15]. In these experiments, intracellular calcium in the neuroblastoma cells was elevated by thapsigargin that influences calcium metabolism in the endoplasmic reticulum (ER) [15]. The results showed that NAIP in conjunction with hippocalcin is important in counteracting cell death induced by calcium elevation caused by different agents. This is of particular significance in the nervous system where NAIP and hippocalcin are co-expressed in the same neuronal population [19]. However, the protection afforded by the NAIP-hippocalcin interaction is not effective in all types of neurons, as shown in cultured sympathetic neurons after nerve growth factor withdrawal [15]. In these neurons, deprivation of nerve growth factor caused cell death that was insensitive to overexpression of NAIP and hippocalcin, although it was blocked by other IAPs, such as XIAP [15,20]. The lack of an effect of NAIP-hippocalcin in the sympathetic neurons is probably related to the fact that other pathways different from calcium are mainly involved in the cell demise caused by deprivation of nerve growth factor.

4. Roles of Hippocalcin in Neuronal Survival

To study in detail the *in vivo* function of hippocalcin in cell death, we cultured neurons from hippocalcin gene-deleted mice generated by homologous recombination and using standard procedures [8] (see also Chapter by Kobayashi and Takamatsu in this book). The results showed that hippocampal neurons lacking hippocalcin were more susceptible to different agents such as thapsigargin (increasing calcium) and staurosporine (activating caspases) and which induced cell death in these neurons [21]. Apart from these compounds, stimulation of the neurons with high concentration of glutamate and thereby activating glutamate receptors also induced cell death that was accentuated in neurons devoid of hippocalcin [21]. This shows that hippocampal neurons lacking hippocalcin are in general more vulnerable towards induction of cell death by glutamate, staurosporine and thapsigargin. In addition, we observed that hippocampal neurons from the gene-deleted animals displayed a loss of viability with time in culture that exceeded that in control cultures from wild-type mice [21]. To substantiate these findings, hippocampal neurons lacking hippocalcin were transfected with cDNA encoding the protein that restored the viability and resistance of these neurons towards different treatments [21]. This demonstrates that there is an increased sensitivity of hippocampal neurons to undergo cell degeneration in cells lacking functional hippocalcin protein. The mechanism for this effect may involve alterations in calcium levels and in the interaction of hippocalcin with NAIP that is expressed in the same hippocampal neurons. The data showed no apparent change in the expression levels of NAIP in the hippocalcin-deficient neurons although a functional deficit may exist in the neurons in some signaling cascade elicited by the two proteins [21]. Interestingly, the results showed no significant change in calcium levels between control and hippocalcin-deficient neurons after different treatments [21]. This suggests that alteration in calcium levels and the calcium buffering capacities in cells expressing or lacking hippocalcin do not directly contribute to the decreased neuronal survival.

To know more about the mechanisms by which hippocalcin can afford neuroprotection, we studied different caspases and their activation in the hippocampal neurons. The results showed that caspase-12, residing in the ER [22] was readily activated in hippocampal neurons lacking hippocalcin [21]. This data was particularly observed using thapsigargin that itself causes ER stress and the elevation of intracellular calcium by blocking calcium uptake into the ER [15]. Apart from caspase-12, calpain [23] was also activated in the hippocampal neurons but there was no clear difference in this respect between wild type and hippocalcin deficient cells [21]. In keeping with an involvement of ER stress in neurons, we observed a slight decrease in the activation of the anti-apoptotic ER chaperone protein, BiP/GRP78 in cells lacking hippocalcin [21]. These results indicate that the hippocalcin-deficient hippocampal neurons are more susceptible to different agents causing cell death, which at least partly is accompanied with the induction of ER stress and the activation of the caspase-12 [21]. However, the signaling pathways and molecular mechanisms involved remain to be studied in more detail.

5. Hippocalcin in Neurodegeneration *in vivo*

To substantiate data obtained in cell culture, we also analyzed hippocampal neurons *in vivo* after kainic acid injections, causing seizures and excitotoxic injury in parts of the hippocampus. The neurons were subsequently analysed by staining for active caspase-3 and using TUNEL labeling, a marker for DNA breaks and apoptosis. The results showed a significant increase in death of hippocampal CA3 neurons in animals lacking hippocalcin [21]. This indicates that hippocalcin has a neuroprotective function in the hippocampal neurons against kainic acid-mediated injury. As mentioned above, hippocampal neurons in NAIP-deficient mice are also more sensitive to excitotoxic injury and kainic acid stimulations [19]. This indicates that hippocalcin and NAIP may affect some common target or function in the hippocampal neurons that contributes to their increased survival.

The *in vivo* data using kainic acid showed that the hippocampal neurons in the hippocalcin gene-deleted mice are particularly sensitive to an excitotoxic injury caused by stimulation of the excitatory glutamate receptors [21]. Prolonged and excessive activation of glutamate receptors can also occurs in the case of high extracellular glutamate levels in the brain affected by epilepsy, trauma, injury and stroke/ischemia.

However, it has been suggested that neuronal viability may also decrease during brain ageing. To study this, we analyzed the brain in wild type and hippocalcin-deficient mice aged more than one year with respect to morphological changes in the hippocampus and cortex. Staining with Fluoro-Jade to label degenerating neurons [24], showed an increase in number of neurons undergoing degeneration in the hilus and dentate gyrus of hippocampus as well as in the cerebral cortex of hippocalcin gene-deleted mice compared with wild-type mice [21]. In agreement with this, staining with cresyl violet revealed an increase in the number of pycnotic neurons in the dentate gyrus of hippocalcin gene-deleted mice indicative of increased cell degeneration [21]. The mechanisms behind the age-dependent enhancement of neurodegeneration in hippocalcin-deficient mice are currently unknown, but may be related to increased excitotoxicity or the induction of ER stress during ageing. We have recently observed alterations in some signaling pathways controlling cell survival and calcium levels in neurons lacking hippocalcin that remain to be studied further.

6. Conclusion

Hippocalcin is a calcium-binding protein that is important in the control of the optimal intracellular calcium levels in neurons and thus may influence a wide variety of cellular activities [1,2]. Hippocalcin can probably have effects on synaptic functions and take part in cellular signaling in the nervous system. Studies reviewed in this chapter disclose still another function of hippocalcin, namely in the control of nerve cell survival and counteraction of calcium-induced cell death. The protective function of hippocalcin against cell death caused by increased calcium involved the protein NAIP that is coexpressed with hippocalcin in specific neuronal populations in the brain. Results obtained so far using hippocalcin-deficient mice have shown an increased vulnerability of hippocampal neurons to increased calcium and excitotoxic injuries. Hippocalcin may also have an influence on age-dependent neuronal degeneration as shown for neurons in the hippocampus and cortex. The mechanisms by which hippocalcin affect neuronal viability after injury and in ageing remain to be studied further. Such studies together with the analysis of the signaling cascades of hippocalcin in neuronal plasticity [8] (see also chapter by Kobayashi and Takamatsu in this book] will lead to novel insights into the function of this protein in the nervous system.

Acknowledgments

Supported by Uppsala University, Cancer Foundation, Sigrid Juselius, and Minerva Foundation. LK received a scholarship from Swedish Brain Foundation. We thank Dr K. Takamatsu and Dr. M. Kobayashi for nice collaboration.

References

[1] Braunewell, K.H., and Gundelfinger, E.D. 1999, *Cell Tissue Res., 295*, 1.

[2] Burgoyne, R.D., and Weiss, J.L. 2001, *Biochem. J., 353*, 1.

[3] Kobayashi, M., Takamatsu, K., Saitoh, S., Miura, M., and Noguchi, T. 1992, *Biochem. Biophys. Res. Commun., 189*, 511.

[4] Paterlini, M., Revilla, V., Grant, A.L., ad Wisden, W. 2000, *Neuroscience, 99*, 205.

[5] Saitoh, S., Takamatsu, K., Kobayashi, M., and Noguchi, T. 1993, *Neurosci. Lett., 157*, 107.

[6] Kobayashi, M., Takamatsu, K., Saitoh, S., and Noguchi, T. 1993, *J. Biol. Chem., 268*, 18898.

[7] O'Callaghan, D.W., Tepikin, A.V., Burgoyne, R.D. 2003, *J. Cell Biol. 163*, 715.

[8] Kobayashi, M., Masaki, T., Hori, K., Masuo, Y., Miyamoto, M., Tsubokawa, H., Noguchi, H., Nomura, M., and Takamatsu K. 2005, *Neuroscience, 133*, 471.
[9] Choi, D.W. 2005, *Nature, 433*, 696.
[10] Mercer, E.A., Korhonen, L., Skoglösa, Y., Olsson, P.A., Kukkonen, J.P., and Lindholm, D. 2000, *EMBO J., 19*, 3597.
[11] Salvesen, G.S., and Duckett, C.S. 2002, *Nat. Rev. Mol. Cell. Biol., 3*, 401.
[12] Riedl, S.J., Renatus, M., Schwarzenbacher, R., Zhou, Q., Sun, C., Fesik, S.W., Liddington, R.C., and Salvesen, G.S. 2001, *Cell, 104*, 791.
[13] Lindholm, D., and Arumae, U. 2004, *J. Cell Biol., 167*, 193.
[14] Maier, J.K., Lahoua, Z., Gendron, N.H., Fetni, R., Johnston, A., Davoodi, J., Rasper, D., Roy, S., Slack, R.S., Nicholson, D.W., and MacKenzie, A.E. 2002, *J. Neurosci., 22*, 2035.
[15] Lindholm, D., Mercer, E.A., Yu, L.Y., Chen, Y., Kukkonen, J., Korhonen, L., and Arumae, U. 2002, *Biochim. Biophys. Acta, 1600*, 138.
[16] Roy, N., Mahadevan, M.S., McLean, M., Shutler, G., Yaraghi, Z., Farahani, R., Baird, S., Besner-Johnston, A., Lefebvre, C., Kang, X. et al. 1995, *Cell, 80*, 167.
[17] Chen, Q., Baird, S.D., Mahadevan, M., Besner-Johnston, A., Farahani, R., Xuan, J., Kang, X., Lefebvre, C., Ikeda, J.E., Korneluk, R.G., and MacKenzie, A.E. 1998, *Genomics, 48*, 121.
[18] Perrelet, D., Ferri, A., Liston, P., Muzzin, P., Korneluk, R.G., and Kato, A.C. 2002, *Nat. Cell. Biol., 4*,175.
[19] Holcik, M., Thompson, C.S., Yaraghi, Z., Lefebvre, C.A., MacKenzie, A.E., and Korneluk, R.G. 2000, *Proc. Natl. Acad. Sci. U S A., 97*, 2286.
[20] Yu, L.Y., Korhonen, L., Martinez, R., Jokitalo, E., Chen, Y., Arumae, U., and Lindholm, D. 2003, *Mol.Cell.Neurosci., 22,* 308.
[21] Korhonen, L., Hansson, I., Kukkonen, J.P., Brannvall, K., Kobayashi, M., Takamatsu, K., and Lindholm, D. 2005, *Mol. Cell. Neurosci., 28*, 85.
[22] Nakagawa, T., Zhu, H., Morishima, N., Li, E., Xu, J., Yankner, B.A., and Yuan, J. 2000, *Nature*, 2000, 403, 98.
[23] Wang, K.K. 2000, *Trends Neurosci., 23*, 20.
[24] Schmued, L.C., Albertson, C., and Slikker, W. Jr. 1997, *Brain Res., 751*, 37.

In: Neuronal Calcium Sensor Proteins
Editors: Pavel Philippov and Karl Koch, pp. 91-113
ISBN 1-59454-978-8

Chapter VI

Constitution and Operational Principles of the Retinal and Odorant-Linked Neurocalcin δ-Dependent Ca2+Modulated ROS-GC Transduction Machinery

Teresa Duda, Venkateswar Venkataraman and Rameshwar K. Sharma*
University of Medicine and Dentistry of New Jersey, USA

Abstract

This chapter is a part of an emerging general Ca^{2+} signal transduction concept proposed to be applicable in all sensory neurons and the sensory-connected secondary neurons. The central theme of the concept is that Ca^{2+} signals through a delicately controlled membrane guanylate cyclase (ROS-GC) transduction machinery. ROS-GC, in turn, generates pulsated levels of cyclic GMP. Cyclic GMP then serves as a Ca^{2+} second messenger. The delicacy and specificity of the transduction machinery is achieved through its unique composition and structural design present in that particular neuron. The present review focuses on these and the operational features of the machinery present in the inner plexifrom layer of the retina and in the cilia of the neuroepithelium, the site of odorant transduction. The illustrations show that in both cases the machinery is operated by the myristoylated form of neurocalcin δ, however, the transducer component of the machinery differs in two systems. In the inner plexiform layer, it is ROS-GC1 and in the olfactory cilia, it is ONE-GC - a variant form of ROS-GC.

* Corresponding author: Rameshkar K. Sharma, The Unit of Regulatory and Molecular Biology, Departments of Cell Biology and Ophthalmology, SOM & NJMS, Stratford, NJ, Email: sharmark@umdnj.edu

1. Introduction

In 1992, Hidaka's group showed that the bovine brain contains four isoforms of 23- to 24-kDa Ca^{2+}-binding proteins [1]. They were collectively named as neurocalcin and their individual isoforms as α, β, γ1, and γ2; the β being the abundant form. The β-form was sequenced for its 185 residues [1] and shown to possess 38 to 45% respective homology with visinin and recoverin, the two previously purified retina-specific proteins [2,3]. Based on Kretsinger's criterion [4], the authors concluded that neurocalcin β contains three potential Ca^{2+}-binding EF-hands [1], a feature identical to that of visinin and recoverin. The authors through calmodulin-specific affinity chromatography and SDS-PAGE further showed that the neurocalcin-like isoform is also present in the retina [1]. Analysis of their figure 3 B shows that the retinal neurocalcin had a slightly higher electrophoretic mobility than that of brain neurocalcins [1]. With these findings the identity of neurocalcin was established. It was concluded that it is a Ca^{2+}-binding protein with three potential Ca^{2+}-binding EF-hands. It is homologous to visinin and recoverin. Finally, its primary residence is in the brain and it, or its orthologue, may also be present in the retinal neurons [1].

In subsequent studies, Hidaka's group raised a polyclonal antibody against the bovine brain neurocalcin [5]. The antibody recognized all four forms of neurocalcin, showed the specific presence of neurocalcin in the bovine cerebellar cortex and in the retinal amacrine and ganglion cells, but not in the photoreceptors [5]. Neurocalcin was not present in bovine lung, heart, liver, spleen, pancreas, intestine, adipose tissue, and testis. These studies affirmed the earlier conclusion of the authors that neurocalcin residence is brain-specific and yet it may also occur in the specified regions of the retinal neurons. However, it does not occur in the photoreceptors. The study also showed that because the antibody was not specific for the neurocalcin isoforms, it was unable to differentiate between the isoforms that existed in the brain or in the retinal neurons. From the physiological perspective, the absence of neurocalcin from the photoreceptors indicated that neurocalcin has no role in phototransduction, yet it may be important for the amacrine and ganglion cell activities. At the time, the authors [1] were under the impression that photoreceptor-specific recoverin is the Ca^{2+}-dependent modulator of a hypothetical photoreceptor guanylate cyclase linked with phototransduction [3,6]. This proposal was later retracted [7]. Apparently influenced by this proposal [3] and by their results that neurocalcin was not present in the photoreceptors, but visinin and recoverin were, it was clear that neurocalcin at the physiological level was distinct from the latter two proteins, yet it was a structural homologue of visinin and recoverin.

Molecular cloning of a bovine brain neurocalcin revealed its entire protein sequence [8] and its crystal analysis at 2.4 Å defined its three-dimensional structure [9]. This form of the neurocalcin has been identified and designated as neurocalcin δ [10].

The present review is focused on neurocalcin δ. A key tool used in these studies is the neurocalcin δ antibody that only recognizes this form of neurocalcin. Thus, the studies discussed in the review are solely reflective of the neurocalcin δ activity towards modulation of the ROS-GC transduction machinery. Some of this material is freely borrowed from a recent review of the investigators [11].

2. Neurocalcin δ is a Member of the Neuronal Ca2+-Sensor Protein Family

Background.

Neurocalcin δ is a family member of the neuronal Ca^{2+} sensor proteins (NCS proteins) [reviewed in refs. 12,13], and NCS proteins, in turn, are members of the Ca^{2+}-binding EF hand super-family of proteins typified by calmodulin (reviewed in ref. 14). A phylogenetic tree showing the relationship of these proteins is presented in Fig. 1. The proteins linked with the ROS-GC transduction machinery are GCAPs, S100B and neurocalcin δ. These proteins have been indicated in red in the phylogenetic tree. The figure shows the S100B and NCS families. The NCS family members are grouped into six subfamilies with the names recoverins, VILIPs, frequenins, KChips, GCAPs and calmodulin. They are involved in diverse regulatory processes ranging from gene expression, ion channel function, modulation of enzyme functions of kinases, guanylate cyclases, adenylate cyclases, membrane traffic through ion channels and receptors to the control of apoptosis. The recent evidence indicates that the NCS proteins GCAPs and neurocalcin δ are specialized and share a common feature with S100B. They are meant to modulate the Ca^{2+}-sensitive ROS-GC transduction machinery linked with the physiological activities of the sensory neurons and of the ones connected with them.

Barring S100B, NCS protein family members share 34 to 45% sequence identity, a conserved myristoylation signal at the amino terminus, two pairs of classical EF-hand Ca^{2+}-binding loop sequences—EF1, EF2, EF3, EF4—and 20-25-kDa subunit mass. In contrast, S100B, a member of the superfamily of S100 proteins, shares only minimal (15%) sequence identity with other NCS proteins and contains only two EF-hand Ca^{2+}-binding sequence motifs. However, it shares a conserved domain with neurocalcin δ: the S100/ICaBP domain. The similarity extends to their three-dimensional structures and functions as discussed below.

The comparison of the Ca^{2+}-bound crystallographic models of S100B and neurocalcin δ shows that despite their overall sequence dissimilarities, the configurational arrangement of their two EF-hand Ca^{2+}-binding sequence motifs are very similar [9,15], and are superimposable (Fig. 2., reproduced from [16], where a detailed discussion on this topic is provided).

The monomeric subunit of Ca^{2+}-bound S100B is a relatively compact and globular structure. It consists of four helices, I-IV. These participate in two EF-hand domains, helix-loop-helix structural motif for Ca^{2+} binding, and are linked by a flexible linker region L1 (Fig. 2A). S100B monomer reveals a splayed helix relationship with neurocalcin δ, involving helices B and C and helices D and E, or helices F and G and helices H and I (Fig. 2B). There are three significant structural similarities between them (Fig. 2C): 1) the S100B monomer superimposes perfectly on either the N-or C-terminal pair of neurocalcin δ with root mean square deviations for main chain atoms being 3.6 and 4.0 Å, respectively; 2) the four helix-packing arrangements, involving either an N-terminal pair or a C-terminal pair of EF hands of neurocalcin, are very similar to those of S100B; Figure 2C shows superimposition of the S100B structure over the C-terminal pair of EF hands of neurocalcin δ; and 3) related to their crystal packing arrangements, the two S100B monomers pack in anti-parallel fashion about its C-terminal helix, and the neurocalcin monomers pack around the N-terminal helix of the EF3 hand.

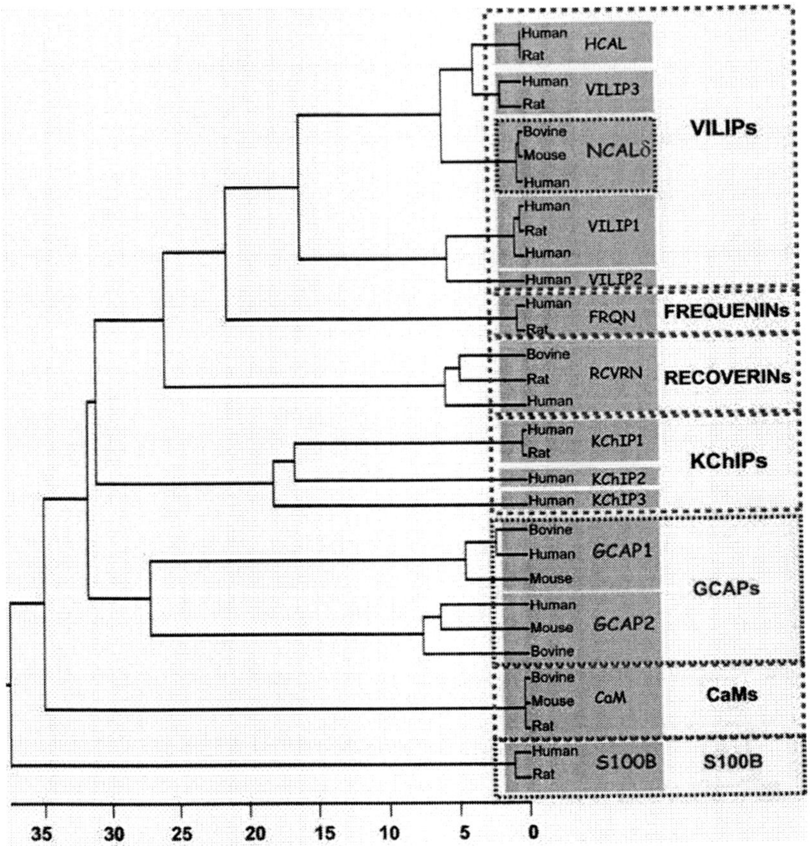

Figure1. Phylogenetic Tree depicting the NCS and S100B families of EF-hand Ca2+-binding proteins. The phylogenetic tree was constructed from the primary sequences of the indicated proteins by the Clustal method of Megalign module of DNASTAR. The bottom line provides the the number of residue substitution events. The sequences and their accession numbers (in parantheses) are: calmodulin (CaM) – bovine (MCBO), mouse (AAH54805) and rat (NP_114175); frequenin (FRQN) – human (AAF01804) and rat (NP_077342); guanylate cyclase activating protein type 1 (GCAP1) – bovine (P46065), human (AAA60542) and mouse (NP_032215); GCAP2 – bovine (P51177), human (Q9UMX6) and mouse (NP_666191); hippocalcin (HCAL) – human (NP_002134) and rat (AAAR14053); KV-channel interacting protein type 1 (KChIP1) – human (AF199597) and mouse (AB075041), KChIP2 – human (AF199598) and KChIP3 – human (AF199599); neurocalcin δ (NCALδ) – bovine (266611), human (NP_114430) and mouse (Q91X97); recoverin (RCVN) – human (JC1227), rat (NM_080901) and bovine (NM_174165); S100B – human (BC001766) and rat (NM_013191); visinin-like protein type 1 (VILIP1) – human (AB001104; BC022012- VSNL1) and rat (NM_012686-VSNL1); VILIP2 – human (AB001105)VILIP3 – human (D16227) and rat (BC088759-VSNL3). The NCS members which serve as Ca2+ sensors for ROS-GC are indicated in bold letters and are colored in red.

It is, thus, concluded that S100B may not appear to be a typical NCS protein family member at the primary amino acid structural level. Yet in reality, at the tertiary structural level it is, and is also at the functional level, as both neurocalcin δ and S100B stimulate Ca^{2+}-dependent ROS-GC1 activity [16,19; reviewed in refs. 11,17,18].

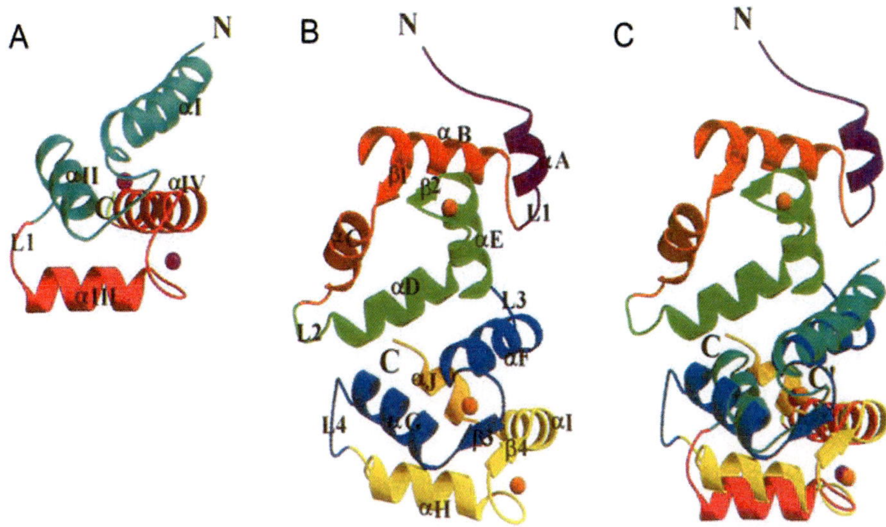

Figure 2. Similarites between neurocalcin δ and S100B. (A) Bovine Ca2+-S100B monomer crystal structure. The bound Ca2+ ions are represented as cyan spheres. The two EF-hand Ca2+-binding domains are color coded: EF1 is in light blue and EF2 is in red. The helices are labeled I-IV, and the linker region connecting the two pairs of EF hands is labeled L1. The C-terminal region is colored in yellow. (B) Bovine Ca2+-neurocalcin monomer structure. The four Ca2+-binding EF-hand domains are color coded: EF1 is in red, EF2 is in green, EF3 is in blue, and EF4 is in yellow. The N- and C-terminal regions are represented in magenta and gold, respectively. The Ca2+ ions that are bound at EF2-, EF3-, and EF4- hand domains are represented as orange spheres. The helices are labeled A-J, and the linker regions connecting EF hands are labeled L1-L4. (C) Best-fit superposition of main chain atoms of the bovine Ca2+-S100B monomer on the C-terminal half of the Ca2+-neurocalcin monomer. The EF1 and EF2 hands of S100B superpose on the EF3 and EF4 hands of Neurocalcin (reproduced from Fig. 6 of [16]).

3. Ca2+-Modulated ROS-GC Transduction Machinery

The discovery of the ROS-GC transduction machinery was made in the year 1994 [20]. Prior to this time the vision investigators had gathered voluminous evidence that cyclic GMP is the direct second messenger of phototransduction, that it directly gates and regulates the channel activities, and, most curiously, it is "inversely-locked in" with the levels of Ca^{2+}, indicating that when there is an increment of Ca^{2+}, there is a concomitant decrement of cyclic GMP and vice versa [for a historical review on these studies, see ref. 21; for earlier reviews covering these studies up to the year 1986, see refs. 22-24; for the original studies see refs. 25- 34].

With the establishment that cyclic GMP is the second messenger of phototransduction, a critical question remained: What is the molecular and biochemical identity of the phototransduction-linked ROS-GC?

Identity of this guanylate cyclase was aided by its two physiological features observed during the physiological investigations in search of the messenger role of cyclic GMP in phototransduction. The guanylate cyclase was of the membrane form and it was inhibited by the submicromolar concentration range of free Ca^{2+} [35-37; for summary see ref. 21]. With the finding that the guanylate cyclase was not a direct recipient of the Ca^{2+} signals [38], an additional important element was added to the feature of the enzyme. The receiver component (soluble protein) of the Ca^{2+} signals was separate from the main enzyme membrane guanylate cyclase ROS-GC transduction component. The receiver detected the Ca^{2+} signals and transmitted them to the ROS-GC and affected its activity. The findings suggested that Ca^{2+} in the nanomolar range stimulated and in the semimicromolar range inhibited hypothetical ROS-GC. These findings became the foundation of the present concept that the phototransduction-linked ROS-GC is a guanylate cyclase activating protein (GCAP)-modulated Ca^{2+} signal transduction machinery. In retrospect, the soluble protein disclosed by the investigators was a GCAP [38] identical to one of the latter two characterized GCAPs: GCAP1 [39-41] or GCAP2 [42]. However, the original soluble protein [38] was mistakenly characterized as recoverin [3,6], and recoverin was mistakenly claimed to be the ROS-GC activator [3,7].

In a strategy to characterize the phototransduction-linked ROS-GC at the biochemical, functional and molecular levels, three criteria were used. Criterion-One, the ROS-GC must reside in the rod outer segments, the site of phototransduction. Criterion-Two, it must mimic the Ca^{2+}-dependent characteristics of the native ROS-GC. Criterion-Three, to differentiate it from the peptide hormone receptor family membrane guanylate cyclases, it must respond only to the intracellular-generated Ca^{2+} signals.

All three criteria were met with the cloned ROS-GC from the bovine retinal cDNA library [20]. These criteria are briefly discussed below.

Criterion-1. A 112 kDa ROS-GC was purified from the bovine rod outer segments [43]. Its guanylate cyclase activity was unresponsive to the natriuretic peptides indicating that it was not a member of the natriuretic peptide receptor family. A 15-amino acid N-terminal segment, and three other internal tryptic peptide fragments were sequenced, and the sequences of these fragments formed the basis for identification of the cloned ROS-GC [20,43]. All sequences matched the sequences of the cloned form, indicating that the native bovine ROS-GC present in the rod outer segments has been cloned [20].

The important aspects of the native ROS-GC sequencing studies were that they experimentally validated the position of the N-terminus amino acid of the mature protein; they showed that the terminus was blocked and thereby protected against proteolytic digestion [43,44].

To date, of all the membrane guanylate cyclases cloned, ROS-GC is the only one cloned on the basis of its protein sequence. Hence, the structure of this ROS-GC serves as the standard for the identity of the other cloned members of the guanylate cyclase family.

Criterion-2 and Criterion-3. At almost the same time that the discovery of ROS-GC was reported [20], the discovery of two forms of GCAP, GCAP1 [39, 41] and GCAP2 [42], were also reported and were shown to be the Ca^{2+}-modulators of a native ROS-GC residing in the bovine rod outer segments. Biochemical and molecular identity of this native ROS-GC, however, was not known. With the availabilities of the cloned bovine ROS-GC1 [20] and the cloned GCAP1 [41], the stage was set to study the direct functional relationship between ROS-GC1 and GCAP1 in Ca^{2+} signal transduction, and to link this transduction mechanism with the phototransduction process [reviewed in ref. 21]. ROS-GC1 expressed in the heterologous system of COS cells responded at 10 nM Ca^{2+} to the GCAP1 stimulation in a dose dependent manner. The GCAP1-stimulated ROS-GC1 activity was incrementally inhibited by free Ca^{2+} with a $K_{1/2}$ of 100 nM [45]. In contrast, under identical conditions, GCAP1 had no effect on the recombinant ANF-RGC, the receptor of the peptide hormone atrial natriuretic factor, ANF [45]. These results demonstrated that (i) recombinant ROS-GC1 mimicked the native ROS-GC present in the rod outer segments in its Ca^{2+}-modulation, and, therefore, on a functional level it was identical to the native enzyme; (ii) ROS-GC received the Ca^{2+} signals through GCAP1; and (iii) ROS-GC differed from the peptide hormone receptor family member in being the transducer of the intracellularly-generated Ca^{2+} signals within the light-sensitive rod outer segments. Thus, ROS-GC belonged to a distinct subfamily of membrane guanylate cyclases that was different from the peptide hormone receptor family and was a transducer of the free Ca^{2+} signals arising in the rod outer segments. It represents the prototype of this form of membrane guanylate cyclase.

These findings also settled an important pending issue. They showed that the previously characterized human photoreceptor ROS-GC, termed retGC [46], was a cloning/ sequencing artifact (for more detailed discussion on this topic consult the recent review [11]). However, reconstitution studies with this retGC clone [46] had reported that GCAP2 is a Ca^{2+}-modulator of the human retGC [47; reviewed in ref. 11].

At the present time, three members of the ROS-GC subfamily have been cloned. All are single transmembrane-spanning proteins. The original ROS-GC has been renamed ROS-GC1, and the names of the other two are ROS-GC2 [48] and ONE-GC (olfactory neuroepithelial membrane guanylate cyclase) [49]. They are alternately called retGC1 and GC-E [46,50], retGC2 and GC-F [50,51] and GC-D [52]. A specialized feature of the ROS-GC subfamily is that it represents a two-component transduction system: the modulator and the ROS-GC. The modulator is a Ca^{2+} sensor protein; it senses and transmits the Ca^{2+} signals to ROS-GC, which in turn, transforms them into the production of cyclic GMP.

4. ROS-GC1 is a Bimodal Ca2+ Signal Transduction Switch

With the disclosure of the biochemical and molecular identity of the physiological form of phototransduction-linked ROS-GC transduction machinery that is inhibited by the Ca^{2+} signals, the question arose: Is there also a ROS-GC form in the retinal neurons that is stimulated by the Ca^{2+} signals?

This question originally arose with an observation made during the purification process of GCAP1 from ROS that was to be used for the ROS-GC reconstitution studies. In a joint collaboration, the groups of Sitaramayya and Sharma observed that one of the chromatographed retinal post-mitochondrial 100,000 *g* supernatant fraction, instead of inhibiting, surprisingly, stimulated the ROS-GC present in the native ROS membranes in a Ca^{2+}-dependent fashion [53]. The stimulatory factor was purified to homogeneity as a 6 to 7 kDa subunit protein, which oligomerized to a functional 40 kDa form [53]. With the concentration of 2 to 3 μM, the factor stimulated the native and the cloned forms of ROS-GC 25-fold. The stimulations were Ca^{2+}-dependent with a $K_{1/2}$ of 2 μM. The factor was not present in outer segments, indicating that it had no role in rod /cone-specific phototransduction. To distinguish it from the newly discovered GCAP that inhibited Ca^{2+} signaling of ROS-GC, the factor was named Ca^{2+}-dependent guanylate cyclase activator protein (CD-GCAP). The factor was cloned, sequenced both in its cloned and the native forms, and reconstituted to show that it was S100B protein [54-56]. The S100B-modulated domain in ROS-GC resided at its C-terminal segment between aa 731-1054 [54]. Because the GCAP1-modulated domain resided in the aa segment 503-786 of ROS-GC1, the investigation showed that the GCAP1- and S100B- modulated domains in ROS-GC1 were separate [54].

These studies established that ROS-GC1 is a dual Ca^{2+} transduction switch. One switch operates within a range of nanomolar to submicromolar Ca^{2+}-concentration and inhibits ROS-GC and the other switch operates at higher Ca^{2+} concentration in a range between semimicromolar to the micromolar Ca^{2+} that stimulates ROS-GC. The switch attains its operational specificity through its Ca^{2+}-sensor partner GCAP or CD-GCAP. Accompanied by a GCAP, it is inhibited, and accompanied by a CD-GCAP, it is stimulated.

5. Neurocalcin δ-modulated Ca2+ Signaling in the Inner Retinal Neurons

With the establishment that ROS-GC1 is a bimodal Ca^{2+} signal transduction switch that, theoretically, empowers it to control diverse Ca^{2+} signaling pathways in retinal neurons, the search to find new Ca^{2+} sensor components of ROS-GC1 began. In this process, neurocalcin δ, a protein initially cloned from the bovine brain [8], was found. The reconstitution experiments with recombinant forms of the ROS-GC1 and neurocalcin δ showed that neurocalcin δ stimulates ROS-GC1 in a Ca^{2+}-dependent manner [16]. The $K_{1/2}$ value for Ca^{2+} is 0.8 μM [57]. The nanomolar to semimicromolar Ca^{2+} range that inhibits the GCAP-linked ROS-GC1 transduction system in phototransduction does not influence the neurocalcin δ-dependent ROS-GC1 activity, indicating that neurocalcin δ, like S100B is a CD-GCAP. This finding was surprising because structurally neurocalcin δ is closer to GCAP yet functionally it behaves like S100B. To explain this paradoxical behavior at a structural level, the crystal structures of Ca^{2+}-bound neurocalcin δ and S100B [9,15] were compared and the details of the comparison have been discussed earlier (see Section 2 above). The comparison shows that the four helix-packing arrangements of the Ca^{2+}-binding EF hands of neurocalcin δ and S100B are very similar. Thus, at the three dimensional level the Ca^{2+}-sensor domains of neurocalcin δ and S100B are structural analogues and their structural identities probably account for their being functional analogues also.

To determine the neurocalcin δ-modulated domain in ROS-GC1, deletion mutagenesis/expression studies were conducted [16]. The domain resides in the aa731-1054 segment of ROS-GC1. The same segment contains the S100B-modulated domain [19,54]. However, the competition experiments involving neurocalcin δ and S100B show that the two modulated domains are distinct [16].

In a recent study, neurocalcin δ-modulated ROS-GC1 Ca^{2+} signal transduction machinery has been characterized at the structural, biochemical and functional level in the inner plexiform layer (IPL) region of the retinal neurons [57]. The significant findings of this study are described below.

Biochemical and Molecular Characterization of Neurocalcin δ

Neurocalcin δ has been purified from the IPL and characterized on the basis of the sequence of its internal fragments. Direct amino acid sequencing failed with the intact native protein, indicating that its N-terminus is blocked, and the protein is acylated in its native state. Mass spectral analysis shows that the acyl group is predominantly in the form of myristoyl group. Consistent with this conclusion, the cloned form of the protein shows that at its N-terminus it contains a myristoylation site motif, MGXXXS. This motif is conserved in all other members of the NCS protein family with the exception of KChIP2, KChIP3, and KChIP4 (three isoforms of Kv-channel interacting proteins, a subfamily of NCS proteins; reviewed in [12,58]). The study shows that it is also conserved in the IPL neurocalcin δ.

The sequence of the cloned IPL neurocalcin δ is identical to the one cloned from the bovine brain [8]. The protein contains four Ca^{2+}-binding EF hands motifs, yet only three (EF2, EF3, and EF4) are predicted to be functional [1,8,9]. The IPL neurocalcin δ is a 20 kDa monomer protein, which, like other Ca^{2+} sensor proteins [39,59,60], exhibits a Ca^{2+}-dependent mobility shift.

Ca^{2+} Myristoyl Switch

In vitro studies show that upon binding with the myristoylated bovine recoverin and brain neurocalcin δ, Ca^{2+} translocates them from the cytosol to membrane [60-67]. This Ca^{2+}-dependent-function of the myristoyl group in the Ca^{2+} sensor proteins has been termed as the "active Ca^{2+} myristoyl switch" [66].

The IPL neurocalcin δ contains an active Ca^{2+} myristoyl switch. In the presence of Ca^{2+}, neurocalcin δ exists as a membrane-bound protein, and in the absence of Ca^{2+}, the majority of it, *but not all*, moves into the cytosol. This last characteristic of neurocalcin δ had functional implications, and, therefore, was investigated further. It was found that even extensive washing of the native IPL membranes with EGTA does not completely dissociate neurocalcin δ from the membranes. This suggested that a residual fraction of neurocalcin δ is always bound with the native membranes. This conclusion was validated with the studies on the recombinant reconstituted system. In the presence of Ca^{2+}, the recombinant form of myristoylated (myr)-neurocalcin δ incubated with native membrane fraction of the IPL remained membrane-bound, and when Ca^{2+} was removed through treatment with EGTA, the majority of it, *but not all*, was no longer membrane-bound. Thus, the recombinant neurocalcin δ in the native membranes of the IPL mimics the Ca^{2+}-dependent mobility and the membrane anchoring profile of the native neurocalcin δ present in the native IPL. There are three physiologically-relevant explanations for the association of neurocalcin δ with the native membranes, and all three are related to the concentration of Ca^{2+} in the resting state of the membranes. First, a cellular Ca^{2+} concentration of 100-200 nM is able to keep neurocalcin δ membrane bound. Second, when present with its target protein at this concentration of Ca^{2+}, neurocalcin δ-target protein complex has a stronger affinity for the membranes than does neurocalcin δ alone. Third, once the Ca^{2+}-dependent complex between neurocalcin δ and ROS-GC1 is formed, existence of the complex is independent of Ca^{2+} concentration. Any or all three of these possibilities suggest that the membrane-bound neurocalcin δ can participate in the Ca^{2+}-dependent events, occurring at millisecond time intervals, the intervals required in the visual transduction events.

Besides recoverin and neurocalcin δ, two other phototransduction-linked Ca^{2+} sensor proteins, GCAP1 and GCAP2 are also myristoylated [68-72]. However, their myristoyl switches function differently from the one in recoverin. In GCAP2, the switch appears to function in a reverse fashion: GCAP2 is membrane-bound in the absence of Ca^{2+} and is in the cytosol in its presence [70]. In GCAP1, the switch has no role in anchoring the protein to the membrane [70]; instead, it controls the Ca^{2+}-dependent regulatory properties of GCAP1 [68,69,73]. Thus, neurocalcin δ Ca^{2+}-myristoyl switch partly works like the one present in recoverin and the evidence to be presented will show that to a certain extent, it works like the one in GCAP1.

Ca^{2+}-Modulated Neurocalcin Δ/ROS-GC1 Transduction Machinery

Guided by the earlier *in vitro* findings of the reconstitution studies that showed that recombinant neurocalcin δ stimulates recombinant ROS-GC1 [16] coupled with the finding that neurocalcin δ is present in the native membrane fraction of the IPL of the retina, natural interaction of neurocalcin δ with ROS-GC1 was envisioned, tested and found. The findings showed that both neurocalcin δ and ROS-GC1 are in close proximity of each other indicating that they interact with each other, and their interaction is possibly functionally productive. The finding that the added presence of the Ca^{2+} in the IPL membranes stimulates native guanylate cyclase activity reveals that the interaction between neurocalcin δ and the membrane guanylate cyclase is, indeed, functionally relevant. It also indicates that these membranes contain a sufficient amount of the residual Ca^{2+} to place neurocalcin in contact with the membranes. Thus, the role of Ca^{2+} appears to be two fold: 1) in its nanomolar (resting) concentration range, to keep neurocalcin δ membrane bound; and 2) in its submicromolar range, to create the physical and functional interaction between neurocalcin δ and the cyclase.

Operational Principles of the Transduction Machinery

These findings established that the myristoylated form of neurocalcin δ is present in the membranes of the IPL and that it physically interacts with ROS-GC1. To have meaningful physiological participation of Ca^{2+} signaling in the neurocalcin δ-modulated ROS-GC1 transduction machinery in the retinal neurons, the machinery must respond to and be operated rapidly by the submicromolar to micromolar range of free Ca^{2+} concentrations. The binding constants determined by surface plasmon resonance spectroscopy show that without Ca^{2+}, neurocalcin δ has no affinity for ROS-GC1. In the presence of Ca^{2+}, it binds ROS-GC1 with high affinity with a K_A of 2.3 x 10^6 M^{-1} and a K_D of 4.6 x 10^{-7} M. This indicates that neurocalcin δ binding to and dissociation from ROS-GC1 are Ca^{2+}-dependent, direct, and of high affinity, and they occur very rapidly in response to the submicromolar to the micromolar Ca^{2+} fluctuations. Thus, once Ca^{2+} binds, neurocalcin δ undergoes conformational change, as evidenced by the mobility shift assay, and stimulates ROS-GC1. The $K_{1/2}$ value of Ca^{2+} for the neurocalcin δ-dependent ROS-GC1 activation is ~0.8 μM, and the $K_{1/2}$ value of neurocalcin δ for ROS-GC1 is also ~0.8 μM. Thus, the stimulation of ROS-GC1 by neurocalcin δ occurs within the physiological concentration range of Ca^{2+} and at the physiological concentration of neurocalcin δ [57].

The study shows that besides anchoring neurocalcin δ to the membrane, the myristoyl switch in neurocalcin δ has another biological function. It raises the V_{max} of ROS-GC1 by 2-fold. Thus, compared to the nonmyristoylated form, myr-neurocalcin δ is a more productive modulator of ROS-GC1. However, the myristoyl group does not change the potency of neurocalcin δ. Its $K_{1/2}$ value for ROS-GC1 activation remains at 0.8 μM.

These findings indicate that myristoyl group in neurocalcin δ has two functions: primary, and secondary. The primary function is to carry neurocalcin δ from the soluble to the membrane portion of the cell. This occurs even in the resting (nanomolar to submicromolar) range of Ca^{2+} concentrations in the neuron. Once membrane bound, its secondary function is to respond effectively and quickly, in the millisecond range, to additional Ca^{2+} fluctuations. The increments in Ca^{2+} concentration enhance the saturation activity of the cyclase, and the decrements proportionately decrease it. Thus, according to these two features, neurocalcin δ functionally is both a recoverin and a GCAP1. And, yet biologically and configurationally it belongs to the class of another CD-GCAP, S100B. The function of both of these CD-GCAPs is to act as Ca^{2+} sensors for ROS-GC1.

The next task of the study was to narrow down the already identified ROS-GC segment where the neurocalcin δ-modulated domain resides [57]. It resides within the segment aa 732-962 of ROS-GC1, and it is distinct from the GCAP1-, GCAP2-, and S100B-modulated domains.

Physiological Implications

With regard to the physiological linkage of the neurocalcin δ/ROS-GC1 transduction pathway with the activity of a specific retinal neuron, it can be categorically stated that the pathway is not present in the outer segments of the rods and cones. Thus, it is not linked with the phototransduction machinery. In the initial survey of the retina, the study shows that both neurocalcin δ and ROS-GC1 are present in the lower strata of the IPL and a subpopulation of ganglion cells [57], and provides a strong argument that the pathway is biologically relevant to the physiology of these retinal compartments.

6. Neurocalcin δ-modulated Ca2+ Signaling in the Cilia of the Olfactory Neuroepithelium

The concept

With the findings that ROS-GC1 transduction machinery is present within and outside of the sensory neurons of the retina, and also in the light-sensitive pinealocytes and in the olfactory bulb, it became a good possibility that the machinery, with the following general principles might be omnipresent in all sensory neurons and the neurons connected with them (this concept is discussed in a recent review [11]). Being present in a particular neuron, it would adapt itself to the workings of that neuron. There, it would function as a precise sensor and transmitter of the Ca^{2+} waves. The neuron phenotype will be faithfully coded by the temporal patterns of the generation, configuration and concentration of the waves. The Ca^{2+}-sensor component of the ROS-GC transduction machinery will sense the waves, and transmit them to the ROS-GC transduction component, which will convert them to the cyclic GMP form. The production of cyclic GMP will be retarded or inhibited. These opposing performances of the operations will be due to the intrinsic bimodal nature of the ROS-GC1 transduction machinery, which can accelerate or retard its operation. In this manner, the machinery is delicately regulated by the parent Ca^{2+} waves. The pulsated level of cyclic GMP through a hypothetical cyclic GMP-gated channel will then control the hyper-or de-polarized state of the neuron membranes. The generated-electrical signal will reflect the original phenotype of the neuron. Such a transmission machine achieves specificity and flexibility through its composition. For instance, when the Ca^{2+} sensor component of the ROS-GC1 transduction machinery is a GCAP, it is inhibited by the Ca^{2+} waves, a situation present in ROS of the retinal neurons. The reverse is true when it is S100B, as is the situation within the pinelocytes and the olfactory bulb.

This concept was tested in the model system of the cilia present in the olfactory epithelium. The choice of this system was based on the following considerations. The cilia are the counter part of the outer segments of the rods (ROS) and cones of the retinal photoreceptors. ROS are the sites of phototransduction, the cilia, of the odorant transduction. Like ROS, the cilia continuously shed from the olfactory receptor neurons and undergo replacement. The ROS contain the entire components of the phototransduction system and the cilia contain the entire components of the odorant transduction system.

Despite these similarities between the two systems, the present evidence indicates that the physiological consequences of the operation of the two systems are diametrically opposite. The phototransduction operation causes hyperpolarization and the odorant transduction causes depolarization in the plasma membranes.

7. ONE-GC, a Variant Form of the ROS-GC Transduction Component in Cilia of the Olfactory Neuroepithelium

The cilia are a part of the olfactory receptor cell, which is a primitive bipolar neuron. Its nucleus resides in the base of the neuroepithelium, and its single dendrite extends toward the base of the epithelial surface. There it widens itself, bearing 10 to 30 motile cilia projecting into the mucous cover (Fig. 3). The cilia contain all the molecular components required for the operation of the odorant transduction machinery.

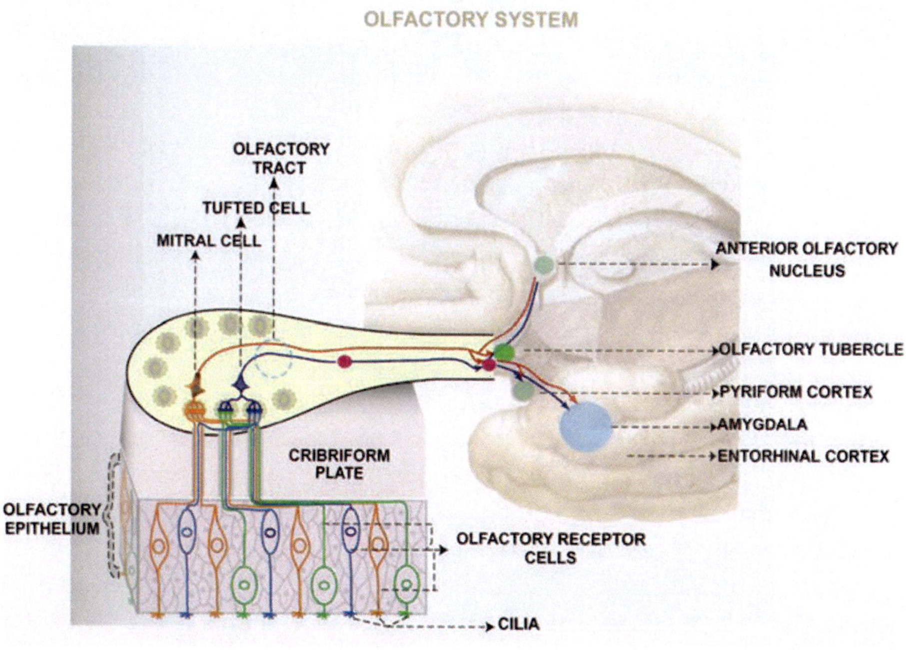

Figure 3. Olfactory system – a schematic illustration. The process of decoding odor begins by the binding of the odorant molecule to its receptors in the cilia of the olfactory receptor cells in the olfactory epithelium. The signal is transmitted through their axons, which traverse the Cribriform Plate and synapse in the glomeruli of the olfactory bulb with the dendrites of the second order neurons – mitral and tufted cells. The efferent axons from the olfactory bulb form the olfactory tract. The olfactory tract carries the processed information from the olfactory bulb to the five regions: anterior olfactory nucleus, olfactory tubercle, pyriform cortex, amygdaloid complex, and entorhinal cortex. From these areas, the signal is further sent out (indicated by dashed lines) to the limbic system (modified from Fig. 1 of [84]; reproduced from Fig. 6 of [11]).

Olfactory signaling is initiated at the ciliated apical border. There the odorant binds to its surface membrane receptor and the binding generates an electrical signal. The biochemical term used for this process is odorant transduction. The present consensus is that the odorant receptor is a member of the superfamily of serpentine receptors and it functions through the cyclic AMP signaling cascade system. The receptor-mediated odorant signal activates G_{olf} [74], a family member of G-proteins (reviewed in refs. 75-79). Activation of G_{olf} stimulates the type III adenylate cyclase (ACIII) [80,81], resulting in the accelerated production of cyclic AMP. Cyclic AMP opens the cyclic AMP-gated channel that causes an influx of Ca^{2+} and Na^{+}. The influx results in the opening of Ca^{2+}-activated Cl^{-} channels [82-85]. The initial influx of Ca^{2+} together with the efflux of Cl^{-} results in the generation of an electrical signal, and depolarization of the ciliary plasma membrane [82-84]. This electrical signal in the form of action potentials is then transmitted through the secondary olfactory neurons of the olfactory bulb and olfactory tract to its final destination, the olfactory cortex neurons [86,87] (see Fig. 3).

With cloning of the membrane guanylate cyclase ONE-GC (originally referred to as GC-D) from the total rat olfactory cDNA library and the demonstration of its presence in a small, selective subpopulation of neuroepithelium neurons via *in situ* hybridization, Garbers's group established the presence of a key component of the membrane guanylate cyclase transduction system in the olfactory neurons [52]. In a subsequent collaborative study with Beavo's group, the authors presented immunocytochemical evidence showing that ONE-GC coexists with its cyclic GMP-PDE partner (PDE2) and these two partners are segregated from the components of the cyclic AMP signaling pathway [88]. Additional immunocytochemical studies from Kaupp's group showed that the subset of olfactory neurons expressing ONE-GC and PDE2 also express a subunit of the cyclic GMP-selective cyclic nucleotide-gated channel, and this subset of neurons contains no (tested) component of the cyclic AMP signaling system [89].

With these immunocytochemical findings, the existence of three important components of the cyclic GMP signaling pathway in the olfactory sensory neurons was settled. It was, thereby, concluded that the cyclic GMP signaling pathway exists in a subset of neurons and these neurons do not contain the cyclic AMP signaling pathway [88,89]. The authors postulated that the cyclic GMP signaling pathway is not linked with the odorant transduction. Instead, it is meant to signal the hypothetical pheromone activity, which regulates sexual behavior of the animals: "These observations suggest the possibility that the function for this group of neurons may be in behavioral responses induced by hormones or pheromones, possibly related to reproduction, rather than a response to specific odorants" [88,89]. Mechanistically, these authors envisioned that the pathway is turned on by the direct binding of the pheromone to the extracellular domain of the ONE-GC [88,89; reviewed in ref. 90]. Because no ligand for the extracellular domain has been found, the authors have coined the term "orphan receptor" for ONE-GC.

In contrast to the conclusion that cyclic GMP signaling pathway is not linked with odorant transduction, the recent biochemical, structural, immunohistochemical and functional studies of Sharma and Duda's group show that the pathway is, indeed, linked with the membrane guanylate cyclase transduction system [49,91]. These studies are briefly documented below.

The subject of investigation in the first study was the isolated rat olfactory neuroepithelium, which house the cilia [49]; and for the second, the isolated cilia, which are the sites of odorant transduction [91]. Five criteria were set forth for the membrane guanylate cyclase transduction machinery to qualify as a genuine transducer of the odorant signal. Criterion 1, in its native state, it must be odorant responsive. Criterion 2, the response should be rapid. Criterion 3, it must reside in the membrane portion of the cilia, the site of odorant transduction. Criterion 4, the physiological levels of Ca^{2+} should mimic the odorant's response. Finally, Criterion 5, with the defined components of the machinery, it should be possible to reconstitute the Ca^{2+} dependency of the machinery.

The discussion below will show that all these five criteria are met by the Ca^{2+}-modulated myr-neurocalcin δ-modulated ONE-GC transduction machinery present in the isolated cilia. Another significant aspect of these studies is the discovery that this transduction machinery is different than that of ROS-GC1 linked with phototransduction [49] and cone pedicles [92], of S100B-modulated ROS-GC1 present in the photoreceptor synaptic region [19], in the pinealocytes [93] and than that of the neurocalcin δ-modulated ROS-GC1 transduction system present in the ganglion cells [57].

Criteria 1 through 4. The cilia were isolated from the neuroepithelium and their purity was verified as in [94]. The intact cilia were then tested for the odorant-responsive membrane guanylate cyclase transduction system [91]. The findings showed that within 30s and up to 8 min the odorant stimulated the cyclic GMP production in a linear fashion. Analysis of the membrane and soluble fractions showed that the membrane guanylate cyclase was the exclusive form present in the cilia and that it was modulated by the physiological concentrations of free Ca^{2+} with a $K_{1/2}$ of 0.70 μM. These findings established that the native olfactory cilia contain the odorant responsive membrane guanylate cyclase transduction machinery, that the machinery rapidly (within seconds) responds to the odorant and that the machinery is Ca^{2+}-modulated.

The identity of the Ca^{2+}-modulated membrane guanylate cyclase transduction system components in the cilia was provided by the molecular cloning and biochemical studies involving Western blot analyses. First, the neuroepithelium of the rat constituted the subject of investigation [49]. From its cDNA library, the ONE-GC was cloned and sequenced. An antibody against its unique sequence was raised in rabbits. It was assured that the antibody only recognizes ONE-GC and not ROS-GC1 and ROS-GC2, the other two members of the Ca^{2+}-modulated ROS-GC. Western-blot analysis of the neuroepithelium membranes showed that they contained ONE-GC, and it was Ca^{2+}-modulated in its native state. Similar studies showed that the membranes also contained neurocalcin δ, and the exogenously supplied neurocalcin δ in the presence of Ca^{2+} stimulated the native membrane guanylate cyclase. Additional analysis of the native membranes of the neuroepithelium through crosslinking studies showed that neurocalcin δ and ONE-GC resided in close proximity to be able to physically interact with each other. Finally, the neurocalcin δ/ONE-GC transduction system was reconstituted in the heterologous cell system of COS cells, using only recombinant neurocalcin δ and recombinant ONE-GC as the sole constituents. The reconstituted transduction machinery faithfully mimicked the native neuroepithelial membrane guanylate cyclase transduction machinery. These findings established the final Criterion 5 set forth for the qualification of the neurocalcin δ/ONE-GC being a genuine transducer of the odorant signal.

In a recent study, using the same strategy, the presence of the neurocalcin δ/ONE-GC transduction machinery at the molecular, biochemical and functional levels has been provided directly in the cilia of the rat neuroepithelium, thereby, bringing the machinery a step closer to its linkage with the odorant transduction system [91]. Immunohistochemical studies validate the conclusions that both neurocalcin δ and ONE-GC are present in apical regions of the cilia, the site of odorant transduction. Two essential features of this study that show neurocalcin δ regulates ONE-GC at the membrane level and functional characterization of the transduction machinery are reproduced below [49].

Presence of Neurocalcin δ in Membranes of the Cilia is Ca^{2+}-Dependent

For the most efficient operation of the odorant linked neurocalcin δ/ONE-GC transduction machinery in the olfactory cilia, the machinery must be able to respond rapidly to the incoming Ca^{2+} signals and, in turn, transduces those instantly into the generation of cyclic GMP. To achieve this end, the most convenient configuration will be where the sensor component, neurocalcin δ, and the transducer component, ONE-GC, are easily accessible. This is made possible by a special structural feature of neurocalcin δ, a myristoyl group at its N-terminus. As in another Ca^{2+} sensor protein, recoverin, this group acts in neurocalcin δ as a "Ca^{2+} myristoyl switch" [61,62,65-67,95,96] (see also Section 4 above). Upon binding Ca^{2+} to its EF hands, the switch signals the migration of neurocalcin δ from cytosol to the membrane. To test if this membrane-anchoring mechanism is operative in cilia, the ciliary membranes were isolated under two Ca^{2+} conditions: Ca^{2+}-treated (1 mM), and Ca^{2+}-depleted (1 mM EGTA). The fractions were analyzed by Western blotting with the neurocalcin δ antibody. A single immunoreactive band corresponding to neurocalcin δ was observed in both fractions. The intensities of the bands were different, however. An intense band was in the Ca^{2+}-treated fraction, whereas in the Ca^{2+}-depleted membranes the band was of significantly lowered intensity. These results showed that 1) there is an active "Ca^{2+} myristoyl switch" in neurocalcin δ, and 2) that there is enough residual Ca^{2+} in the cilia to keep some amount of neurocalcin δ always membrane-bound.

8. Functional Characterization of the Neurocalcin δ-Modulated ONE-GC Signal Transduction Machinery in the Cilia

To decipher the operational principles of the neurocalcin δ/ONE-GC transduction machinery of the cilia, two reconstituted systems were used. Reconstitution was carried out with recombinant myr-neurocalcin δ. In the first, the native ciliary membranes were used, and in the second, recombinant ONE-GC mimicking the native system was used.

The EGTA-washed ciliary membranes were exposed to the incremental concentrations of myr-neurocalcin δ in the presence of saturated amounts of Ca^{2+}. Myr-neurocalcin δ caused a dose-dependent stimulation of the cyclase with a $K_{1/2}$ of 0.8 μM and saturation at 4 μM. The stimulation was Ca^{2+}-dependent with a $K_{1/2}$ of ~0.75 μM. These results showed that ONE-GC present in the native membranes of the cilia responds to the physiological levels of Ca^{2+} and neurocalcin δ; hence, it is functional in receiving and transmitting the Ca^{2+} signals.

To determine if the properties of the native system can be mimicked by the in vitro reconstituted system, membranes of COS cells expressing recombinant ONE-GC were treated identically as the native ciliary membranes. In the presence of the saturating concentrations of Ca^{2+}, myr-neurocalcin δ stimulated ONE-GC in a concentration-dependent fashion with an $K_{1/2}$ of 0.8 μM. The total stimulation was 8-fold over the basal value. At saturated (4 μM) concentration of myr-neurocalcin δ, Ca^{2+} stimulated ONE-GC in a dose-dependent fashion with an EC_{50} of ~0.8 μM. These results showed that the reconstituted system with its sole recombinant components faithfully mimics the native ciliary system of ONE-GC in its Ca^{2+}-dependent modulation [91].

Through these studies the constitution and operational principles of the machinery have been established as stated below.

Constitution

The machinery is composed of a core ONE-GC transduction component and its Ca^{2+}-sensor component, myr-neurocalcin δ. It is noteworthy that the soluble form of the guanylate cyclase is not present in the cilia. Thus, the odorant-responsive neurocalcin δ/ONE-GC transduction machinery is cilia-specific and it is the only one linked with odorant transduction. There, in the resting state of the cilia, with a concentration range of 60-100 nM of free Ca^{2+}, it is partially active and ever ready to receive and transduce the odorant-induced Ca^{2+} signals into the production of cyclic GMP.

Operational Principles

The operation of the machinery occurs through the Ca^{2+}-dependent direct binding of myr-neurocalcin δ with its site in the ONE-GC segment, aa 836-1028. The binding is rapid and of high affinity with a K_D of 0.34 μM with fast association (k_{on} of 5.7 x $10^3 M^{-1} s^{-1}$) and dissociation (k_{off} of 1.56 x $10^{-3} s^{-1}$) constants [91]. Thus, the machinery is finely tuned to respond to the continuous changes of Ca^{2+} beats.

A critical functional feature of neurocalcin δ resides in its active "Ca^{2+} myristoyl switch". The switch enables neurocalcin δ to move from cytosol to the membrane compartment of the cilia. The switch is partially active even in the resting state of the cilia where the free Ca^{2+} range is between 60 to 100 nM. Therefore, a fraction of neurocalcin δ in the cilia is always membrane bound, is in physical contact with ONE-GC and, in this manner, keeps the transduction machinery ever ready to receive and transmit the Ca^{2+} signals. The odorant signal causes an increment in free Ca^{2+} and this, in turn, strengthens further interaction between neurocalcin δ and ONE-GC. The interaction results in enhanced stimulation of ONE-GC and the enhanced production of cyclic GMP.

An important piece of information from these comprehensive studies is that expression of the Ca^{2+}-modulated neurocalcin δ/ONE-GC transduction machinery is more general in the apical region of the cilia [49,91], than it was initially proposed [88,89; reviewed in ref. 90]. Thus, the machinery is abundantly expressed in the sensory neurons and it is linked with odorant transduction.

9. Odorant Transduction Models

A critical question remains in understanding the mechanism of linkage between neurocalcin δ/ONE-GC transduction machinery and odorant transduction. Is the cyclic GMP signaling pathway independent of the cyclic AMP signaling pathway? Depending upon the answer, additional questions arise. If the two pathways are independent of each other, do they exist in the same or different neurons? And, if they are interdependent, do they exist in all or only some of these neurons? Fully aware of these problems, yet with the present consensus that the direct linkage of the odorant receptor exists only with the adenylate cyclase system, the authors of this review propose two mechanisms. In mechanism 1, the cyclic AMP and cyclic GMP pathways are linked [91]. The mechanism 2 is independent of the cyclic AMP signaling. The central theme of both mechanisms is that the three pivotal signaling components—Ca^{2+}, myr-neurocalcin δ and ONE-GC—of the transduction machinery are locked to control odorant-responsive cyclic GMP levels in the olfactory cilia.

Mechanism 1

It is proposed that the odorant binds to a serpentine receptor, activates G_{olf}, which, in turn, stimulates the type III adenylate cyclase. The resultant rise in cyclic AMP opens the cyclic AMP/cyclic GMP-gated channel. This results in an influx of Ca^{2+} and Na^{+}. The Ca^{2+} influx has two consequences. One, Ca^{2+} regulates the Cl^{-} channels; two, *via* myr-neurocalcin δ, it activates ONE-GC, and the resultant rise in cyclic GMP via the cyclic GMP-gated channel causes further influx of Ca^{2+}. This depolarizes the ciliary membranes. The generated action potentials carry the information from the cilia to the soma. The model is simple; it is testable and paves the way for explaining the precise role of cyclic GMP in odorant transduction. It is in agreement with a recent study [97]. This study, through Ca^{2+} imaging, shows that the odorant generates a wave like Ca^{2+} elevation in the olfactory cilia. The elevation begins at its apical region and spreads through the dendrite toward the soma [97].

Mechanism 2

The odorant signal causes an influx of Ca^{2+} possibly through a voltage-gated channel. This will then activate the neurocalcin δ-modulated ONE-GC transduction system. The elevation in Ca^{2+} is first sensed by neurocalcin δ, which, then undergoes a conformational change. This change, in turn, enhances its binding to the cyclase and is transduced to the cyclase catalytic domain via the neurocalcin δ site on the cyclase. This cascade results in accelerated synthesis of cyclic GMP, which controls the activity of cyclic GMP-gated channels, and the channels then regulate the depolarization state of the ciliary membranes.

10. Concluding Remarks

From its inception to its present age, the field of membrane guanylate cyclase has had a turbulent history: euphoria, disappointment, and finally euphoria once again. In 1963, cyclic GMP was discovered from rat urine. Six years later it and its catalyst, membrane guanylate cyclase, were found in all the tested mammalian tissues, and like the cyclic AMP signaling system, cyclic GMP was proposed as a second messenger of certain hormonal signals. However, during the period covering 1974 to 1984, barring a few (including the author's), the major laboratories denied even the existence of the membrane guanylate cyclase in nature and proposed that the sole functional form of the guanylate cyclase was the soluble form. The field was rekindled with the first purification of a membrane guanylate cyclase, which surprisingly was also found to be a surface receptor for the peptide hormone ANF. Some of the corner stone developments of the field are cited below (adapted from ref. 17).

The first, and foremost, is the discovery of ANF-RGC. This discovery showed the independent existence of a membrane guanylate cyclase signal transduction system. It also revealed an unusual nature of the membrane guanylate cyclase enzyme. The enzyme was also a hormone receptor. This discovery dispelled the belief that the guanylate cyclase existed only in its soluble form and that it transduced only the intracellularly generated nitric oxide signals.

The second is the discovery of ROS-GC from the rod outer segments of the retina. The discovery linked ROS-GC with the molecular machinery of phototransduction. At a biochemical level it showed the existence of a membrane guanylate cyclase class, which transduced the Ca^{2+} signals generated inside the sensory neurons of the retina, and was not involved in the transduction of the peptide hormone signals. The discovery established that the membrane guanylate cyclase family consists of two subfamilies: the peptide hormone receptor and Ca^{2+}-modulated ROS-GC subfamily. Thus, the discovery negated the concept that all members of the membrane guanylate cyclase are also peptide hormone receptors.

The third development is the discovery that the ROS-GC transduction system is not uniquely expressed in the rod outer segments of the retina. Its existence is broad and complex, and in these variant forms, it is linked or has the potential to be linked with the neurosensory processes of sight, smell, taste and the light-regulated processes of the pinealocytes.

The fourth is the observation that ROS-GC transduction machinery is also linked with the signal mediated by the $\alpha_{2D/A}$-adrenoreceptor. This forms a potential link of the ROS-GC transduction machinery with the biological clock where the neurotransmitter norepinephrine may be an important player. The $\alpha_{2D/A}$-adrenoreceptor is one of the norepinephrine receptors.

Now that the Ca^{2+}-modulated ROS-GC transduction field has been established and is shown to be very broadly linked with the Ca^{2+} spikes generated in the sensor neurons, many new problems and challenges have come to the forefront. The first group of the problems has to do with the membrane guanylate cyclases themselves. Have all their subtypes been discovered? Have all the Ca^{2+} sensor components of each of the subtypes been disclosed? What is the tissue distribution of the paired Ca^{2+} sensor protein and its ROS-GC transduction component? What is their function and mechanism of regulation in these specific tissues?

A second group of problems concerns the increasing diversity of the Ca^{2+}-modulated pathways controlled by each of the ROS-GCs. What are the molecular mechanisms involved in those controls? What do they mean in physiological terms?

The third group of problems deals with elucidation of the precise biochemical mechanisms involved in each of the membrane guanylate cyclase regulations. All membrane guanylate cyclases are modular in nature and each module individually exhibits a tight control over the catalytic module of ROS-GC. The challenge is to explain that control in molecular terms. These explanations will result in developing therapies against any of the pathological conditions of the modules.

Finally, to resolve the biochemical mechanistic pathway in absolute terms, availability of the crystal structure of each of the membrane guanylate cyclase and its Ca^{2+} sensor partners is essential. To obtain these crystal structures is a formidable future task.

Acknowledgements

This research was supported by USPHS awards: DC 005349 (R.K.S.), HL 070015 (T.D.) and a grant from the UMDNJ foundation (V.V.). The authors thank Joan Sharma for editorial assistance.

References

[1] Terasawa, M., Nakano, A., Kobayashi, R., and Hidaka, H. 1992, *J. Biol. Chem.* 267, 19596.
[2] Yamagata, K., Goto, K., Kuo, C.H., Kondo, H., and Miki, N. 1990, *Neuron*, 4, 469.
[3] Dizhoor, A.M., Ray, S., Kumar, S., Niemi, G., Spencer, M., Brolley, D., Walsh, K.A., Philipov, P.P., Hurley, J.B., and Stryer, L. 1991, *Science* 251, 915.
[4] Kretsinger, R.H. *Ann. N. Y. Acad. Sci.*, 1980, 356,14.
[5] Nakano, A., Terasawa, M., Watanabe, M., Usuda, N., Morita, T., and Hidaka, H. 1992, *Biochem. Biophys. Res. Commun.*, 186, 1207.
[6] Lambrecht, H.G. and Koch, K.W. 1991, *EMBO J.* 10, 793.
[7] Hurley, J.B., Dizhoor, A.M., Ray, S., and Stryer, L. 1993, *Science* 260, 740.
[8] Okazaki, K., Watanabe, M., Ando, Y., Hagiwara, M., Terasawa, M., and Hidaka, H. 1992, *Biochem. Biophys. Res. Commun.* 185, 147.
[9] Vijay-Kumar, S. and Kumar, V.D. 1999, *Nat. Struct. Biol.* 6, 80.
[10] Kumar, V.D., Hidaka, H., Okazaki, K., Vijay-Kumar, S. 1996, *Proteins*, 25, 261.
[11] Sharma, R. K., Duda, T., Venkataraman, V., and Koch, K-W. 2004, *Curr. Topics Biochem. Res.*, 6, 111.
[12] Braunewell, K.-H. and Gundelfinger, E.D. 1999, *Cell Tissue Res.* 295, 1.
[13] Burgoyne, R. D., O'Callaghan, D. W., Hasdemir, B., Haynes, L.P., and Tepikin, A.V. 2004, *Trends Neurosci.*, 27, 203.
[14] Cheung, W.Y. 1980, *Science*, 207, 19.

[15] Matsumura, H., Shiba, T., Inoue, T., Harada, S., and Kai, Y. 1998, *Structure* 6, 233.

[16] Kumar, V.D., Vijay-Kumar, S., Krishnan, A., Duda, T., and Sharma R.K. 1999, *Biochemistry* 38, 12614.

[17] Sharma, RK. 2002, *Mol. Cell. Biochem.* 230, 3.

[18] Koch, K.-W., Duda, T., and Sharma, R.K. 2002, *Mol.Cell.Biochem.* 230, 97.

[19] Duda, T., Koch, K.-W., Venkataraman, V., Lange, C., Beyermann, M., and Sharma, RK. 2002, *EMBO J.* 21, 2547.

[20] Goraczniak, RM., Duda, T., Sitaramayya, A., and Sharma, RK. 1994, *Biochem. J.,* 30, 455.

[21] Pugh, E.N. Jr, Duda, T., Sitaramayya, A., and Sharma, R.K. 1997, *Biosci. Rep.*, 17, 29.

[22] Lamb, T.D. 1986, *Trends Neurosci.*, 9, 224.

[23] Pugh, E.N. Jr. and Cobbs, W.H. 1986, *Vis. Res.*, 26, 1613.

[24] Stryer, L. 1986, *Ann. Rev. Neurosci.*, 9, 87.

[25] Cohen, A.I., Hall, I.A., and Ferrendelli, J.A. 1978, *J. Gen. Physiol.*, 71, 595.

[26] Cote, R.H., Biernbaum, M.S., Nicol, G.D., and Bownds, M.D. 1984, *J. Biol. Chem.*, 259, 9635.

[27] Lolley, R.N. and Racz, E. 1982, *Vis. Res.*, 22, 1481.

[28] Woodruff, M.L. and Bownds, M.D. 1979, *J. Gen. Physiol.*, 73, 629.

[29] Fesenko, E.E., Kolesnikov, S.S., and Lyubarsky, A.L. 1985, *Nature*. 313, 310.

[30] Cook, N.J., Hanke, W., and Kaupp, U.B. 1987, *Proc Natl Acad Sci USA*. 84, 585.

[31] Shinozawa, T., Sokabe, M., Terada, S., Matsusaka, H., and Yoshizawa, T. 1987, *J. Biochem* (Tokyo), 102, 281.

[32] Lamb, T.D., Matthews, H.R., and Torre, V. 1986, *J. Physiol.*, 372, 315.

[33] Matthews, H.R., Torre V., and Lamb, T.D. 1985, *Nature* 313, 582.

[34] Kaupp, U.B., Niidome, T., Tanabe, T., Terada, S., Bönigk, W., Stühmer, W., Cook, N.J., Kangawa, K., Matsuo, H., Hirose, T., Miyata, T., and Numa, S. 1989, *Nature* 342, 762.

[35] Hodgkin, A.L. and Nunn, B.J. 1988, *J. Physiol.* 403, 439.

[36] Kawamura, S. and Murakami, M. 1989, *J. Gen. Physiol.* 94, 649.

[37] Pepe, I.M. Panfoli, I., and Cugnoli, C. 1986,. 203, 73.

[38] Koch, K.-W. and Stryer, L. 1988, *Nature* 334, 64.

[39] Frins, S., Bonigk, W., Muller, F., Kellner, R., and Koch, KW. 1996, *J. Biol. Chem.*, 271, 8022.

[40] Gorczyca, W.A., Gray-Keller, M.P., Detwiler, P.B., and Palczewski, K. 1994, *Proc. Natl. Acad. Sci. USA*. 91, 4014.

[41] Palczewski, K., Subbaraya, I., Gorczyca, W.A., Helekar, B.S., Ruiz, C.C., Ohguro, H., Huang, J., Zhao, X., Crabb, J.W., Johnson, R.S., and Baehr, W. 1994, *Neuron* 13, 395.

[42] Dizhoor, A.M., Olshevskaya, E.V., Henzel, W.J., Wong, S.C., Stults, J.T., Ankoudinova, I., and Hurley, J.B. 1995, *J. Biol. Chem..* 270, 25400.

[43] Margulis, A., Goraczniak, R.M., Duda, T., Sharma, R.K., and Sitaramayya, A. 1993, *Biochem. Biophys. Res. Commun.* 194, 855.

[44] Koch, K.-W., Stecher, P., and Kellner, R. 1994, *Eur. J. Biochem.* 222, 589.

[45] Duda, T., Goraczniak, R., Surgucheva, I., Rudnicka-Nawrot, M., Gorczyca, W.A., Palczewski, K., Sitaramayya, A., Baehr, W., and Sharma, R.K. 1996, *Biochemistry* 35, 8478.

[46] Shyjan, A.W., de Sauvage, F.J., Gillett, N.A., Goeddel, D.V., and Lowe, D.G. 1992, *Neuron* 9, 727.

[47] Dizhoor, A.M., Lowe, D.G., Olshevskaya, E.V., Laura, R.P., and Hurley, J.B. 1994, *Neuron* 12, 1345.

[48] Goraczniak, R.M., Duda, T., and Sharma, R.K. 1998, Biochem. Biophys. Res. Commun. 245, 447-453.

[49] Duda, T., Jankowska, A., Venkataraman, V., Nagele, R.G., and Sharma, R.K. 2001, *Biochemistry* 40, 12067.

[50] Yang, R.B., Foster, D.C., Garbers, D.L., and Fulle, H.J. 1995, *Proc. Natl. Acad. Sci.USA*. 92, 602.

[51] Lowe, D.G., Dizhoor, A.M., Liu, K., Gu, Q., Spencer, M., Laura, R., Lu, L., and Hurley, J.B. 1995, *Proc. Natl. Acad. Sci. USA*., 92, 5535.

[52] Fulle, H.J., Vassar, R., Foster, D.C., Yang, R.B., Axel, R., and Garbers, D.L. 1995, *Proc. Natl. Acad. Sci. USA*., 92, 3571.

[53] Pozdnyakov, N., Yoshida, A., Cooper, N.G., Margulis, A., Duda, T., Sharma, R.K., and Sitaramayya, A. 1995, *Biochemistry* 34, 14279.

[54] Duda, T., Goraczniak, R.M., and Sharma, R.K. 1996, Biochemistry 35, 6263.

[55] Margulis, A., Pozdnyakov, N., and Sitaramayya, A. 1996, *Biochem. Biophys. Res. Commun.* 218, 243.

[56] Pozdnyakov, N., Goraczniak, R., Margulis, A., Duda, T., Sharma, R.K., Yoshida, A., and Sitaramayya A. 1997, *Biochemistry* 36, 14159.

[57] Krishnan, A., Venkataraman, V., Fik-Rymarkiewicz, E., Duda, T., and Sharma, R.K. 2004, *Biochemistry* 43, 2708.

[58] Burgoyne, R.D. and Weiss, J.L. 2001, Biochem J. 353, 1 (Erratum, *Biochem J.* 354, 727).

[59] Burgess, W.H., Jemiolo, D.K., and Kretsinger, R.H. 1980, *Biochim. Biophys. Acta*. 623, 257.

[60] Ladant, D. 1995, *J. Biol. Chem*. 270, 3179.

[61] Ames, J.B., Tanaka, T., Stryer, L., and Ikura, M. 1996, *Curr. Opin. Struct. Biol.* 6, 432.

[62] Beven, L., Adenier, H., Kichenama, R., Homand, J., Redeker, V., Le Caer, J.P., Ladant, D., and Chopineau, J. 2001, *Biochemistry* 40, 8154.

[63] Hughes, R.E., Brzovic, P.S., Klevit, R.E., and Hurley, J.B. 1995, *Biochemistry* 34, 11410.

[64] Senin, I.I., Vaganova, S.A., Weiergraber, O.H., Ergorov, N.S., Philippov, P.P., and Koch, K.-W. 2003, *J. Mol. Biol.* 330, 409.

[65] Valentine, K.G., Mesleh, M.F., Opella, SJ., Ikura, M., and Ames, J.B. 2003, *Biochemistry* 42, 6333.

[66] Zozulya, S. and Stryer, L. 1992, *Proc. Natl. Acad. Sci. USA* 89, 11569.

[67] Tanaka, T., Ames, J.B., Harvey, T.S., Stryer, L., and Ikura M. 1995, *Nature* 376, 444.

[68] Hwang, J. Y. and Koch, K.-W. 2002, *Biochim. Biophys. Acta* 1600, 111.

[69] Hwang, J. Y. and Koch, K.-W. 2002, *Biochemistry* 41, 13021.

[70] Olshevskaya, E.V., Ermilov, A.N., and Dizhoor, A.M. 1999, *J.Biol Chem.* 274, 25583.

[71] Otto-Bruc, A., Buczylko, J., Surgucheva, I., Subbaraya, I., Rudnicka-Nawrot, M., Crabb, J.W., Arendt, A., Hargrave, P.A., Baehr, W,. and Palczewski, K. 1997, *Biochemistry* 36, 4295.

[72] Schrem, A., Lange, C., Beyermann, M. and Koch, K.-W. 1999, *J.Biol.Chem.* 274, 6244.

[73] Hwang, J.-Y., Lange, C., Helten, A., Höppner-Heitmann, D., Duda, T., Sharma, R.K., and Koch, K.-W. 2003, *Eur.J.Biochem.* 270, 3814.

[74] Belluscio, L., Gold, G.H , Nemes, A., and Axel, R. 1998, *Neuron* 20, 69.

[75] Breer, H. 2003, *Anal. Bioanal. Chem.* 377, 427.

[76] Buck, L.B. 1995, *Cell* 83, 349.

[77] Buck, L.B. 2000, *Cell* 100, 611.

[78] Crasto, C. Singer, M.S., and Shepherd, G.M. 2001, *Genome Biol. 2*, 1027.1.

[79] Touhara, K. 2002, *Microsc. Res. Tech.* 58, 135.

[80] Bakalyar, H. A. and Reed, R. R. 1990, *Science* 250, 1403.

[81] Wong, S.T., Trinh, K., Hacker, B., Chan, G.C., Lowe, G., Gaggar, A., Xia, Z.,Gold, G.H., and Storm, D.R. 2000, *Neuron* 27, 487.

[82] Firestein, S. and Werblin, F. 1989, *Science* 244, 79.

[83] Firestein, S., Picco, C., and Menini, A. 1993, *J. Physiol.* 468, 1.

[84] Kurahashi, T. 1989, J. Physiol. 419,177-192 (*Erratum, J Physiol* 424,561-562).

[85] Nakamura, T. and Gold, G.H. 1987, *Nature* 325, 442.

[86] Duda, T., Venkataraman, V., Krishnan, A., Nagele, R.G., and Sharma, R.K. 2001, *Biochemistry* 40, 4654.

[87] Firestein, S. 2001, *Nature* 413, 211.

[88] Juilfs, D.M., Fulle, H.-J., Zhao, A. Z., Houslay, M.D., Garbers, D.L., and Beavo, J.A. 1997, *Proc. Natl. Acad. Sci. USA.* 94, 3388.

[89] Meyer, M.R., Angele, A., Kremmer, E., Kaupp, U.B., and Muller, F. 2000, *Proc. Natl. Acad. Sci. USA.* 97, 10595.

[90] Kaupp, U. B. and Seifert, R. 2002, *Physiol. Rev.* 82, 769.

[91] Duda, T., Fik-Rymarkiewicz, E., Venkataraman, V., Krishnan, A., and Sharma, R.K. 2004, *Mol. Cell. Biochem.* 267, 107.

[92] Venkataraman, V., Duda, T., Vardi, N., Koch, K.-W., and Sharma, R.K. 2003, *Biochemistry* 42, 5640.

[93] Venkataraman, V., Nagele, R., Duda, T., and Sharma, R.K. 2000, *Biochemistry* 39, 6042.

[94] Anholt, R.R., Aebi, U., and Snyder, S.H. 1986, *J. Neurosci.*, 6, 1962.

[95] O'Callaghan, D.W., and Burgoyne, R.D. 2003, *Biochem. Soc. Trans.*, 31, 963.

[96] Spilker, C., and Braunewell, K.H. 2003, *Mol. Cell. Neurosci.*, 24, 766.

[97] Zufall, F., Leinders-Zufall,T., and Greer, C.A. 2000, *J. Neurophysiol.* 83, 501.

In: Neuronal Calcium Sensor Proteins
Editors: Pavel Philippov and Karl Koch, pp. 115-135
ISBN 1-59454-978-8

Chapter VII

STRUCTURE, FUNCTION AND EXPRESSION OF MEMBERS OF THE VILIP (VISININ-LIKE PROTEIN) SUBFAMILY OF NEURONAL CA^{2+} SENSOR PROTEINS

***Marian Brackmann*[1], *Andreas Hofmann*[2] *and Karl-Heinz Braunewell*[1,*]**

[1]Signal Transduction Research Group, Neuroscience Research Center/Institute for Neurophysiology, Charité Berlin, Tucholskystr. 2, D-10117 Berlin, Germany;
[2]Institute of Structural & Molecular Biology, School of Biological Sciences, The University of Edinburgh, Mayfield Road, Edinburgh EH9 3JR, Scotland.

Abstract

The subfamily of VILIPs, with its members VILIP-1, -2, -3, hippocalcin and neurocalcin δ, has been implicated in the modulation of a variety of intracellular signaling cascades. This chapter will compare the members of the subfamily and particularly focus on two proteins, VILIP-1 and VILIP-3, which have recently been characterized in more detail. Comparative studies have shown that the proteins are predominantly expressed in the brain but show restricted expression patterns in different subsets of neurons. Moreover, VILIP-1 and -3 show differences in calcium affinities, membrane binding kinetics and in their intracellular targets to which they associate after calcium binding. Although both proteins use a similar calcium-myristoyl switch mechanism, the proteins show different calcium-dependent localization to subcellular compartments when expressed in the same neuron. These distinct myristoyl switch properties enable the VILIPs to take part in different signaling cascades in a calcium-dependent manner. For VILIP-1, an

* Correspondence concerning this article should be addressed to Dr. K.-H. Braunewell Signal Transduction Research Group, Neuroscience Research Center/Institute for Neurophysiology, Charité Berlin, Tucholskystr. 2, D-10117 Berlin, Germany.

involvement in cyclic nucleotide signaling has been demonstrated, whereas VILIP-3 most likely modulates MAPK activity. Thus, the highly homologous proteins VILIP-1 and -3 seem to have evolved to fulfil specialized functions in the neuronal signaling of subsets of neurons in the nervous system.

Keywords: adenylyl cyclase, calcium-myristoyl switch, guanylyl cyclase, intracellular signaling pathways, MAPK pathways, neuronal calcium sensors, nicotinic acetylcholine receptor, Visinin-like proteins.

Abbreviations

CNP	C-type natriuretic peptide
ERK	extracellular signal-regulated kinase
EST	expressed sequence tag
GC	guanylyl cyclase
MAPK	mitogen activated protein kinase
MLK	mixed lineage kinase
nAChR	nicotinic acetylcholine receptor
NCS	neuronal calcium sensor
PLD	phospholipase D
retGC	retinal guanylyl cyclase
RT-PCR	reverse transcriptase polymerase chain reaction
VILIPs	visinin-like proteins

1. Introduction

A variety of calcium-sensing proteins have been identified in the nervous system, in addition to the ubiquitous calcium sensor protein calmodulin, reflecting the importance of calcium and its regulative function in nerve cells. These molecules have been termed neuronal calcium sensor (NCS) proteins [1,2]. To date, 14 NCS proteins are known to exist in the human genome with orthologs in various species. This chapter will focus on one of the subfamilies, the VILIPs (Fig. 1), particularly on VILIP-1 and VILIP-3 and compare them with hippocalcin and neurocalcin δ. Proteins in this subfamily show a high degree of amino acid identity, between 67 and 94% [2-4].

2. Distribution of VILIPs in the Central Nervous System and in the Periphery

In the nervous system, VILIPs show a widespread but distinct expression pattern primarily in nerve cells. The expression patterns of NCS proteins and their transcripts have been reviewed previously [2]. More recently, several comparative expression studies for VILIP subfamily members have been performed. These include an analysis of the differential protein expression of neurocalcin isoforms α and δ in the rat cerebellum [5]; of VILIP-1 and VILIP-3 in the rat cerebellum and hippocampus [6]; and of VILIP-1 and VILIP-3 in the human brain [7]. A detailed differential mRNA expression study of the VILIP subfamily describes the expression on VILIP-1, -2, -3 and hippocalcin in the rat brain [8]. VILIP-1 mRNA shows a widespread distribution in virtually all brain areas except the caudate-putamen. Hippocalcin and VILIP-2 exhibit overlapping expression patterns in the forebrain including neocortex, hippocampus and caudate-putamen. Finally, VILIP-3 has the most restricted expression pattern in the cerebellum where it localizes to Purkinje and granule cells [8].

Immunohistochemical data, in general, have to be taken with some caution because antibodies might recognize several NCS proteins due to the high similarity of members of the subfamily. These studies should be supported by *in situ* hybridisation or RT-PCR data when highly specific antibodies are not available. Many of the detailed studies on the distribution of VILIPs in the brain have focussed on the hippocampus and the cerebellum. For VILIP-1 and VILIP-2, early studies in the hippocampus of rat and gerbil [9,10] indicated that the two proteins are differentially distributed. Whereas VILIP-2 expression is most prominent in CA1/CA2 regions and the dentate gyrus of the rat hippocampus, VILIP-1 immunoreactivity is strongest in the CA3 region of the gerbil. Hippocalcin is well expressed, with varying intensities, in different parts of the rat hippocampus and in some other brain regions including the cerebellum [11,12]. VILIP-3 expression has been mainly localized in the cerebellum but there is also weak expression in other brain regions including the hippocampus [6,13]. At the mRNA level regional differences in the expression levels of hippocalcin and VILIP-3 can be observed in the rat hippocampus. Compared to other hippocampal subregions, a relatively high expression level of VILIP-3 and a relatively low level of hippocalcin are observed in the dentate gyrus [6]. VILIP-1 and -3 are colocalized in neurons in the hippocampus, showing a strong expression for VILIP-1 in many neurons and weaker expression of VILIP-3 in a subset of neurons [6]. Co-immunolocalization studies of neurocalcin in the rat hippocampus have suggested that all neurocalcin immunoreactive neurons (neurocalcin antibodies reacted with neurocalcins α, β, γ, δ) are GABAergic. Neurocalcin-positive neurons comprise about 19% of GABAergic neurons, suggesting that neurocalcins/VILIPs are marker proteins for a subpopulation of GABAergic interneurons [14].

The second brain structure which has been studied in more detail is the cerebellum. Northern blot data indicate that VILIP-2 is not expressed, whereas VILIP-3 is prominently expressed in the rat cerebellum [15]. Immunohistochemical and *in situ* hybridisation data revealed that at the cellular level, VILIP-3 is mainly expressed in Purkinje cells, but granule cells also show weak VILIP-3 immunoreactivity [6,16]. In comparison, VILIP-1 transcripts are primarily found in the granule cell layer but are absent from Purkinje cells in chicken and rat cerebellum [17,5,6]. VILIP-1 protein and its ortholog neurocalcin α are consistently absent from Purkinje cells. Immunoreactivity is restricted to granule cells and to the molecular layer of the chicken and rat cerebellum [18,5,6]. Studies in the human brain have been performed with antibodies against VILIP-1 and VILIP-3. Generally both proteins are expressed in subsets of neurons in virtually all brain regions. However, VILIP-1 displays a much more intense immunoreactivity than VILIP-3 [7].

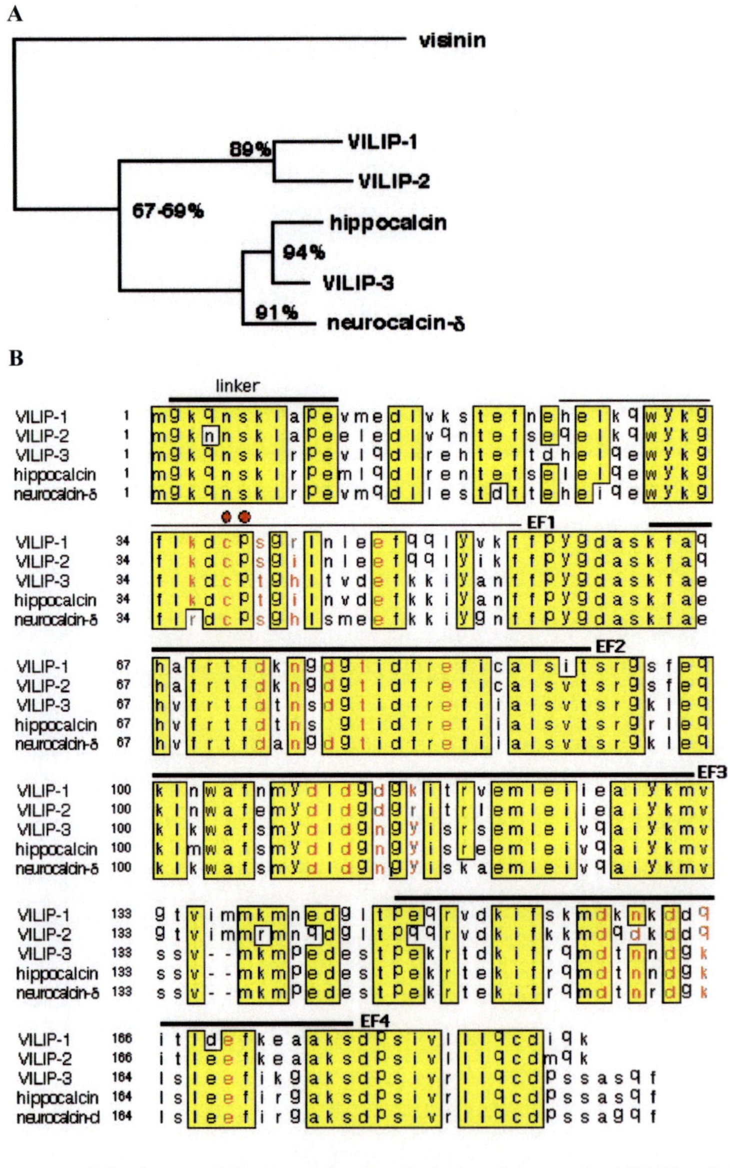

Fig. 1. A. Dendrogram of the human VILIP-subfamily. Relationships in the VILIP-subfamily based on amino acid similarity. The dendrogram was generated using human sequences from the SwissProt database and the clustalX and NJplot program. Data base accession numbers: VILIP-1 (VSNL1, U14747), VILIP-2 (HLP4, AB00105), VILIP-3 (BDR-1, D16227), neurocalcin δ (NCALD, AF251061), hippocalcin (BDR-2, D16593). Visinin/recoverin, the founder of the NCS protein family, is used as a reference. The identity of amino acids in % is indicated. B. Alignment of visinin-like proteins. Sequences of the VILIP subfamily members VILIP-1, VILIP-2, VILIP-3, hippocalcin and neurocalcin δ, were taken from the SwissProt database. EF-hands 1-4 are marked with a black line. The conserved calcium-binding motifs of the four EF-hands are indicated in red. The conserved Cys-Pro amino acid pair in EF-hand 1 which disables calcium binding is marked with red dots.

Members of the VILIP-group of NCS proteins are also expressed in neurons of sensory pathways including the retina and olfactory system. In the retina of various species, VILIP-1 and neurocalcin δ are not found in photoreceptor cells but are expressed in subsets of bipolar, amacrine and retinal ganglion cells [17,19-21]. Moreover, a subset of olfactory receptor neurons in the rat olfactory epithelium were shown to express VILIP-1 [22,23,2], neurocalcin δ [24,25] and hippocalcin [26].

Although NCS proteins have been described as a nervous system-specific gene family, expression outside the nervous system has been observed as well [2]. The family member VILIP-3, also known as REM-1, has not only been described in cells of the haematopoietic system and in the gut [27], but also in kidney, spleen and testis [6]. For VILIP-1, major expression has been detected in heart, liver, lung and testis in human and rat and also in stomach and skin of rat [28]. At the mRNA level, some expression of VILIP-1 in heart and lung had already been implied by Northern blot data [15]. Furthermore, the presence of VILIP-1 in heart and colon was recently shown by RT-PCR [29], and its presence in skin was shown at the protein and mRNA level [30]. In contrast to the other VILIPs, hippocalcin shows a very restricted protein expression pattern exclusively in brain tissues. Although expressed sequence tag (EST) sequences were found in a few additional peripheral tissues, no protein expression could be detected [28]. The peripheral distribution of VILIPs in relation to brain expression was much higher in embryonic tissue compared to adult rat tissues, indicating a developmental restriction of expression. During development changes in the strength of protein expression of VILIPs are observed, e.g. a developmental decrease of expression in liver, lung, kidney, spleen, pancreas and colon occurs. These results might additionally hint to a specific function of VILIP-1 and -3 during the development of distinct organs.

It was concluded from the foregoing data that hippocalcin may have adapted to specific brain functions particularly in the hippocampus, whereas VILIP-1 and particularly VILIP-3 still fulfill the specific functions of some peripheral organs. In addition, EST clones containing VILIP sequences were also found in disease-related tissues especially in cancer tissues such as nervous tumor, hepatocellular carcinoma, adenocarcinoma, oligodendroglioma, lung squamous cell carcinoma and head/neck tumors. Thus, besides playing a role in the development and physiology of peripheral organs, it is likely that VILIPs are also involved in the pathophysiology of cancer. This assumption has recently been confirmed for VILIP-1 as it has been shown to be involved in the growth and invasiveness of head/neck tumors [30,31].

3. Structure of VILIPs

3.1. Relationship of VILIPs

VILIP subfamily members, including VILIP-1, -2, -3, hippocalcin and neurocalcin δ, have been cloned and sequenced from various species. Sequence analysis of the NCS protein family illustrates only 30-60% sequence similarity with other NCS subfamilies. Nevertheless, subfamily members share between 67 to 94 % identical amino acids, forming a closely related subfamily (Fig. 1A). Of the two VILIP branches, VILIP-1 and -2 are 89% homologous. VILIP-3 and hippocalcin share 94% identity, and the latter two proteins are 91% identical to neurocalcin δ. There is 67-69% homology between the VILIP-1 and -3 branches. VILIPs are typically 191-193 amino acid residues long and have a M-G-X_3-S consensus sequence for N-terminal myristoylation. Besides the EF-hand calcium-binding motifs which identify them as members of the EF-hand superfamily of Ca^{2+}-binding proteins, no other domains have been identified yet.

3.2. The Calcium-Myristoyl Switch

Eleven out of the 14 mammalian NCS proteins are myristoylated at their N-termini. Thus, besides KChIP2-4, which are not myristoylated, and NCS-1, GCAP-1 and –2, which do not show calcium-dependent membrane association, many NCS proteins can translocate to subcellular membrane compartments by a molecular mechanism termed calcium-myristoyl switch, which is dependent on calcium-binding and N-terminal myristoylation of the proteins [32]. The molecular mechanism of the calcium-myristoyl switch has been analyzed in detail from tertiary structure data of NCS proteins. These were first available for recoverin which was analyzed in its calcium-bound and calcium-free myristoylated forms [33,34]. The four EF-hand motifs in recoverin are arranged in a compact tandem array, with the myristoyl side chain buried in a hydrophobic pocket [35]. Binding of calcium to recoverin induces a conformational change leading to the surface exposure of hydrophobic parts of the polypeptide and of the myristoyl side chain, thereby making these structures available for interaction with cellular membranes and/or target proteins. Biochemically, the existence of the calcium-myristoyl switch mechanism has been shown for all members of the VILIP-subfamily [6,36-38]. It has also been observed in living cells after increasing intracellular calcium concentration [39-41,13]. The calcium-myristoyl-switch mechanism seems to apply to all members of the VILIP subfamily as expected by biochemical and cell biological data.

3.3. General Structure of NCS Proteins

Like all NCS proteins, VILIPs possess four EF-hand motifs, EF1 – EF4 (Fig. 1B). Since EF1 also forms the most variable part in the sequence of NCS proteins (Fig. 1B, EF1 thin line) it is therefore discussed as a possible interaction site with target proteins. Parts of EF1 and EF2 in GCAP1 have already been shown to interact with retinal guanylyl cyclase [42]. However, small stretches of hydrophobic amino acids following EF3 and EF4 might also participate in the target interaction (Fig. 1B).

Canonical calcium binding in EF-hands occurs by seven oxygen ligands in a pentagonal bipyramidal geometry [43]. The amino acid sequence motif for calcium binding in NCS proteins is D-X-D/N-X-D/N-X-Y-X_4-E (X is any amino acid) where the side chain oxygens of the Asp and Asn residues occupy three mono-dentate coordination sites, and the side chain oxygens of the Glu residue provides a bi-dentate coordination. Residue Y coordinates the calcium ion by its carbonyl oxygen and the apical coordination position is occupied by a water molecule. Therefore, EF1 in most NCS proteins is rendered dysfunctional as the amino acid sequence is changed to K/R-X-C-P-S/T-X-R/H/I-X_4-E where most residues lack the required side chain oxygens and the positive charge of Lys/Arg prevents attraction of a metal ion. In recoverin or KChIP1, the calcium binding sequence in the fourth EF-hand motif is also dysfunctional, and these proteins bind only two calcium ions.

A number of NCS protein structures have been solved by X-ray crystallography or NMR spectroscopy in the presence and absence of calcium, including recoverin [34,35,44-46], GCAP-2 [47], NCS-1 [48,49], neurocalcin δ [50], and KChIP1 [51,52]. For recoverin, structures of the myristoylated protein in the apo- [44,35], partially [45,46] and fully [34] loaded calcium forms are known. NCS proteins appear as a highly compact structure with a protein fold constituted by two pairs of EF-hand calcium-binding motifs.

3.4. Modeling of Tertiary Structure of VILIP-1 and -3

As mentioned above, at the level of primary structure, VILIP-1 and VILIP-3 are most similar to neurocalcin δ and recoverin. Notably, the highest similarity is shared with neurocalcin δ. Both VILIPs have the required amino acid sequence for calcium binding in the fourth EF-hand, further supporting the close relationship with neurocalcin δ. The degree of similarity between VILIPs and NCS proteins of known structure is the basis for predicting the structures of VILIPs by the homology modeling approach. In short, the three-dimensional structures of apo-recoverin (PDB entry 1IKU, [35]) and calcium-bound neurocalcin δ (PDB entry 1BJF, [50]) were used as a template. For the apo-form, the conformational difference between different NCS proteins is assumed to be negligible, thus, in the absence of a structure of apo-neurocalcin δ, the recoverin structure can be used. However, for the calcium-bound form, recoverin and neurocalcin δ differ in their conformations (root-mean-square deviation: 8.9 Å) and in the calcium binding site EF4. Therefore, the structure of neurocalcin δ is the preferred template for the modeling approach. Alignments of amino acid sequences were generated with MUSCLE [53]. Twenty independent homology models were calculated with the programme Modeller 4 [54] and the one with the lowest energy was selected. The selected model was visually inspected with O [55] and its overall geometry was scrutinized using PROCHECK [56]. Neurocalcin δ is the preferred template structure for VILIP modeling due to the higher amino acid sequence similarity. Nevertheless, for the apo-form a structure is only available for recoverin.

The homology models for VILIP-1 and VILIP-3 in the apo- and calcium-bound state are shown in Fig.2. In the apo-form, the myristoyl group is harbored in the hydrophobic cleft constituted by residues Ala9, Met13, Lys27, Trp30, Tyr31, Phe55, and Phe56 from EF1, Phe82, Phe85, Ile86, and Leu89 from EF2, and Trp103 and Met107 from EF3 (VILIP-1 numbering). All of these residues are highly conserved in the NCS family of proteins and the structure of the hydrophobic cleft is most likely similar in all members of this family. The binding of calcium in EF2, EF3, and EF4 provokes a conformational change of the protein whereby several residues of the hydrophobic cleft are being exposed and the myristoyl group is being extruded.

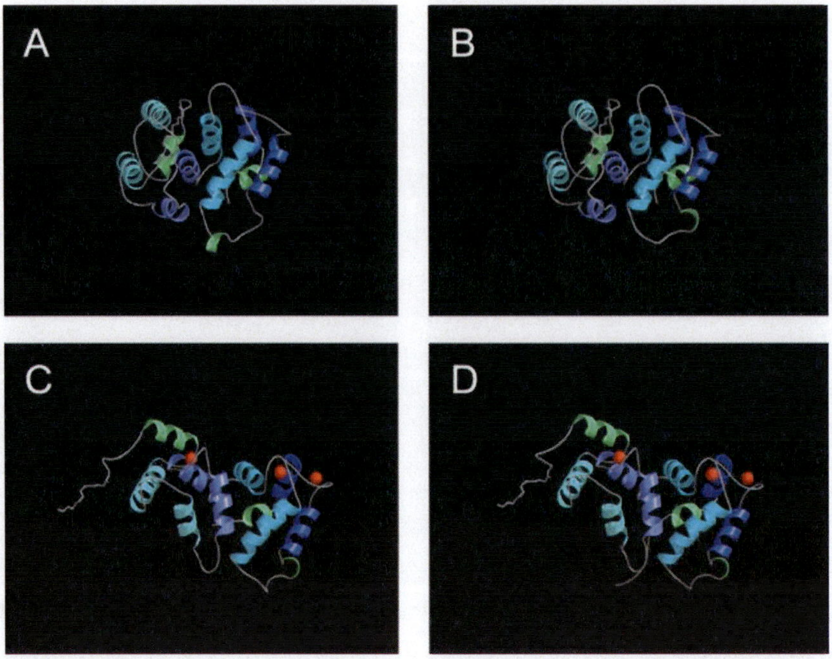

Fig. 2. Structural models of VILIP-1 and VILIP-3. The structures were modeled using the coordinates of calcium-free recoverin (1IKU) and calcium-bound neurocalcin δ (1BJF). Models A and B illustrate the calcium-free forms of VILIP-1 and VILIP-3, respectively. Models C and D show their calcium-bound forms. Colouring of the EF-hand motifs follows the scheme EF1 (light blue), EF2 (light purple), EF3 (dark blue), EF4 (dark purple). The N-terminal helix A, the C-terminal helix J and the helical turn before the fourth EF-hand motif are shown in green. The figure was prepared with BOBSCRIPT (Esnouf, 1999) and rendered with POVRay (POVRay - The Persistence of Vision Raytracer. version 3.5, Persistence of Vision Development Team, 2002, http://www.povray.org/).

4. Calcium Myristoyl Switch as a Mechanism of Signal Transduction: Differences for VILIP-1 and VILIP-3

The reversible localization of signaling proteins and calcium sensors to distinct membrane compartments or to associated cytoskeletal components in living neurons via the molecular mechanism of the calcium-myristoyl switch, has been postulated to be a signal transduction mechanism for the selective activation of downstream signaling cascades, such as receptors, receptor signaling complexes and signal effector molecules [13,39-41,57]. Like most NCS proteins, VILIP-1 and -3 have been shown in biochemical fractionation experiments to translocate from the cytosolic to the particulate fraction in a calcium-dependent manner (Fig. 3A). Similarly, when co-expressed in neural cell lines, VILIP-1 and -3 exhibit biochemically defined calcium-myristoyl switch properties. However, differences in the degree and/or mechanism of calcium-dependent membrane association became evident in biochemical experiments following repeated EGTA extraction of the membrane fraction. Under these conditions, VILIP-1 is less extractable from the membrane fraction compared to

VILIP-3 [4]. Moreover, neurocalcin α and δ, the cow counterparts of VILIPs, have different calcium affinities and the degree of the calcium-induced conformational change differs as shown by calcium overlay, gel migration assay and tryptophan fluorescence assay [4,5]. In line with these biochemical data, the membrane association of VILIP-1 and VILIP-3 also appears to vary. VILIP-1 displays a tight membrane association in granule cells, whereas VILIP-3 is evenly distributed in Purkinje cells [4]. An equivalent subcellular distribution has been observed in hippocampal neurons co-expressing VILIP-1 and -3, which strongly indicates that these differences reflect the intrinsic properties of the proteins [13]. The calcium-dependent subcellular membrane localization differed substantially when endogenous VILIP-1 and VILIP-3 in the same hippocampal neuron was observed. Endogenous VILIP-1 shows cell surface membrane association, including membranes of axons and dendrites, which is in line with the described function of VILIP-1 as modulator of cell surface associated proteins [30,58-61]. In addition, VILIP-1 only affiliates with trans-Golgi membranes following a calcium stimulus in hippocampal neurons [39], while VILIP-3 showed a weak calcium-independent Golgi localization that was only gradually enhanced following stimulation of hippocampal neurons [13]. Furthermore, VILIP-3 interacts with intracellular juxtanuclear membranes and granular structures in the whole cytosol, which fits to a possible function as a modulator of the intracellular MAP kinases [4]. Interestingly, an enhanced juxtanuclear localization of VILIPs also exists under conditions of disturbed calcium homeostasis in Alzheimer`s disease [62,63].

It has been suggested that the *in vitro* differences in calcium affinities for NCS proteins might increase the dynamic range over which calcium and NCS proteins can regulate neuronal activities [3]. The observed *in vitro* differences in calcium affinities of VILIPs have been found to be paralleled by those in the subcellular localization and activity-dependent translocation to other cellular signaling compartments, when co-expressed [4,5,13]. Thus, each NCS protein might respond to different calcium concentrations, have a high specificity for certain cell-types and specificity towards distinct receptors and signaling pathways. Further comparative analyses of the calcium affinities of myristoylated VILIPs are necessary, along with a comparison of their functional activities.

5. VILIPs Modulate Different Signaling Cascades

A variety of extracellular signals including light, odorants, hormones, growth factors, electrical activity and neurotransmitters evoke highly specific changes of intracellular calcium levels. There are several recent observations indicating that VILIP-1 and -3 act as signaling modulators in different signaling compartments. The VILIPs can serve as effectors to transduce these very different calcium signals, which is in line with the view of NCS proteins as "pleiotropic" signal transducers (Table 1). In this review, we will summarize the effects of VILIP subfamily proteins, concentrating on the effects of VILIP-1 and -3 on various signaling components and pathways.

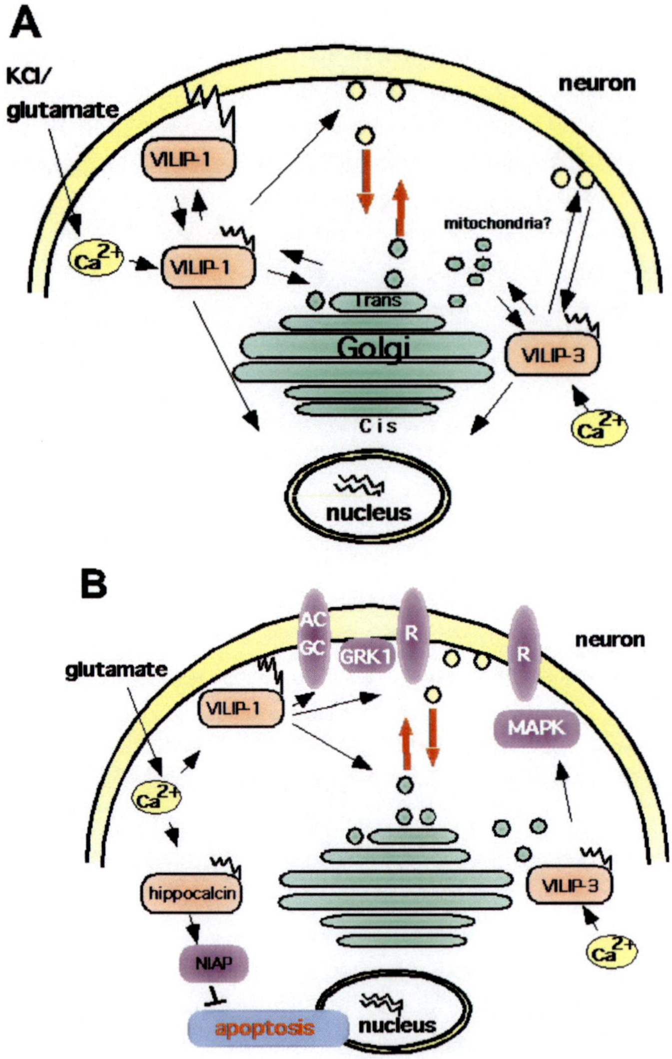

Fig. 3. A. The calcium-myristoyl switch and activity-dependent localization of VILIP-1 and -3 to cellular membranes. Following binding of calcium, a calcium-induced conformational change of cytosolic VILIP-1 and -3 occurs leading to the exposure of the hydrophobic myristoyl side chain. The NCS proteins can then interact with cellular membranes, and are able to shuttle from the cytosol to cellular membranes and *vice versa*. Both VILIPs can associate with clathrin-coated vesicles involved in endocytotic processes, the cell surface membrane, the trans-Golgi and under certain pathological conditions juxtanuclear membranes. VILIP-3 might also localize to microsomal and mitochondrial membranes as indicated by interaction with cytochrome oxidase B. The calcium- and myristoylation-dependent interaction of VILIPs with cell membranes may be a prerequisite for the interaction with membrane-bound target molecules from various cellular signaling cascades in different cellular compartments. B. NCS proteins from the VILIP-subfamily are calcium-dependent modulators of intracellular signaling cascades. VILIP-1 has been identified as a modulator of adenylyl cyclase (AC), guanylyl cyclase (GC) and nicotinic acetylcholine receptor (nAChR) signaling. In the case of AC, VILIP-1 expression leads to the enhancement of cAMP accumulation in various transfected cell lines and tumor cells. Thereby, VILIP-1 affects rhoA signaling in tumor cells. Similarly, VILIP-1-expression increases cGMP-accumulation in transfected cell lines and acts as modulator of receptor GC-B in primary neurons. Expression of membrane localized VILIP-1 increases the surface expression of the

receptor GC-B in primary neurons by influencing clathrin-dependent receptor recycling. In contrast, the cytosolically localized myristoylation mutant of VILIP-1 influences the NO-stimulated soluble GC. In addition, VILIP-1-expression enhances surface expression of nAChR in frog oocytes and increases the sensitivity of the receptor for its ligand ACh. VILIP-3 was identified as modulator of MAPK signaling in transfected cell lines. The NCS protein affects expression level and activity of the MAPKinase ERK1/2 in PC12 cells. Similar to VILIP-1, neurocalcin δ and hippocalcin have been identified as modulators of adenylyl (olfactory adenylyl cyclase) und guanylyl cyclase (olfactory and retinal guanylyl cyclase) enzymes. Similar to VILIP-3, hippocalcin increases MAPKinase activity. It also interacts with the neuronal inhibitor of apoptosis NAIP and protects neurons against apoptotic cell death. This is the opposite effect of VILIP-1 and -3 which appear to be neurotoxic when overexpressed in PC12 cells.

Table 1. NCS proteins as "pleiotropic" signal transducers.

NCS protein	Functional interaction partners	Reference
VILIP-1	adenylyl cyclase (type III/VI)	[22,30,31,58,61]
	guanylyl cyclase (GC-A/-B, soluble GC)	[59,67]
	GRK1, 2	[5,90]
	acetylcholine receptor (α4ß2)	[60]
	trkb mRNA	[91]
	actin, cortical cytoskeleton	[18,92]
VILIP-2	Cav2.1 calcium channel (P/Q-type)	[73]
VILIP-3	ERK1/2	[4]
	Cytochrome b5	[93]
	Na/Pi cotransporter (NaPi-IIa)	[74]
Neurocalcin δ	GRK2	[5]
	Retinal guanylyl cyclase	[21,66]
	olfactory guanylyl cyclase	[24,25]
	actin, clathrin and tubulin	[80,81]
Hippocalcin	adenylyl cyclase (type III),	
	retinal guanylyl cyclase	[26]
	mixed linage kinase MLK2/3	
	mitogen activated protein kinases (MAPK)	[75,76]
	PLD (cdc42-dependent)	[78]
	neuronal apoptosis inhibitory protein (NAIP)	[94-96]

5.1. VILIP-1

Previous results suggested that VILIPs are modulators of multiple intracellular targets showing a 'pleiotropy' of actions similar to calmodulin (Table 1). Indeed, VILIP-1 (see Fig. 3B) has been implicated in modulation of adenylyl cyclase activity [30,58,61,64] and related cyclase enzymes, including the soluble and the membrane-localized guanylyl cyclases (GCs) [4,59,65], nicotinic acetylcholine receptor [60] and integrin receptors [32].

5.1.1. Influence on Adenylyl Cyclase Signaling

Regulation of cAMP levels by VILIP-1 has first been observed in VILIP-1-transfected C6 glioma cells, which most likely express the adenylyl cyclase type VI [58,64]. After stimulation of ß-adrenergic receptors by isoproterenol or of adenylyl cyclase type VI by Forskolin, the cAMP level in these cells were increased several-fold in wild-type VILIP-1 transfected cells compared to untransfected C6 cells. A myristoylation-deficient mutant of VILIP-1, which does not show membrane association in these cells, does not exhibit this effect. Interestingly, the basic cAMP levels appear to be elevated in VILIP-1-transfected C6 cells, which induces differentiation processes in the glioma cells [64]. The effect of VILIP-1 on adenylyl cyclase was not observed in other neural cell lines tested, such as PC12 or Neuro2A cells. These results point to an isoform-specific effect of VILIP-1, e.g. on adenylyl cyclase type VI. In addition, at high μM concentrations, VILIP-1 directly inhibits olfactory adenylyl cyclase type III activity in a calcium-dependent manner in olfactory membrane assays following odor stimulation. The protein is proposed as an additional modulator of olfactory adaptation processes [22]. These findings fit the strong VILIP-1 immunoreactivity in the olfactory epithelium especially in the olfactory knobs containing the olfactory signaling machinery [2].

Further support by more recent results indicate that VILIP-1 influences adenylyl cyclase activity in other cell types including human embryonic kidney cells [61] and various skin tumor cell lines [30]. High-level expression of VILIP-1 occurs in less aggressive squamous cell carcinoma and papilloma cell lines. The expression is lost in more aggressive tumor cells. The reduced expression correlates with decreased cAMP levels, which increased the expression of the metalloproteinase MMP-9 and rhoA activity. The loss of VILIP-1 and its effect on cAMP-dependent signaling intensifies the invasiveness of carcinoma cells, which can be reversed by forcing the expression of VILIP-1 and increasing the VILIP-1-induced cAMP levels [30].

Thus, data obtained in different cellular systems suggest a modulatory role for VILIP-1 on the activity of distinct adenylyl cyclase isoforms either by direct physical interaction or by involving an additional, yet unknown, mechanism. Moreover, in the periphery, the VILIP-1 effect on cAMP-signaling has important pathophysiological implications and identifies the protein as a putative tumor metastasis suppressor gene.

5.1.2. Influence on Guanylyl Cyclase Signaling

In addition to the effect on adenylyl cyclases, an influence of VILIP-1 on non-retinal guanylyl cyclases, namely the receptor GC-A and -B and soluble cyclases, has been observed *in vitro* [59]. In several transfected neural cell lines including C6, PC12 and Neuro2A cells, VILIP-1 affects cGMP-levels. Following the stimulation of receptor GCs with natriuretic peptides or the soluble GC with nitric oxide, an enhanced cGMP-accumulation was observed in all three cell lines when transfected with VILIP-1. Interestingly, the effect on soluble and particulate GCs depends on the subcellular localization of VILIP-1. The membrane-localized wild type VILIP-1 mainly affects the particulate receptor cyclases GC-A and GC-B, whereas the myristoylation mutant of VILIP-1, localized in the cytosol, influences the soluble cyclase enzyme. Thus, membrane-localized VILIP-1 was shown to modulate the membrane-localized natriuretic peptide receptor guanylyl cyclase GC-A and -B [4,59]. Similarly, the highly homologous protein neurocalcin δ modulates the activity of homologous membrane guanylyl cyclases, such as retGC1 expressed in retinal neurons [21,66] and ONE-GC expressed in the olfactory neuroepithelium [24,25].

The effect of VILIP-1 on cGMP-signaling was also found in primary cerebellar cultures, where the protein specifically affected neuronal GC-B but not glial GC-A [59]. Additionally, the relevance of the effect of VILIP-1 on GC-B activity in primary hippocampal neurons was further investigated. VILIP-1 co-localized with GC-B in many, but not all, cultured hippocampal neurons, and also significantly enhanced C-type natriuretic peptide (CNP)-stimulated cGMP accumulation following transient transfection with the VILIP-1 cDNA. Moreover, heterologous VILIP-1-expression led to an increase in the surface expression and attenuation of GC-B internalization in hippocampal neurons. The higher level of VILIP-1 expression thereby enhances basal, as well as CNP-stimulated, cGMP-accumulation. Interestingly, VILIP-1 also affected the recycling of the unrelated transferrin receptor, indicating that VILIP-1 is a general modulator of membrane trafficking. On the other hand, the effects on GC are not observed with VILIP-3 [67].

In summary, it is likely that VILIP-1 is a physiological modulator of the neuronal CNP signaling pathway, the receptor GC-B and the second messenger cGMP. Interestingly, it was demonstrated recently that other proteins of the VILIP subfamily also influence cyclases. Neurocalcin δ was found to interact with and directly influence related GCs, namely the retinal and the olfactory GCs [21,24,25,66]. Moreover, hippocalcin inhibits olfactory adenylyl cyclase but activates olfactory GC [26]. Thus, a major function of VILIPs might be the modulation of cyclase enzymes, similar to the related GCAP family which regulates retinal GCs [68,69]. The co-localization of VILIP-1 with GC-B in the soma and dendritic membranes of hippocampal neurons suggests that VILIP-1 might influence GC activity under physiological conditions in the hippocampus *in vivo*. Regarding the Ca^{2+}-dependent localization of VILIP-1 at the trans-Golgi network and dendritic membrane specializations in hippocampal neurons [39], it is conceivable that GC-B and VILIP-1 may influence neuronal signaling in the dendrites of hippocampal neurons in a Ca^{2+}- and cGMP-dependent manner. Thus VILIP-1 might influence cGMP-dependent neuronal processes, including different forms of synaptic plasticity [70,71] and learning and memory [72].

5.1.3. Influence of VILIP-1 on nAChR Signaling

Similar to its effect on GC receptors, VILIP-1 also affects ionotropic receptors, such as nicotinic acetylcholine receptors (nAChRs). VILIP-1 interacts with $\alpha_4\beta_2$ nAChR and is able to modulate acetylcholine receptor function [60]. Co-expression of VILIP-1 with recombinant $\alpha_4\beta_2$ nAChR up-regulated the surface expression levels by two-fold and increased the agonist-sensitivity to acetylcholine by three-fold. The VILIP-1 myristoylation mutant or mutants not able to bind calcium are found to attenuate the modulation of $\alpha_4\beta_2$ nAChR by VILIP-1. Furthermore, VILIP-1 and $\alpha_4\beta_2$ nAChR are co-localized in pyramidal cells and interneurons in the hippocampus [13,39, Brackmann, Gierke, Bernstein & Braunewell, unpublished]. These results suggest that VILIP-1 represents a novel modulator of $\alpha_4\beta_2$ nAChR, which increases the surface expression level and agonist sensitivity of the receptor in response to changes in the intracellular levels of calcium [60]. An effect of VILIP-1 on nAChR phosphorylation, which is involved in the sensitization/desensitization process of nAChR, has yet to be examined.

5.2. The Related Protein VILIP-2

For VILIP-2, which is 89% homologous to VILIP-1, there are virtually no functional data available. It was recently reported that VILIP-2 binds to the presynaptic P/Q-type calcium channel (Cav2.1). When co-expressing VILIP-2 with Cav2.1, an effect on the Cav2.1 inactivation rate was observed, which is dependent on N-terminal myristoylation [73]. Although it has not been tested whether VILIP-2 influences the signaling pathways affected by VILIP-1, from the differential distribution of the two proteins in the cerebellum, where VILIP-1 but not VILIP-2 is present [15], and in the hippocampus, where they show differential distribution [9], it can be assumed that the two proteins possess different cellular functions despite their high homology.

5.3. VILIP-3 in Comparison with the Related Proteins Neurocalcin δ and Hippocalcin

5.3.1. VILIP-3

The second branch of VILIPs comprises VILIP-3, hippocalcin and neurocalcin δ with homologies between 91 and 94%. Similar to VILIP-1 and -2, they show differential expression patterns, e.g. hippocalcin showing high abundance in the hippocampus and VILIP-3 in the cerebellum. In contrast to hippocalcin and neurocalcin δ, VILIP-3 is less well examined with respect to its function. VILIP-3 is mainly expressed in cerebellar Purkinje cells and might be specialized to fulfill the signaling requirements of this neuronal cell type [4,6]. However, expression in hippocampal neurons [13] and in the periphery [28,74] also point to more general functions. Since VILIP-3 is highly homologous to hippocalcin, the effect of VILIP-3 on MAPK activation seems likely. Hippocalcin specifically interacts with MAP kinases and influences phosphorylation of the MAP-kinase ERK2 [75,76]. The overexpression of VILIP-3 should lead to increased ERK activation since a reduced ERK2-activation was demonstrated in hippocalcin-deficient mice. This was investigated in transfected PC12 cells where following VILIP-3-transfection, a difference in the degree of ERK1/2-phosphorylation was detected in comparison to non-transfected cells. In addition, an effect on ERK2 protein expression was observed in VILIP-3-expressing cells, indicating that VILIP-3 might also affect the expression patterns of signaling proteins. As mentioned previously, in contrast to VILIP-1, VILIP-3 did not influence cGMP-signaling [4]. Differences in the membrane association properties of VILIP-1 and VILIP-3 [13], together with the fact that both proteins interact with a different set of binding partners [4], make it quite likely that VILIP-3 activates a distinct set of signaling cascades. The fact that VILIP-3 associates with intracellular juxtanuclear membranes and granular structures in the whole cytosol fits to a possible function of the protein as a putative modulator of the intracellular ERK1/2 MAP kinases [4,13].

5.3.2. Hippocalcin

As stated before, hippocalcin was identified as a possible interaction partner of MLK2/3 (mixed-lineage Ser/Thr kinases) following an interaction screen of a yeast two-hybrid cDNA library [75]. MLKs are closely related to the mitogen-activated protein kinase kinase kinase (MAPKKK) family [77]. MLK2 and 3 interact with the small GTPases Rac and Cdc42, with motor proteins of the kinesin superfamily and colocalize with the tubulin cytoskeleton, suggesting that they are involved in regulation of cytoskeleton dynamics in cells. MLK2/3 activate the JNK (c-Jun N-terminal kinase), ERK (extracellular signal-regulated kinase) and p38 [77]. However, to date no influence on kinase activity has been observed. Therefore, the biological significance of the MLK-hippocalcin interaction remains unclear. Interestingly, in this context, expression of hippocalcin in conjunction with the small rho-like GTPase cdc42 leads to a 1.8 fold increase in the calcium-dependent but PKC-independent PLD activation. However, hippocalcin alone did not activate PLD activity [78], which probably indicates MAPK-dependent activation of PLD. Another involvement of hippocalcin in MAPK-signaling was suggested by studies in hippocalcin-deficient knock-out mice, which show deficits in learning behavior [76]. At the molecular level, the ionomycin- or N-methyl-D-aspartate-induced phosphorylation of the MAP-kinase ERK2 was reduced in hippocampal slice cultures from hippocalcin-deficient mice compared to wild-type mice. Therefore, several lines of evidence indicate that hippocalcin has a modulatory influence on MAP-kinase-dependent signaling [76].

Hippocalcin might also have additional functions. It was recently found that at high calcium concentrations, the protein activates olfactory adenylyl cyclase but inhibits olfactory guanylyl cyclase ONE-GC (GC-D) in membrane assays [26]. Notably, this is the opposite effect of VILIP-1, which inhibits olfactory adenylyl cyclase [22] but activates GC-B [59,65], a natriuretic receptor which seems to be abundantly expressed in the brain including the hippocampus and the olfactory epithelium [78]. Thus, VILIP-1 and hippocalcin are discussed to be involved in the fine-tuning of stimulus detection and odor adaptation mediated by adenylyl cyclases and GCs in the olfactory neuroepithelium [26].

5.3.3. Neurocalcin δ

Similar to VILIP-1 [59,65], neurocalcin δ can influence cGMP-signaling [21,24,25,66]. Neurocalcin δ is a calcium-dependent regulator of the retinal GC, retGC1 (GC-E). RetGC1 colocalizes with neurocalcin δ in neurons in the inner plexiform layer of the retina, e.g. in amacrine and ganglion cells and is discussed to influence synaptic signaling processes [21]. The effect is specific for retGC1 since no influence on retGC2 or GC-A was observed [66], the latter one being a target for VILIP-1 [59]. Neurocalcin δ interacts with retGC1 and modulates its function at 1 μM calcium concentration with a halfmaximal activity of neurocalcin δ at 0.8 μM. In addition, neurocalcin δ which is expressed in the olfactory epithelium can interact and activate olfactory GC with similar high sensitivity. In contrast, GCAPs which can activate retGC1 and 2 failed to activate olfactory GC-D. Thus, besides the GCAPs, VILIP-1 and hippocalcin, neurocalcin δ is the fourth NCS protein to affect one of the GC family members. Finally, an interaction of neurocalcin δ with actin [80, 81], the clathrin heavy chain, alpha- and beta-tubulin [81] as well as of hippocalcin with β-2-adaptin [82] was observed, indicating a possible involvement in endocytotic membrane transport.

6. New Hypothesis for NCS Protein Signaling

NCS proteins like the KChIPs and NCS-1 have been shown to influence surface expression of Ca^{2+} and potassium channels [83-85]. KChIP1 affects post ER trafficking of Kv4 potassium channels [83,41]. NCS-1 interacts with ARF1 to control trans-Golgi network-plasma membrane trafficking of the Kv4 potassium channel [86]. Various other NCS proteins can interact with molecules connected to membrane trafficking, including neurocalcin which interacts with α and β-clathrin and β-2-adaptin [81, 82]. Furthermore, in an approach to find proteins which interact with clathrin-coated vesicles isolated from rat brain and therefore linked to clathrin-mediated trafficking, VILIP-1 and -3 were identified. Vesicles which were enriched in clathrin heavy and light chain, the α, β, γ, μ - adaptin subunits of the AP-2 complex, which functions at the plasma membrane, and the AP-1 complex, which functions at the trans-Golgi network, where also associated with VILIP-1 and -3. These data implicate the two VILIPs in clathrin-dependent membrane trafficking processes such as endocytosis [87].

A role for VILIP-1 in membrane transport and receptor trafficking has been suggested earlier by its localization to raft [88] and Golgi membranes [39], the compartments known to be involved in endocytosis [89]. VILIP-1 influences rhoA activity, a modulator of the cortical cytoskeleton [30], and it can interact with the cortical actin cytoskeleton itself [38,39], which is also involved in the endocytosis process. These results are in line with a proposed role of VILIP-1 as regulator of post-Golgi membrane trafficking [67]. The protein is not only able to increase cell surface expression of the $\alpha_4\beta_2$ nicotinic acetylcholine receptor [60], but also influences receptor recycling of transferrin and GC-B receptor [67]. As shown in detail for the example of GC-B, VILIP-1 appears as a general regulator of clathrin-mediated events throughout the cell. The effects of VILIP-1 include modulation of the surface expression of the nAChR [60], receptor recycling of GC-B and the transferrin receptor [67], regulation of integrin αV and 5 surface expression [31], and most likely modulation of a set of other neuronal receptors. Through this global mechanism on receptor trafficking, VILIP-1 might contribute to processes of synaptic plasticity under physiological and pathological conditions in the hippocampus.

7. Summary

Members of the VILIP subfamily of NCS proteins are involved in a variety of calcium-dependent signaling cascades in nerve cells. The calcium-myristoyl switch mechanism may enable NCS proteins to shuttle information from the cytoplasm to the cell membrane and *vice versa* in a calcium-dependent manner. NCS proteins may therefore act as mediators to confer calcium-dependence to key players of intracellular signaling pathways including G-protein-coupled receptor kinases, the MAP-kinase pathway and cyclic nucleotide signaling. Furthermore, recent reports indicate that NCS proteins might regulate membrane trafficking processes on different vesicular pathways from endoplasmatic reticulum to Golgi apparatus, from Golgi apparatus to cell surface as well as regulating the recycling pathway for receptors.

8. Acknowledgements

Work in the authors' laboratories was supported by grants from Deutsche Forschungsgemeinschaft, Deutsche Krebshilfe, Charité Berlin, Kultusministerium des Landes Sachsen-Anhalt, BBSRC and the Bial Foundation. The authors would like to thank Adlina Mohd Yusof for critically reading the manuscript.

References

[1] Nef, P. 1996, *Guidebook to the calcium-binding proteins,* Celio, MR, (ed) Oxford University Press, New York, Oxford, pp 94.

[2] Braunewell, K.-H., and Gundelfinger, E.D. 1999, *Cell Tissue Res., 299*, 1.

[3] Burgoyne, R.D., and Weiss, J.L. 2001, *Biochem. J., 353*, 1.

[4] Spilker, C., Gundelfinger, E. D., and Braunewell, K.-H. 2002b, *BBA-proteins and proteomics, 1600*, 118.

[5] Kato, M., Watanabe, Y., Iino, S., Takaoka, Y., Kobayashi, S., Haga, T., Hidaka, H. 1998, *Biochem. J., 331*, 871.

[6] Spilker, C., Richter, K., Smalla, K.-H., Manahan-Vaughan, D., Gundelfinger, E.D., and Braunewell, K.-H. 2000, *Neurosci. 96*, 121.

[7] Bernstein, H.-G., Baumann, B., Danos, P., Diekmann, S., Bogerts, B., Gundelfinger, E.D. and Braunewell K.-H. 1999, *J. Neurocytol., 28*, 655

[8] Paterlini, M., Revilla, V., Grant, A.L., Wisden, W. 2000, *Neuroscience, 99*, 205.

[9] Saitoh, S., Kobayashi, M., Kuroki, T., Noguchi, T., Takamatsu, K, 1995, *Neurosci. Res., 23*, 383.

[10] Lenz, S.E., Zuschratter, W., and Gundelfinger, E.D., 1996b, *Neurosci. Lett., 206*, 133.

[11] Saitoh, S., Takamatsu, K., Kobayashi, M., Noguchi, T., 1994, *Brain Res. Dev. Brain Res., 80*, 199.

[12] Grant, A.L., Jones, A., Thomas, K.L., and Wisden, W., 1996, *Neuroscience, 75*, 1099.

[13] Spilker, C., and Braunewell, K.-H., 2003, *Mol. Cell. Neurosci., 24*, 766.

[14] Martinez-Guijarro, F.J., Brinon, J.G., Blasco-Ibanez, J.M., Okazaki, K., Hidaka, H., and Alonso, J.R., 1998, *Hippocampus. 8*, 2.

[15] Kajimoto, Y., Shirai, Y., Mukai, H., Kuno, T., and Tanaka, C., 1993, *J. Neurochem., 61*, 1091.

[16] Hamashima, H., Tamaru, T., Noguchi, H., Kobayashi, M., and Takamatsu, K., 2001, *Neurosci. Res., 39*, 133.
[17] Lenz, S.E., Henschel, Y., Zopf, D., Voss, B., and Gundelfinger, E.D., 1992, *Brain Res Mol. Brain Res., 15*, 133.
[18] Lenz, S.E., Jiang, S., Braun, K., Gundelfinger, E.D., 1996, *Cell Tissue Res., 283*, 413.
[19] Nakano, A., Terasawa, M., Watanabe, M., Usuda, N., Morita, T., and Hidaka, H, 1992, *Biochem. Biophys. Res. Commun., 186*, 1207.
[20] De Raad, S., Comte, M., Nef, P., Lenz, S.E., Gundelfinger, E.D., Cox, J.A., 1995, *Histochem. J., 27*, 524.
[21] Krishnan, A., Venkataraman, V., Fik-Rymarkiewicz, E., Duda, T., and Sharma R.K., 2004, *Biochemistry, 43*, 2708.
[22] Boekhoff, I., Braunewell, K.-H., Andreini, I., Breer, H., and Gundelfinger, E.D., 1997, *Eur. J. Cell Biol., 72*, 151.
[23] Bastianelli, E., Polans, A.S., Hidaka, H., and Pochet, R., 1995, *J. Comp. Neurol., 354*, 395.
[24] Duda, T., Jankowska, A., Venkataraman, V., Nagele, R.G., Sharma, R.K., 2001, *Biochemistry. 40*, 12067.
[25] Duda, T., Fik-Rymarkiewicz, E., Venkataraman, V., Krishnan, A., Sharma, R.K., 2004, *Mol. Cell. Biochem., 267*, 107.
[26] Mammen, A., Simpson, P.J., Nighorn, A., Imanishi, Y., Palczewski, K., Ronnett, G.V., and Moon, C., 2004, *Biochem. Biophys. Res. Commun., 322*, 1131.
[27] Kraut, N., Frampton, J., and Graf, T., 1995, *Oncogene, 10*, 1027.
[28] Gierke, P., Zhao, C., Linke, B., Brackmann, M., Heinemann, U. and Braunewell, K.-H., 2004, *Biochem. Biophys. Res. Commun., 323*, 38.
[29] Ohya, S., and Horowitz, B., 2002, *Am. J. Physiol. Gastrointest. Liver Physiol., 283*, 1290.
[30] Mahloogi, H., Gonzalez-Guerrico, A.M., De Cicco, R.L., Bassi, D.E., Goodrow, T., Braunewell, K.H., and Klein-Szanto, A. J.P., 2003, *Cancer Res., 63*, 4997.
[31] Gonzalez Guerrico, A.M., Jaffer, Z.M., Page, R.E., Braunewell, K.-H., Chernoff, J. and Klein-Szanto, A.J.P., 2005, *Oncogene, 24,* 2307.
[32] Zozulya, S., and Stryer, L.,1992, *Proc. Natl. Acad. Sci. USA, 89*, 11569.
[33] Ames, J.B., Tanaka, T, Stryer, L, Ikura, M , 1996, *Curr. Opin. Struct. Biol., 6*, 432.
[34] Ames, J.B., Ishima, R., Tanaka, T., Gordon, J.I., Stryer, L., and Ikura, M. 1997, *Nature, 389*, 198.
[35] Tanaka, T., Ames, J.B., Harvey, T.S., Stryer, L. & Ikura, M., 1995, *Nature, 376*, 444.
[36] Kobayashi, M., Takamatsu, K., Saitoh, S., and Nogushi, T., 1993, *J. Biol. Chem., 268*, 18898.
[37] Ladant, D, 1995, *J. Biol. Chem., 270*, 3179.
[38] Lenz, S.E., Braunewell, K.H., Weise, C., Nedlina-Chittka, A., Gundelfinger, E.D., 1996, *Biochem. Biophys. Res. Commun., 225*, 1078.
[39] Spilker, C., Dresbach, T. and Braunewell, K.-H., 2002a, *J. Neurosci., 22*, 7331.
[40] O'Callaghan, D. W., Ivings, L., Weiss, J. L., Ashby, M. C., Tepikin, A.V. and Burgoyne R.D., 2002, *J. Biol. Chem., 277*, 14227.

[41] O'Callaghan, D.W., Hasdemir, B., Leighton, M., Burgoyne, R.D., 2003, *J. Cell Sci., 116*, 4833.

[42] Krylov, D.M., and Hurley, J.B., 2001, *J. Biol. Chem., 276*, 30648.

[43] Kretsinger, R.H., 1980, *Crit. Rev. Biochem., 8*, 119.

[44] Flaherty, K.M., Zozulya, S., Stryer, L., and McKay, D.B., 1993, *Cell, 75*, 709.

[45] Ames, J.B., Hamasaki, N., and Molchanova, T., 2002, *Biochemistry, 41*, 5776.

[46] Weiergraber, O.H., Senin, II, Philippov, P.P., Granzin, J., and Koch, K.W., 2003, *J. Biol. Chem., 278*, 22972.

[47] Ames, J.B., Dizhoor, A.M., Ikura, M., Palczewski, K., and Stryer, L., 1999, *J. Biol. Chem., 274*, 19329.

[48] Ames, J.B., Hendricks, K.B., Strahl, T., Huttner, I.G., Hamasaki, N., and Thorner, J., 2000, *Biochemistry, 39*, 12149.

[49] Bourne, Y., Dannenberg, J., Pollmann, V., Marchot, P., and Pongs, O., 2001, *J. Biol. Chem., 276*, 11949.

[50] Vijay-Kumar, S., and Kumar, V.D., 1999, *Nat. Struct. Biol., 6*, 80.

[51] Scannevin, R.H., Wang, K., Jow, F., Megules, J., Kopsco, D.C., Edris, W., Carroll, K.C., Lu, Q., Xu, W., Xu, Z., Katz, A.H., Olland, S., Lin, L., Taylor, M., Stahl, M., Malakian, K., Somers, W., Mosyak, L., Bowlby, M.R., Chanda, P., and Rhodes, K.J., 2004, *Neuron, 41*, 587.

[52] Zhou, W., Qian, Y., Kunjilwar, K., Pfaffinger, P.J., and Choe, S., 2004, *Neuron, 41*, 573.

[53] Edgar, R.C., 2004, *Nucl. Acids Res., 32*, 1792.

[54] Sali, A., and Blundell, T.L., 1993, *J. Mol. Biol., 234*, 779.

[55] Jones, T., Zou, J., Cowan, S., and Kjeldgaard, M., 1991, *Acta Crystallogr. A Found. Cryst., 47*, 110.

[56] Laskowski, R., MacArthur, M., Moss, D., and Thornton, J., 1993, *J. Appl. Cryst., 26*, 283.

[57] Teruel, M.N, and Meyer, T., 2000, *Cell, 103*, 181.

[58] Braunewell, K.-H., Spilker, C., Behnisch, T., and Gundelfinger, E.D., 1997, *J Neurochem, 68*, 2129.

[59] Braunewell, K.-H., Brackmann, M., Schaupp, M., Spilker, C. Anand, R., and Gundelfinger, E. D., 2001, *J. Neurochem., 78*, 1277.

[60] Lin, L., Jeanclos, E.M., Treuil, M., Braunewell, K.H., Gundelfinger, E.D., and Anand, R., 2002b, *J. Biol. Chem., 277*, 41872.

[61] Lin, L., Braunewell, K.H., Gundelfinger, E.D., and Anand, R., 2002a, *Biochem. Biophys. Res. Commun., 296*, 827.

[62] Blandini, F., Braunewell, K.-H., Manahan-Vaughan, D., Orzi, F., and Sarti, P., 2004, *Cell Death Differ., 11*, 479.

[63] Braunewell, K.-H., Riederer, P., Spilker, C., Gundelfinger, E.D., Bogerts, B., and Bernstein, H.G., 2001, *Dement. Geriatr. Cogn. Disord., 2*, 110.

[64] Braunewell, K.-H., Gundelfinger, ED, 1997. Low level expression of calcium-sensor protein VILIP induces cAMP- dependent differentiation in rat C6 glioma cells]. *Neurosci. Lett., 234*, 139.

[65] Brackmann, M., Zhao, C., Kuhl, D. Manahan-Vaughan., D., and Braunewell, K.-H., 2004, *Biochem. Biophys. Res. Commun., 322*, 1073.
[66] Kumar, V.D., Vijay-Kumar, S., Krishnan, A., Duda, T., and Sharma, R.K, 1999, *Biochemistry, 38*, 12614.
[67] Brackmann, M., Schuchmann, S., Anand, R., and Braunewell, K.-H., 2005, *J. Cell Sci., 118*, 2495-2505.
[68] Polans, A., Baehr, W., Palczewski, K., 1996, *Trends Neurosci., 19*, 547.
[69] Palczewski, K., Sokal, I., and Baehr, W., 2004, *Biochem. Biophys. Res. Commun., 322*, 1123.
[70] Schuman, E.M., and Madison, D.V., 1991, *Science, 254*, 1503.
[71] Monfort, P., Munoz, M.D., Kosenko, E., Felipo, V., 2002, *J. Neurosci., 22*, 10116.
[72] Telegdy, G., 1994, *Rev. Neurosci., 5*, 309
[73] Lautermilch, N.J., Few, A.P., Scheuer, T., and Catterall, W.A., 2005, *J. Neurosci. 25*, 7062.
[74] Pribanic, S., Gisler, S., Bacic, D., Kocher, O., Braunewell, K.-H., Nakadai, T., Murer, H., and Biber, J., 2003, *Nephron - Physiology., 95*, 76.
[75] Nagata, K., Puls, A., Futter, C., Aspenstrom, P., Schaefer, E., Nakata, T., Hirokawa, N., and Hall, A., 1998, *EMBO J., 17*, 149.
[76] Kobayashi, M., Masaki, T., Hori, K., Masuo, Y., Miyamoto, M., Tsubokawa, H., Noguchi, H., Nomura, M., and Takamatsu, K., 2005, *Neuroscience, 133*, 471.
[77] Tibbles, L A, and Woodgett, JR, 1999, *Cell Mol. Life Sci., 55*, 1230.
[78] Hyun, J.K, Yon, C., Kim, Y.S., Noh, D.Y., Lee, K.H., and Han, J.S., 2000 D. *Mol. Cells., 10*, 669.
[79] Simpson, P.J., Miller, I., Moon, C., Hanlon, A.L., Liebl, D.J., and Ronnett, G.V., 2002, *J. Neurosci., 22*, 5536.
[80] Mornet, D., and Bonet-Kerrache, A., *Biochim. Biophys. Acta.*, 2001, 1549,197.
[81] Ivings, L., Pennington, S.R., Jenkins, R., Weiss, J.L., and Burgoyne, R.D., 2002, *Biochem. J., 363*, 599.
[82] Palmer, C. L., Burbidge, S. A. and Henley, J. M., 2001, *Abstr. Soc. Neurosci. Vol 27.*
[83] An, W.F., Bowlby, M.R., Betty, M., Cao, J., Ling, H.P., Mendoza, G., Hinson, J.W., Mattsson, K.I., Strassle, B.W., Trimmer, J.S., and Rhodes, K.J., 2000, *Nature, 403*, 553.
[84] Weiss, J. L., Archer, D. A., and Burgoyne, R. D., 2000, *J. Biol. Chem., 275*, 40082.
[85] Zhao, X., Varnai, P., Tuymetova, G., Balla, A., Toth, Z. E., Oker-Blom, C., Roder, J., Jeromin, A., and Balla, T., 2001, *J. Biol. Chem., 276*, 40183.
[86] Haynes, L.P., Thomas, G.M., and Burgoyne, R.D., 2005, *J. Biol. Chem., 280*, 6047.
[87] Blondeau, F., Ritter, B., Allaire, P. D., Wasiak, S., Girard, M., Hussain, N. K., Angers, A., Legendre-Guillemin, V., Roy, L., Boismenu, D., Kearney, R. E., Bell, A. W., Bergeron, J. J., and McPherson, P. S., 2004, *Proc. Natl. Acad. Sci. USA, 101*, 3833.
[88] Orito, A., Kumanogoh, H., Yasaka, K., Sokawa, J., Hidaka, H., Sokawa, Y., and Maekaw, S., 2001, *J. Neurosci. Res., 64*, 235.
[89] Herreros, J., Ng, T., and Schiavo, G., 2001, *Mol. Biol. Cell., 12*, 2947.
[90] Sallese, M., Iacovelli, L., Cumashi, A., Capobianco, L., Cuomo, L., De Blasi, A., 2000, *Biochim. Biophys. Acta., 1498*, 112.

[91] Mathisen, P.M., Johnson, J.M., Kawczak, J.A., and Tuohy, V.K., 1999, *J. Biol. Chem., 274*, 31571.

[92] Spilker, C, Gundelfinger, ED, and Braunewell, K-H., 1997, *Neurosci. Lett., 225*, 126.

[93] Oikawa, K., Kimura, S., Aoki, N., Atsuta, Y., Takiyama, Y., Nagato, T., Yanai, M., Kobayashi, H., Sato, K., Sasajima, T., and Tateno, M., 2004, *J. Biol. Chem., 279*, 15142.

[94] Mercer, E.A., Korhonen, L., Skoglosa, Y., Olsson, P.A., Kukkonen, J.P., and Lindholm, D., 2000, *EMBO J., 19*, 3597.

[95] Lindholm, D., Mercer, E.A., Yu, L.Y., Chen, Y., Kukkonen, J., Korhonen, L., and Arumae, U., 2002, *Biochim. Biophys. Acta., 1600*, 138.

[96] Korhonen, L., Hansson, I., Kukkonen, J.P., Brannvall, K., Kobayashi, M., Takamatsu, K., and Lindholm, D., 2005, *Mol Cell Neurosci., 28*, 85.

PART III: RECOVERIN

In: Neuronal Calcium Sensor Proteins
Editors: Pavel Philippov and Karl Koch, pp. 139-151

ISBN 1-59454-978-8

Chapter VIII

Recoverin: A Calcium-Dependent Regulator of the Visual Transduction

Pavel P. Philippov[1,*], Ivan I. Senin[1] and Karl-Wilhelm Koch[2]
[1]A.N. Belozersky Institute of Physico-Chemical Biology,
Moscow State University, 119992 Moscow, Russia;
[2]AG Biochemistry, Institute of Biology and Environmental Sciences, Faculty V,
Carl von Ossietzky University Oldenburg, D-26111 Oldenburg, Germany.

Abstract

The visual cascade operating in vertebrate photoreceptor cells consists of the visual receptor rhodopsin, responsible for photon capture, a G-protein named transducin (G_t), and an effector enzyme cGMP-phosphodiesterase. Activation of the cascade causes a fall in the cytoplasmic concentration of the second messenger cGMP and in turn closure of cationic channels in the photoreceptor plasma membrane, which is the first step of information transmission on the way from the photoexcited visual cell – via the optic nerve – to the brain. After excitation, the visual cell should be recovered to the unstimulated state. This process is controlled by several Ca^{2+}-binding proteins one of which is recoverin, a Ca^{2+}-sensor of rhodopsin kinase (GRK-1) responsible for phosphorylation and thus desensitization of the visual receptor. This chapter is focused on structural and functional characteristics of recoverin, one of the first NCS proteins to be discovered.

Keywords: recoverin, neuronal calcium sensor, calcium-binding protein, rhodopsin kinase, protein phosphorylation, visual transduction, surface plasmon resonance spectroscopy.

---

* Correspondence concerning this article should be addressed to Pavel P. Philippov A.N. Belozersky Institute of Physico-Chemical Biology, Moscow State University, 119992 Moscow, Russia. tel.: +7(495)939-50-17; fax: +7(495)939-09-78; e-mail: ppph@belozersky.msu.ru.

Abbreviations

EF	EF-hand Ca^{2+}-binding motif
EF1, EF2, EF3, and EF4	the 1st, 2nd, 3rd, and 4th Ca^{2+}-binding site of recoverin, respectively
E85Q	recoverin mutant with substitution E85Q in the 2nd Ca^{2+}-binding site of recoverin
NCS	neuronal calcium sensor
Rho	nonbleached rhodopsin
Rho*	bleached rhodopsin
ROS	rod outer segments
SPR-spectroscopy	surface plasmon resonance spectroscopy
wt-recoverin	recoverin of wild type
$[Ca^{2+}]_f$	free Ca^{2+}concentration

1. Introduction

Recoverin, one of the first described member of the neuronal calcium sensor (NCS) protein family, was almost simultaneously discovered, by several groups in Russia, USA and Germany, as a protein with an apparent molecular mass of 26 kDa present in bovine rod outer segments (ROS). In 1989, P.Philippov's group from Russia used a column with delipidated visual rhodopsin immobilized on Concanavalin A Sepharose for purification of the visual G-protein, transducin from bovine ROS and G_i/G_o-proteins from bovine brain tissue [1]. A preparation of transducin thus obtained was contaminated with cGMP-phosphodiesterase and an unknown protein with an apparent molecular weight of 26 K. Then the latter protein, preliminary named as "p26" [2], was investigated in more detail [3]. Thus, it has been demonstrated that p26 is highly specific for retina, present in photoreceptors (both in rods and cones), and contains several calcium binding sites of the EF-hand type. Concurrently, K.-W.Koch's group from Germany also isolated "a 26 kd protein" from bovine ROS [4]. P26 has also been suggested [3,4] to be a Ca^{2+}-specific regulator of photoreceptor guanylate cyclase, a key enzyme of photoreceptor recovery, and finally christened as recoverin [3]. That recoverin is capable of activating guanylate cyclase was, however, disproved in subsequent works [5-7]. Instead rhodopsin kinase (G-protein-coupled receptor kinase 1, GRK 1), which is responsible for phosphorylation and thus desensitization of the visual receptor rhodopsin, was supposed to be a recoverin intracellular target [5,8-10]. Now the name "recoverin" is usually used to designate various orthologs of bovine recoverin with the exception of the chicken and frog orthologs which bear the names "visinin" [11] and "S-modulin" [12], respectively. It should be added that in 1987 Thirkill et al. [13] found autoantibodies against an antigen having a molecular weight of 23 K in sera of patients with cancer-associated retinopathy (CAR) and named that antigen as "CAR-antigen". Later CAR-antigen was purified from bovine ROS by K.Palczewski's group and shown to be identical to recoverin [14].

In this chapter, we discuss different structural and functional aspects of recoverin as a visual signaling protein. Recoverin as CAR-antigen and a principal cause of paraneoplastic retinal degeneration is reviewed in this issue by G.Adamus.

2. Tissue and Cellular Distribution

Immunocytochemically, recoverin was detected in adult retina of all species investigated so far: man [15-17], bull [2,3,18] monkey [16], mouse [19], rat [19], rabbit [19], frog [19], chameleon [19], and newt [20] with the only exception presented by chicken retina [19]. (In the works cited, antibodies against bovine recoverin were used to visualize recoverin immunocytochemically). In addition to retina, recoverin immunoreactivity was observed in ocular ciliary epitelium [21], pineal organ [3,15,22], and the rat olfactory epithelium [23]. Within retina, recoverin-positive reaction was found in photoreceptor cells [3], both rods and cones, as well as in higher order neurons (bipolar and ganglion cells of a number of species [16,18,19,24]) and in amacrine cells of lamprey *Lampreta fluviatilis* [25]. Within photoreceptor cells, recoverin is present in ROS, inner segments, cell bodies, and synaptic pedicles of both rods and cones [3].

3. Recoverin Gene

The gene encoding recoverin was cloned and localized to chromosome 17p13.1 in human [26] and to chromosome 11.36 in mouse [27] in proximity to the p53 tumor suppressor gene. The gene of 9 to 10 kb length is presented by a single copy and contains at least 3 exons and 2 introns. The 602-bp open reading frame of the human gene corresponds to a 23 kDa protein which is 89% identical to bovine and mouse recoverin, and 59% identical to chicken visinin. All EF-hand binding motives present in recoverin are encoded in the first exon. The arrangement of introns relative to these motives is unique within the superfamily of EF-hand proteins, suggesting that recoverin constitutes a new subgroup [28].

The 5' untranslated promoter region of recoverin includes the supressor sequence between the positions -1200 and -900 and the enhancer sequence between the positions -1440 and -1200. The positions -900 – -213 and -2400 – -2000 in the untranslated 5' promoter region of the gene contain several consensus *cis*-elements which are involved both in the activation and suppression of the gene transcription. The sequences of the *cis*-elements corresponds to AP-1, Oct-1, BCO9, Sp-1, Ret-1/PCE-1, CREB, and Crx (cone rod homebox). Of these, an important regulatory role has been demonstrated for the Crx transcription factor, the binding site for which is present in the position -40 – -46 upstream of the recoverin gene promoter [29]. CRX is shown to play a crucial role in the differentiation of photoreceptors and maintenance of their normal functions [30]. A conserved consensus site (TAATCC/A) for binding of CRX is detected in the upstream promoter regions of several photoreceptor-specific genes, including those of rhodopsin [31,32], cone opsin [33], arrestin [34], and interphotoreceptor retinoid-binding protein [35]. The AP-1 motif is localized in the 5' flanking region of the human recoverin gene in the position from -642 to -652 [29]. The AP-1 binding site is a target for transcription factors of the c-Fos and c-Jun family. In photoreceptors, the AP-1 element is identified to be present in the 5' flanking region of the gene for the human rod β-subunit of cGMP phosphodiesterase [36]. The photoreceptor specific *cis*-element PCE 1 is detected more than 2 kb before the site of the transcription start in the human recoverin gene [29]. The PCE 1 core consensus sequence is localized in the promoter regions of numerous photoreceptor genes, including those of opsin, arrestin, and interphotoreceptor retinoid-binding protein in different vertebrates and *Drosophila* [34,37].

4. Molecular Properties and Ca^{2+}-Myristoyl switch of Recoverin

The electrophoretic mobility of human and bovine recoverin on SDS polyacrylamide gel corresponds to a 23 and 26 kDa protein, respectively [3,4,38]. The complete amino acid sequence of recoverin deduced from the cDNA contains 202 amino acid residues and has a calculated mass of 23.3 kDa [3,39]. During posttranslational modification, a consensus sequence for myristoylation present at the N-terminus of recoverin, undergoes cleavage of the N-terminal methionine residue, and the resulting N-terminal glycine is acylated by N-myristoyl transferase. Finally, N-acylated retinal recoverin contains 201 residues and its molecular mass is determined to be 23.4 kDa. The N-terminal acylation of recoverin is heterogeneous: the electrospray mass spectrometry demonstrated that it is modified by myristic acid (14:1) and three other fatty acids (14:0, 14:2 and 12:0) [40]. These fatty acyl modifications are essential for Ca^{2+}-dependent binding of recoverin to ROS membranes and phospholipid vesicles (see below).

Recoverin contains four potential Ca^{2+}-binding sites of an EF-hand type, which include the following amino acid residues: 36-48 for EF1, 73-85 (EF2), 109-121 (EF3), and 159-170 (EF4). Of the four potential Ca^{2+}-binding sites, only EF2 and EF3 are capable of binding Ca^{2+}, whereas EF1 and EF4 do not possess this ability. Non-myristoylated recombinant recoverin exhibits heterogeneous and uncooperative binding of Ca^{2+} ions with dissociation constants of 0.11 and 6.9 μM. Myristoylated recombinant recoverin binds cooperatively two Ca^{2+} with a Hill coefficient of 1.75 and an apparent dissociation constant of 17 μM [41]. The binding of Ca^{2+} to recoverin is a sequential process with EF3 being filled first [41-43].

Most NCS proteins are covalently acylated at the N-terminus, and for some of these, including recoverin, the predominant modification is myristoylation. Recoverin's myristoyl group operates like a flexible anchor. It is extruded when Ca^{2+} is bound to EF2 and EF3 and it is buried within a hydrophobic pocket when recoverin is Ca^{2+}-free. This Ca^{2+}-dependent change of conformation has been named Ca^{2+}-myristoyl switch [44,45]. The main feature of this switch is that the exposed myristoyl group facilitates the binding of recoverin to native or artificial membranes, whereas at low $[Ca^{2+}]_f$, recoverin is released into the soluble phase. This operational principle has been verified in a series of publications dealing with the three-dimensional structure of Ca^{2+}-bound and Ca^{2+}-free recoverin [46-51]. (The three-dimensional structure of recoverin is covered in more detail elsewhere in this book).

Equilibrium centrifugation assays were performed to determine the $[Ca^{2+}]_f$ at which the binding of myristoylated recoverin to membranes is halfmaximal [43,44]. EC_{50} values fall in the range 2-5 μM and are slightly out of the physiological range of cytoplasmic $[Ca^{2+}]_f$. However, extrapolation to the cellular conditions in ROS bearing stacks of densely packed membranes reveales that the apparent affinity of Ca^{2+}-loaded recoverin for membranes is in the submicromolar (i.e. physiological) range [44]. In addition, Senin et al. [52] have recently shown that the binding of recoverin to membranes and also the inhibition of rhodopsin kinase by recoverin is facilitated when the cholesterol content of membranes is increased. The cholesterol content in disk membranes differs longitudinally and is highest at the base of an outer segment [53-55]. Thus, recoverin would be heterogeneously distributed along the axis of a photoreceptor (for a more detailed discussion see ref. [52]) and might even possess different Ca^{2+}-sensitivities depending on the membrane composition it is attached to.

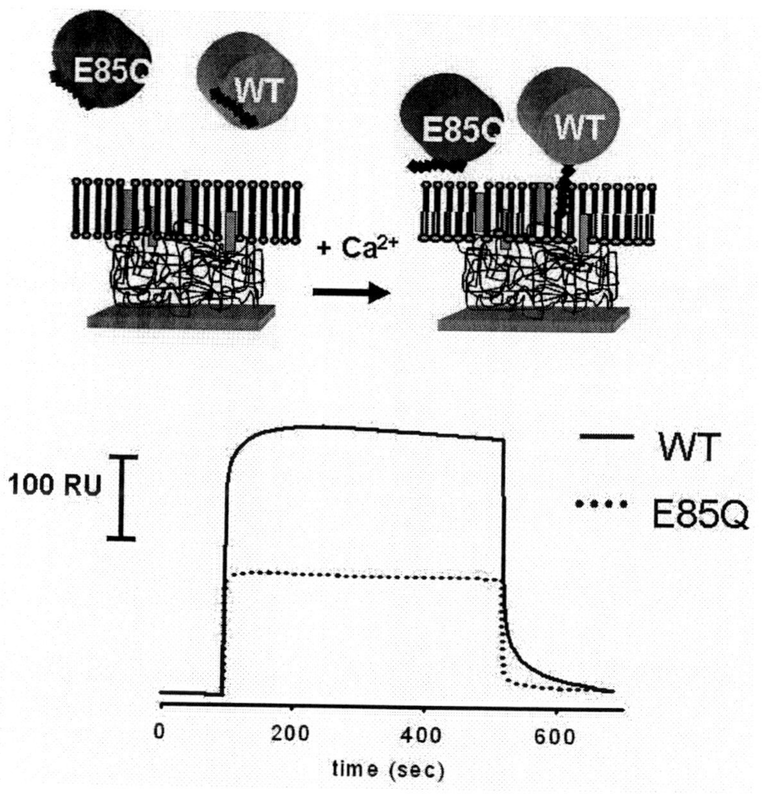

Figure 1. A lipid bilayer is immobilized on a commercially available hydrophobic sensor chip (L1, BIAcore). Flushing of wild-type recoverin (WT) and recoverin mutants (example E85Q) over the surface results in binding to the membranes when the Ca^{2+}-concentration is raised (*upper part*). Both recoverin forms are myristoylated, but the myristoyl group is only fully exposed in the case of wildtype recoverin. Myristoylated E85Q represents an intermediate state during the conformational change. Binding to membranes is less pronounced and results in lower amplitude (measured in resonance units, RU) than in the case of wild-type recoverin (*lower part*, see text for more explanation).

Surface plasmon resonance (SPR) spectroscopy was recently employed to investigate dynamic aspects of recoverin´s association with membranes [43,52,56]. In general, SPR-spectroscopy monitors the interaction of biomolecules on specifically modified sensor chip surfaces. Changes in the refractive index in close proximity to the sensor surface are recorded in a time-dependent manner and provide raw data to determine kinetic and affintiy constants of biomolecular interaction processes. Details of the measuring principles and the wide range of applications has been reviewed elsewhere [57,58]. A specific approach using immobilized membranes on sensor surfaces is illustrated in Figure 1 and was used to study the Ca^{2+}-myristoyl switch of wt-recoverin and its mutants [43]. The sensor chip consists of a thin gold layer that is coated with carboxy dextran. Small organic hydrophobic molecules are embedded within the dextran layer allowing the assembling of phospholipid bilayers as illustrated. Proteins like wt-recoverin and its mutant E85Q with the disabled EF2 are flushed over the surface, increase of $[Ca^{2+}]_f$ leads to an extrusion of the myristoyl group in wt-recoverin and subsequent binding to the immobilized membrane. Any binding event on the surface results in a positive resonance signal (expressed in resonance units, RU). The myristoylated form of the mutant E85Q represents a transition state during the Ca^{2+}-induced conformational change [43,50]. Since the myristoyl group is only partially exposed, binding to the membranes is reduced. Interestingly, the dissociation phase of the mutant E85Q does only show a fast phase without the prominent slower phase visible in the sensorgram obtained with wt-recoverin (see Figure 1 and [43]). Apparently, the mean residence time of wt-recoverin on the membrane is controlled by EF2. If this site is nonfunctional, the transient binding of the corresponding mutant is shortened at least by a factor of 10 in comparsion to wt-recoverin [43]. In fact, the mean residence time is too short to allow efficient inhibition of rhodopsin kinase. These conclusions were supported by subsequent work on recoverin mutants in which EF4 was functionally restored [59].

5. Recoverin as a Ca^{2+}-Dependent Inhibitor of Rhodopsin Kinase *in vitro*

By now a great body of the *in vitro* data have been accumulated on the ability of recoverin to be a Ca^{2+}-dependent inhibitor of rhodopsin phosphorylation catalyzed by rhodopsin kinase. The first indications that recoverin can perform this function have been described in ref. [5,8]. Gorodovikova and Philippov [5] suggested a Ca^{2+}-dependent complex between recoverin and rhodopsin kinase to exist in bovine ROS. Simultaneously, Kawamura et al. [8] described an ability of recoverin to inhibit cGMP hydrolysis in a suspension of frog ROS in a Ca^{2+}-dependent manner and suggested this effect to be a result of the recoverin inhibitory action upon rhodopsin kinase. More recently a Ca^{2+}-dependent inhibitory effect of recoverin upon light-dependent rhodopsin phosphorylation was directly demonstrated in bovine ROS suspension [9]. Finally, the same effect of recoverin was reproduced in a reconstituted system consisting of rhodopsin (in the content of urea-washed photoreceptor membranes) and two other purified retinal proteins: rhodopsin kinase and recoverin [10]. These data were then confirmed by other authors [60,61] in similar systems, in one of which retinal rhodopsin kinase was substituted for recombinant rhodopsin kinase [60].

Similar to retinal recoverin which is heterogeneously acylated [40] recombinant non-acylated recoverin is capable of inhibiting light-dependent rhodopsin phosphorylation in a Ca^{2+}-dependent manner [60,62,63]. However, N-terminal acylation of recoverin enhances its inhibitory efficiency with respect to rhodopsin kinase [63]. At saturating $[Ca^{2+}]$, inhibition of rhodopsin phosphorylation can be described as a hyperbolic function of the recoverin concentration yielding IC_{50} of 6.5 [63]- 8 μM [60] for its non-myristoylated form, 0.4 - 0.5 μM [64] for the recombinant myristoylated form and 1 - 6 μM [10,61,63] for retinal recoverin. However, acylated and non-acylated forms of recoverin have different Hill coefficients estimated from the Ca^{2+}-dependence of the recoverin inhibitory effects upon rhodopsin phosphorylation: 0.7 - 0.9 in the case of the non-acylated recoverin [60,63] and 1.5 - 1.7 in the case of both retinal [63] and recombinant myristoylated [60,63] recoverin forms.

Though the mechanism of the recoverin inhibitory action upon rhodopsin kinase is poorly investigated, one may suggest as a first approximation two hypothetic mechanisms [64]. The 1st type is not genuine inhibition in the kinetic sense and can be named as "inhibition by compartmentalization". According to this mechanism, in the presence of Ca^{2+}, recoverin forms a complex with rhodopsin kinase and targets the complex - via its exposed myristoylated N-terminus - to photoreceptor membranes. As a result, rhodopsin kinase stays away from its substrate, photoexcited rhodopsin (Rho*), which prevents Rho* phosphorylation. The 2nd type involves a genuine inhibition of the phosphorylation by direct interference of recoverin with the complex between Rho* and rhodopsin kinase. In this case, several subtypes of the inhibitory mechanism could be considered, but all of these imply a two-step scheme of Rho* phosphorylation by rhodopsin kinase. According to this scheme [65-67], before rhodopsin kinase phosphorylates the C-terminus of Rho* (the 2nd step), it is activated by interaction with the 3rd cytoplasmic loop of Rho* (the 1st step). This two-step mechanism can be devided into the following events: (i) inactive rhodopsin kinase binds to the 3rd cytoplasmic loop of Rho*, and (ii) due to this binding, conformational changes occur leading to the activation of rhodopsin kinase; (iii) the C-terminal part of Rho* comes into contact with the substrate-binding site of rhodopsin kinase that allows the C-terminus to occupy the proper position in the catalytic site of the enzyme; (iv) rhodopsin kinase starts to catalyze the phosphate incorporation into the C-terminus of Rho*. In principle, any of these events (or their combinations) could be sensitive to a Ca^{2+}-dependent inhibition by recoverin. In addition, recoverin could interfere with the binding of ATP (the second substrate of rhodopsin kinase) to the enzyme and/or with the dissociation of ADP (the product) from rhodopsin kinase. So far the choice between above-mentioned options cannot be made.

6. Is Recoverin a Ca^{2+}-Sensor of Rhodopsin Kinase and/or Ca^{2+}-Buffer in ROS *in vivo*?

Since the first *in vitro* data on recoverin as a Ca^{2+}-sensor of rhodopsin kinase have been obtained, the question of whether it functions in the same manner under physiological conditions have been discussed over many years. The discussion was fuelled by a number of *in vitro* experiments and a limited amount of works performed on *ex vivo* and *in vivo* models (for a review, see ref. 64). However, they did not provide a clear answer to the aforementioned question. Moreover, a minimal hypothesis on recoverin function *in vivo* proposed that it could simply play the role of a Ca^{2+}-buffer in ROS [68], a role that has been envisioned already in one of the earliest publications on recoverin [4]. At the same time, there existed a compromise opinion which did not exclude two actions of recoverin under physiological conditions: the inhibition of rhodopsin kinase and calcium buffering [69]. (For a critical review of the arguments for and against recoverin as a Ca^{2+}-sensor of rhodopsin kinase *in vivo* see ref. 64).

Recently Makino et al. [70] obtained mice in which one or both copies of the recoverin gene were disrupted. For some experiments, recoverin knockout (rec – / –) mice were crosssed with mice in which guanylate cyclase activating proteins 1 and 2 were knocked out (GCAPs – / –) to obtain rods in which both sets of proteins were absent (rec – / –, GCAPs – / –). The comparison of the photoresponse of rods in rec – / – mice with that of rods in normal mice indicated that in the presence of Ca^{2+}, recoverin prolonged the dark-adapted flash response and increased the rod's sensitivity to dim steady light. Knockout rods had faster Ca^{2+} dynamics, indicating that recoverin was a significant Ca^{2+}-buffer in ROS. Imporant was, however, that incorporation of the exogenous buffer did not restore the wild-type photoresponse. Recoverin's action was still present in GCAPs – / – rods, whose cyclase activity did not vary with a background light level. Thus, the shorted incremental response of rec – / – rods arose rather from reduced activity of cGMP-phospodiesterase than from an effect on the activity of guanylate cyclase. The authors suggested that in addition to Ca^{2+}-buffering, recoverin potentiates light-triggered phosphodiesterase activity in a Ca^{2+}-dependent manner probably by prolonging the catalytic activity of photoexcited rhodopsin. (These experiments are described in more detail elsewhere in this book).

Kennedy et al. [71] used mutant zebrafish lacking the alpha-subunit of cone transducin and normal zebrafish to test whether phosphorylation of cone opsin is sensitive to the level of intracellular Ca^{2+} during illumination. Using such an *in vivo* model, the authors proceeded from the assumption that while the level of intracellular Ca^{2+} in mutant cones is not sensitive to light, it falls in wild-type cones upon illumination. Thus, such an approach opened a way to compare phosphorylation of cone opsin in intact cells at different concentrations of intracellular calcium. *In vivo* measurements of the steady-state level of the phosphorylation in the mutant and wild-type fishes after 4 min of exposure to steady bright light revealed approximately 7-fold reduction of cone opsin phosphorylation in the mutant cones in comparison with those of wild-type. A set of control experiments confirmed that this reduction results precisely from the effect of Ca^{2+} upon light-dependent phosphorylation of opsin in zebrafish cones.

Though Makino et al. [70] did not identify the target of the Ca^{2+}-dependent action of recoverin upon activity of cGMP-phosphodiesterase, their *in vivo* data, together with the *in vivo* data of Kennedy et al. [71] and a number of the *in vitro* data described in the preceding Section, provide strong evidence in favour of the dual function of recoverin as a Ca^{2+}-sensor of rhodopsin kinase and Ca^{2+}-buffer in photoreceptor outer segments. In Figure 2, recoverin is placed into the general scheme of visual transduction.

7. Recoverin and High-Gain Rhodopsin Phosphorylation

One more functional aspect of recoverin is its ability to operate as a Ca^{2+}-dependent inhibitor of phosphorylation of non-bleached rhodopsin (Rho). One rhodopsin molecule contains 9 potential sites for phosphorylation, and at a high bleaching level of ROS, almost all these sites can be phosphorylated by rhodopsin kinase *in vitro* [72]. However, at a low bleaching level, the phosphate incorporation into total rhodopsin, calculated per Rho*, can reach tens in reconstituted systems [60,73-75] or even hundreds in suspension of electropermeabilized ROS [76]. This phenomenon named "high-gain rhodopsin phosphorylation" [76] can be explained by incorporation of ATP phosphoryl groups not only into Rho* but into Rho as well. Direct measurements of phosphorylated forms of Rho* and Rho in the system, containing photoreceptor membranes and rhodopsin kinase, confirm that phosphorylated Rho is present in such a system, thereby the lower the level of rhodopsin bleaching is, the higher is the relative amount of phosphorylated Rho [73].

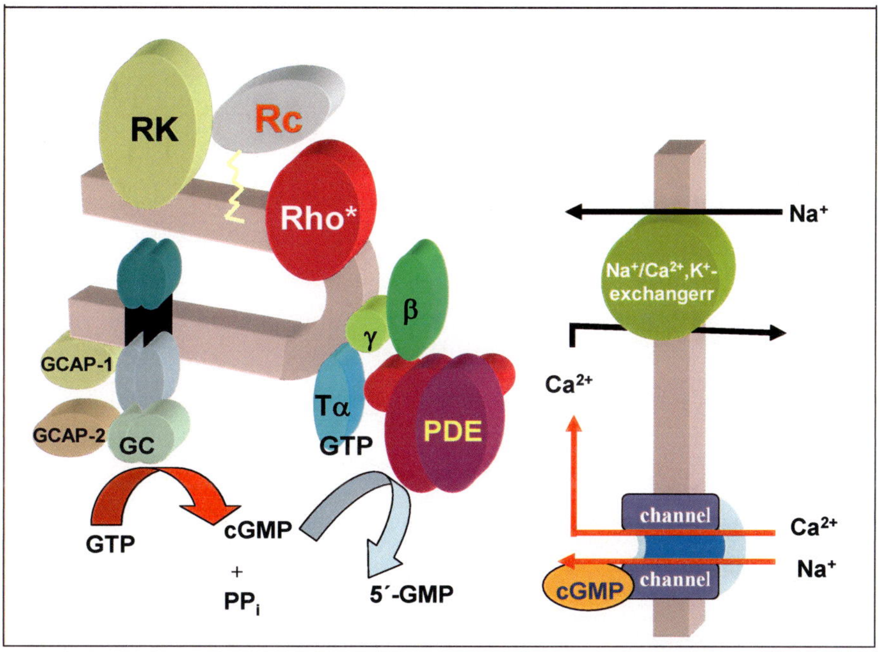

Figure 2. General scheme of phototransduction. *In the dark*: the level of free Ca^{2+} in ROS cytoplasm is high and rhodopsin kinase (RK) stays in an inactive complex with recoverin (Rc); the photoreceptor second messenger cGMP keeps cationic channels localized in ROS plasma membranes in the open state. *In light*: photoexcited rhodopsin (Rho*) catalyses the transit of α-subunit (Tα) of transducin (Gt) to the GTP-bound form which is capable of activating cGMP-phosphodiesterase (PDE) and thus stimulating hydrolysis of cGMP. A decrease of the cytoplasmic level of cGMP causes the closure of the channels and hyperpolarization of the ROS plasma membrane, that is the first step of the electrophysiological response of the photoreceptor cell. Though external Ca^{2+} stops entering into ROS via the channels, light-independent Na^+/Ca^{2+}, K^+-exchanger continues pumping cytoplasmic Ca^{2+} out, resulting in a fall of free Ca^{2+} in ROS cytoplasm. As a consequence, a complex between RK and Rc dissociates, RK becomes active and phophorylates Rho* the catalytic activity of which decreases, thus preventing the signal transduction in the cascade Rho – Gt – PDE. Also, a fall of cytoplasmic Ca^{2+} allows guanylate cyclase-activating proteins (GCAPs) to activate guanylyl cyclase (GC) which synthesizes cGMP and restores its dark level.

What might be the biological sense of high-gain rhodopsin phosphorylation? To answer this question, it has been suggested [64,74,75] that high-gain rhodopsin phosphorylation is a consequence of an unwanted side-reaction, that occurs because a high excess of Rho over Rho* exists in ROS membranes, i.e. in the conditions which correspond to the normal working regime of retinal rods. In such a regime, rhodopsin kinase needs to find and phosphorylate a few molecules of Rho* in the presence of an exceeding number (an order of 10^6) of Rho molecules. Thus, the possibility exists that rhodopsin kinase, despite its preference for Rho* as a substrate [73] may phosphorylate Rho. If so, a reasonable mechanism would be to prevent rhodopsin kinase from the wasteful phosphorylation of Rho *in vivo* thereby directing the kinase to fulfill its "correct" function of Rho* phosphorylation. Indeed, it was demonstrated in *in vitro* experiments [60,74,75] that recoverin inhibited Rho phosphorylation more effectively than Rho* phosphorylation. Furthermore, the recent data [71] demonstrate an ability of Ca^{2+} to regulate the extent of phosphorylation of non-bleached pigments in zebrafish cones *in vivo*. As it is very probable that the only mediator of the Ca^{2+} effect upon visual pigment(s) phosphorylation is recoverin, one may believe that this protein is capable of quenching non-bleached rhodopsin phosphorylation under physiological conditions.

In conclusion, it may be said that there are yet no experimental data or even experimentally justified hypotheses concerning recoverin's function(s) in cells or cell compartments different from ROS. As mentioned above, these structures include other parts of retinal rods, higher order neurons of retina, ocular ciliary epitelium, and the pineal organ. It remains to be seen whether recoverin is involved in control of protein phosphorylation in these structures or whether it fulfils quite different function(s).

Acknowledgements

Cited work of the authors was supported by grants from the Ludwig Institute for Cancer Research (P.P.Ph), Russian Foundation for Basic Research NN 06-04-48018a and 04-04-04001 (P.P.Ph), N 06-04-48761 (I.I.S.), and by grants from the Deutsche Forschungsgemeinschaft (K.-W.K.).

References

[1] Dizhoor, A.M., Nekrasova, E.R., and Philippov, P.P. 1989, *Biochem. Biophys. Res Commun., 162*, 544.

[2] Dizhoor, A.M.,Nekrasova, E.R., and Philippov, P.P. 1991, *Biokhimia, 56*, 225.

[3] Dizhoor, A.M., Ray, S., Kumar, S., Niemi, G., Spencer, M., Brolley, D., Walsh, K.A., Philipov, P.P., Hurley, J.B., and Stryer, L. 1991, *Science, 251*, 915.

[4] Lambrecht, H.-G., and Koch, K.-W. 1991, *EMBO J., 10*, 793.

[5] Gorodovikova, E.N., and Philippov, P.P. 1993, *FEBS Lett., 335*, 277.

[6] Gray-Keller, M,P., Polans, A.S., Palczewski, K., and Detwiler, P.B. 1993, *Neuron, 10*, 523.

[7] Hurley, J.B., Dizhoor, A.M., Ray, S., and Stryer, L. 1993, *Science, 260*, 740.

[8] Kawamura, S., Hisatomi, O., Kayada, S., Tokunaga, F., and Kuo, C.-H. 1993, *J. Biol. Chem., 268*, 14579.

[9] Gorodovikova, E.N., Gimelbrant, A.A., Senin, I.I., and Philippov, P.P. 1994, *FEBS Lett., 349*, 187.

[10] Gorodovikova, E.N., Senin, I.I., and Philippov, P.P. 1994, *FEBS Lett., 353*, 171.

[11] Yamagata, K., Goto, K., Kuo, C.H., Kondo, H., and Miki, N. 1990, *Neuron, 4*, 469.

[12] Kawamura, S., Takamatsu, K., and Kitamura, K. 1992, *Biochem. Biophys. Res. Commun., 18*, 411.

[13] Thirkill, C.E., Roth, A.M., and Keltner, J.L. 1987, *Arch. Ophtalmol., 105*, 372.

[14] Polans, A.S., Buczylko, J., Crabb, J., and Palczewcki, K. 1991, *J. Cell Biol.*, 112, 981.

[15] Korf, HW., White, B.H., Schaad, N.C., Klein, D.C. 1992, *Brain Research, 595*, 57.

[16] Milam, A.H., Dacey, D.M., and Dizhoor A. 1993, *Vis., Neurosci., 10*, 1.

[17] Polans, A., Burton, M.D., Haley, T.L., Crabb, J.W., and Palczewski, K. 1993, *Invest. Ophtalmol. Vis. Sci., 34*, 81.

[18] De Raad, S., Comte, M., Nef, P. Lenz, S.E., Gundelfinger, E.D., and Cox, J. 1995, *Histochem. J., 27*, 524.

[19] McGinnis, J.F., Stepanik, P.L., Jariangprasert, S., Lerious V. 1997, *J. Neurosci. Res., 50*, 487.

[20] Bazhin, A.V., Grigorian, E.N., Tikhomirova, N.K., Poplinskaia, V.A., and Filippov, P.P. 2002, *Izv. Akad. Nauk. Ser. Biol., 4*, 427.

[21] Bertazolli-Filho, R., Ghosh, S., Huang, W., Wollmann, G., and Coca-Prados, M. 2001, *Biochem. Biophys. Res. Commun., 284*, 317.

[22] Bastianelli, E., and Pochet, R. 1994, *J. Pineal. Res., 17*, 103.

[23] Bastianelli, E., Polans, A.S., Hidaka, H., and Pochet, R. 1995, *J. Comp. Neurol. 354*, 395.

[24] Wiechmann, A.F., and Hammarback, J.A. 1993, *Exp. Eye. Res., 57*, 763.

[25] Dalil-Thiney, N., Bastianelli, E., Pochet, R., Reperant, J., and Versaux-Botteri, C. 1998, *Neurosci. Lett., 247*, 163.

[26] Murakami, A., Yajima, T., and Inana, G. 1992, *Biochem. Biophys. Res. Commun., 187*, 234.

[27] McGinnis, J. F., Lerious, V., Pazik, J., and Elliott, R.W. 1993, *Mamm Genome, 4,* 43.

[28] Freund, C. L., and Valle, D. 1992, *Am. J. Hum. Genet., 51 (suppl.),* A6.

[29] Wiechmann, A., and Howard, E. 2003, *Curr. Eye. Res., 26*, 25.

[30] Akagi, T., Mandai, M., Ooto, S., Hirami, Y., Osakada, F., Kageyama, R., Yoshimura, N., and Takahashi, M. 2004, *Invest. Ophthal. Vis. Sci., 45*, 4570.

[31] Zack, D.J., Bennett, J., Wang, Y., Davenport, C,. Klaunberg, B., Gearhard, J., and Nathans, J. 1991, *Neuron, 6*, 187.

[32] Kumar, R., Chen, S., Scheurer, D., Wang, Q., Duh, E., Sung, C., Rehemtulla, A., Swaroop, A., Adler, R., and Zack, D.J. 1996, *J. Biol. Chem., 271*, 29612.

[33] Saha, M.S., Servetnick, M., and Grainger, R.M. 1992, *Curr. Opin. Genet. Dev., 2*, 582.

[34] Kikuchi, T., Raju, K., Breitman, M.L., and Shinohara T. 1993, *Mol. Cell. Biol., 13*, 4400.

[35] Liou, G.I., Matragoon, S., Yang, J., Geng, L., Overbeek, P.A., and Ma, D.P. 1991, *Biochem. Biophys. Res. Commun., 181*, 159.

[36] Polo, A.D., and Farber, D.B. 1995, *Proc Natl Acad Sci USA, 92*, 4016.

[37] Yu, X., Leconte, L., Martinez, J.A., and Barnstable, C.J. 1996, *J. Neurochem., 67*, 2494.

[38] Stepanik, P.L., Lerious, V, and McGinnis, J.F. 1993, *Exp. Eye. Res., 57*, 189.

[39] Ray, S., Zozulya, S., Niemi, G.A., Flaherty, K.M., Brolley, D., Dizhoor, A.M., McKay, D.B., Hurley, J., and Stryer, L. 1992, *Proc. Natl. Acad. Sci. USA, 89*, 5705.

[40] Dizhoor, A.M, Ericsson, L.H., Johnson, R.S., Kumar, S., Olshevskaya, E., Zozulya, S., Neubert ,T.A., Stryer, L., Hurley, J.B., and Walsh, K.A. 1992, J. Biol. Chem., 267, 16033.

[41] Ames, J.B., Porumb, T., Tanaka, T., Ikura, M., and Stryer, L. 1995, *J. Biol. Chem., 270*, 4526.

[42] Permyakov, S.E., Cherskaya, A.M., Senin, I.I., Zargarov, A.A., Shulga-Morskoy, S.V., Alekseev, A.M., Zinchenko, D.V., Lipkin, V.M., Philippov, P.P., Uversky, V.N., and Permyakov, E.A. 2000, *Protein Eng., 13*, 783.

[43] Senin, I.I., Fisher, T., Komolov, K.E., Zinchenko, D.V., Philippov, P.P., and Koch, K.-W. 2002, *J. Biol. Chem., 52*, 50365.

[44] Zozulya, S., and Stryer, L. 1992, *Proc. Natl. Acad. Sci. USA, 89*, 11569.

[45] Ames, J.B., Tanaka, T., Stryer, L., and Ikura, M. 1996, *Curr. Opin. Struct. Biol. 6*, 432.

[46] Flaherty, K.M., Zozulya, S., Stryer, L., and McKay, D.B. 1993, *Cell, 75*, 709.

[47] Ames, J.B., Tanaka, T., Ikura, M., and Stryer, L. 1995, *J. Biol. Chem., 270*, 30909.

[48] Tanaka, T., Ames, J.B., Harvey, T.S., Stryer, L., and Ikura, M. 1995, *Nature, 376*, 444.

[49] Ames, J.B., Ishima, R., Tanaka, T., Gordon, J.I., Stryer, L., and Ikura, M. 1997, *Nature, 389*, 198.

[50] Ames, J.B., Hamasaki, N., and Molchanova, T. 2002, *Biochemistry, 41*, 5776.

[51] Weiergräber, O.H., Senin, I.I., Philippov, P.P., Granzin, J., and Koch, K.-W. 2003, *J. Biol. Chem., 278*, 22972.

[52] Senin, I.I., Hoeppner-Heitmann, D., Polkovnikova, O.O., Churumova, V.A., Tikhomirova, N.K., Philippov, P.P., and Koch, K.-W. 2004, *J. Biol. Chem., 279*, 48647.

[53] Boesze-Battaglia, K., Hennessey, T., and Albert, A.D. 1989, *J. Biol. Chem., 264*, 8151.

[54] Boesze-Battaglia, K., Fliesler, S.J., and Albert, A.D. 1990, *J. Biol. Chem., 265*, 18867.

[55] Boesze-Battaglia, K., and Albert, A.D. 1990, *J. Biol. Chem., 265*, 20727.

[56] Lange, C., and Koch, K.-W. 1997, *Biochemistry, 36*, 12019.

[57] Schuck, P. 1997, *Annu. Rev. Biophys. Biomol. Struct., 26*, 541.

[58] Koch, K.-W. 2000, *Methods Enzymol., 315*, 785.

[59] Senin, I.I., Vaganova, S.A., Weiergräber, O.H., Ergorov, N.S., Philippov, P.P., and Koch, K.-W. 2003, *J. Mol. Biol., 330*, 409.

[60] Chen, C.-K., Inglese, J., Lefkowitz, R.J. and Hurley, J.B. 1995, *J. Biol. Chem., 270*, 18060.

[61] Klenchin, V.A., Calvert, P.D., and Bownds, M.D. 1995, *J. Biol. Chem., 270*, 16147.

[62] Kawamura, S., Cox, J.A., and Nef, P. 1994, *Biochem. Biophys. Res. Commun., 203*, 121.

[63] Senin, I.I., Zargarov, A.A., Alekseev, A.M., Gorodovikova, E.N., Lipkin, V.M., and Philippov, P.P. 1995, *FEBS Lett., 376*, 87.

[64] Senin, I.I., Koch, K.-W., Akhtar, M., and Philippov, P.P. 2002, Photoreceptor and Calcium, W.Baehr and K.Palczewski (Eds.), Landes Bioscience, Georgetown, 69.
[65] Fowles, C., Sharma, R., and Akhtar, M. 1988, *FEBS Lett., 238*, 56.
[66] Brown, N.G., Fowles, C., Sharma, R., and Akhtar, M. 1992, *Eur. J. Biochem., 208*, 659.
[67] Palczewski, K., Buczylko, J., Kaplan, M.W., Polans, A.S., and Crabb, J.W. 1991, J. *Biol. Chem, 266*, 12949.
[68] Otto-Bruc, A.E., Fariss, R.N., Van Hooser, J.P., and Palczewski, K. 1998, *Proc. Natl. Acad. Sci. USA, 95*, 15014.
[69] Nikonov, S., Lamb, T.D., and Pugh, E.N. 2000, *J. Gen. Physiol., 116*, 795.
[70] Makino, C.L., Dodd, R.L., Chen, J., Burns, M.E., Roca, A., Simon, M.I., and Baylor D.A. 2004, *J. Gen. Physiol., 123*, 729.
[71] Kennedy, M.J., Dunn, F.A., and Hurley, J.B. 2004, *Neuron, 41*, 915.
[72] 72.Wilden, U., and Kühn, H. 1982, *Biochemistry, 2*, 3014.
[73] Dean, K.R., and Akhtar, M. 1993, *Eur. J. Biochem., 21*, 881.
[74] Senin, I.I., Dean, K.R., Zargarov, A.A., Akhtar, M, and Philippov, P.P. 1997. *Biochem. J. 32*, 551.
[75] Senin, I.I., Zargarov, A.A., Akhtar, M, and Philippov, P.P. 1997, *FEBS Lett., 408*, 251.
[76] Binder, B.M., Biernbaum, M.S., and Bownds, M.D. 1990, *J. Biol. Chem., 265*, 15333.

In: Neuronal Calcium Sensor Proteins
Editors: Pavel Philippov and Karl Koch, pp. 153-180
ISBN 1-59454-978-8

Chapter IX

RECOVERIN SHAPES THE PHOTORESPONSE OF RETINAL RODS

Ronald A. Bush[1] and Clint L. Makino[2,*]

[1]Section for Translational Research on Retinal and Macular Degeneration, National Institute on Deafness and Other Communication Disorders, National Institutes of Health, 50 South Drive MSC 8021, Bethesda, MD 20892;
[2]Department of Ophthalmology, Massachusetts Eye and Ear Infirmary and Harvard Medical School, 243 Charles Street, Boston, MA 02114.

Abstract

Recoverin is a member of the EF-hand family of Ca^{2+}-binding proteins that is found in selected neurons, including retinal rods, cones and some types of bipolar cells. This chapter reviews physiological and behavioral studies aimed at understanding recoverin's role in visual signaling. Recoverin has two roles in phototransduction: it confers Ca^{2+}-dependence to the shutoff of photoexcited rhodopsin and it buffers Ca^{2+}, delaying changes in Ca^{2+}-dependent guanylate cyclase activity. The consequence of both actions is to prolong the photocurrent flash response and to enhance step sensitivity. However, by mechanisms that have not yet been elucidated, recoverin prolongs the flash response of second order neurons *prior* to the changes in the photocurrent response of the rods. The light-driven activities of amacrine and ganglion cells are also affected indirectly by recoverin. At the organismal level, recoverin improves visual performance near absolute threshold.

Keywords: Ca^{2+}-binding protein, Ca^{2+} buffer, electroretinogram, knockout mice, light adaptation, photoreceptor, phototransduction, recoverin, retinal bipolar cell, single cell recording, visual sensitivity

* Correspondence concerning this article should be addressed to Clint L. Makino, Department of Ophthalmology, Massachusetts Eye and Ear Infirmary and Harvard Medical School, 243 Charles Street, Boston, MA 02114. tel.: 617-573-4462; fax: 617-573-4290; e-mail: cmakino@meei.harvard.edu.

Abbreviations

ERG	electroretinogram
GCAP	guanylate cyclase activating protein
GRK	G protein-coupled receptor kinase
PDE	phosphodiesterase
STR	scotopic threshold response
pSTR	positive scotopic threshold response
nSTR	negative scotopic threshold response

1. Introduction

Working under exceedingly dim lighting conditions, retinal rods encode sufficient information about the outside world to support vision. Such performance is possible because rods have evolved the capacity to signal the presence of single photons. Yet despite their sensitivity, rods continue to operate over many log units of intensity. Highly sophisticated biochemical machinery powers the conversion of a photon into an electrical, neural signal. Amplification is achieved with a cascade of biochemical reactions, in which each activation stage has been optimized for speed. Numerous mechanisms underlie the shutoff and recovery in order to limit the response amplitude and duration, minimize noise and extend the rod's dynamic range. Neuronal calcium sensors provide for important regulatory controls. Recoverin is one such sensor that regulates the shutoff of rhodopsin. Because recoverin holds a position at the top of a cascade where small changes in activity have large effects on the events that follow, this chapter will explore the impact of recoverin on phototransduction in some detail. The rod signal must then be transmitted to second order neurons, processed by the retina and sent to other areas of the brain. Despite the massive convergence in the rod pathway, a high signal to noise is maintained. Here again, neuronal calcium sensors, including recoverin, are likely to play a role. This chapter will also describe early studies showing how recoverin affects the photon signal as it journeys through the inner retina. The molecular structure of recoverin, its interactions with Ca^{2+}, membranes and rhodopsin kinase and its involvement in pathological conditions are discussed in other chapters of this issue.

2. Overview of Phototransduction

Rods extend a specialized ciliated structure, the outer segment, whose purpose is to convert photons into an electrical signal [reviewed in refs. 1-3]. It may contain up to a billion rhodopsin molecules in order to ensure a high probability of photon capture. When photons are scarce, the voice of each individual gets heard. One rhodopsin molecule, excited by a single photon, activates more than a hundred transducins, each of which activates a cGMP phosphodiesterase (PDE) catalytic subunit. PDE hydrolyzes cGMP, leading to the closure of cGMP-gated channels. The continuous entry of Na^{+} and Ca^{2+} into the outer segment in darkness is terminated and the rod hyperpolarizes. At the synapse, the hyperpolarization closes voltage-sensitive Ca^{2+} channels and diminishes the release of the neurotransmitter glutamate onto bipolar and horizontal cells.

The size of the photon response depends on how long the cascade remains in operation. Thermal shutoff of rhodopsin is slow, stochastic and incomplete, so the rod takes proactive measures. Multiple phosphorylations of rhodopsin progressively diminish its activity. Full quench follows the binding of arrestin. The intrinsic shutoff of transducin is also slow, but it becomes accelerated by a regulator of G protein signaling molecule type 9 which exists as a complex with a G protein β5 subunit and a membrane anchoring protein. PDE shutoff occurs concomitant with transducin shutoff. Cyclic GMP-gated channels reopen as cGMP accumulates to steady-state levels and cation influx resumes.

Overlaid on the system are regulatory mechanisms that assist in photon response recovery and adapt the rod to steady light. These mechanisms monitor the progress of phototransduction activity through changes in the levels of Ca^{2+}. The bulk of the current entering through the cGMP-gated channel is carried by Na^{+}, but there is also a trickle of Ca^{2+}. Ca^{2+} is removed by a Na^{+}/ Ca^{2+},K^{+} exchanger. By closing channels and preventing Ca^{2+} influx light allows the exchanger to lower intracellular Ca^{2+}. Three sensors detect the change and set in motion processes that oppose the effects of phototransduction. In lowered Ca^{2+}, guanylate cyclase activating proteins (GCAPs) stimulate guanylate cyclases to synthesize cGMP at a faster rate. Ca^{2+}/calmodulin (or Ca^{2+} bound to a related protein) decreases the apparent affinity of the channel for cGMP, so in low Ca^{2+}, less cGMP is needed to open the channel. Recoverin inhibits rhodopsin kinase and slows rhodopsin shutoff at high Ca^{2+}, but allows for more rapid quenching of rhodopsin when Ca^{2+} is low.

3. Recoverin Slows Flash Response Recovery and Enhances Step Sensitivity in Rods

To explore the operation of recoverin *in situ*, including the time course and magnitude of its effect, Chen and Simon prevented recoverin expression in mice by targeted genetic disruption. Photocurrent responses of single rods from these recoverin knockout (rec-/-) mice were recorded with a suction electrode. Flash responses of rec-/- rods rise normally, but they are brief and there is an undulation in the recovery phase that is most pronounced at midrange (Figure 1) [4,5]. Integration time, taken as the integral of the dim flash response normalized by its amplitude, reduces from 250 ms in normal rods to 157 ms in rec-/- rods. While abbreviated flash responses are to be expected with a faster shutoff of rhodopsin, the rec-/- response is notable for the absence of other features. Shutoff of rhodopsin normally begins within 100 ms after photon absorption, i.e., before the peak of the response [6,7], yet neither the amplitude nor the time to peak of the rec-/- rod single photon response decrease [4,5]. An unexpected increase in rec-/- rod sensitivity was reported in one study [4]. It likely arose from a small increase in outer segment size and quantal catch, because quantal response size was unchanged. The disparity vanishes when much greater care is taken to control for genetic background and uniformity of rearing conditions [5].

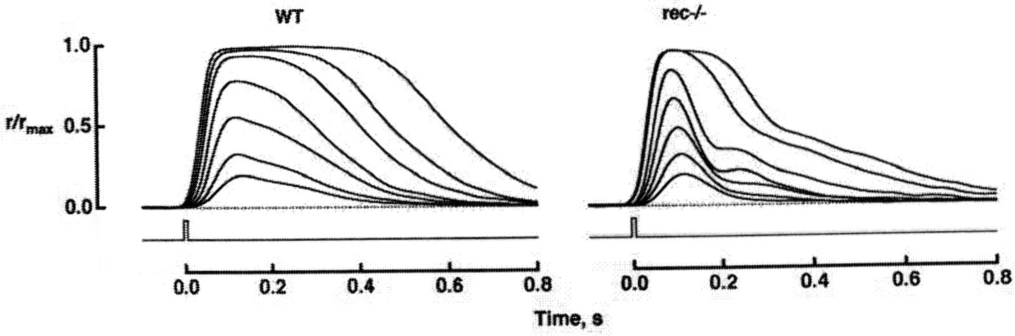

Figure 1. Rapid recovery of a rec-/- mouse rod response to flashes. After reaching a peak, the rec-/-rod begins to recover at a faster than normal rate. Recovery then slows and exhibits a "hump" in some responses ~0.25 s after the flash. Adapted from C.L. Makino, R.L. Dodd, J. Chen, M.E. Burns, A. Roca, M.I. Simon and D.A. Baylor, The Journal of General Physiology 2004, 123, 729-741 by copyright permission of The Rockefeller University Press.

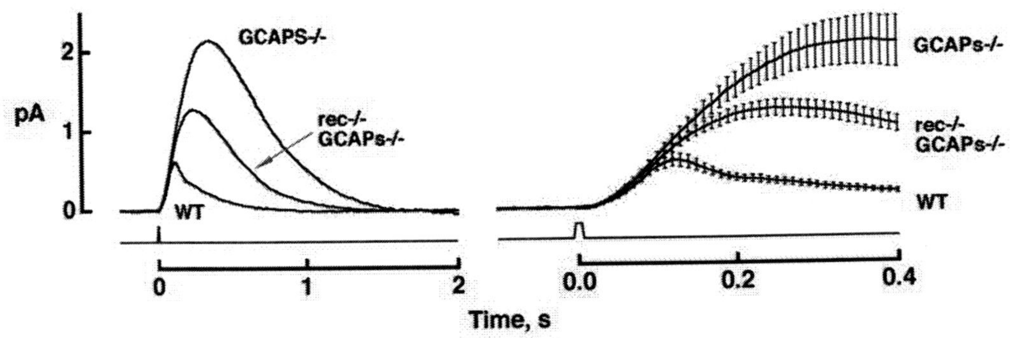

Figure 2. Shorter time to peak and diminished single photon responses of GCAPs-/- mouse rods after deletion of recoverin. Compared to wild type rods, GCAPs-/- photon responses are large and prolonged. Knockout of recoverin reduces the GCAPs-/- response size and accelerates the recovery. Responses are shown on two time scales with inclusion of the wild type response for comparison. Error bars show SEM. Adapted from C.L. Makino, R.L. Dodd, J. Chen, M.E. Burns, A. Roca, M.I. Simon and D.A. Baylor, The Journal of General Physiology 2004, 123, 729-741 by copyright permission of The Rockefeller University Press and from A. Mendez, M.E. Burns, I. Sokal, A.M. Dizhoor, W. Baehr, K. Palczewski, D.A. Baylor and J. Chen, Proceedings of the National Academy of Sciences of the USA 2001, 98, 9948-9953 copyright 2001 National Academy of Sciences of the USA.

To test whether compensatory Ca^{2+} feedback onto guanylate cyclase activity might preserve the rising phase of the rec-/- response, Chen and Simon further deleted GCAPs from rec-/- mice [4]. Coincidence of the photon responses of rec-/- GCAPs-/-, GCAPs-/- rods and for that matter, wild type rods, over the first ~100 ms (Figure 2) is consistent with the idea that shutoff and recovery mechanisms do not yet attenuate the rising phase in this interval. It further suggests that recoverin does not lengthen the latency to rhodopsin shutoff. But soon after, the rec-/- GCAPs-/- rod response rises more slowly and then peaks earlier at a lower amplitude than the GCAPs-/- rod response. Recovery is faster with no signs of undulation. Thus Ca^{2+} feedback regulation of cGMP synthesis provided by GCAPs prevents the absence of recoverin from lowering photon sensitivity, but introduces an undulation in the response recovery. As will be seen below, the undulation has consequences downstream from the rod outer segment.

Since the response of a rod to dim, steady light summates the effects of single photons, it depends upon the rate of photon absorption and the integration time of the quantal response. Hence the 1.6-fold lower integration time of rec-/- rods results in a proportional fall in step sensitivity (Figure 3). The intensity of 250 photons μm^{-2} s^{-1} at 500 nm required for a half-maximal response for wild type rods rises to 410 photons μm^{-2} s^{-1} for rec-/- rods. With time in steady light, the response recovers partially as the rod adapts to light. Adaptation is faster in rec-/- rods, due to an indirect effect on guanylate cyclase activity (see below). Eventually though, step response amplitudes of rec-/- rods match those of wild type many seconds after light onset, i.e., when both types of rods are light adapted (C.L. Makino, R.L. Dodd, J. Chen, M.E. Burns, A. Roca, M.I. Simon and D.A. Baylor, unpublished result).

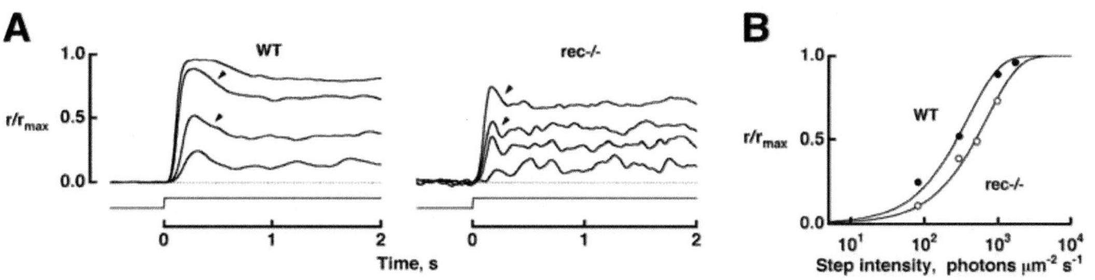

Figure 3. Reduced sensitivity of a rec-/- rod to stepped changes in illumination. A. Averaged photocurrent responses of wild type and rec-/- mouse rods to steps of 500 nm light. The reduction in response that occurs as the rod adapts (arrowheads) is more rapid in the rec-/- rod. B. Stimulus-response relations for the rods in A. The wild type rod required 273 photons μm^{-2} for a half-maximal response while the rec-/- rod required 495 photons μm^{-2}. Adapted from C.L. Makino, R.L. Dodd, J. Chen, M.E. Burns, A. Roca, M.I. Simon and D.A. Baylor, The Journal of General Physiology 2004, 123, 729-741 by copyright permission of The Rockefeller University Press.

Mutant rods over-expressing recoverin have not yet been produced, but Gray-Keller and coworkers [8] succeeded in dialyzing extra recoverin into isolated rod outer segments with a patch pipette (Figure 4). Outer segments lack mitochondria, so GTP and ATP were included in the pipette for photoresponse recordings under whole cell voltage clamp. The presence of recoverin in the pipette extends the rising phase of the flash response causing it to reach a greater amplitude and to take longer to recover (Figure 4B, left column). Peak responses to steps of light are also greater in outer segments dialyzed with recoverin (Dr. Peter Detwiler, personal communication). The effects are only partially reproduced by substituting a much higher concentration of the Ca^{2+} buffer BAPTA for recoverin in the pipette. However, similar effects on flash responses are obtained by filling the pipette with sangivamycin to suppress rhodopsin phosphorylation or with phytic acid to inhibit arrestin binding to phosphorylated rhodopsin [9].

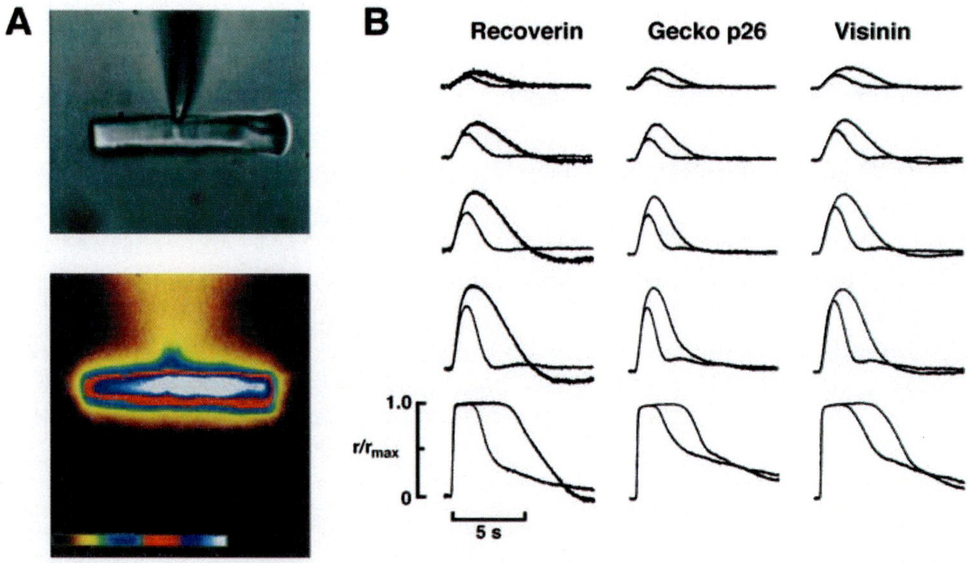

Figure 4. Flash response prolongation after the addition of extra recoverin orthologues to isolated outer segments. A. Method for introducing recoverin or other agents into the outer segment and recording photoresponses. The upper panel shows a patch pipette sealed onto an isolated rod outer segment from Gecko. Disruption of the underlying membrane allows the contents of the pipette to enter the cell. The lower panel shows the diffusion of fluorescently labeled dextran, mw = 10,000, from the pipette into the outer segment. Courtesy, Dr. Peter Detwiler, University of Washington. B. Larger, slower flash responses in outer segments dialyzed with recoverin, Gecko p26 or visinin at nominal concentrations of 16, 30 and 10 μM, respectively. Adapted from Neuron, 10, M.P. Gray-Keller, A.S. Polans, K. Palczewski and P.B. Detwiler, The effect of recoverin-like calcium-binding proteins on the photoresponse of retinal rods, 523-531, copyright 1993, with permission from Elsevier.

4. Prolongation of Rhodopsin Lifetime by Recoverin

Kawamura and Murakami [10] discovered that recoverin mediates Ca^{2+}-dependent prolongation of light-activated PDE activity. The mechanism, discussed in greater depth in other chapters, consists of inhibition of rhodopsin phosphorylation as shown in vitro [11,12], in broken rod outer segments [13] and in intact rods of isolated retinas [5 although cf. 14]. Two lines of electrophysiological evidence show that recoverin slows the quenching of rhodopsin's catalytic activity *in situ.*

In response to bright flashes that close all of the cGMP-gated channels, rec-/- rods emerge from saturation sooner than wild type rods (Figure 5) [4,5,15]. Time in saturation increases in proportion to the natural logarithm of the flash strength and the slope of the relation estimates the time constant for the slowest step in the response recovery (τ_c) [16,17]. τ_c for wild type rods is 205 ms, while that for rec-/- rods is 50 ms faster (Figure 5). Work on other types of transgenic mice is relevant to these observations. τ_c is reduced for mutant rods expressing a palmitoylation deficient rhodopsin mutant that supports faster rhodopsin phosphorylation [18] and photon response recovery is slowed in rods made to express half the normal amount of rhodopsin kinase [7]. One interpretation is that in mouse rods, (i) the timing of rhodopsin shutoff imposes a limit on how fast the photoresponse recovers and (ii) rhodopsin shutoff is faster in the absence of recoverin. GCAPs contribute to the change in rec-/- rods as evidenced by the smaller difference in mean τ_c after knockout of recoverin from GCAPs-/- rods and the lack of statistical significance in the difference between the τ_c values of GCAPs-/- and rec-/- GCAPs-/- rods.

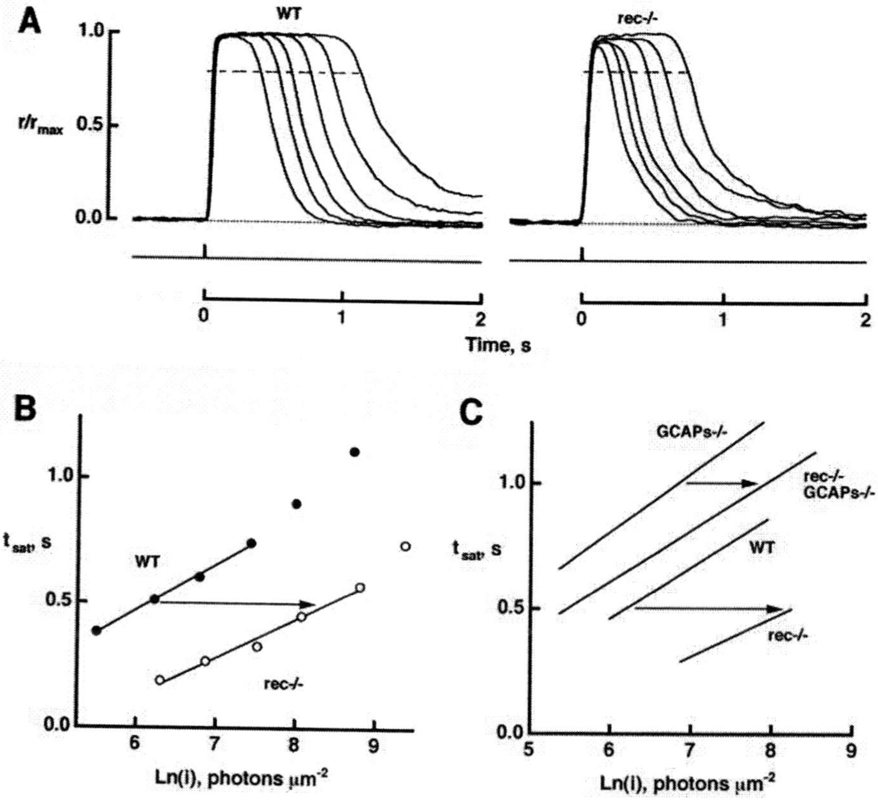

Figure 5. Faster recovery of rec-/- rods to saturating flashes. A. Lengthening of response saturation with increasing flash strength. At each flash strength, saturation time is shorter for the rec-/- rod. B. Pepperberg plots of the rods in A, with saturation time measured from midflash to 20% recovery (dashed lines in A). The relation for the rec-/- rod has a reduced slope and shifts to higher flash strengths. The 2.3 ln unit increase in flash strength necessary to obtain a criterion saturation time of 0.5 s in the rec-/- rod (arrow) translates to a 9.7-fold reduction in transduction gain during the recovery period. Panels A and B adapted from C.L. Makino, R.L. Dodd, J. Chen, M.E. Burns, A. Roca, M.I. Simon and D.A. Baylor, The Journal of General Physiology 2004, 123, 729-741 by copyright permission of The Rockefeller University Press. C. Pepperberg plots for GCAPs-/-, rec-/- GCAPs-/-, wild type and rec-/- rods. The curves depict the averaged behavior of 23 GCAPs-/-, 11 rec-/- GCAPs-/-, 21 wild type and 20 rec-/- rods with slopes of: 0.234, 0.206, 0.205, 0.153 s, respectively. Slopes were estimated using an unstructured covariance model with random slopes and intercepts (MIXED procedure of SAS, version 8; C.L. Makino, R.L. Dodd, J. Chen, M.E. Burns, A. Roca, M.I. Simon and D.A. Baylor, unpublished result). The gain reduction in the recovery phase of the saturated response was 2.9-fold in rec-/- GCAPs rods, evaluated at a saturation time of 1.0 s, and 7.9-fold in rec-/- rods evaluated at 0.5 s (arrows).

Interestingly, rhodopsin shutoff does not appear to slow the photoresponse recovery in intact amphibian rods because τ_c does not change in background light [16] or when intracellular Ca^{2+} is kept constant [19]. Moreover, the susceptibility window for Ca^{2+} regulation of rhodopsin lifetime closes with a time constant of ~0.5 s (see below). This time constant, which may describe rhodopsin lifetime, is considerably faster than the τ_c for amphibian rods which is ~2.5 s [16,19,20].

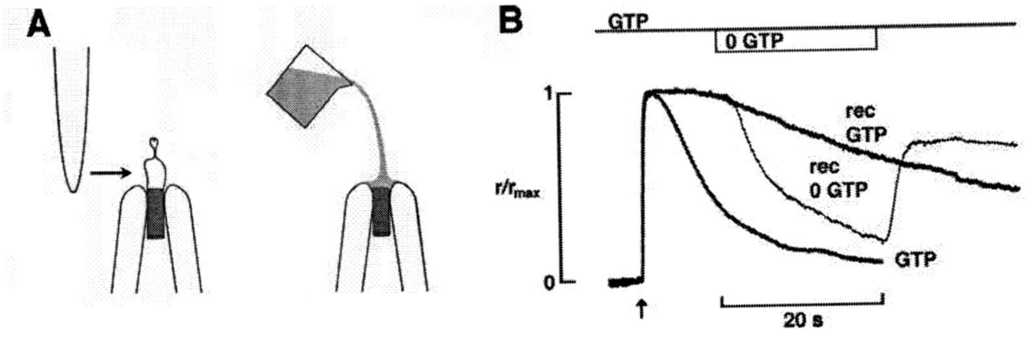

Figure 6. Extension of activated rhodopsin lifetime with recoverin. A. Isolation of a rod outer segment from salamander in a suction electrode. Photoresponses are elicited from the truncated outer segment when it is perfused with a pseudo-intracellular medium supplemented with ATP, GTP and cGMP. B. Photocurrent assay for measuring rhodopsin activity. The presence of 20 μM myristoylated recoverin prolongs the response to a saturating flash of 97 photons μm^{-2} (arrow). Removal of GTP curtails activation of transducins and causes the response to decline. Addition of GTP back to the perfusion medium reinstates the response to the approximate level where it would be with GTP present throughout. Thereafter, the response resumes its slow course of recovery. Adapted from M.A. Erickson, L. Lagnado, S. Zozulya, T.A. Neubert, L. Stryer and D.A. Baylor, The effect of recombinant recoverin on the photoresponse of truncated rod photoreceptors, Proceedings of the National Academy of Sciences of the USA 1998, 95, 6474-6479, copyright 1998 National Academy of Sciences of the USA.

A second demonstration of recoverin's control over rhodopsin catalytic lifetime utilizes rod outer segments as cellular test tubes [21]. When a rod outer segment is positioned inside a suction electrode and its inner segment removed (Figure 6A), continuous "pouring" of ATP, GTP and cGMP in a pseudo-intracellular medium into the open end of the truncated outer segment enables it to generate photocurrent responses to flashes. The response recovers as the decline in light-activated PDE activity permits cGMP to accumulate within the outer segment and reopen channels. Ca^{2+} feedback onto guanylate cyclase is largely inoperative because the perfusion arrangement clamps internal Ca^{2+} and because the GTP concentration is too low to support any significant amount of cGMP production. Addition of myristoylated recoverin prolongs the response, but only in high Ca^{2+}. High Ca^{2+} added alone or added with unmyristoylated recoverin has no effect. In the retina, recoverin is heterogeneously acylated [22,23]. In perfused truncated outer segments, recoverins bearing laurate, myristate or unsaturated forms of myristate work with equal efficacy. Transient removal of GTP during the photoresponse interrupts transducin activation and quickens the recovery (Figure 6B). Restoration of GTP causes the response to jump to the level observed with GTP present throughout thus revealing that some rhodopsins are still active. The premature recovery of the photoresponse after removal of GTP in the presence of recoverin follows a time course that is similar to the recovery of the response in the absence of recoverin, arguing against any effect of recoverin on the shutoff of cascade components downstream from rhodopsin.

5. Recoverin Boosts Phototransduction Gain at Late Times

The reduction in saturation time after bright flashes in rec-/- rods provides a means for evaluating the gain of transduction during the response recovery period. Disregarding the slight variation in the slope of the saturation function, rec-/- rods require as much as eight times more light to produce a response with a criterion saturation time (see Figure 5C). A gain reduction of nine times in the absence of recoverin was reported earlier for a smaller number of cells [4]. Part of the effect is mediated by differences in dynamic feedback onto cGMP synthetic activity as demonstrated by a step-flash paradigm [24]. Exposure of the rod to steady light closes a fraction of the cGMP-gated channels, which reduces Ca^{2+} influx and lowers the steady-state level of internal Ca^{2+} inside the outer segment. A saturating test flash is then given. The flash must be sufficiently bright so that the duration of the saturation period allows Ca^{2+} to fall to a minimum and stimulate guanylate cyclase activity maximally. Then Ca^{2+} feedback onto guanylate cyclase is taken out of the equation. The only difference between responses elicited in darkness and those following a conditioning step is the degree to which recoverin inhibits rhodopsin shutoff at the time of the flash. In wild type mice, inhibition is greatest for the test flash given in darkness when Ca^{2+} is highest, and least with the flash preceded by the conditioning step (Figures 7A, B, D). The decrease in flash response saturation time caused by the conditioning step corresponds to a gain reduction of up to 1.8 ± 0.1 fold (mean $\pm$ SEM, n=8). In "control" experiments on rec-/- rods, conditioning steps fail to decrease saturation time of the test flash response (Figures 7C, D). Another means of determining the effect of recoverin separately from dynamic guanylate cyclase activity is to examine the effect of recoverin deletion on the flash response saturation behavior in rods that lack Ca^{2+} feedback onto guanylate cyclase, e.g. in GCAPs-/- rods. The saturation function of rec-/- GCAPs-/- rods shifts to flash strengths three-fold higher than that of the GCAPs-/- rods (see Figure 5C). Thus by prolonging the activity of rhodopsin, recoverin boosts the effective gain of transduction at late times after the flash in mammalian rods by nearly an order of magnitude. A two- to three-fold gain reduction arises from the slower rhodopsin shutoff and the remainder, from an effect on guanylate cyclase activity, to be discussed below.

Recoverin may be more powerful in amphibian rods than in mouse rods because light-induced Ca^{2+} changes are more extensive. In mouse rods, the Ca^{2+} falls from 250 nM in darkness to 20 nM in saturating light [25]. But in salamander rods, Ca^{2+} may be as high as 670 nM in darkness and fall to 30 nM in the light [26]. Indeed, the gain reduction in toad and salamander rods determined from step-flash or analogous paired flash (described below) experiments is 7- to 14-fold [27-29], several-fold higher than the ~2-fold change in mouse.

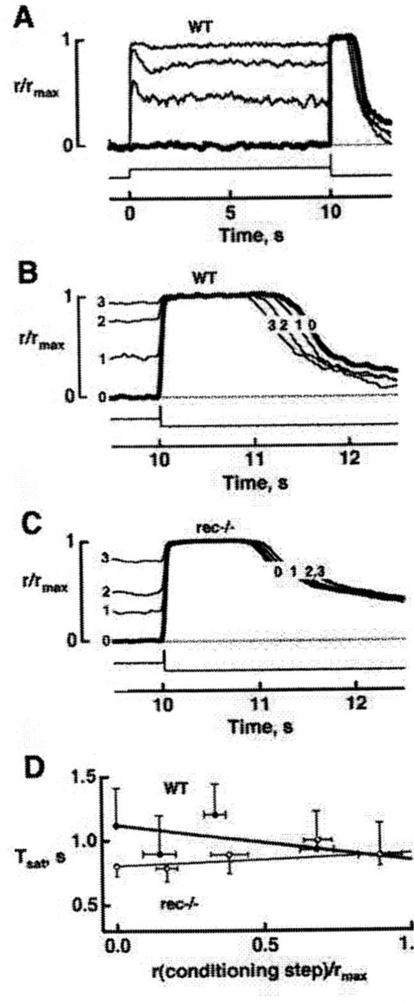

Figure 7. Dependence of saturation time after a bright flash on internal Ca^{2+} in wild type but not in rec-/- rods. A. Reduced saturation time of the response to a flash given immediately after termination of a step of light. The flash responses are shown on an expanded time scale in B. C. Lack of a reduction in flash response saturation time after conditioning steps in a rec-/- rod. D. Convergence in saturation times for wild type and rec-/- rods at the Ca^{2+} levels achieved after elimination of almost all of the circulating current. Error bars show SEM. Adapted from C.L. Makino, R.L. Dodd, J. Chen, M.E. Burns, A. Roca, M.I. Simon and D.A. Baylor, The Journal of General Physiology 2004, 123, 729-741 by copyright permission of The Rockefeller University Press.

6. Buffering of Ca^{2+} by Recoverin

How does recoverin affect guanylate cyclase activity? As an abundant Ca^{2+}-binding protein, recoverin perturbs Ca^{2+} dynamics. When light closes the cGMP-gated channels and terminates Ca^{2+} influx, the continuous operation of the Na^+/ Ca^{2+},K^+ exchanger causes internal Ca^{2+} to plummet. The action of the Na^+/ Ca^{2+},K^+ exchanger is electrogenic, coupling the entry of 4 Na^+ and the release of 1 K^+ to the removal of one Ca^{2+} [30]. So after nearly all of the circulating current is rapidly abolished by closure of the cGMP-gated channels, a small component continues to disappear slowly as the outer segment Ca^{2+} gets depleted by the exchanger (Figure 8). The amplitude of the exchange current in rec-/- rods is normal indicating that the magnitude of the light-induced change in Ca^{2+} is unaffected by recoverin, however, the time constant for Ca^{2+} removal drops from 113 ms in wild type rods to 86 ms in rec-/- rods [4]. The faster light-induced Ca^{2+} decline in rec-/- rods quickens flash response recovery kinetics (see Figure 1) and the sag in the step response (see Figure 3) by accelerating feedback onto cGMP synthesis. Genetic deletion of GCAPs in mice has no significant effect on the rate of Na^+/ Ca^{2+},K^+ exchange [31], therefore, GCAPs contributes little to Ca^{2+} buffering in rod outer segments.

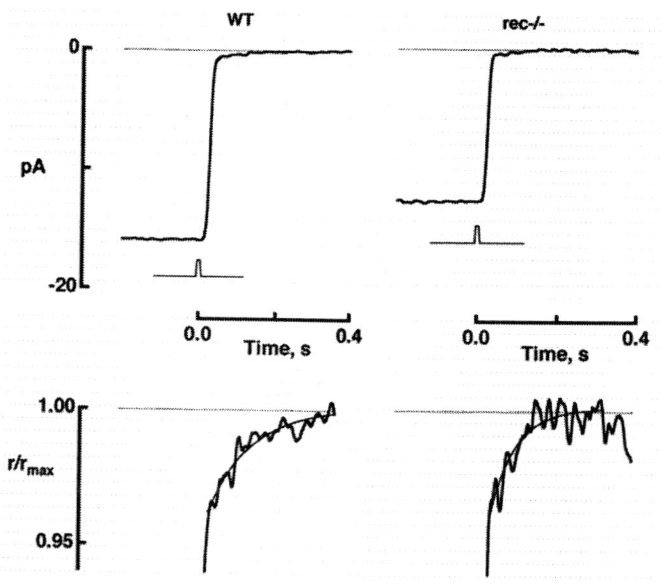

Figure 8. Faster Ca^{2+} removal from a rec-/- rod. Closure of the cGMP channels with a bright flash reveals a small Na^+/ Ca^{2+},K^+ exchange current (upper panels) that declines exponentially with a time constant of 112 ms for the wild type rod and 61 ms for the rec-/- rod (lower panels). The initial amplitudes are –0.7 and –0.6 pA, respectively. Adapted from C.L. Makino, R.L. Dodd, J. Chen, M.E. Burns, A. Roca, M.I. Simon and D.A. Baylor, The Journal of General Physiology 2004, 123, 729-741 by copyright permission of The Rockefeller University Press.

Physiological experiments reveal that under normal conditions, most of the Ca^{2+} in a salamander outer segment is bound to a soluble buffer with K_{buffer} ~0.7 μM [32]. Myristoylated recoverin behaves with an apparent dissociation constant for Ca^{2+} in the outer segment in the same range (reviewed in a separate chapter). If each molecule of buffer binds 2 Ca^{2+}, then its concentration would exceed a hundred μM, in agreement with the physiological estimate for recoverin in mouse rod outer segments [4] and the highest biochemical estimates in frog [33,34]. Thus recoverin is likely to be the main Ca^{2+} buffer in the outer segment. By slowing the rate with which free Ca^{2+} changes, recoverin modulates all Ca^{2+}-dependent processes in rod outer segments.

7. Recoverin Confers Limited Dynamic Feedback over Rhodopsin Shutoff

In order for recoverin to provide dynamic feedback regulation over rhodopsin lifetime, it must be capable of acting within that lifetime. Using a paired flash paradigm, Murnick and Lamb [29] determined the magnitude of recoverin's effect and the rapidity with which it could be removed after a bright flash in toad rods (Figure 9).

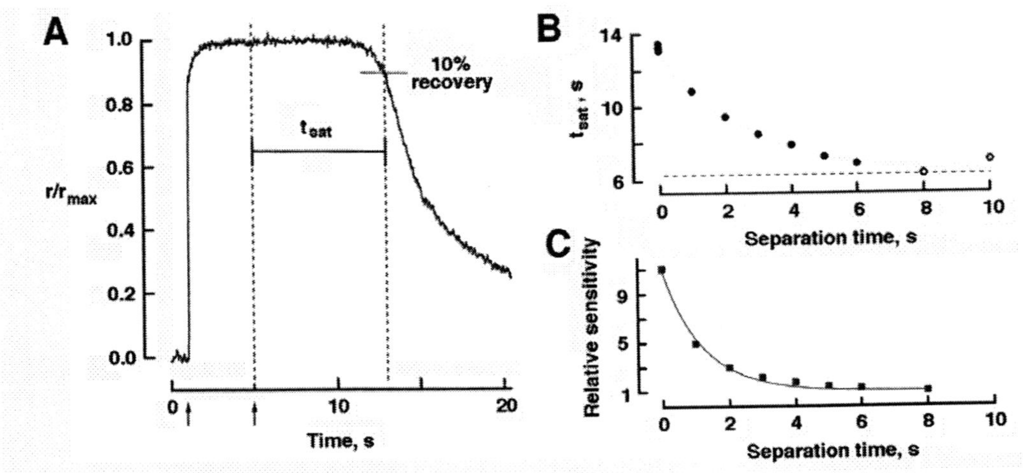

Figure 9. Time course over which there is inhibition of rhodopsin shutoff in a toad rod. A. Paired flash protocol. A bright flash saturates the rod and lowers intracellular Ca^{2+} (arrow on the left). After a variable delay, a brighter, test flash is given (second arrow) and the saturation time measured as the time required for 10% recovery of the test flash response. B. Decrease in saturation time from 12.96 to 6.28 s with increasing inter-flash interval (separation time). C. Decline in sensitivity with inter-flash interval. For a τ_c of 2.88 s, the relative sensitivity falls 11-fold with a time constant of 1.3 s. Adapted with permission from J.G. Murnick and T.D. Lamb, Kinetics of desensitization induced by saturating flashes in toad and salamander rods, Journal of Physiology 1996, 495, 1-13, copyright 1996 Blackwell Publishing Ltd and from Calvert PD, Makino CL, The time course of light adaptation in vertebrate retinal rods. In: Baehr W, Palczewski K. Photoreceptors and Calcium. Georgetown: Landes Bioscience, New York: Kluwer Academic/ Plenum Publishers, 2003: 37-60.

In their method, a bright "priming" flash saturates the rod and begins to lower Ca^{2+}. The rod gradually reaches a state of minimal inhibition by recoverin. The saturation time of the response to a test flash is measured as a function of time after the priming flash. The test flash is so much brighter than the priming flash that simultaneous presentation of both flashes elicits a response no different than the response to the test flash alone. Saturation time of the test flash response declines with increasing inter-flash interval, i.e., as Ca^{2+} falls and recoverin loses its efficacy. Since the response to the test flash maximally activates guanylate cyclase activity, any reduction in test flash response duration caused by priming may be attributed solely to removal of the inhibition of rhodopsin shutoff by recoverin. τ_c in amphibian rods does not change with Ca^{2+}, so the reduction in saturation time, Δt_{sat}, can be equated with a decrease in S, the sensitivity relative to that in darkness according to [19,35]:

$$S = \exp(\Delta t_{sat}/\tau_c) \qquad (1)$$

Relative sensitivity falls tenfold with a time constant of 1.2 s (calculated from the results of [29]). The time constant describes the time course of the portion of the fall in Ca^{2+} that affects rhodopsin shutoff.

Matthews [28] pursued the question of whether that time course is fast relative to the period in which rhodopsin shutoff can be modulated by "clamping" outer segment Ca^{2+} in a salamander rod at the level present in darkness for periods extending into the response to a bright flash and then returning the rod to Ringer's (Figure 10A). The flash elicits a response with saturation time long enough for Ca^{2+} to fall and maximally stimulate guanylate cyclase activity. Maintaining outer segment Ca^{2+} at the dark level for only a few seconds after the flash extends saturation time maximally, marking the end of the period in which rhodopsin shutoff is subject to feedback regulation (Figures 10A, B). The fact that saturation time extends at all indicates that recoverin does provide some feedback. Yet the magnitude of gain reduction is only about 1.4-fold (Figure 10C), a small fraction of the 7-14-fold range possible [27-29]. The reason is that susceptibility of rhodopsin shutoff to regulation fades with a time constant of 0.5 s that is faster than the 1.2 s time constant for the effective removal of Ca^{2+}, measured with paired flashes [29]. In contrast to bright flashes, which close all of the cGMP-gated channels in less than 100 ms, single photons in amphibian rods gradually close channels for up to 0.6 to 4 s [36]. Intracellular Ca^{2+} changes are smaller and slower than with bright flashes, so any regulation of phototransduction gain would be considerably less than 1.4-fold.

Could feedback regulation of rhodopsin lifetime be more important in mammalian rods, where Ca^{2+} removal is faster? It is extremely difficult to manipulate outer segment Ca^{2+} in mammalian rods while recording their photoresponses. A prediction can be made, though. Measurements made with Ca^{2+}-sensitive dyes show that upon exposure of a mouse rod to saturating light, 40% of the internal free Ca^{2+} declines with a time constant of 150 ms, while 60% declines with a time constant of 540 ms [25]. If the lifetime of rhodopsin in a mouse rod is 200 ms, then the timing of Ca^{2+} decline is more favorable for regulation during the response to bright flashes. But the single photon response peaks in 140 ms, recovers completely in ~500 ms and gives rise to a small, slow decline in Ca^{2+}. Moreover, the full magnitude of regulation by recoverin in mouse rods is several-fold less than in amphibian rods. Recoverin then cannot act dynamically in this regime. Similar conclusions obtain from molecular modeling of the rod photoresponse [37]. As a test, addition of the Ca^{2+} buffer BAPTA to mouse rods lacking GCAPs delays the onset of any effect mediated by recoverin. It should prolong the dim flash response if recoverin were to provide feedback regulation. Prolongation does not occur - dim flash response kinetics are unchanged [38]. Therefore, the main function of recoverin is not to provide negative feedback during responses to isolated flashes. Instead, recoverin's dynamic regulation applies to repetitive stimulation and to steady light; it is a mechanism of light adaptation [4,27,29,39]. The beauty of the system is that prolongation of photon responses in darkness improves step sensitivity without detracting from temporal resolution under bright ambient lighting.

On a longer time scale, recoverin expression undergoes a diurnal cycle [40]. Levels of message and protein are highest just before and throughout the daylight hours. The ~3-fold fluctuation in protein level is remarkably high. Since rods only shed ~15% of their outer segment each day [41], most of the recoverin must be lost by other means. It remains to be seen whether outer segment concentration of recoverin is affected, but the pattern seems opposite that which would promote maximal phototransduction sensitivity during the night. One might speculate that the rhythm of recoverin expression serves purposes in the inner segment and synapse (see below) rather than in the outer segment.

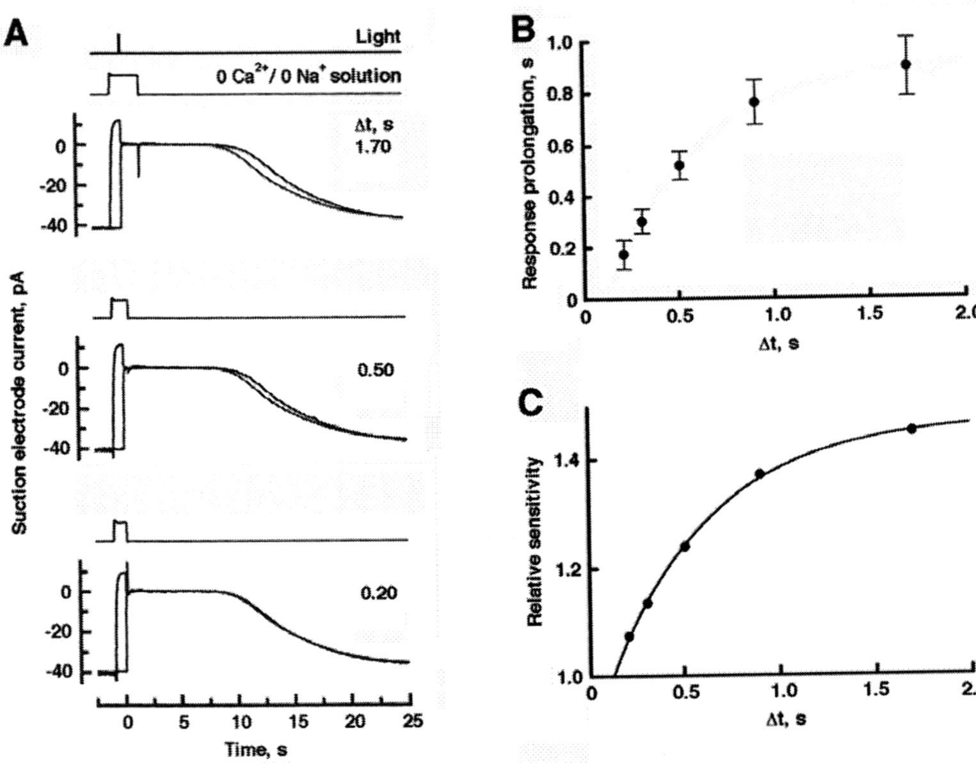

Figure 10. Closure of the susceptibility window for rhodopsin shutoff. A. Flash responses of a rod subjected transiently to "Ca^{2+} clamping." Bathing a salamander rod in a zero sodium, zero calcium solution eliminates calcium influx and disables the exchanger, preventing efflux as well. After the flash, intracellular calcium essentially remains at the dark adapted level until the rod is returned to Ringer's. In each trial shown, the response stays saturated longer when the rod is treated briefly with the zero sodium, zero calcium solution than when bathed in Ringer's from the start. B. Increase in saturation time with Δt, the duration of exposure to zero sodium, zero calcium after a test flash. C. Conversion of saturation time to relative sensitivity according to eq. 1 assuming a τ_c of 2.5 s (the mean value for salamander rods from [16,19,20]. Sensitivity declines 1.4-fold with a time constant of 0.5 s. Panels A and B adapted from H.R. Matthews, The Journal of General Physiology 1997, 109, 141-146 by copyright permission of The Rockefeller University Press.

8. Regulation of Visual Pigment Lifetime in Cones

The phosphorylation rate of photoactivated visual pigment in cones is also Ca^{2+}-dependent [42]. Ca^{2+}-dependent regulation may be more powerful in cones than in rods because the cone cGMP channel and Na^{+}/ Ca^{2+},K^{+} exchanger support a higher Ca^{2+} traffic through the outer segment [reviewed in ref. 43]. In salamander, Ca^{2+} changes 80-fold in red-sensitive cones after exposure to saturating light, compared to 20-fold in rods, and the time constants for the light-induced decline are three- to six-fold faster in cones than in rods [26,44]. Since the resting level of Ca^{2+} in darkness in rods and cones is less than the $K_{1/2}$ of recoverin for Ca^{2+}, the relative magnitude of recoverin's inhibition should increase with the greater dynamic range of Ca^{2+} in cones.

Recoverin is present in cones of primate, cow, rat, squirrel, chameleon and salamander [45-50] but a distinct cone orthologue was discovered frog (s26 [34]) and chick (visinin [51,52]). Retinal photoreceptors of human [although cf. ref. 48], cat, chick, frog and carp immunolabel for these "cone recoverins" but those of cow, mouse and rat do not [34,51,52]. The requirement for recoverins with different amino acid sequences in cones of some species might depend upon the type of pigment kinase that they express. Although rhodopsin kinase or GRK1 is critical for the timely recovery of the cone photoresponse in mouse [53], cones in e.g., dog, pig and medaka lack GRK1 and utilize GRK7 instead [54-56]. The use of both GRKs in primate cones [56-58] may justify their expression of both types of recoverin. Chick rods and cones use a hybrid kinase that bears closer sequence homology to GRK1. But instead of being farnesylated like other members of the GRK1 family [59], it is geranylgeranylated [60], a feature of the GRK7 family [54,55]. Functionally, there is an equivalence in the actions of bovine recoverin, gecko p26 or visinin in dialyzed gecko rods (see Figure 4B), however, subtle differences in the behavior of the different orthologues may get obscured by the presence of the endogenous protein in those experiments [8].

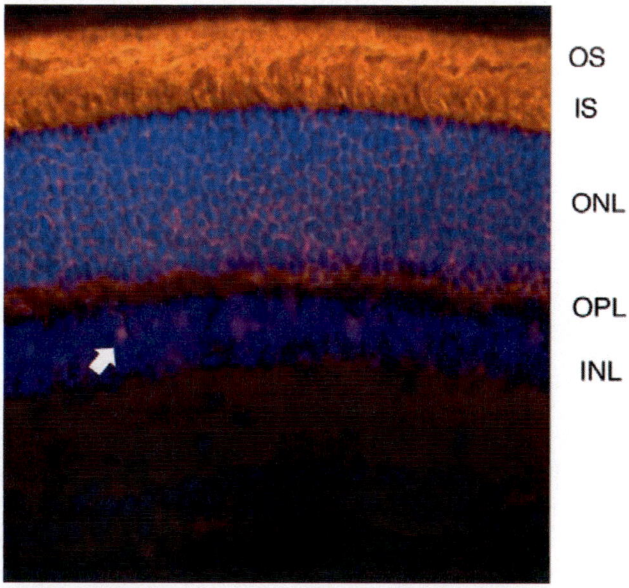

Figure 11. Localization of recoverin (orange, labeled with anti-recoverin antibodies [45]) throughout the photoreceptor, from the outer segment to the synapse in the outer plexiform layer, and in a subset of cone bipolar cells (arrow) in mouse retina. Nuclei are stained blue with Hoechst dye 33342. OS, outer segment; IS, inner segment; ONL outer nuclear layer; OPL, outer plexiform layer; INL, inner nuclear layer. Courtesy, Dr. Tiansen Li, Massachusetts Eye and Ear Infirmary and Harvard Medical School.

9. Beyond the Outer Segment: ERG Findings

Recoverin's presence is not limited to rod and cone outer segments [45]; it pervades the entire photoreceptor (Figure 11). Recoverin is also expressed in some types of bipolar cells [46] and in some neurons in the ganglion cell layer [61,62]. At least in rat, the recoverin immunopositive neurons in the ganglion cell layer appear to be "mislocalized" bipolar cells [63]. We have begun to explore the significance of recoverin to signaling beyond the outer segment by recording the electroretinogram (ERG) of rec-/- mice (Figure 12).

The ERG summates the massed voltage changes across several layers of retinal neurons and glia [for reviews on murine ERG, see refs. 64,65]. With bright flashes, the earliest component is the negative-going a-wave. It accurately reflects changes in the circulating dark current of the rods. The signal from cone photoreceptors, which constitute only about 3% of the photoreceptors in mice [66], is relatively weak and makes a negligible contribution to the a-wave even at the high end of the ERG range [e.g. see ref. 67]. The a-wave amplitude and stimulus-response relation of the rec-/- mice were similar to those of wild type (Figures 12A,B, 13A). These results differ slightly from an earlier study in which a reduced rec-/- a-wave amplitude was reported [15]. The basis for the discrepancy is not known, but could have arisen from variations in genetic background, light-rearing conditions and characteristics of the stimulus.

Excitation of second order, ON-bipolar cells gives rise to the corneal positive b-wave. Due to signal amplification and convergence at the rod to bipolar cell synapse, the b-wave is present at flash strengths too dim to observe an a-wave and reaches half saturation at about 2 rhodopsin photoexcitations per rod [68]. In rec-/- mice, the maximal amplitude and intensity-response relationship of the b-wave were not altered (Figures 12, 13B) nor was there a change in the ratio of b- to a-wave amplitude (Figure 13C). The time to peak of the b-wave, or implicit time, diminished with flash strength for both types of rods (Figure 13D). Implicit times of rec-/- mice were normal for responses to the dimmer flashes but preceded those of wild type animals over a certain range of bright flashes. The b-wave duration was decreased (Figure 13E) due to a faster rate of decline from peak amplitude. These features were especially pronounced for very bright flashes that were expected to recruit cone-driven activity. For our brightest flashes, b-wave width at half-maximal amplitude was 131 $\pm$ 4 ms (mean $\pm$ SEM) in rec-/- mice and 173 $\pm$ 9 ms in wild type mice, $p<0.001$. Diminution of implicit time and narrowing of b-wave width could have resulted from the later, negative-going, components generated in the inner retinal rod [68] or from cone [69] pathways becoming more vigorous in the absence of recoverin. In addition, a subset of cone hyperpolarizing bipolar cells express recoverin [11]. These cells may be responsible for some of the differences in b-wave kinetics seen in figures 13D and E because they arose at intensities bright enough to activate the cone pathway. Though the contribution of hyperpolarizing bipolar cells to the rat ERG is minor [69], their role in mice has not been fully explored. In primates, they contribute substantially to the cone a- [70] and b-waves [71]. Therefore, our ERG results give no evidence for any deficit in rod to bipolar transmission in rec-/- mice.

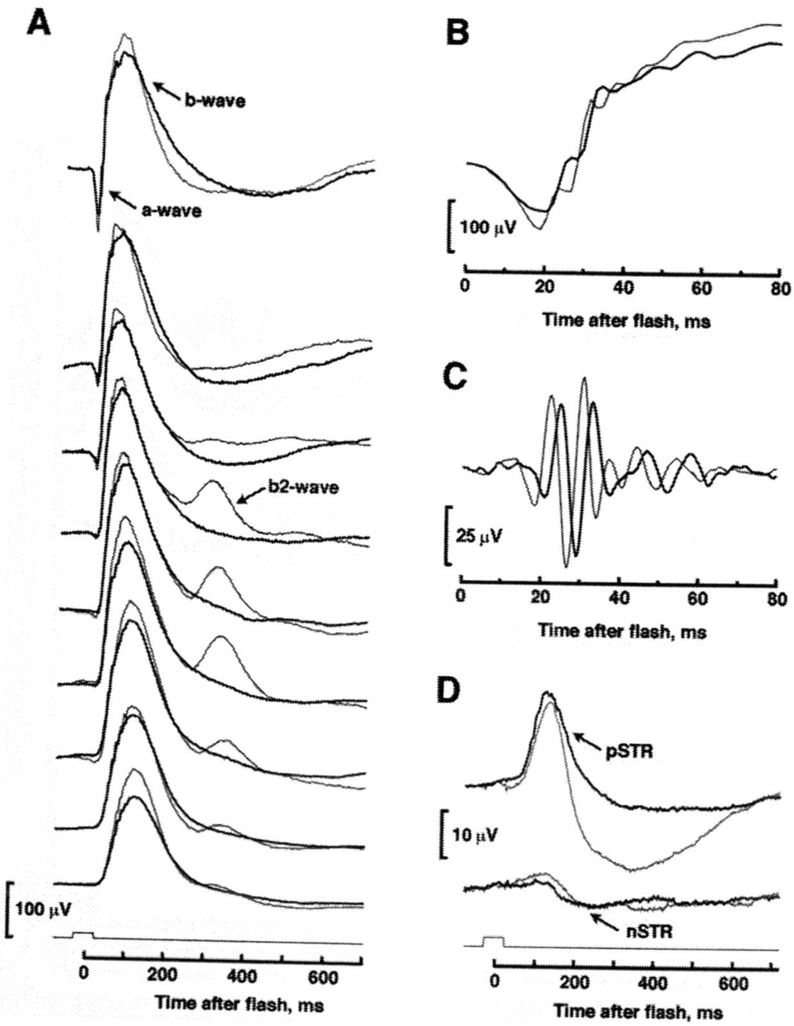

Figure 12. ERGs of 6 dark adapted wild type (heavy lines) and 6 rec-/- mice (thin lines) obtained using methods described previously [95]. A. Averaged responses to 50 ms exposures to a 2700 K source. The b-waves in rec-/- mice were similar in waveform and timing to those of wild type mice except that rec-/- responses peaked slightly sooner after some bright flashes and recovered faster at the highest flash strengths. The b-waves in rec-/- mice were followed by a second peak (b2-wave) along the trailing edge for midrange flashes. An a-wave became visible with brighter flashes. Intensities were (bottom to top): 0.14, 0.27, 0.54, 1.36, 2.72, 5.43, 13.6, 27.2 and 272 scotopic cd m^{-2}. B. Expanded view of the a-waves. The kinetics and amplitude of the a-wave in response to the 272 cd m^{-2} exposure (7000 rhodopsin photoexcitations per rod for the 50 ms flash, assuming that 1 scotopic cd m^{-2} leads to the excitation of 516 rhodopsins per rod each second [96]) were largely normal in rec-/- mice, however, the oscillatory potentials appeared earlier. C. Extraction of the oscillatory potentials from the traces in B using a digital bandpass filter (73-300 Hz). There was a phase advancement of the oscillatory potentials in rec-/- mice of 2 ms. Due to the brightness of the flash, time on the abscissa is plotted from flash onset in panels B and C. D. Scotopic threshold responses. Negative scotopic threshold response (nSTR) thresholds were comparable in wild type and rec-/- mice, ~1 rhodopsin photoexcitation for every 300 rods. The positive component (pSTR) became visible with slightly stronger flashes. Stronger flashes also caused nSTR amplitude growth in rec-/- but not in wild type mice. Intensities were (bottom to top): 2.7 x 10^{-4} and 2.7 x 10^{-3} cd m^{-2}.

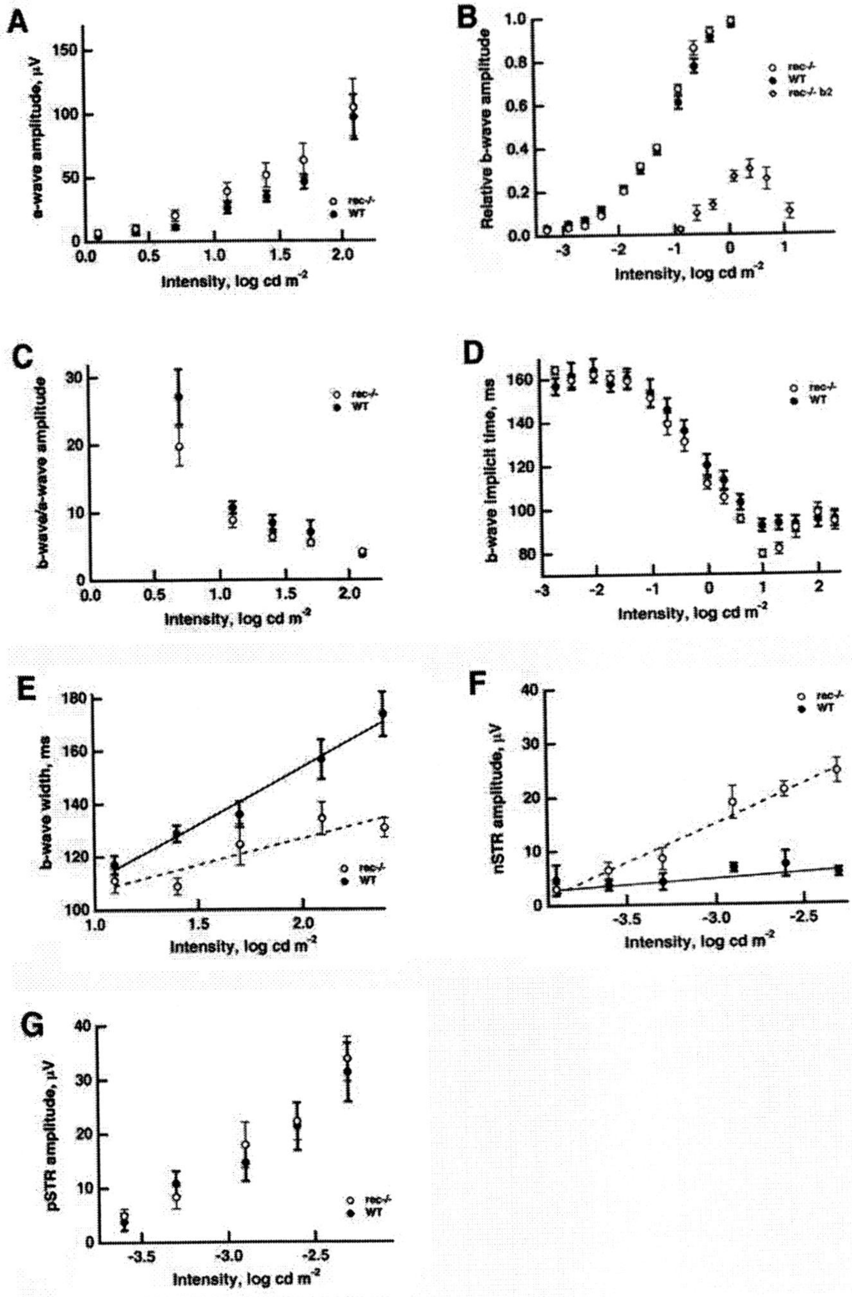

Figure 13. ERG parameters of wild type and rec -/- mice, mean ± SEM. Intensities are expressed as log scotopic cd m^{-2}. A. Stimulus-response relation for the a-wave. Although some a-wave amplitudes of rec-/- mice appeared to be larger than those of wild type over a restricted stimulus range, the differences were not significant at any intensity. The amplitude of the a-wave was measured from the trough of the a-wave to the baseline. B. Matching b-wave stimulus-response relations for rec-/- mice and normal mice. The b- and b2-waves were normalized to the b-wave amplitude at 1 scotopic cd m^{-2} (26 rhodopsin photoexcitations per rod), the approximate saturating amplitude. Half saturation occurred at 2.1 and 1.8 rhodopsin photoexcitations per rod for wild type and rec-/- mice, respectively. The amplitude of the b-wave was measured from baseline or from the trough of the a-wave to response peak, while the b2-

wave amplitude was measured from the trough just preceding it on the trailing edge of the b-wave. Maximal b-wave amplitude was 250 $\pm$ 49 for wild type and 324 $\pm$ 61 for rec-/- mice, a difference that was not significant. C. Ratio of b- to a-wave amplitudes reflecting the relationship between photoreceptor potentials and the corresponding responses of second order neurons. There were no significant differences across intensities. D. Decrease in implicit time with flash strength. The decrease was somewhat greater in rec-/- mice, $p<0.0001$ (2-way ANOVA). E. Stimulus-response relations for b-wave width. Measured at half-maximal amplitude, the bright flash b-wave width in rec-/- mice was low, $p<0.005$, and broadened with flash strength at an attenuated rate, 20 ms (log cd $m^{-2})^{-1}$, compared to wild type mice, 43 ms (log cd $m^{-2})^{-1}$, $p<0.005$. F. Stimulus-response relation for the nSTR. nSTR amplitude increased with flash strength at a rate of 14.4 μV (log cd $m^{-2})^{-1}$ in rec-/- mice whereas it changed little in wild type animals, 2.1 μV (log cd $m^{-2})^{-1}$, $p<0.0001$. G. Normal stimulus-response relation for the pSTR in rec-/- mice.

Rec-/- mice exhibit a peculiar "bump" on the trailing edge of the b-wave in their response to midrange flashes (Figure 12; see also ref. 15). This "b2-wave" followed the b-wave by 233 $\pm$ 9 ms and attained a maximal amplitude of 30% that of the b-wave at flash strengths that just saturated the b-wave. The b2-wave was probably driven by the undulation in the photocurrent recovery of rec-/- rods (see Figure 1) because both were present over the same range of flash strengths. In neither case did the peak to peak interval vary systematically with flash strength, though the peak to peak interval in the rod photocurrent responses was shorter, 156 $\pm$ 5 ms (n = 24 rods, C.L. Makino, R.L. Dodd, J. Chen, M.E. Burns, A. Roca, M.I. Simon and D.A. Baylor, unpublished result). The reason for the increased delay to the second peak in the ERG, also seen in the study by Hurley and Chen [15], is not known. One possibility is that even though transmission through the rod pathway is not delayed for the initial flash response, there are distortions in the bipolar cell response to the recovery phase of the rod response (cf. Figure 14). In addition, the b2-wave is influenced by the same potentials that shape the b-wave, mentioned above. In response to the stimulus light being turned off, hyperpolarizing bipolar cells give rise to a large corneal positive potential in the cone ERG of primates [71] and a smaller positive potential in dark- and light-adapted rats at flash strengths stimulating the rod and cone pathways [72]. However, the b2-wave was not an "OFF" response since its timing was unaffected by varying the duration of the light exposure from 10 to 300 ms (data not shown).

With bright flashes, oscillations intrude on the rising phase of the b-wave. These oscillatory potentials, which are thought to arise from the activity of bipolar-amacrine-ganglion cell circuits, are phase advanced in rec-/- by ~2 ms (see Figure 12C). This phase advance was seen in each of the four major peaks of the oscillatory potentials, beginning shortly after the a-wave peak. Inner retinal signaling at the dimmest stimulus intensities was monitored using the scotopic threshold response (STR) which dominates the ERG near visual threshold [73]. Arising from third order amacrine cells and ganglion cells, the corneal negative STR has the lowest threshold of any ERG component. The positive STR has a slightly higher threshold. It reflects predominantly ganglion cell activity and is eventually overshadowed by the b-wave with increasing flash strength [74]. These slow components appear at intensities even dimmer than those required for observing a b-wave because of further signal convergence and amplification during processing of the dim flash response in the retina downstream from bipolar cells. STR threshold was similar for rec -/- and wild type mice (see Figures 12D, 13F, G) indicating that the retinas of both mice were sensitive to light flashes exciting only a single rhodopsin molecule in one rod out of several hundred. Since the single photon response amplitude was normal in rec -/- mice [4,5], these results imply that exceedingly weak signals are faithfully transmitted at all levels of the retina. Development of the pSTR with increasing flash strength was similar in both types of mice. For reasons unknown, we did not detect the growth in nSTR amplitude with flash strength described in other wild type mice [74], however, we did observe a positive correlation in rec-/- mice (see Figure 13F).

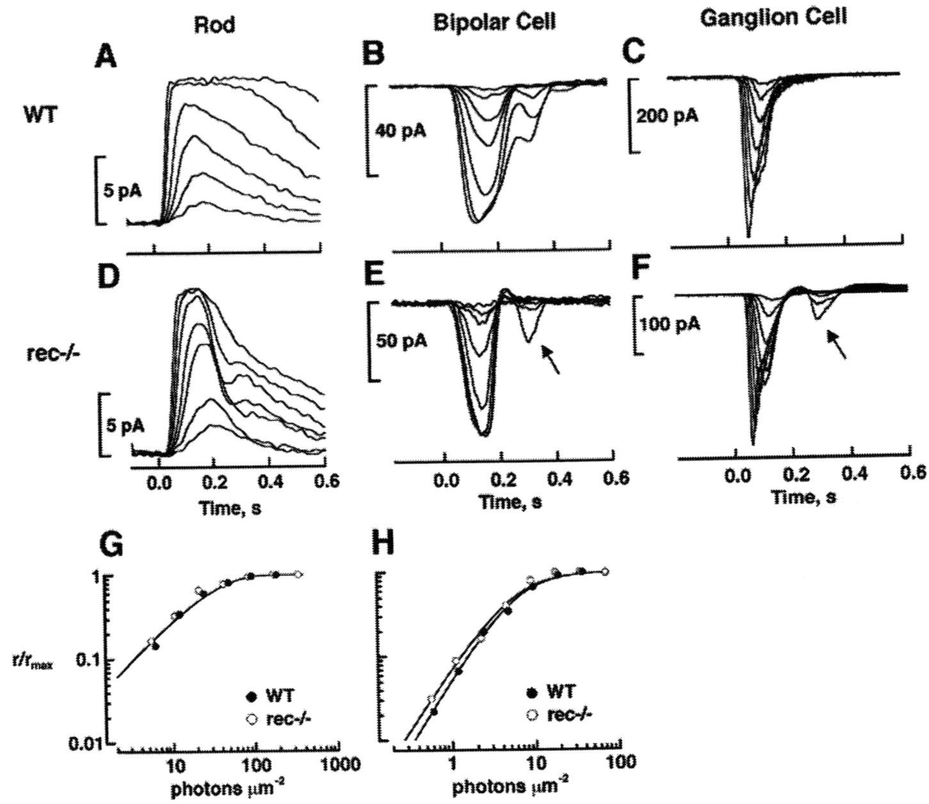

Figure 14. Faster flash responses of rod, bipolar and ON-ganglion cells of rec-/- mice compared to those of wild type mice in whole-cell voltage clamp recordings from retinal slices or from whole-mounted retinas. In rec-/- retinas, there is a prominent secondary peak (arrows) in the bipolar and ganglion cell responses of cells. The sensitivities of rods and bipolar cells to flashes were unchanged by deletion of recoverin. Adapted from Neuron, 46, A.P. Sampath, K.J. Strissel, R. Elias, V.Y. Arshavsky, J.F. McGinnis, J. Chen, S. Kawamura, F. Rieke and J.B. Hurley, Recoverin improves rod-mediated vision by enhancing signal transmission in the mouse retina, 413-420, copyright 2005, with permission from Elsevier.

Surprisingly, b-wave flash responses of rec-/- and rec+/- mice are larger than normal under light adapted conditions (Dr. Neal Peachey, personal communication). In preliminary experiments, we found that background light attenuated the b-wave to a greater extent than the b2-wave in rec-/- mice. At least part of the explanation was that a trough developed between the b- and b2-waves that deepened with increasing background light (not shown). The trough caused the b2-wave amplitude to grow relative to that of the b-wave so that at flash strengths most favorable for observing the b2-wave, its amplitude nearly matched that of the b-wave. These effects on the ERG cannot be attributed to differences in the photocurrent responses of rec-/- and wild type rods. Step sensitivity as measured at the peak of the photocurrent response was greater for wild type rods (see Figure 3), but the response levels equalized over time in steady light as rods became fully light adapted. Furthermore, the sensitivity of rec-/- rods to flashes superimposed on background light was not different from that of wild type rods, based on observations of a limited number of cells [4]. Enhancement of the light adapted rec-/- ERG must therefore originate downstream from the outer segment. A likely site of action is the photoreceptor synapse, but there could be involvement of higher order neurons since recoverin is present in some bipolar cells (e.g. see Figure 11). The overall conclusion from these ERG recordings is that recoverin alters the retinal processing of flashes with manifestations in the oscillatory potentials, b-wave and b2-wave. Further study is required for an understanding of the nature of these changes.

10. Recoverin Prolongs the Photon Response of Rod Bipolar Cells

Single cell recordings of proximal retinal neurons provide additional insight into the function of recoverin in the processing of rod mediated signals [5]. Flash responses of rod bipolar cells in rec-/- retinas include a secondary "response" over the same range of flashes that give undulating response recoveries in rods and the b2-wave in the ERG (see Figures 1, 12, 13, and 14). Transmission of the undulation may be particularly efficient due to the nonlinearity and high-pass filtering characteristics of the bipolar cell response [reviewed in ref. 75]. The presence of a secondary response, albeit less distinct, in wild type bipolar cells may be explained by the "kink" in the recovery that is sometimes observed in wild type rods [cf. 31].

The nonlinearity in the bipolar cell response precludes a rigorous evaluation of the rising phase in rec-/- mice compared to those of wild type. Nevertheless, it is clear that the rec-/- bipolar cell response "peels away" from the wild type response at or before 115 ms post-flash [5]. The rec-/- response peaks at 130 ms, compared to 150 ms for the wild type response, and integration time is reduced by 20% due to a faster time course of recovery. The disparity in rec-/- and wild type bipolar cell responses within this time frame is remarkable because it precedes any difference in the rod photoresponses (Figure 15).

With dim flashes, the faster peak and recovery of the rec-/- bipolar cell response were not apparent in our ERG recordings. Undoubtedly, these changes were overwhelmed by the contributions of other b-wave components. The presence of recoverin at the rod synapse (outer plexiform layer in Figure 11) raises the possibility that faster bipolar cell response kinetics in rec-/- mice arise from disruption of a mechanism that modulates neurotransmitter release. Alternatively, recoverin deletion could change the intrinsic properties of the bipolar cell that directly truncate the response or do so through inhibitory feedback circuitry. Although rod bipolar cells of normal mice do not appear to contain recoverin [76], rabbit rod bipolar cells do show weak recoverin immunofluorescence under optimized conditions [77], suggestive of low levels of expression. As was inferred by ERG recordings, quicker rec-/- rod photoresponse recovery kinetics do not affect the stimulus-response relations of rod bipolar cells (see Figure 14).

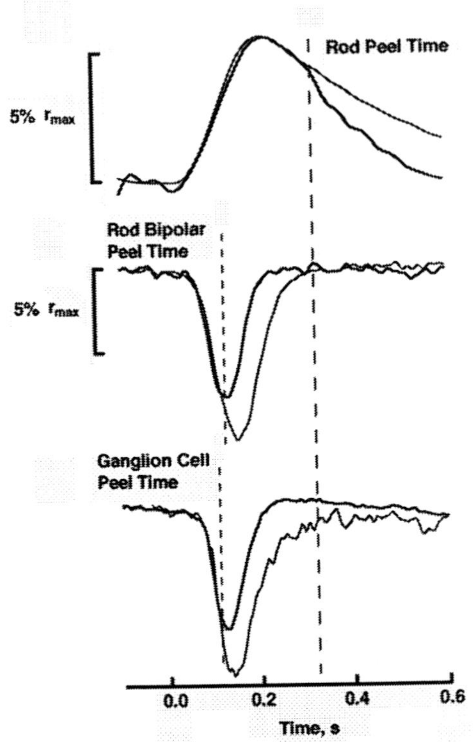

Figure 15. Early "peeling away" of rec-/- rod bipolar and ganglion cell responses from those of the wild type cells at ~115 and 110 ms, respectively, almost 200 ms before any divergence in the rod response. Adapted from Neuron, 46, A.P. Sampath, K.J. Strissel, R. Elias, V.Y. Arshavsky, J.F. McGinnis, J. Chen, S. Kawamura, F. Rieke and J.B. Hurley, Recoverin improves rod-mediated vision by enhancing signal transmission in the mouse retina, 413-420, copyright 2005, with permission from Elsevier.

The accelerated flash response recovery and the secondary responses of bipolar cells are passed along to ON-ganglion cells in rec-/- mice (see Figures 14C,F). For responses to very dim flashes, ganglion cell time to peak is ~20 ms faster and the integration time is 40% less in rec-/- mice than in wild type (see Figure 15). ON-ganglion cell sensitivity is normal, however. Therefore in mammals, (i) recoverin prolongs the signal at the level of the bipolar cell and (ii) signaling of dim flashes through the ON-pathway of the retina does not depend upon the integration time of the rod or rod bipolar cell responses. In the future, additional pharmacological studies and single cell recordings of other retinal neurons will be necessary to evaluate all of the rec-/- ERG findings and to determine the full significance of recoverin in retinal processing.

11. Recoverin Improves Visual Sensitivity

What is the importance of recoverin to vision? Rec-/- mice are less sensitive than wild type mice in behavioral tests of visual threshold (Figure 16) [5]. At these intensities, only a small fraction of the rods absorbs a photon so sensitivity must reflect the integration time of the bipolar cell. But the ~4-fold loss in behavioral sensitivity in rec-/- mice does not follow simply from the 1.3-fold lower bipolar cell integration time. Neural processing must amplify the relatively small difference in rod bipolar cell integration time or incorporate information from OFF-pathways. A prediction is that rec-/- mice will exhibit a threshold minimum at 4-6 Hz, due to the undulation in the rod photoresponse. It would be interesting to test the performance of other mutant mice with brief rod photoresponses and normal recoverin expression, to determine the contribution of recoverin apart from its effect on rod integration time. Recoverin could also be knocked out of mice lacking scotopic vision (e.g. rod transducin knockout mice [78]) to see whether there would be an even larger impact on photopic vision.

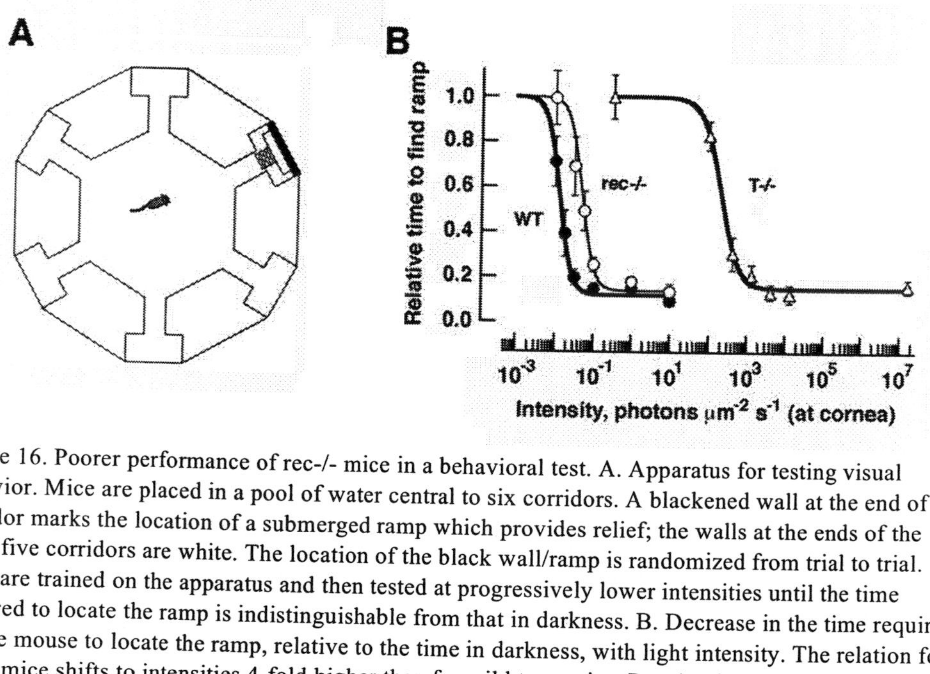

Figure 16. Poorer performance of rec-/- mice in a behavioral test. A. Apparatus for testing visual behavior. Mice are placed in a pool of water central to six corridors. A blackened wall at the end of one corridor marks the location of a submerged ramp which provides relief; the walls at the ends of the other five corridors are white. The location of the black wall/ramp is randomized from trial to trial. Mice are trained on the apparatus and then tested at progressively lower intensities until the time required to locate the ramp is indistinguishable from that in darkness. B. Decrease in the time required for the mouse to locate the ramp, relative to the time in darkness, with light intensity. The relation for rec-/- mice shifts to intensities 4-fold higher than for wild type mice. Rec-/- mice still outperform mice lacking rod transducin (T-/-), which rely entirely on cones [78]. At comparable intensities, pupil diameter is the same for wild type and rec-/- mice. Error bars show SEM. Adapted from Neuron, 46, A.P. Sampath, K.J. Strissel, R. Elias, V.Y. Arshavsky, J.F. McGinnis, J. Chen, S. Kawamura, F. Rieke and J.B. Hurley, Recoverin improves rod-mediated vision by enhancing signal transmission in the mouse retina, 413-420, copyright 2005, with permission from Elsevier.

12. Novel Targets for Recoverin?

The presence of rod and cone recoverins in pinealocytes [52,79-81] and rod recoverin in ocular ciliary epithelial cells [82] may not be too surprising, since these cells also express opsin-based pigments [reviewed in refs. 83,84] and rhodopsin kinase [82,85-87]. Substantial rhodopsin kinase activity is demonstrable in the pineal [88]. In these tissues, recoverins may regulate pigment phosphorylation, as in photoreceptor outer segments, and play a role in nonvisual photic signaling. Recoverin expression in inner retinal neurons [46,61,62], olfactory receptor neurons [89] and certain carcinomas [90,91] is more difficult to explain because rhodopsin kinase is not co-expressed. If recoverin does more than just buffer Ca^{2+} in these cells, then it must possess an alternate target(s). Such a target remains to be identified. An intriguing finding is that under conditions where photoreceptors are strongly labeled with a recoverin antibody, immunoreactivity of bipolar cells varies considerably across and within vertebrate classes and sometimes even within the same order [46,49,50,76,77,92-94]. One might suspect that recoverin serves a redundant function and that it finely tunes the response characteristics of certain neurons. Consistent with this notion, bipolar cells that express recoverin also express other Ca^{2+}-binding proteins [e.g. see ref. 77]. In the future, it will be important to find out whether these species-specific patterns of recoverin expression help endow an organism with the proper sensitivity and adaptational capacity to suit its particular visual ecology.

Acknowledgments

Drs. J. Chen and M. Simon deserve special credit for creating the recoverin knockout mouse used in many of the studies described herein. We thank Drs. D. Gerwick and P. Sieving for assistance with the ERG recording, Ms. C. Mfalila for consultations on the statistical analysis and numerous colleagues for sharing preprints and unpublished work and for stimulating discussions. Supported by Foundation Fighting Blindness, the Intramural Research Program of the National Institute on Deafness and Other Communication Disorders and the National Eye Institute: EY06094 (Dr. Paul Sieving), EY07003 (core grant to the Kellogg Eye Center), EY12944 (CLM) and EY14104 (core grant to the Massachusetts Eye and Ear Infirmary).

References

[1] Burns, M.E., and Lamb, T.D. 2003, *Visual Neurosciences*, L.M. Chalupa and J.H. Werner (Eds.), MIT Press, Cambridge, 215.

[2] Makino, C.L., Wen, X.-H., and Lem, J. 2003, *Curr. Op. Neurobiol., 13*, 404.

[3] Pugh, E.N. Jr., and Lamb, T.D. 2000, *Handbook of Biological Physics Vol. 3, Molecular Mechanisms of Visual Transduction*, D.G. Stavenga, W.J. DeGrip, and E.N. Pugh Jr. (Eds.), Elsevier, Amsterdam, 183.

[4] Makino, C.L., Dodd, R.L., Chen, J., Burns, M.E., Roca, A., Simon, M.I., and Baylor, D.A. 2004, *J. Gen. Physiol., 123*, 729.
[5] Sampath, A.P., Strissel, K.J., Elias, R., Arshavsky, V.Y., McGinnis, J.F., Chen, J., Kawamura, S., Rieke, F., and Hurley, J.B. 2005, *Neuron, 413.*
[6] Chen, J., Makino, C.L., Peachey, N.S., Baylor, D.A., and Simon, M.I. 1995, *Science, 267*, 374.
[7] Chen, C.-K., Burns, M.E., Spencer, M., Niemi, G.A., Chen, J., Hurley, J.B., Baylor, D.A., and Simon, M.I. 1999, *Proc. Natl. Acad. Sci. USA, 96*, 3718.
[8] Gray-Keller, M.P., Polans, A.S., Palczewski, K., and Detwiler, P.B. 1993, *Neuron, 10*, 523.
[9] Palczewski, K., Rispoli, G., and Detwiler, P.B. 1992, *Neuron, 8*, 117.
[10] Kawamura, S., and Murakami, M. 1991, *Nature, 349*, 420.
[11] Chen, C.-K., Inglese, J., Lefkowitz, R.J., and Hurley, J.B. 1995, *J. Biol. Chem., 270*, 18060.
[12] Klenchin, V.A., Calvert, P.D., and Bownds, M.D. 1995, *J. Biol. Chem., 270*, 16147.
[13] Kawamura, S. 1993, *Nature, 362*, 855.
[14] Otto-Bruc, A.E., Fariss, R.N., van Hooser, J.P., and Palczewski, K. 1998, *Proc. Natl. Acad. Sci. USA, 95*, 15014.
[15] Hurley, J.B., and Chen J. 2001, *Progr. Brain Res., 131*, 395.
[16] Pepperberg, D.R., Cornwall, M.C., Kahlert, M., Hofmann, K.P., Jin, J., Jones, G.J., and Ripps, H. 1992, *Vis. Neurosci., 8*, 9.
[17] Nikonov, S., Engheta, N., and Pugh, E.N. Jr. 1998, *J. Gen. Physiol., 111*, 7.
[18] Wang, Z., Wen, X.-H., Ablonczy, Z., Crouch, R.K., Makino, C.L., and Lem, J. 2005, *J. Biol. Chem., 280*, 24293.
[19] Lyubarsky, A., Nikonov, S., and Pugh, E.N. Jr. 1996, *J. Gen. Physiol., 107*, 19.
[20] Calvert, P.D., Govardovskii, V.I., Arshavsky, V.Y., and Makino, C.L. 2002, *J. Gen. Physiol., 119*, 129.
[21] Erickson, M.A., Lagnado, L., Zozulya, S., Neubert, T.A., Stryer, L., and Baylor, D.A. 1998, *Proc. Natl. Acad. Sci. USA, 95*, 6474.
[22] Dizhoor, A.M., Ericsson, L.H., Johnson, R.S., Kumar, S., Olshevskaya, E., Zozulya, S., Neubert, T.A., Stryer, L., Hurley, J.B., and Walsh, K.A. 1992, *J. Biol. Chem., 267*, 16033.
[23] Johnson, R.S., Ohguro, H., Palczewski, K., Hurley, J.B., Walsh, K.A., and Neubert, T.A. 1994, *J. Biol. Chem., 269*, 21067.
[24] Fain, G.L., Lamb, T.D., Matthews, H.R., and Murphy, R.L.W. 1989, *J. Physiol., 416*, 215.
[25] Woodruff, M.L., Sampath, A.P., Matthews, H.R., Krasnoperova, N.V., Lem, J., and Fain, G.L. 2002, *J. Physiol., 542*, 843.
[26] Sampath, A.P., Matthews, H.R., Cornwall, M.C., and Fain, G.L. 1998, *J. Gen. Physiol., 111*, 53.
[27] Matthews, H.R. 1996, *J. Physiol., 490*, 1.
[28] Matthews, H.R. 1997, *J. Gen. Physiol., 109*, 141.
[29] Murnick, J.G., and Lamb, T.D. 1996, *J. Physiol., 495*, 1.

[30] Cervetto, L., Lagnado, L., Perry, R.J., Robinson, D.W., and McNaughton, P.A. 1989, *Nature, 337*, 740.

[31] Mendez, A., Burns, M.E., Sokal, I., Dizhoor, A.M., Baehr, W., Palczewski, K., Baylor, D.A., and Chen, J. 2001, *Proc. Natl. Acad. Sci. USA, 98*, 9948.

[32] Lagnado, L., Cervetto, L., and McNaughton, P.A. 1992, *J. Physiol., 455*, 111.

[33] Kawamura, S., Hisatomi, O., Kayada, S., Tokunaga, F., and Kuo, C.-H. 1993, *J. Biol. Chem., 268*, 14579.

[34] Kawamura, S., Kuwata, O., Yamada, M., Matsuda, S., Hisatomi, O., and Tokunaga, F. 1996, *J. Biol. Chem., 271*, 21359.

[35] Pepperberg, D.R., Jin, J., and Jones, G.J. 1994, *Vis. Neurosci., 11*, 53.

[36] Baylor, D.A., Lamb, T.D., and Yau, K.-W. 1979, *J. Physiol., 288*, 589.

[37] Hamer, R.D., Nicholas, S.C., Tranchina, D., Lamb, T.D., and Jarvinen, J.L.P. 2005, *Vis. Neurosci., 22*, 417.

[38] Burns, M.E., Mendez, A., Chen, J., and Baylor, D.A. 2002, *Neuron, 36*, 81.

[39] Koutalos, Y., Nakatani, K., and Yau, K.-W. 1995, *J. Gen. Physiol., 106*, 891.

[40] Weichmann, A.F., and Sinacola, M.K. 1997, *Mol. Brain Res., 45*, 321.

[41] Young, R.W. 1967, *J. Cell Biol., 33*, 61.

[42] Kennedy, M.J., Dunn, F.A., and Hurley, J. B. 2004, *Neuron, 41*: 915.

[43] Korenbrot J.I., and Rebrik, T.I. 2002, *Adv. Exp. Med. Biol., 514*, 179.

[44] Sampath, A.P., Matthews, H.R., Cornwall, M.C., Bandarchi, J., and Fain, G.L. 1999, *J. Gen. Physiol., 113*, 267.

[45] Dizhoor, A.M., Ray, S., Kumar, S., Niemi, G., Spencer, M., Brolley, D., Walsh, K.A., Philipov, P.P., Hurley, J.B., and Stryer, L. 1991, *Science, 251*, 915.

[46] Milam, A.H., Dacey, D.M., and Dizhoor, A.M. 1993, *Vis. Neurosci., 10*, 1.

[47] Von Schantz, M., Szel, A., van Veen, T., and Farber, D.B. 1994, *Invest. Ophthalmol. Vis. Sci., 35*, 3922.

[48] Polans, A.S., Burton, M.D., Haley, T.L., Crabb, J.W., and Palczewski, K.. 1993, *Invest. Ophthalmol. Vis. Sci., 34*, 81.

[49] McGinnis, J.F., Stepanik, P.L., Jariangprasert, S., and Lerious, V. 1997, *J. Neurosci. Res., 50*, 487.

[50] Zhang, J., and Wu, S.M. 2003, *J. Comp. Neurol., 461*, 276.

[51] Hatakenaka, S., Kiyama, H., Tohyama, M., and Miki, N. 1985, *Brain Res., 331*, 209.

[52] Yamagata, K., Goto, K., Kuo, C.-H., Kondo, H., and Miki, N. 1990, *Neuron, 2*, 469.

[53] Lyubarsky, A.L., Chen, C.-K., Simon, M.I., and Pugh, E.N. Jr. 2000, *J. Neurosci., 20*, 2209.

[54] Hisatomi, O., Matsuda, S., Satoh, T., Kotaka, S., Imanishi, Y., and Tokunaga, F. 1998, *FEBS Lett., 424*, 159.

[55] Weiss, E.R., Raman, D., Shirakawa, S., Ducceschi, M.H., Bertram, P.T., Wong, F., Kraft, T.W., and Osawa, S. 1998, *Mol. Vision, 4*, 27.

[56] Weiss E.R., Ducceschi, M.H., Horner, T.J., Li, A., Craft, C.M., and Osawa, S. 2001, *J. Neurosci., 21*, 9175.

[57] Cideciyan, A.V., Zhao, X., Nielsen, L., Khani, S.C., Jacobson, S.G., and Palczewski, K. 1998, *Proc. Natl. Acad. Sci. USA, 95*, 328.

[58] Zhao, X., Huang, J., Khani, S.C., and Palczewski, K. 1998, *J. Biol. Chem., 273*, 5124.

[59] Anant, J.S., and Fung, B.K.-K. 1992, *Biochem. Biophys. Res. Comm., 183*, 468.
[60] Zhao, X., Yokoyama, K., Whitten, M.E., Huang, J., Gelb, M.H., and Palczewski, K. 1999, *FEBS Lett., 454*, 115.
[61] Wiechmann, A.F., and Hammarback, J.A. 1993, *Exp. Eye Res., 57*, 763.
[62] Dalil-Thiney, N., Bastianelli, E., Pochet, R., Reperant, J., and Versaux-Botteri, C. 1998, *Neurosci. Lett., 247*, 163.
[63] Gunhan, E., van der List, D., and Chalupa, L.M. 2003, *J. Neurosci., 23*, 1383.
[64] Peachey, N.S., and Ball, S.L. 2003, *Doc. Ophthalmol., 107*, 13.
[65] Nusinowitz, S., Ridder, W.H. III, and Heckenlively, J.R. 2002, *Systematic Evaluation of the Mouse Eye: Anatomy, Pathology, and Biomethods*. R.S. Smith, S.W.M. John, P.M. Nishina, J.P. Sundberg (Eds.), CRC Press LLC, Boca Raton, 320.
[66] Carter-Dawson, L.D., and LaVail, M.M. 1979, *J. Comp. Neurol., 188*, 245.
[67] Lyubarsky, A.L., Lem, J., Chen, J., Falsini, B., Iannaccone, A., and Pugh, E.N. Jr. 2002, *Vision Res., 42*, 401.
[68] Robson, J.G., Maeda, H., Saszik, S.M., and Frishman, L.J. 2004, *Vision Res., 44*, 3253.
[69] Xu, L., Ball, S.L., Alexander, K.R., and Peachey, N.S. 2003, *Vis. Neurosci., 20*, 297.
[70] Bush, R.A., and Sieving, P.A. 1994, *Invest. Ophthalmol. Vis. Sci., 35*, 635.
[71] Sieving, P.A., Murayama, K., and Naarendorp, F. 1994, *Vis. Neurosci., 11*, 519.
[72] Naarendorp, F., and Williams, G.E. 1999, *Vis. Neurosci., 16*, 91.
[73] Sieving, P.A., Frishman, L.J., and Steinberg, R.H. 1986. *J. Neurophysiol., 56*, 1049.
[74] Saszik, S.M., Robson, J.G., and Frishman, L.J. 2002, *J. Physiol., 543*, 899.
[75] Field, G.D., Sampath, A.P., and Rieke, F. 2005, *Annu. Rev. Physiol., 67*, 491.
[76] Haverkamp, S., and Wassle, H. 2000, *J. Comp. Neurol., 424*, 1.
[77] Massey, S.C. and Mills, S.L. 1996, *J. Comp. Neurol., 366*, 15.
[78] Calvert, P.D., Krasnoperova, N.V., Lyubarsky, A.L., Isayama, T., Nicolo, M., Kosaras, B., Wong, G., Gannon, K.S., Margolskee, R.F., Sidman, R.L., Pugh, E.N. Jr., Makino, C.L., and Lem, J. 2000, *Proc. Natl. Acad. Sci. USA, 97*, 13913.
[79] Goto, K., Yamagata, K., Miki, N., and Kondo, H. 1990, *Cell Tiss. Res., 262*, 501.
[80] Korf, H.-W., White, B.H., Schaad, N.C., and Klein, D.C. 1992, *Brain Res., 595*, 57.
[81] Blackshaw, S., and Snyder, S.H. 1997, *J. Neurosci., 17*, 8074.
[82] Bertazolli-Filho, R., Ghosh, S., Huang, W., Wollmann, G., and Coca-Prados, M. 2001, *Biochem. Biophys. Res. Comm., 284*, 317.
[83] Foster, R.G., and Hankins, M.W. 2002, *Progr. Ret. Eye Res., 21*, 507.
[84] Vigh, B., Manzano, M.J., Zadori, A., Frank, C.L., Lukats, A., Rohlich, P., Szel, A., and David, C. 2002, *Histol. Histopathol., 17*, 555.
[85] Palczewski, K., Carruth, M.E., Adamus, G., McDowell, J.H., and Hargrave, P.A. 1990, *Vision Res., 30*, 1129.
[86] Lorenz, W., Inglese, J., Palczewski, K., Onorato, J.J., Caron, M.G., and Lefkowitz, R.J. 1991, *Proc. Natl. Acad. Sci. USA, 88*, 8715.
[87] Zhao, X., Haeseleer, F., Fariss, R.N., Huang, J., Baehr, W., Milam, A.H., and Palczewski, K. 1997, *Vis. Neurosci., 14*, 225.
[88] Somers, R.L., and Klein, D.C. 1984, *Science, 226*, 182.
[89] Bastianelli, E., Polans, A.S., Hidaka, H., and Pochet, R. 1995, *J. Comp. Neurol., 354*, 395.

[90] Polans, A.S., Witkowska, D., Haley, T.L., Amundson, D., Baizer, L., and Adamus, G. 1995, *Proc. Natl. Acad. Sci. USA, 92*, 9176.

[91] Maeda, A., Ohguro, H., Maeda, T., Wada, I., Sato, N., Kuroki, Y., and Nakagawa, T. 2000, *Cancer Res., 60*, 1914.

[92] Euler, T., and Wassle, H. 1995, *J. Comp. Neurol., 361*, 461.

[93] Chan, T.L., Martin, P.R., Clunas, N., and Grunert, U. 2001, *J. Comp. Neurol., 437*, 219.

[94] Cuenca, N., Deng, P., Linberg, K.A., Lewis, G.P., Fisher, S.K., and Kolb, H. 2002, *J. Neurocytol., 31*, 649.

[95] Humphries, M.M., Rancourt, D., Farrar, G.J., Kenna, P., Hazel, M., Bush, R.A., Sieving, P.A., Sheils, D.M., McNally, N., Creighton, P., Erven, A., Boros, A., Gulya, K., Capecchi, M.R., and Humphries, P. 1997, *Nature Genet., 15*, 216.

[96] Lyubarsky, A.L., Daniele, L.L., and Pugh, E.N. Jr. 2004, *Vision Res., 44*, 3235.

In: Neuronal Calcium Sensor Proteins
Editors: Pavel Philippov and Karl Koch, pp. 181-199

ISBN 1-59454-978-8

Chapter X

THE ROLE OF RECOVERIN IN AUTOIMMUNITY

*Grazyna Adamus**
Neurological Sciences Institute, Oregon Health & Science University,
Beaverton, Oregon, USA.

Abstract

Visual paraneoplastic syndrome is a group of degenerating diseases of the retina that are associated with extraocular cancer and anti-retinal autoantibodies (Anti-Rc AAbs). Recoverin, a calcium-binding protein typically found in retina, has been the most intensively studied paraneoplastic antigen in cancer-associated retinopathy (CAR). It has been postulated that autoimmunity plays a major role in the pathogenicity of CAR. Recoverin has potent immunogenic and pathogenic properties, and it induces strong antibody and cytotoxic T cell responses. Recoverin is also expressed in tumors in patients with and without symptoms of CAR, and in both cases, recoverin is believed to trigger anti-Rc immune responses, resulting in effective tumor protection but not always in autoimmunity. Anti-Rc AAbs are cytotoxic to retinal cells and affect only those cells that express the recoverin autoantigen. The entry of AAbs into the retina and into the photoreceptor cells triggers antibody-mediated retinal cell death through the activation of apoptotic transduction pathways. In this process, AAbs induce an increase in the level of intracellular calcium, which activates the mitochondria-dependent and caspase 9-dependent activation of caspase 3. Activation of the apoptotic pathway ultimately leads to DNA fragmentation and cell death. In this review, we examine the role of recoverin and anti-Rc AAbs in the etiology of CAR and propose a molecular mechanism model of autoimmune retinopathy.

Keywords: recoverin, CAR sybdrome, paraneoplastic, anti-recoverin antinodies, autoimmune retinopathy, apoptosis, photoreceptor, cancer, CTL

* Correspondence concerning this article should be addressed to Grazyna Adamus, Neurological Sciences Institute, Oregon Health & Science University, Beaverton, Oregon, USA. Tel: 503-418-2540; Fax 503-418-2541; email: adamusg@ohsu.edu.

Abbreviations

Anti-Rc	anti-recoverin
Abs	antibodies, AAbs - autoantibodies
MAb	monoclonal antibody
CAR	cancer-associated retinopathy
CTLs	cytotoxic T cells
SCCL	small-cell carcinoma of the lung
NSCLC	non-small-cell lung carcinoma.

1. Introduction

Retinal degeneration is one of the leading causes of visual loss in the world, but the pathogenesis of retinal diseases is not fully understood. There is evidence that a subset of patients with retinopathy may have an autoimmune component to the disease. Some patients with retinal dysfunction and retinal degeneration contain circulating AAbs against retinal proteins. This finding has led to the hypothesis that anti-retinal AAbs contribute to the development of retinal disease. Basic immunological studies support the pathogenic role of these AAbs, which may function through the activation of a cell apoptotic response after prolonged exposure of the retina to such AAbs. The AAbs against recoverin was one of the first AAbs reported in association with visual loss [1-3].

In this review, we describe pathogenic mechanisms that underlie anti-Rc antibody-associated retinal dysfunction and propose a model for cancer-associated retinopathy (CAR).

2. Recoverin and Cancer-Associated Retinopathy

Anti-Rc AAbs were first found in cancer patients with retinal degeneration [1-4]. This type of retinal degeneration, associated with AAbs and extraocular tumors, was named cancer-associated retinopathy syndrome (CAR syndrome) [5,6]. CAR is a heterogeneous group of paraneoplastic autoimmune diseases associated with several antigens expressed by rod, cone, and ganglion cells. Two major autoantigens that have been most intensively studied are recoverin and α-enolase. Recoverin is a calcium-binding protein that is present mostly in retinal photoreceptor cells and also in bipolar cells. The calcium-bound form of recoverin plays a role in regulating rhodopsin phosphorylation in photoreceptor cells. The biochemical and biophysical properties of recoverin are discussed in other chapters of this book.

The presence of AAbs in human blood is usually an indication of an autoimmune disease. AAbs against recoverin (the so-called "CAR antigens") were first reported in association with small-cell carcinoma of the lung [7]. Clinically, CAR is characterized by sudden and progressive visual loss, suggesting retinal dysfunction with ring scotoma, photopsia, attenuated retinal arterioles, and abnormalities of the a- and b-waves from electroretinograms (ERG) [8,9]. Eye examination and histopathology revealed only minimal or no evidence of eye inflammation. Pathology of CAR is not widely available, but when available, it confirms the loss of retinal cells, particularly the loss of cells in the photoreceptor cell layer [3,8] (Figure 1). Severe degeneration of photoreceptors has been associated with melanophagic activities [3].

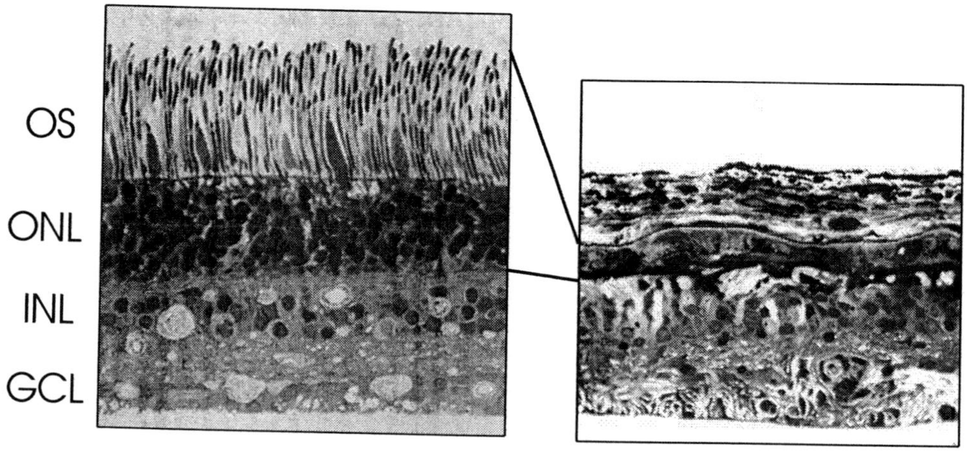

Figure 1. Pathology of CAR. Photomicrographs of normal retina (left) and retina from a representative CAR patient showing an absence of photoreceptor cells (right). The patient had anti-recoverin antibodies and small-cell carcinoma of the lung. OS – outer segments of photoreceptor cells, ONL – outer nuclear layer, INL – inner nuclear layer, GCL – ganglion cell layer. Magnification x5000.

A correlation has been found between antibody titers and visual symptoms [10,11]. An increase in antibody titers was observed before vision deterioration or the decrease in antibody titers after treatment of clinical symptoms. In contrast, T cell specific responses to recoverin in CAR patients were detected, but they were not correlated with the progression of visual loss [10]. In some cases, serum anti-Rc AAbs were detected several months before a diagnosis of cancer, suggesting that anti-Rc AAbs might have a diagnostic value in the early detection of cancer. The question whether anti-Rc Abs can be used as a marker for CAR has emerged early in the study of CAR and is still unresolved. From our studies of AAbs occurring in retinopathy patients with cancer, only about 25% of the patients have serum anti-Rc AAbs [12]. Importantly, anti-Rc AAbs have not been detected in the sera of healthy subjects. Our laboratory has analyzed over 1500 samples of sera from patients experiencing visual loss, and of these patients, anti-Rc AAbs were detected only in the sera from patients with cancer-related retinopathies. The Thirkill laboratory has also analyzed about 3000 serum samples from patients with retinopathy, with similar results [13]. Moreover, until just recently, there have not been reports indicating a correlation between the antibody against recoverin and other forms of retinopathy, including retinitis pigmentosa, diabetic retinopathy, and age-related macula degeneration. The occurrence of anti-Rc AAbs has been recently reported in a patient with retinopathy but without underlying neoplasm at the time of testing [14] and in patients with idiopathic retinopathies with a history of abnormal ERG [15]. Clinically, these patients present visual symptoms associated with paraneoplastic retinopathy.

The presence of anti-Rc AAbs in patients with sudden and progressive loss of vision but who do not have cancer at the time of testing may be explained by the presence of a very small tumor, too small to be detected, but which might be detected in years to come. The presence of anti-Rc AAbs in these patients might indicate a tumor in an early, otherwise non-detectable, stage of development. For example, in one of our patients with CAR, there was a 15-month gap between the detection of anti-Rc AAbs and the discovery a secondary malignant tumor (Adamus, unpublished study). Therefore, it is essential to perform follow-up clinical diagnostic tests for the detection of cancer, but the length of time for such follow-up tests has not been established.

It is important to note that not all patients with CAR have anti-Rc AAbs. AAbs to other retinal antigens have been detected by Western blot analysis, including α-enolase [16], heat shock cognate (hsc) protein 70 [17], 40-kDa protein [18], 45-kDa protein [19], PTB-like protein [20], neurofilament proteins [21], collapsin response-mediated protein 5 (CRMP-5-IgG, 62kDa [22]), and other retinal proteins [7,23,24]. AAbs against the ganglion cells of the retina, which share antigens with small-cell carcinoma cells, were found in the cerebrospinal fluid of CAR patients, and immune deposits were found in their retinas [25]. In those patients, tissue damage with the specific loss of retinal ganglion cells and their processes were found only in the retina and optic nerve. In addition, AAbs against a number of retinal proteins have been found in the sera of patients with various retinopathies, as well as in individuals with no known retinal disease [12]. Thus, it is hypothesized that circulating anti-retinal AAbs contribute to retinal pathology. The role of these AAbs in pathogenicity needs to be further evaluated. Since paraneoplastic anti-retinal AAbs also bind to tumor cells [26], this finding suggests that some paraneoplastic syndromes, presenting without identifiable tumors, may result in the elimination of tumor by immune-mediated responses.

3. Immunological Properties of Recoverin

Occurrences of eye diseases that are of an autoimmune nature, as well as experimental models for these diseases, have been attributed to the sequestration of ocular antigens from the immune system – a sequestration that prevents the development of tolerance against these antigens [27]. However, transcripts of certain ocular antigens, including recoverin, are constitutively expressed in the thymus, the site of central tolerance induction [28]. Resistance to an organ-specific autoimmune disease, such as retinopathy, may be regulated at least in part by the capacity of the immune system to establish central tolerance to the relevant autoantigen.

Recoverin is a potent antigen and pathogen. When injected into susceptible animals, recoverin causes antibody and T cell responses, and it induces intraocular inflammation [29,30]. The recoverin sequence within amino acid residues 64-70 (KAYAQHV) has been found to be a major pathogenic epitope, causing complete photoreceptor degeneration when Lewis rats were immunized with the corresponding peptide [31]. An additional recoverin T cell and a uveitogenic epitope have been found within the recoverin sequence 136-167 [32].

In patients with CAR, anti-Rc AAbs that were found to be restricted to the IgG1 subclass [33] as naturally occurring normal antibodies belong largely to the IgM class. Two major epitopes of human AAbs have been identified within the sequence residues 64-70 (KAYAQHV), in proximity to the calcium-binding domain EF-hand 2 and residues 48-52 (QFQSI) [33]. Autoantibody binding was found to depend on recoverin calcium-binding properties, which induces conformational changes. Such changes induced by bound calcium were found to enhance the binding of Abs to recoverin.

4. Expression of Recoverin in Tumors

Recoverin is typically expressed in the retina but can also be found in tumors during the development of a neoplasm. This unusual expression of recoverin in tumors could be related to the location of the recoverin gene. The gene that regulates human recoverin expression is located at chromosome 17p13.1 in proximity to the tumor suppressor gene p53 [34]. A single mutation may possibly activate the recoverin gene and deactivate the p53 gene, in effect causing the aberrant expression of recoverin in tumor cells and later eliciting the production of anti-Rc immune responses. Not only recoverin but also other autoantigens implicated in neurological paraneoplastic syndromes have been found in tumor cells during tumor transformation [35,36]. AAbs made against those paraneoplastic antigens are essential to diagnosing paraneoplastic disorders and related cancers.

The expression of recoverin in tumors has led to the hypothesis that CAR results, in part, from the immunological response of the host to the tumor. Early studies to identify tumor antigens in CAR have shown that recoverin is expressed by the neuroendocrine-related tumor SCCL in patients with CAR [10,31,37]. Moreover, similar tumors and cell lines obtained from lung tumors of individuals without retinopathy did not express recoverin [31]. In those studies, the expression of recoverin was tested in human lung cancer cell lines by reverse transcription polymerase chain reaction (RT-PCR), Northern blot analysis, and Western blotting. Results showed that recoverin was expressed only in SCCL cell lines developed from patients with CAR [31,37]. The sequence of recoverin cDNA from the cells of lung tumors was identical to the human recoverin sequence obtained from the retina. These significant findings corroborate the hypothesis that atypical expression of recoverin in SCCL induces immunological responses, leading to visual loss in CAR. Moreover, a sudden loss of vision and this "cancer-retina" immunological connection through the elicitation of anti-Rc AAbs may be two of the most apparent indications of an extraocular tumor. Testing for anti-Rc AAbs has become a tool in diagnosing CAR and in medical investigations for cancer.

The localization of recoverin to lung and endometrial tumors and to primary cultured CAR-related tumor cells [10,31,37] has prompted a further investigation of recoverin in cancers not associated with visual loss. Surprisingly, many tumors, not just those associated with the clinical profile of CAR, have been shown to express recoverin, including SCCL, lung adenocarcinoma, non-small-cell lung carcinoma (NSCLC), gastric cancer, pancreatic cancer, breast cancer, uterine cervical cancer, endometrial cancer, leukemia, and thymoma [38-40]. Recoverin has been found in more than 50% of gastric cancer cases, mostly in early clinical stages I and II [41]. However, such a high rate of recoverin expression is inconsistent with the low occurrence of CAR. Moreover, patients with recoverin-positive tumors had a similar rate of survival, as did patients with recoverin-negative tumors, and the expression of recoverin in tumors did not always correlate with the production of anti-Rc AAbs [38]. In those cases, lack of retinopathy could be explained by the low production of AAbs. As a result, anti-Rc AAbs may act only on easily accessible tumor cells, not on the retina behind the blood-retinal barrier (BRB). Consequently, the early presence of anti-tumor recoverin immune responses may indicate a favorable prognosis for primary cancer in CAR patients. Together, the anti-Rc antibody response and the development of CAR are not just the result of an aberrant expression of recoverin by cancer cells, but may depend on other factors, such as genetic control of immune responses.

The role of recoverin in retinal cells other than photoreceptors cells, such as a subpopulation of cone bipolar and ganglion cells that express recoverin, is not apparent [42,43]. In addition, an abundant expression of recoverin in tumors suggests that recoverin might also have a different but significant role in the physiology of the cell than in phototransduction (see another chapter in this book). From recent studies, there is evidence that tumor-expressed recoverin can be a growth regulator of the cell and an indicator of tumor progression [44]. Recoverin from cell lysates of 293 or SSTW cells immunoprecipitated with G-protein-coupled receptor kinases (GRKs) and caveolin-1, suggesting a possible functional connection. Additionally, immunocytochemistry has revealed that immunolabeling of recoverin overlaps with the immunolabeling of GRKs and caveolin-1. Based on these recent findings, recoverin may function in a calcium-dependent manner in tumors via GRKs – a function that would be homologous to the role of rhodopsin kinase in photoreceptor cells. The fact that cells transfected with the recoverin gene proliferate much more slowly than do cells without the recoverin gene imply a regulatory role for recoverin in tumors [38].

In the retina, the rate of recoverin transcription may occur maximally during the period of greatest exposure to light, presumably when recoverin is most needed by photoreceptors to perform their role in phototransduction. Then recoverin decreases at night when high levels of expression are not required [45]. The rate of recoverin transcription in tumor cells is unknown. However, recent studies have suggested that the expression of recoverin might be manipulated within the cell. Human retinoblastoma cell line Y79 possesses an mRNA transcript for recoverin, but the recoverin protein is not always expressed. The addition of differentiating agents, such as 2 mM butyrate or 4 mM dbcAMP, to the cell culture was found to cause an increased expression of the recoverin protein associated with morphological differentiation to a neuronal phenotype [46]. Two SCCL cell lines, NCI-H69 and NCI-H82, became recoverin-positive after treatment with butyrate [47]. In both cases, the recoverin protein in butyrate-treated cells was confirmed by Western blotting using Abs against recoverin and by analyzing the expression of recoverin mRNA. The induction of the recoverin expression in cancer cell lines in vitro suggests the possibility of investigating mechanisms underlying the aberrant expression of recoverin as well as other paraneoplastic antigens in tumor cells.

Not only have anti-Rc AAbs been detected in cancer patients' sera, but the recoverin protein has also been found in their sera. Recently, in a search for useful tumor markers, recoverin was measured in the sera of cancer patients at different stages of gliomas [48]. Levels of recoverin in the sera were 10-fold higher (>30 pmol/l) in patients with recurrent glioma compared with sera from healthy control subjects (<4 pmol/l). Only a 3- to 5-fold increase was observed in the sera from patients with low-grade glioma, anaplastic glioma, and gliomas with no evidence of recurrence. The physiological significance of the presence of recoverin in the serum is unknown. However, these findings imply that tumor released-recoverin may have applications for tumor diagnosis, surveillance, and treatment response.

5. Anti-Rc AAbs in Cancer Patients without CAR Symptoms

The relationship among extraocular cancer, anti-Rc AAbs, and retinal dysfunction is important in understanding the pathogenesis of CAR. The fact that recoverin was found in tumors of patients with CAR implies that this aberrant expression may trigger anti-Rc autoimmune responses, which, in effect, result in retinopathy. However, the expression of recoverin in tumors of subjects without CAR and the presence of anti-Rc AAbs in cancer patients without loss of vision complicate the use of CAR as a sole biomarker for CAR. Nevertheless, the finding of anti-Rc AAbs in patients with sudden and unexplained loss of vision could indicate underlying cancer.

The "tumor-antibody" connection of recoverin has been investigated by searching for anti-Rc AAbs in cancer patients with no visual symptoms of CAR. The most common cancer that is associated with CAR and other neurological paraneoplastic syndromes is SCCL. In the study of neurological paraneoplastic syndromes, SCCL patients were examined for serum anti-neuronal AAbs, and the occurrence of these AAbs was compared to the serum Abs in other patients with other types of lung cancer [49]; for example, SCCL patients were examined for anti-Hu AAbs. A significant proportion of SCCL patients (20%) had paraneoplastic AAbs, but only in low titers. These patients had better prognoses and had no neurological disease [50].

In visual paraneoplastic syndrome, it is not unusual to find AAbs at low frequencies. In fact, some patients with SCCL and NSCLC were found to have low titers of anti-Rc AAbs [47,51-53]. Surprisingly, recoverin was expressed in SCCL tumors, as frequently as in NSCLC tumors, but a correlation between recoverin-positive tumors and serum anti-Rc AAbs was not always found. In one study, low titers of AAbs (below 1:100) were detected in 10% of patients with SCCL without CAR [51]. The examination of 279 patients without CAR visual symptoms for serum AAbs and recoverin-tumor expression showed AAbs in 15% of the SCCL patients and 20% of the NSCL patients [40]. However, an examination of SCCL and NSCLC tumor sections from these patients revealed much higher frequencies of recoverin in tumors compared to frequencies of circulating serum AAbs: 68% of the SCCL and 85% of the NSCLC were recoverin-positive. Only 2 (of 86) patients with non-malignant pulmonary disorders and no healthy control subjects showed anti-Rc AAbs. In these studies, a gradation in the frequency of CAR, the presence of AAbs, and recoverin-positive lung tumors were apparent as follows: in SCCL, $<1\%$, 15%, and 68%; and in NSCLC, $<1\%$, 20%, and 85%, respectively [40]. In spite of these findings, the reported percentage of sero-positive patients has been found to differ, depending on the patient population. As an example, in studies of malignant versus non-malignant samples, 72 of NSCLC patients and 29 of SCLC patients were compared with 27 patients with sarcoidosis (non-malignant lung disease) and 32 healthy controls. Only 7% of the samples from the SCCL patients and 1% of the samples from the NSCLC patients were recoverin-positive, and none in non-malignant sera [53]. Although the percentage of recoverin-positive samples was different in both those studies, it was remarkable that normal sera did not contain anti-Rc AAbs. Such a quantitatively lower response, although adequate for the development of tumor immunity, may not be sufficient to cause autoimmune disease.

6. Anti-Tumor Cytotoxic T Cells Specific to Recoverin

Aberrantly expressed recoverin in tumor cells can become a target not only of AAbs but also of cytotoxic T cells (CTLs). Many putative tumor antigens are intracellular proteins, and CTLs respond to antigenic peptides presented in the context of MHC class I, which are most often derived from the intracellular proteins [54]. CTLs can directly kill tumors; therefore, patients with tumor specific CTLs have a favorable prognosis. Recoverin-specific CTL responses were found in some CAR patients of HLA-A24 (MHC class I) haplotypes and in healthy control subjects [55,56]. The frequency of CTL precursors (CTLp) was assessed in 20 HLA-A24^{+} cancer patients (10 colon, 6 stomach, 4 breast cancers) and 7 healthy control subjects by the application of an HLA-A24-recoverin peptide tetramer [55]. Seven cancer patients and 2 healthy individuals showed anti-Rc cytotoxic responses in an HLA-A24-restricted manner after they were stimulated in vitro with the recoverin peptide. Higher frequencies of CTLp were found in cancer patients with and without CAR than in healthy control subjects.

In a study of recoverin-derived HLA class I-A24-binding peptides to generate anti-tumor-recoverin CTLs, 3 recoverin epitopes - R49 (QFQSIYAKF), R49.2 (QFQSIYAKFF), and R64 (AYAQHVFRSF) - were found to induce peptide-specific CTLs. In part, these CTL epitope sequences (QFQS and AYAQHV) correspond to the major binding sites for anti-Rc AAbs in humans [33,57], implying the importance of these sequences in the generation of autoimmune responses, both humoral and cellular. Recoverin-specific CTLs recognized recoverin-positive tumor cells from patients with CAR, suggesting that the activation of recoverin-specific, anti-tumor CTL responses might lead to a better prognosis for patients with CAR. Therefore, recoverin might be the target of epitope-based immunotherapy in treating HLA-A24^{+} cancer patients with a recoverin-expressing tumor.

The recoverin peptide containing the major immunogenic site R64 (AYAQHVFRSF) induced anti-tumor effects in BALB/c mice transplanted with recoverin-expressing fibrosarcoma [58]. In this mouse model of CAR, the administration of the recoverin peptide simultaneously produced retinal dysfunction and tumor cell regression. Anti-tumor CTLs and anti-Rc AAbs were produced, and higher antibody titers correlated with more severe ERG reduction, indicating an effect of Abs on retinal function. Although these are remarkable findings, it is important to point out that a cancer vaccine for CAR patients may be possible only if a specific epitope can be isolated to induce an anti-tumor effect without triggering an anti-retinal response. So far, this goal has not been achieved.

7. CAR AAbs Entering the Retina

Potentially autoimmune antibodies may need to be in sufficiently high titers to enter the eye and cause retinopathy. Anti-Rc AAbs were found not only in the blood but also in the aqueous humor [59], which strongly suggests that anti-Rc AAbs penetrate the aqueous humor and vitreous cavity beyond the BRB.

The entry of serum AAbs into the retina has been investigated following intravenous administration of purified, biotinylated anti-Rc IgG (MAb Rec-1) or IgM (MAb Rec-2) in the rat in our laboratory. Intravenously injected rat IgG showed significant penetration through a compromised BRB, and diffusion of these Abs in the retinal tissue through retinal vessels was evident (Figure 2).

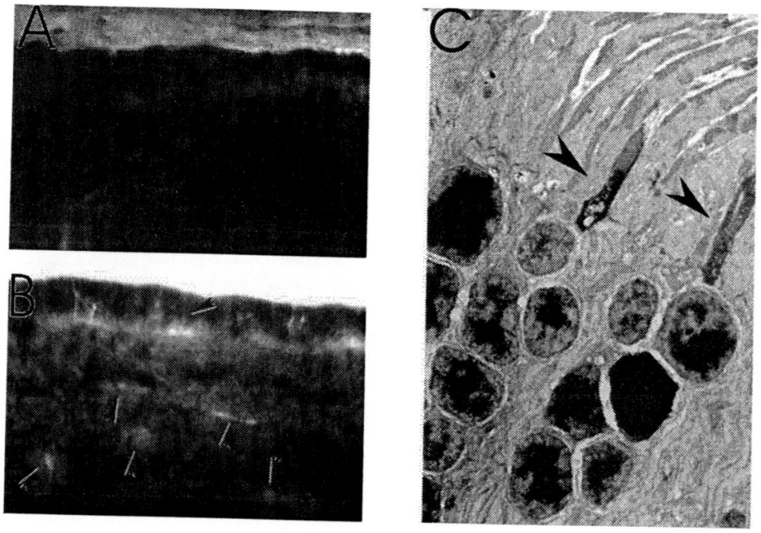

Figure 2. Effect of anti-recoverin antibodies on the rat retina. (A) and (B) Micrographs representing cryosections from the Lewis rat retina intravenously injected with biotinylated antirecoverin antibodies after immunostaining with Texas red-streptavidin. (A) Control rat received saline instead of anti-recoverin antibodies. (B) Diffusion of anti-recoverin antibodies through retinal vessels seven days after injection of biotinylated rat anti-recoverin MAb Rec-1. In addition, rats received 1000 U IL-2 intraperitoneally at the time of antibody injection. Arrows indicate anti-recoverin antibodies leaking from retinal vessels. (C) Electron micrograph of the photoreceptor cells from the Lewis rat injected intravenously with a CAR patient's antibodies against recoverin. The rat received pertussis toxin intraperitoneally at the time of antibody injection. Note pyknotic nuclei and dying outer segments (arrows). The eye was collected two weeks after the initial immunization with CAR patient's autoantibodies of IgG class.

Anti-retinal AAbs persisted in the blood for a few days after intravenous injection, and considerable levels of IgG were in the retina even 7 days later (Figure 3). With the aid of cytokines, which are often detected during cancerous disease, significantly more AAbs can enter the retina. It has been shown that some cytokines – such as VEGF, TNFα, IL-2, IL-6, and IL-1β – cause a breakdown of the BRB by opening tight junctions between retinal and brain endothelium cells [60-62]. In Lewis rats, administration of AAbs in the presence of IL-2 increased significantly the access of AAbs to the retina, supporting the ability of IL-2 to increase the permeability of the BRB to IgG. The IgM class of anti-Rc AAbs was not effective in penetrating the retina, possibly due to the molecular size of IgM (Figure 3). Also, pertussis toxin, which has been shown to facilitate the entry of immune cells into the eye by increasing permeability across the BRB [63], improved the entry of anti-Rc AAbs into the retina (Figure 3). The intravenous administration of purified CAR anti-Rc AAbs in the rat induced a degeneration of rods, which was represented by a large number of pyknotic nuclei in the outer nuclear layer and a few degenerating outer segments. Dying photoreceptors were detected among normal-looking neighboring cells (Figure 2C). These observations lead to the conclusion that during the CAR disease process, anti-Rc AAbs reach the retina from the serum.

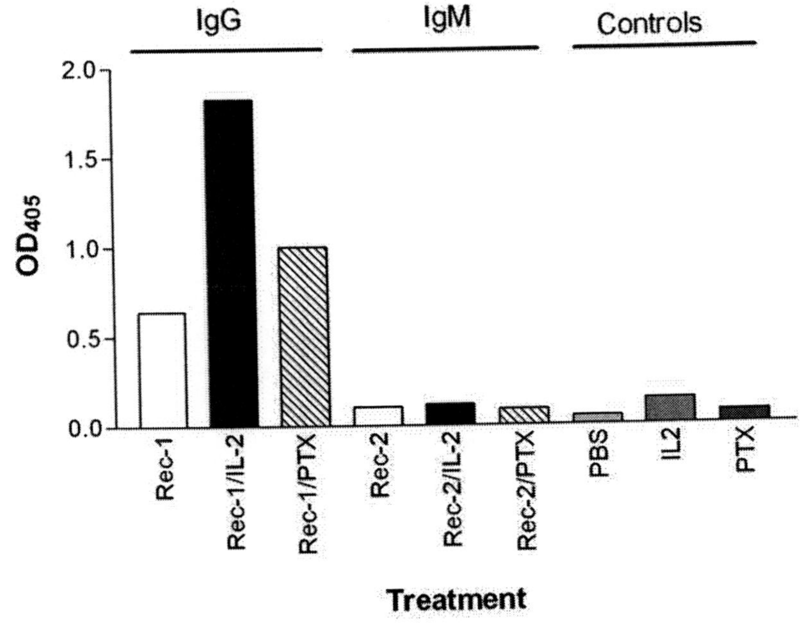

Figure 3. Presence of anti-recoverin antibodies in the retinas of Lewis rats injected with antibodies and measured by ELISA using Streptavidin-peroxidase conjugate. The rats received intravenously two doses of biotinylated MAbs against recoverin of different immunoglobulin classes. PTX or IL-2 was used to increase permeability of BRB at the time of antibody injection. To assess the penetration of serum antibodies into the eyes, control rats obtained saline instead of MAb and an equal amount of PTX, IL-2, or PBS. A 10 μg sample of the total retinal proteins obtained from saline-perfused eyes was used for ELISA. Note that only anti-recoverin IgG was found in the retinas of injected rats.

8. ERG and Anti-Retinal AAbs

It is generally accepted that CAR, in association with anti-Rc AAbs, mostly affects photoreceptors and that in CAR, rod and cone ERG a- and b-waves are markedly reduced in amplitude, suggesting both rod and cone dysfunction. When a diagnosis of CAR is made late in the course of the disease, paraneoplastic retinopathies are often associated with a profoundly abnormal or extinguished ERG [3,64,65].

The limited studies on the physiological role of AAbs, which focused on measuring ERG changes, support the hypothesis that anti-retinal AAbs contribute to retinal cell dysfunction [66,67]. For example, anti-Rc Abs, which target photoreceptor cells when injected into the vitreous of a rat eye, were found to cause a marked suppression of the ERG [68]. Interestingly, the ERG amplitude was reduced to a large extent in the presence of anti hsc70 Abs, which, along with anti-Rc Abs, are occasionally found in some CAR patients [66].

9. Apoptotic Cell Death

Anti-Rc AAbs are cytotoxic to retinal cells. They seem to affect only those cells that express the autoantigen protein [69,70]. This antibody-mediated destruction of retinal cells was independent of complement. Normal Abs did not influence cell physiology, and cells lacking the target antigen were not affected by the entry of AAbs. Not only anti-Rc IgG but also its Fab fragment can penetrate retinal cells by an active process of endocytosis [71]. Exposing E1A.NR3 rat retinal cells to anti-Rc Abs in vitro caused a significant transient increase in intracellular Ca^{2+}, while non-specific Abs did not show such an effect [72]. Moreover, the increase in intracellular Ca^{2+} was found to activate the mitochondrial apoptotic pathway [73]. The increase in intracellular Ca^{2+} was blocked with nifedipine, thus inhibiting apoptotic processes and cell death.

Elevated extracellular K^+ protects retinal cells from anti-Rc antibody-mediated cell death [70]. The protective effects of K^+ were shown to be time- and dose-dependent events, suggesting that the mobilization of internal Ca^{2+} stores rescues the cells by interfering with apoptotic signal transduction pathways.

The apoptotic cell death that is induced by anti-Rc Abs has been shown in in vitro studies using cultured retinal cells [69-71 and in vivo studies using a Lewis rat CAR model [74]. Living retinal cells grown in the presence of anti-Rc Abs displayed morphological changes that are characteristic of apoptotic death, such as shrinking cell bodies, blebs, retraction of processes and detachment from a culture dish, followed by chromatin condensation and internucleosomal DNA degradation [69], and these changes occurred within 18 to 24 hours after treatment with anti-Rc Abs. In studies of the molecular mechanism of apoptotic death, anti-Rc Abs induced a decrease in anti-apoptotic bcl-x_L, while pro-apoptotic bcl-x_S and bax increased during a 1-hour period [71]. In addition, increases in cytosolic cytochrome c and apaf-1 were apparent, and their presence was consistent with the initiation of formation of the caspase 9/apaf-1 complex followed by the sequential activation of caspase 3 and the cleavage of the caspase 3 substrate, PARP (poly (ADP-ribose) polymerase). Similar to other cells, caspase 3 plays an effector role in retinal cells, which is essential for nuclear changes, including chromatin condensation. Inhibiting caspase-3 by the irreversible caspase-3 inhibitor z-DEVD-fmk protected cells from death, thus providing evidence that caspase-3 activation promotes the death of retinal cells [69]. This sequence of events observed in the retinal cells in vitro strongly supports the hypothesis that anti-Rc AAbs induce apoptosis.

A similar order of intracellular events was detected in vivo when anti-retinal Abs were injected into the vitreous of rat eyes. After intravitreal injection, anti-Rc Abs penetrated toward the photoreceptor cell layer within 12 to 24 hours and their presence could be measured in the photoreceptor cell layer for 48 to 72 hours [66,74]. Histopathology revealed a loss of 25%-30% of cells in the outer nuclear layer and in the inner nuclear layer in the affected retina, in comparison to the retinal layers in control rats. Apoptotic cells were present throughout the outer and inner nuclear layers [74]. In the antibody-treated retinas, caspase 9 and 3 were sequentially activated after the treatment of Abs. Eight hours after injecting the Abs, the enzymatic activity of caspase 9 peaked, followed by the maximum activation of caspase 3 at 20 hours (Figure 4). The activity of PARP was found to correlate with the activity of caspase 3 in the diseased retina. There was no upregulation of caspases or PARP activities in the control retinas (sham, PBS injected). These studies provide clear evidence that anti-Rc Abs are capable of penetrating in vivo photoreceptor cells and of inducing apoptotic cell death. Thus, a similar mechanism of anti-Rc antibody penetration may occur in patients with CAR, which may lead to visual loss and blindness.

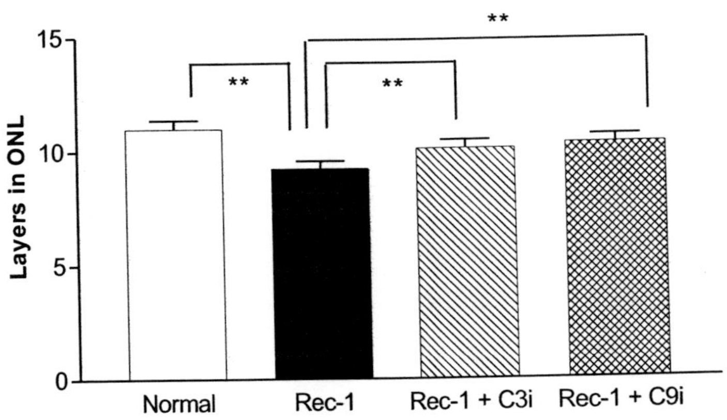

Figure 4. Effect of caspase 9 and 3 inhibitors on protection of retina from degeneration after injection with anti-recoverin antibodies. Layers of nuclei in the outer and inner nuclear layers were counted 7 days after injection. The bars represent means of rows in the outer nuclear layer, calculated from 5 animals ±SD. Asterisks indicate values of statistical significance: *, $p<0.05$, **, $p<0.001$.

10. Experimental Approaches to Anti-Recoverin Treatment

Some paraneoplastic syndromes respond to treatment of the underlying cancer or to immunosuppression; however, for most paraneoplastic syndromes, no effective treatment exists. In addition, there is no established protocol for treating CAR. A review of the medical literature has shown no spontaneous improvement of vision. Presently, anti-Rc-associated treatment for CAR involves modulating the immune system to reduce the autoimmune response. Corticosteroids, plasmapheresis, or intravenous immunoglobulin infusion may be considered for treatment of patients because anti-Rc retinopathy frequently responds to steroid or immunomodulation therapy – if the disorder is treated early in its course [11,75]. Without treatment, however, retinopathy associated with anti-Rc AAbs almost always progresses rapidly to severe vision loss and blindness.

Understanding the pathogenesis of CAR and other paraneoplastic syndromes is important for the development of new treatment strategies. For retinopathy, there are several possible targets: tumor recoverin, anti-Rc AAbs, intracellular Ca^{2+}, apoptosis/cell death and degeneration, and anti-Rc cellular responses. The most challenging treatment strategy is to control anti-Rc AAbs in subjects with malignant tumors because of the "double sword effect" of anti-Rc AAbs, in which AAbs simultaneously play a positive and negative role: positive in controlling the tumor growth but negative in inducing retinopathy. Therefore, any treatment leading to the suppression of AAbs in the presence of cancer has to be carefully considered. Immunotherapy may be possible if a specific epitope of recoverin contains the antigenic site for anti-tumor AAbs and does not cross-react with retinal antigens.

Caspase inhibitors and calcium antagonists have been used in animal models to block or to suppress the effects of anti-Rc Abs on the retina, with promising results. The involvement of caspases in antibody-induced retinal damage has been evaluated in a rat model of CAR (Adamus, unpublished studies). Intravitreal treatment with inhibitors for caspase 9 and 3 in the presence of anti-Rc Abs protected photoreceptor cells from degeneration in Lewis rats (Figure 5). In another study [76], injection of caspase inhibitors with anti-Rc Abs into the eyes of rats inhibited ERG impairment and significantly suppressed the antibody-induced enhancement of rhodopsin phosphorylation.

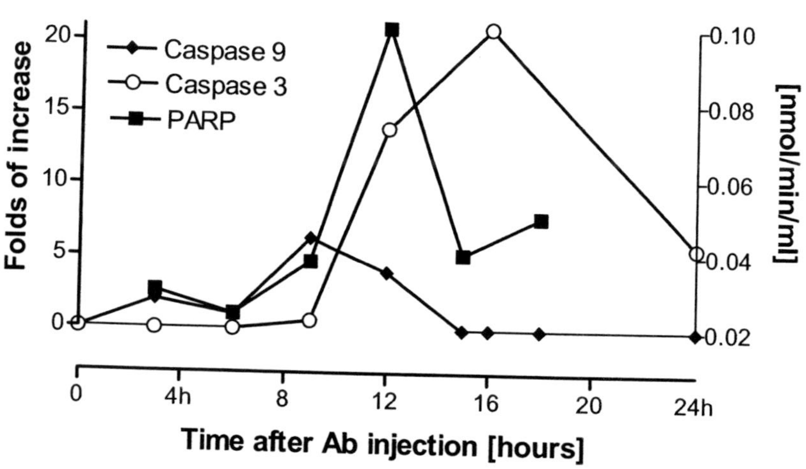

Figure 5. Sequential activation of caspase 9, caspase 3, and PARP in Lewis rat retina after an injection of anti-recoverin antibody Rec-1. PARP activity correlates with caspase 3 activity (in folds of increase). PARP activity was determined by an Apoptosis Detection Kit that determines DNA damage present in cell extracts by measuring PARP enzymatic activity (in units).

Elevated calcium may lead to retinal dysfunction and degeneration. Because intracellular Ca^{2+} is elevated in the early stages of apoptotic death, controlling Ca^{2+} could be a key candidate for treating CAR [77]. The application of Ca^{2+} blockers as a strategy for treating antibody-induced retinal dysfunction is just starting to be explored, and with positive results thus far. In a Lewis rat model, the administration of nilvadipine, a Ca^{2+} antagonist, to anti-Rc-antibody-treated rats significantly improved their worsening ERG [68]. Therefore, blockers and/or environmental changes that mobilize intracellular Ca^{2+} could be used to prevent the death of photoreceptor cells occurring in CAR.

Neurotrophic factors have protective properties for dying photoreceptors. They can be delivered directly into the eye or via a virus construct, such as an adeno-associated virus. Virus-mediated delivery of a gene encoding for a neurotrophic factor was shown to protect the retina from different forms of retinal degeneration [78,79]. In a rat model of CAR, a virally delivered transgene of a ciliary neurotrophic factor (CNTF) protected photoreceptor cells from death that was induced with anti-Rc Abs [80]. CNTF was found to produce photoreceptor-protected cells through the activation of STAT3 and the suppression of caspase 3 activities. Because CNTF is a critical survival factor for photoreceptors, CNTF receptor (CNTFRα) gene delivery was also tested as an additional treatment for retinal degeneration, using a rat model of CAR [81]. The enrichment of retinal cells in the CNTFRα protected the rat retina from apoptotic death that was induced by anti-Rc Abs. Thus, delivery of CNTF or CNTFRα may be useful to prevent retinal degeneration in human antibody-mediated retinal degeneration.

11. The Molecular Mechanism Model of Autoimmune Retinopathy

As earlier noted in this chapter, anti-Rc AAbs in the sera of patients who progressively lose vision have recently been implicated as a major cause of retinal death. A model of antibody-induced apoptosis of photoreceptor cells has been proposed, based on this new evidence [69-71,74,76,80]. The model consists of a two-phase mechanism: an anti-tumor response and an anti-retinal response.

In the anti-tumor response phase (Figure 6A), during normal tumor cell turnover, tumor cells release recoverin, which triggers anti-tumor immune responses. Dendritic cells in the periphery, take up the aberrantly expressed recoverin, which, in effect, activates cellular and humoral responses that recognize recoverin as a nonself. Recoverin-specific T cells and B cells are then produced. After maturing, B cells produce AAbs against recoverin expressed in tumors. AAbs, in addition to CTLs, slow tumor growth. Once the immune system is activated, such as in the proposed anti-tumor response phase, serum AAbs persist for many years, even after just one activation [82]. This persistent antibody production may be due to repeated stimulation of antigen-specific B-cells with antigen, or may result from the slow release of antigens stored in lymphoid tissue over a long period, or may result from an occasional contact of the immune cells with metastatic cells containing the recoverin antigen.

Anti-Rc AAbs also induce retinopathy when they are produced in high titers. In the anti-retinal response phase (Figure 6B), circulating AAbs cross the BRB and penetrate into retinal layers, where anti-Rc AAbs attack photoreceptors that express recoverin. Anti-Rc AAbs then penetrate into retinal cells by an active process of endocytosis. Once in the cell, anti-Rc AAbs block the recoverin function in phototransduction, which results in the enhancement of rhodopsin phosphorylation and an increase in intracellular Ca^{2+} levels. The increase in intracellular Ca^{2+} activates the mitochondria-dependent and caspase-9-dependent activation of caspase 3, ultimately leading to DNA fragmentation and cell death. Massive death of photoreceptor cells leads to retinal dysfunction, degeneration, and blindness.

This model of autoimmune CAR proposed for anti-Rc AAbs could be applied to other anti-retinal AAbs associated with the syndrome, the molecular roles of which have not been fully evaluated. The AAbs may be associated with a specific clinical phenotype of retinopathy and may serve as a biomarker for the diagnosis of cancer and retinal disease. The study of immune responses in patients with cancer expressing paraneoplastic antigens, as well as in animal models, will provide better understanding of ocular immunity of the syndrome.

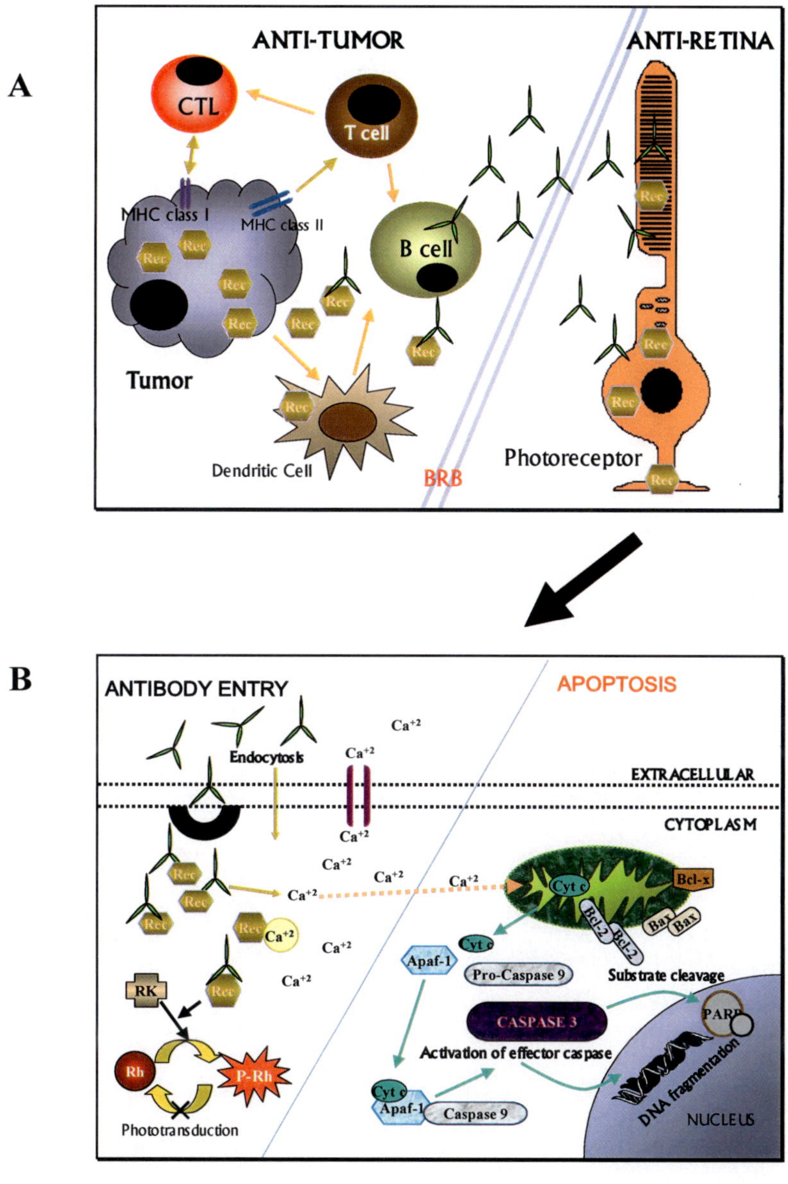

Figure 6. Diagrams representing the molecular mechanism model of cancer-associated retinopathy (see text for description).

Acknowledgements

This work was supported by the grant from the National Eye Institute EY13053.

References

[1] Thirkill, C.E., Tait, R.C., Tyler, N.K., Roth, A.M., Keltner, J.L. 1992, *Invest. Ophthalmol. Vis. Sci., 33*, 2768.
[2] Polans, A.S., Burton, M.D., Haley, T.L., Crabb, J.W., Palczewski, K. 1993, *Invest. Ophthalmol. Vis. Sci., 34*, 81.
[3] Adamus, G., Guy, J., Schmied, J.L., Arendt, A., Hargrave, P.A. 1993, *Invest. Ophthalmol. Vis. Sci., 34*, 2626.
[4] Polans, A.S., Buczylko, J., Crabb, J., Palczewski, K. 1991, *J. Cell Biol., 112*, 981.
[5] Thirkill, C.E., Roth, A.M., Keltner, J.L. 1987, *Arch. Ophthalmol., 105*, 372.
[6] Thirkill, C. 1994, *Neuro-ophthalmology, 14*, 297.
[7] Thirkill, C.E., FitzGerald, P., Sergott, R.C., Roth, A.M., Tyler, N.K., Keltner, J.L. 1989, *New England Journal of Medicine, 321*, 1589.
[8] Sawyer, R.A., Selhorst, J.B., Zimmerman, L.E., Hoyt, W.F. 1976, *Am. J. Ophthalmol., 81*, 606.
[9] Jacobson, D.M. 1996, *Curr. Opin. Ophthalmol., 7*, 30.
[10] Adamus, G., Amundson, D., MacKay, C., Gouras, P. 1998, *Arch. Ophthalmol., 116*, 251.
[11] Keltner, J.L., Thirkill, C.E., Tyler, N.K., Roth, A.M. 1992, *Arch. Ophthalmol., 110*, 48.
[12] Adamus, G., Ren, G., Weleber, R.G. 2004, *BMC Ophthalmol., 4*, 5.
[13] Keltner, J.L., Thirkill, C.E. 1998, *Am. J. Ophthalmol., 126*, 296.
[14] Whitcup, S.M., Vistica, B.P., Milam, A.H., Nussenblatt, R.B., Gery. I. 1998, *Am. J. Ophthalmol., 126*, 230.
[15] Heckenlively, J.R., Jordan, B.L., Aptsiauri, N. 1999, *Am. J. Ophthalmol., 127*, 565.
[16] Adamus, G., Aptsiauri, N., Guy, J., Heckenlively, J., Flannery, J., Hargrave, P.A. 1996, *Clin. Immunol. Immunopath., 78*, 120.
[17] Ohguro, H., Ogawa, K., Nakagawa, T. 1999, *Invest. Ophthalmol. Vis. Sci., 40*, 82.
[18] Peek, R., Dijkstra, B.G., Meek, B., Kuijpers, R.W. 2002, *Clin. Exp. Immunol., 128*, 498.
[19] Yoon, Y.H., Cho, E.H., Sohn, J., Thirkill, C.E. 1999, *Korean J. Ophthalmol., 13*, 43.
[20] Kikuchi, T., Arai, J., Shibuki, H., Kawashima, H., Yoshimura, N. 2000, *J. Neuroimmunol., 103*, 26.
[21] Kornguth, S.E., Kalinke. T., Grunwald, G.B., Schutta, H., Dahl, D. 1986, *Cancer Res., 46*, 2588.
[22] Cross, S.A., Salomao, D., Parisi, J., Kryzer, T., et al. 2003, *Ann. Neurol., 54*, 38.
[23] Grunwald, G.B., Simmonds, M.A., Klein, R., Kornguth, S.E. 1985, *Lancet., March, 23*, 658.
[24] Murphy, MA, Thirkill, C.E., Hart, W.M. 1997, *J. Neuroophthalmol., 17*, 77.

[25] Grunwald, G.B., Kornguth, S.E., Towfighi, J., Sassani, J., Simmonds, M.A., et al. 1987, *Cancer, 60*, 780.
[26] Tora, M., Graus, F., Bolos, C., Real, F.X. 1997, *Neurology, 48*, 735.
[27] Gery, I., Egwuagu, C.E. 2002, *Int. Rev. Immunol., 21*, 89.
[28] Charukamnoetkanok, P., Fukushima, A., Whitcup, S.M., Gery, I., Egwuagu, C.E. 1998, *Curr. Eye Res., 17*, 788.
[29] Adamus, G., Ortega, H., Witkowska, D., Polans, A. 1994, *Exp. Eye Res., 59*, 447.
[30] Gery, I, Chanaud, N.P. 3rd, Anglade, E. 1994, *Invest. Ophthalmol. Vis. Sci., 35*, 3342.
[31] Polans, A., Witkowska, D., Haley, T., Amundson, D., Baizer, L., Adamus, G. 1995, *Proc. Natl. Acad. Sci. USA, 92*, 9176.
[32] Ohkoshi, M., Abe, T., Satoh, N., Nakajima, A., Sakuragi, S. 2001, *Curr. Eye Res., 22*, 304.
[33] Adamus, G., Amundson, D. 1996, *J. Neurosci. Res., 45*, 863.
[34] McGinnis, J.F., Austin, B., Klisak, I., Heinzmann, C., Kojis, T., et al. 1995, *J. Neurosci. Res., 40*, 165.
[35] Gomm, S.A., Keevil, B.G., Thatcher, N., Hasleton, P.S., Swindell, R.S. 1988, *Br. J. Cancer, 58*, 797.
[36] Albert, M.L., Darnell, R.B. 2004, *Nat. Rev. Cancer, 4*, 36.
[37] Matsubara, S., Yamaji, Y., Soto M., Fujita, J, Takahara, J. 1996, *Br. J. Cancer, 74*, 1419.
[38] Maeda, A., Ohguro, H., Maeda, T., Wada, I., Sato, N., et al. 2000, *Cancer Res., 60*, 1914.
[39] Yamada, G., Ohguro, H., Aketa, K., Itoh, T., Shijubo, N., et al. 2003, *Hum. Pathol., 34*, 717.
[40] Bazhin, A.V., Savchenko, M.S., Shifrina, O.N., Demoura, S.A., Chikina, S.Y., et al. 2004, *Lung Cancer, 44*, 193.
[41] Ohguro, H., Odagiri, H., Miyagawa, Y., Ohguro, I., Sasaki, M., Nakazawa, M. 2004, *Tohoku J. Exp. Med., 202*, 213.
[42] Milam, A.H., Dacey, D.M., Dizhoor, A.M. 1992, *Vis. Neurosci., 10*, 1.
[43] Wiechmann, A.F., Hammarback, J.A., 1993, *Exp. Eye Res., 57*, 763.
[44] Miyagawa, Y., Ohguro, H., Odagiri, H, Maruyama, I., Maeda, T., et al. 2003, *Biochem. Biophys. Res. Commun., 300*, 669.
[45] Wiechmann, A.F., Sinacola, M.K. 1997, *Brain Res. Mol. Brain Res., 45*, 321.
[46] Wiechmann, A.F. 1996, *J. Neurochem., 67*, 105.
[47] Bazhin, A.V., Savchenko, M.S., Belousov, E.V., Jaques, G., Philippov, P.P. 2004, *Lung Cancer, 45*, 299.
[48] Sampath, P., Weaver, C.E., Sungarian, A., Cortez, S., Alderson, L., Stopa, E.G. 2004, *Cancer Control, 11*, 174.
[49] Dalmau, J., Furneaux, H.M., Cordon-Cardo, C., Posner, J.B. 1992, *Am. J. Pathol., 141*, 881.
[50] Roberts, W.K., Darnell, R.B. 2004, *Curr. Opin. Immunol., 16*, 616.
[51] Bazhin, A.V., Shifrina, O.N., Savchenko, M.S., Tikhomirova, N.K., Goncharskaia, M.A., et al. 2001, *Lung Cancer, 34*, 99.

[52] Savchenko, M., Bazhin, A., Shifrina, O., Demoura, S., Kogan, E., et al. 2003, *Lung Cancer, 41*, 363.

[53] Jankowska, R., Witkowska, D., Porebska, I., Kuropatwa, M., Kurowska, E., Gorczyca, W.A. 2004, *Pathobiology, 71*, 323.

[54] Knutson, K.L., Disis, M.L. 2005, *Cancer Immunol Immunother., 54*, 721.

[55] Rong, S., Ikeda, H., Sato, Y., Hirohashi, Y., Takamura, Y., et al. 2002, *Cancer Immunol. Immunother., 51*, 282.

[56] Maeda, A., Ohguro, H., Nabeta, Y., Hirohashi, Y., Sahara, H., et al. 2001, *Eur. J. Immunol., 31*, 563.

[57] Polans, A.S., Adamus G. 1995, *Behav. Brain. Sci., 18*, 483.

[58] Maeda, A., Maeda, T., Ohguro, H., Palczewski, K., Sato, N. 2002, *Eur. J. Immunol., 32*, 2300.

[59] Ohguro, H., Maruyama, I., Nakazawa, M., Oohira, A. 2002, *Am. J. Ophthalmol., 134*, 605.

[60] Luna, J.D., Chan, C.-C., Derevjanik, N.L., Mahlow, J., Chiu, C., et al. 1997, *J. Neurosci. Res., 49*, 268.

[61] Saija, A., Princi, P., Lanza, M., Scalese, M., Aramnejad, E., De Sarro, A. 1995, *Life Sci., 56*, 775.

[62] Ellison, M.D., Krieg, R.J., Povlishock, J.T. 1990, *J. Neuroimmunol., 28*, 249.

[63] Agarwal, R.K., Sun, S.H., Su, S.B., Chan, C.C., Caspi, R.R. 2002, *J. Neuroimmunol., 129*, 133.

[64] Klingele, T.G., Burde, R.M., Rappazzo, J.A., Isserman, M.J., Burgess, D., Kantor, O. 1984, *J. Clin. Neuroophthalmol., 4*, 239.

[65] Jacobson, D.M., Thirkill, C.E., Tipping, S.J. 1990, *Ann. Neurol., 28*, 162.

[66] Ohguro, H., Ogawa, K., Maeda, T., Maeda, A., Maruyama, I. 1999, *Invest. Ophthalmol. Vis. Sci., 40*, 3160.

[67] Lei, B., Bush, R.A., Milam, A.H., Sieving, P.A. 2000, *Invest. Ophthalmol. Vis. Sci., 41*, 262.

[68] Ohguro, H., Ogawa, K., Maeda, T., Maruyama, I., Maeda, A., et al. 2001, *Invest. Ophthalmol. Vis. Sci., 42*, 2589.

[69] Adamus, G., Machnicki, M., Seigel, G.M. 1997, *Invest. Ophthalmol. Vis. Sci., 38*, 283.

[70] Chen, W., Elias, R.V., Cao, W., Lerious, V., McGinnis, J.F. 1999, *J. Neurosci. Res., 57*, 706.

[71] Shiraga, S., Adamus, G. 2002, *J. Neuroimmunol., 132*, 72.

[72] Adamus, G., Webb, S., Shiraga, S., Duvoisin, R. 2003, *FASEB J., 17*, C186.

[73] Adamus, G., Webb, S., Shiraga, S., Duvoisin, R.M. 2005, In submission.

[74] Adamus, G., Machnicki, M., Elerding, H., Sugden, B., Blocker, Y.S., Fox, D.A. 1998, *J. Autoimmun., 11*, 523.

[75] Guy, J., Aptsiauri, N. 1999, *Arch. Ophthalmol., 117*, 471.

[76] Maeda, T., Maeda, A., Maruyama, I., Ogawa, K.I., Kuroki, Y., et al. 2001, *Invest. Ophthalmol. Vis. Sci., 42*, 705.

[77] Wenzel, A., Grimm, C., Samardzija, M., Reme, C.E. 2005, *Prog. Retin. Eye Res., 24*, 275.

[78] LaVail, M.M., Unoki, K., Yasumura, D., Matthes, M.T., Yancopoulos, G.D., Steinberg, R.H. 1992, *Proc. Natl. Acad. Sci. USA, 89*, 11249.

[79] Liang, F.Q., Aleman, T.S., Dejneka, N.S., Dudus, L., Fisher, K.J., et al. 2001, *Mol. Ther., 4*, 461.

[80] Adamus, G., Sugden, B., Shiraga, S., Timmers, A.M., Hauswirth, W.W. 2003, *J. Autoimmun., 21*, 121.

[81] Adamus, G., Sugden, B., Timmers, A.M., Hauswirth, W.W. 2005, in submission.

[82] Zinkernagel, R.M., Ehl, S., Aichele, P., Oehen, S., Kundig, T., Hengartner, H. 1997, *Immunol. Rev., 156*, 199.

Part IV: GCAPs

In: Neuronal Calcium Sensor Proteins
Editors: Pavel Philippov and Karl Koch, pp. 203-219
ISBN 1-59454-978-8

Chapter XI

CALCIUM SENSITIVITY OF PHOTORECEPTOR GUANYLYL CYCLASE (RETGC) AND CONGENITAL PHOTORECEPTOR DEGENERATION: MODELING *IN VITRO* AND *IN VIVO*.

Alexander M. Dizhoor**, *Elena V. Olshevskaya and Igor V. Peshenko
Hafter Research Laboratories, Pennsylvania College of Optometry,
Elkins Park, PA 19027, USA.

Abstract

Regulation of retinal guanylyl cyclase (retGC) plays critical role in shaping photoresponses of vertebrate rods and cones. The activity of retGC is sharply reactive to the free Ca^{2+} variations that occur in photoreceptors between light and dark. Ca^{2+}-sensor proteins, GCAPs, can activate retGC when they are in Mg^{2+}-bound form (in the light), but they inhibit retGC in the dark, when Ca^{2+} concentrations rise and it replaces Mg^{2+} in their EF-hand domains. There are several mutations in GCAPs and retGC that have been linked to severe forms of congenital blindness and result in anomalous regulation of the cyclase *in vitro* and *in vivo*. The abnormal Ca^{2+}-sensitivity of retGC regulation in these cases can be attributed to a change in the Ca^{2+}-sensitivity of GCAP per se (such as produced by Tyr99Cys substitution in GCAP1), as well as to the altered GCAP/retGC interaction (such as produced by mutations in dimerization domain of retGC). The biochemical and biological effects of mutations in GCAPs and retGC related to human congenital retinal dystrophy– have been successfully studied both *in vitro*, by using

* Correspondence concerning this article should be addressed to Alexander M. Dizhoor, Pennsylvania College of Optometry, 8360 Old York Road, Elkins Park, PA 19027, USA. Tel.: +1 (215) 780-1468; e-mail: adizhoor@pco.edu.

recombinant isoforms of GCAP and retGC, and *in vivo*, by utilizing a transgenic mouse model for the Tyr99Cys substitution in GCAP1.

Keywords: photoreceptor, guanylyl cyclase, phototransduction, retinal degeneration

Abbreviations

CNG	cyclic nucleotide-gated channel
GCAP	guanylyl cyclase activating protein
ECD	extracellular domain
KHD	kinase homology domain
PDE	cyclic nucleotide phosphodiesterase
R*	metarhodopsin II
retGC	retinal photoreceptor guanylyl cyclase.

1. Guanylyl Cyclase (retGC) and Guanylyl Cyclase Activating Proteins (GCAPs) in Photoreceptor Function

Cyclic GMP serves as the intracellular messenger in phototransduction in vertebrate rods and cones through its ability to regulate the inward current of Na^+ and Ca^{2+} via cGMP-gated channels, CNG, in the plasma membrane of photoreceptor cells. Two key processes control the level of free cGMP in photoreceptor outer segments – hydrolysis by light-activated phosphodiesterase (PDE6) and synthesis by membrane guanylyl cyclase (retGC) (see ref. [1-5] for review). Even in the dark, most of cGMP in photoreceptor remains tightly bound, primarily to non-catalytic domains of PDE6 [6-8], and only a few percent of total cGMP (~2-4 μM) remains free in the dark at equilibrium with the channels to regulate their activity. At this free cGMP level, only a few percent of the CNG channels remain open, but this small fraction of the channels is sufficient to change the membrane potential of the photoreceptor in response to light thus providing normal physiological responses of rods and cones. The changes in the rate of the hydrolysis and synthesis of cGMP in the light versus dark result in (1) closure of the channels via G-protein-mediated hydrolysis of cGMP by PDE6, thus causing hyperpolarization of the cellular membrane, or (2) opening of the CNG channels and partial depolarization of the photoreceptor during recovery and light adaptation, by accelerating cGMP synthesis by retGC (Figure 1).

Photoexcited rhodopsin triggers the hydrolysis of cGMP by activating first a G protein, transducin, which then releases PDE6 from self-inhibition and thus causes cGMP-gated channels, CNG, to close (see ref. [1-4] for review). Rods and cones rapidly recover to their resting potential after excitation induced by a non-saturating flash of light, a property essential for their ability to repeatedly respond to light stimuli. Another process, light adaptation, partially restores light sensitivity of the cells exposed to a constant background illumination. This process is especially effective in cones and defines their function as a photoreceptor for the daylight conditions. Compared to a dark-adapted retina, light-adapted photoreceptors decrease the amplitude and accelerate the kinetics of their electrical responses to a test flash of standard strength, and the degree of such change in their behavior depends on the intensity of the background light. Both recovery and light adaptation require reopening of the CNG channels in order to restore photoreceptor responses to light. Reopening of the CNG channels occurs as a result of multiple reactions [9] that include (1) suppression of cGMP hydrolysis by inactivating bleached rhodopsin and turning off the active form of transducin and (2) fast resynthesis of cGMP by retGC to compensate for the drop in free cGMP concentration caused by light. Processes of quenching light-dependent PDE6 cascade are described in detail in many comprehensive articles (for example, refs. [1-5]), and in this chapter, we will only give a brief overview of the processes related to cGMP synthesis by retGC.

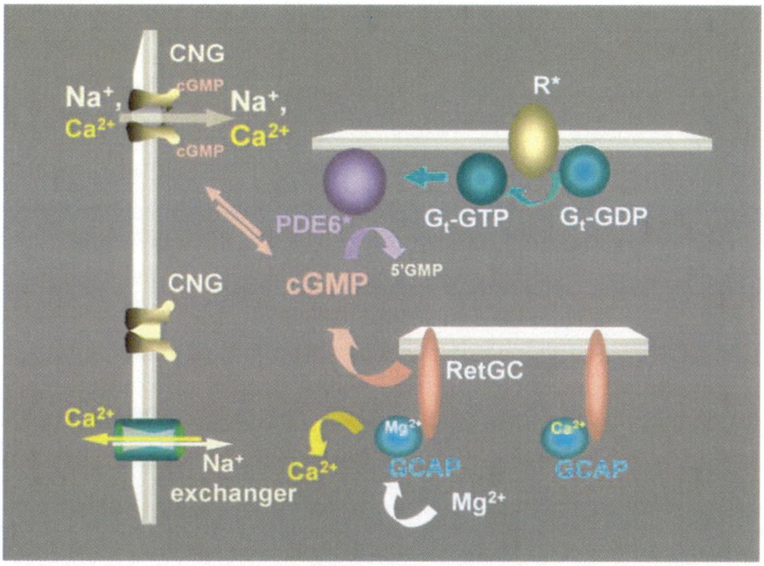

Figure 1. Phototransduction and recovery in photoreceptors. Free cGMP level in photoreceptors maintains a small fraction of cGMP-gated channels (CNG), in the open state thus sustaining inward current of Na^+ and Ca^{2+}. Light activated rhodopsin R*, catalyzes conversion of G-protein, transducin (Gt), into GTP-bound form, which, in its turn, stimulates activity of PDE6 and hydrolysis of cGMP to 5'GMP, concequently, the CNG channels close. In the absence of Ca^{2+} influx, its intracellular concentration is rapidly reduced by Na^+/Ca^{2+}, K^+ exchanger, thus causing a Ca^{2+}-sensor protein, GCAP, to convert from its Ca^{2+}-bound state into a Mg^{2+}-bound state and begin to stimulate the activity of a membrane guanylyl cyclase (retGC) to replenish cGMP in the cytoplasm of the outer segment and accelerate the recovery.

The activity of retGC changes inversly with free Ca^{2+} in the outer segments of photoreceptors in the light versus dark [10-13]. In the dark, Ca^{2+} extruded from the photoreceptor outer segment by a Na^+/K^+, Ca^{2+}-exchanger re-enters the outer segment through the open CNG channels, and the average free Ca^{2+} levels in outer segment reaches ~300-450 nM in lower vertebrates and ~220-250 nM in mammals [14-16]. Closure of cGMP gated channels caused by light-activated cGMP hydrolysis blocks the influx of Na^+ and Ca^{2+}

from the extracellular medium, while the exchanger continues to extrude Ca^{2+} ions from the photoreceptor outer segment, thus leading the free Ca^{2+} to drop to as low as ten-fold after illumination [14-17]. The fall in the intracellular Ca^{2+} concentrations can contribute to several processes, such as stimulation of cGMP binding to the CNG channels and inactivation of photoexcited rhodopsin [1-5], but most importantly, it accelerates cGMP synthesis by retGC [10-13].

Two isozymes of retGC, named retGC1 and retGC2, have been identified in human photoreceptors [18]. Their homologs in other species are also known as GC-E and GCE-F or ROSGC1 and ROSGC2, respectively [19,20]. Anti-retGC1 antibodies produce strong immunofluorescence in cones, but much weaker immunofluorescence in rods [21,22], while retGC2 is strongly immunoreactive only in rods [22]. Consistently with that, disruption of retGC1 (GC-E) gene results in complete loss of cone responses and rapid degeneration of cones in GC-E null mice, but has little effect on viability and function of rods [23].

As typical membrane guanylyl cyclases, both retGC1 and -2 are single transmembrane domain-containing proteins and they have three homology regions, also found in receptor guanylyl cyclases [24,25]. The first homology region is the extracellular domain (ECD) that in receptor cyclases is responsible for binding extracellular ligands, such as cyclic peptides. This region shows rather little resemblance between retGC and other members of this family. There is much more similarity between the intracellular segment of retGC and other membrane guanylyl cyclases [18,24]. This portion of the enzyme includes a catalytic domain, proximal to its C-terminus, and a kinase homology domain (KHD), located upstream of the catalytic domain and downstream of a single transmembrane domain of retGC (reviewed in refs. [26,27]). The actual consensus sequence implicated in protein kinase activity has not been found in KHD and, although some protein kinase activity was found in preparations of retGC [28], it remains unclear whether or not KHD in retGC can catalyze the transfer of the phosphate residue from ATP. Yet, retGC is apparently able to bind ATP, and this binding both stabilizes the structure of the enzyme and enhances its stimulation by GCAPs [29-31].

There is also a short stretch of amino acid residues located between the KHD and the catalytic domain that forms a putative "dimerization domain", essential for the ability of the cyclase to form a homodimer [24]. Although there is no direct structural evidence that it actually creates a direct contact between two retGC molecules, the peptide fragments corresponding to the dimerization domain of retGC1 has been shown to form a coiled-coil dimer in solution [32]. As described further, several mutations in this short domain have been linked to inherited forms of severe retinal pathology [32].

Unlike other members of the membrane guanylyl cyclases family who become active only when extracellular peptide ligand is bound to their ECD, both isozymes of retGC are regulated by Ca^{2+}-binding proteins, GCAPs (guanylyl cyclase activator proteins - see [22,33] for review), that interact with the cytoplasmic portion of the cyclase [29].

Two homologous 24-kDa proteins, GCAP1 and GCAP2, that impart Ca^{2+}-sensitivity to retGC, were isolated from a bovine retina [21,34], and their homologs are present in all vertebrate species [35].

GCAPs are recoverin-like proteins, a separate group within the EF-hand superfamily of calcium-binding proteins that include proteins remotely homologous to calmodulin. However, unlike calmodulin, all these proteins are N-fatty acylated (see refs. [26,35] for review). Recoverin-like proteins (frequently also referred to as neuronal calcium sensor group, NCS) have four EF-hand related structures, usually named EF-1 through EF-4, but only three, or sometimes just two of them form true Ca^{2+}-binding domains. The N-terminal EF hand-related motif, EF-1, deviates from the consensus amino acid sequence and lacks several key side chain residues required for coordinating Ca^{2+} ions. Instead of binding Ca^{2+}, in GCAPs this domain is required for interaction with the target enzyme, retGC [36,37]. Three-dimensional structure of the Ca^{2+}-bound GCAP2 established by Ames at al. [38] (Figure 2) is strikingly different from calmodulin and closely resembles that of recoverin and neurocalcin, thus demonstrating the distinctiveness of the molecular design for recoverin family. GCAP molecule has three cation-binding EF-hands and consists of two globular halves connected by a flexible region between the EF hands 2 and 3, sensitive to Ca^{2+}-dependent conformational changes [38-40].

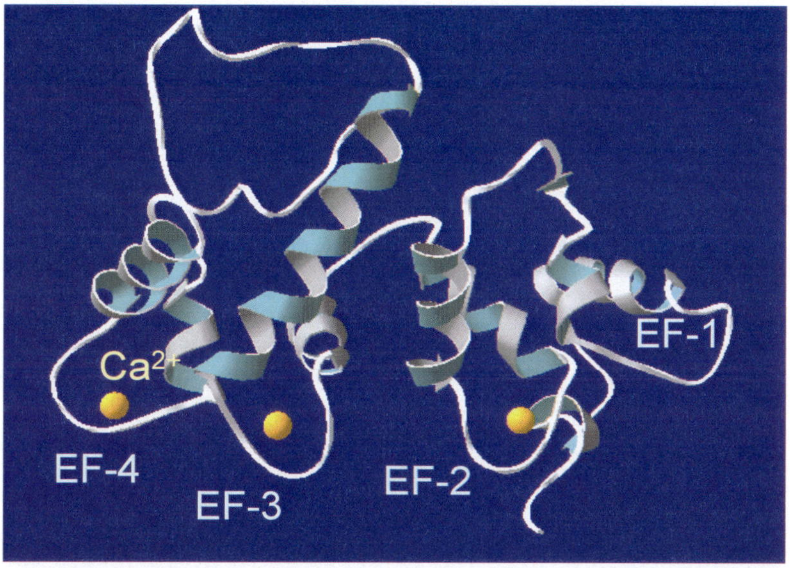

Figure 2. Three-dimensional structure of GCAP2 [38]. GCAPs consist of four EF-hand helix-loop-helix domains (EF-1 through EF-4). The first EF-hand deviates from the Ca^{2+}-binding consensus motif and does not bind Ca^{2+}, but is required for interaction with its target, retGC. Three other EF-hands are required for Ca^{2+} and, at least some of them, for Mg^{2+} binding.

Although both GCAPs can bind three Ca^{2+} ions, these are not just Ca^{2+}-binding, but rather Ca^{2+}/Mg^{2+}-binding proteins [41]. The importance of Mg^{2+} binding to EF hands of GCAP is two-fold. First, at physiological concentrations of free Mg^{2+} (~0.9 mM [42]), GCAP undergoes conformational change that can be monitored *in vitro* by a sharp decrease in its tryptophan fluorescence. Importantly, this is also the form in which GCAPs are *activator*s of retGC. Substitution of Mg^{2+} for Ca^{2+} makes GCAPs undergo an "activator-to-inhibitor" transition, and either fully Ca^{2+} loaded or mixed Mg^{2+}/Ca^{2+} form of GCAP (this is still remains a subject for more detailed analysis) becomes an inhibitor of retGC at the Ca^{2+} concentrations, similar to those found in dark-adapted rods and cones. Second, competitive binding of Mg^{2+} in GCAP EF hands, through its competition with Ca^{2+} under physiological concentrations, sets the Ca^{2}-sensitivity of retGC regulation precisely in the range within which free Ca^{2+} levels change between light and dark. If GCAPs were unable to bind Mg^{2+} at physiological concentrations, they would be effectively converted into inhibitor form at free Ca^{2+} concentrations below the light-adapted level of free Ca^{2+} in mammalian rods [14] (Figure 3).

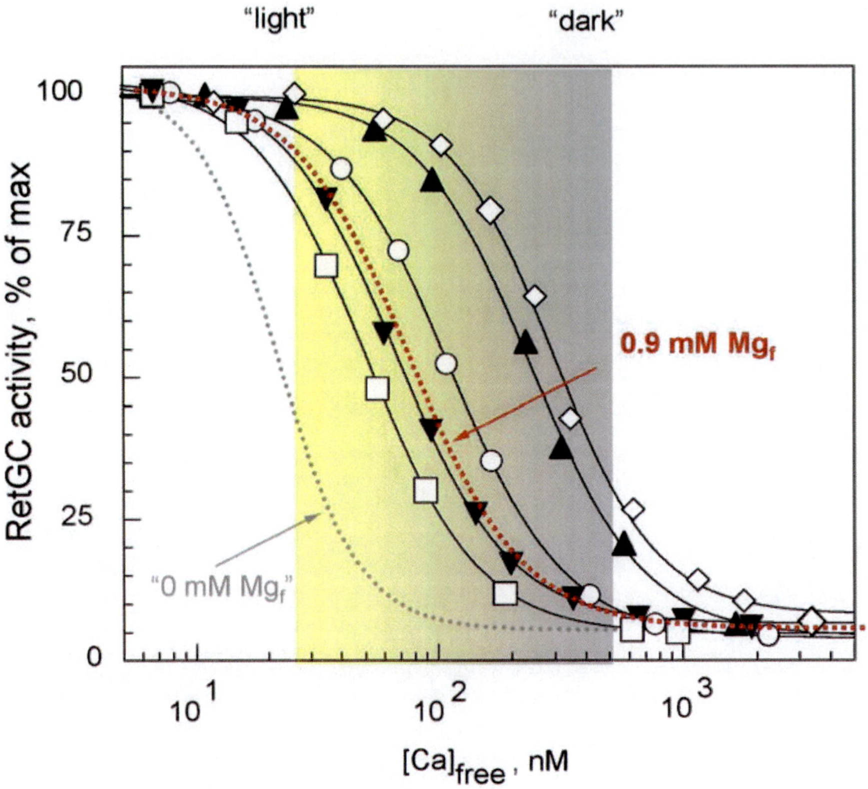

Figure 3. GCAPs are Mg^{2+}/Ca^{2+} binding proteins [41], therefore, the competition between Mg^{2+} and Ca^{2+} shifts Ca^{2+}-sensitivity of retGC inhibition depending on free Mg^{2+} in the assay. In the absence of Mg^{2+}, Ca^{2+}-sensitivity of retGC inhibition would become below the lowest free Ca^{2+} levels found in photoreceptors in the light (gray dashed line). At the free Mg^{2+} concentrations near its actual intracellular value (red dashed line), Ca^{2+}-sensitivity of GCAP•retGC lies within the physiological free Ca^{2+} interval in photoreceptor, changing between the light- and the dark-adapted conditions [14-16].

The exact mechanism of cyclase regulation by GCAPs is still unclear, although some evidence point at GCAP-mediated dimerization of the cyclase, in which GCAP dimer may serve as an adaptor for dimerization of the two subunits of retGC1, and perhaps retGC2, to tighten the complex [43-45], which is necessary for the catalytic activity of the enzyme [46] (Figure 4). The identity of the site(s) for the interaction between retGC and GCAPs is ambiguous due to the variance between the results from different laboratories, but there are several regions in the intracellular portion of retGC1 that are able to participate in GCAP/cyclase interaction *in vitro* and some of them can even discriminate between different isoforms of GCAPs [47-51].

Inactivation of a tail-to-tail gene array for GCAP1 and GCAP2 in a mouse model delays the recovery and creates abnormally high light sensitivity in their rods [52,53]. The rate, at which rods recover in these mice, can be restored by transgenic expression of individual GCAPs, either GCAP1 or GCAP2, which is consistent with the ability of both GCAPs to stimulate retGC *in vitro* in a Ca^{2+}-sensitive manner. However, the response kinetics in mice overexpressing either GCAP1 or GCAP2 in the double GCAP knockout background depend on mosaicism of the promoter activity in the transgenic constructs, producing cell-to-cell variations of the expression levels, as well as potential alterations in Ca^{2+} buffering introduced by overexpression of the GCAP [53,54]. Therefore, the exact physiological role of GCAP1 and GCAP2 in shaping light responses in rods and cones remains to be determined by using individual GCAP1 and GCAP2 gene knockout models. *In vitro*, GCAP1 and GCAP2 make retGC operate in slightly different ranges of free Ca^{2+} concentrations [41,55], and their expression levels are seemingly different in rods versus cones [22,56-58]. This may result in GCAP1 and GCAP2 being able to regulate different phases of recovery after dim or strong flashes of light, or contribute differently to rod versus cone response kinetics.

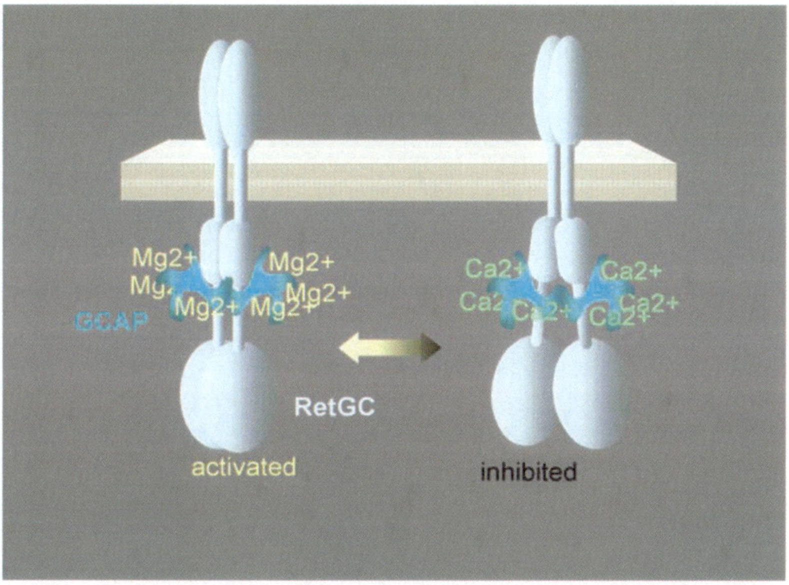

Figure 4. GCAP promotes tighter dimerization of retGC, but Ca^{2+} binding to GCAPs affects stability of the dimer [43].

2. Factors that Determine Ca^{2+}-Sensitivity of retGC

Given that retGC itself is not sensitive to Ca^{2+} within the physiologically relevant range of concentrations, its sensitivity to Ca^{2+} should dependent entirely on the presence of GCAPs. Therefore, the main factor that determines Ca^{2+}-sensitivity of the cyclase (defined as $EC50_{Ca}$) must be the affinity of GCAPs for Ca^{2+} or, more precisely, their relative affinities for Mg^{2+} and Ca^{2+}. However, two additional factors can potentially contribute to the overall Ca^{2+}-sensitivity of the retGC/GCAP complex (Figure 5). First, binding to the cyclase may affect conformation of GCAPs so that their affinity for Mg^{2+} and Ca^{2+} significantly changes within the complex GCAP•retGC (i.e., Kd_{2Ca} and/or Kd_{2Mg} substantially change from Kd_{1Ca} and Kd_{1Mg} in Figure 5). Second, if the complex, retGC•GCAP, is at equilibrium with GCAP in solution or on the membrane, then the Ca^{2+}-bound (inhibitor) and the Mg^{2+}-bound (activator) forms of GCAP can compete against each other at any given free Ca^{2+} concentration. In such a case, the affinity of retGC for the activator (Kd_A) versus the inhibitor (Kd_I) form of GCAP can make Ca^{2+}-sensitivity of the enzyme be different from the Ca^{2+}-sensitivity of GCAP itself.

As described further, naturally occurring mutations in GCAP and retGC may affect the Ca^{2+}-sensitivity of retGC by changing both the Ca^{2+}-binding properties of GCAP and the affinity of the cylcase for the activator versus the inhibitor form of GCAP.

In addition, GCAP1 and GCAP2 display different Ca^{2+}-sensitivity *in vitro* [41,55]. If the properties of both recombinant GCAPs adequately reflect their behavior *in vivo*, then relative concentrations of GCAP1 versus GCAP2, as well as retGC1 versus retGC2, may contribute differently to the overall Ca^{2+} -sensitivity of cGMP production in rods and cones. However, it would require comparing their individual gene knockout models in order to address this possibility.

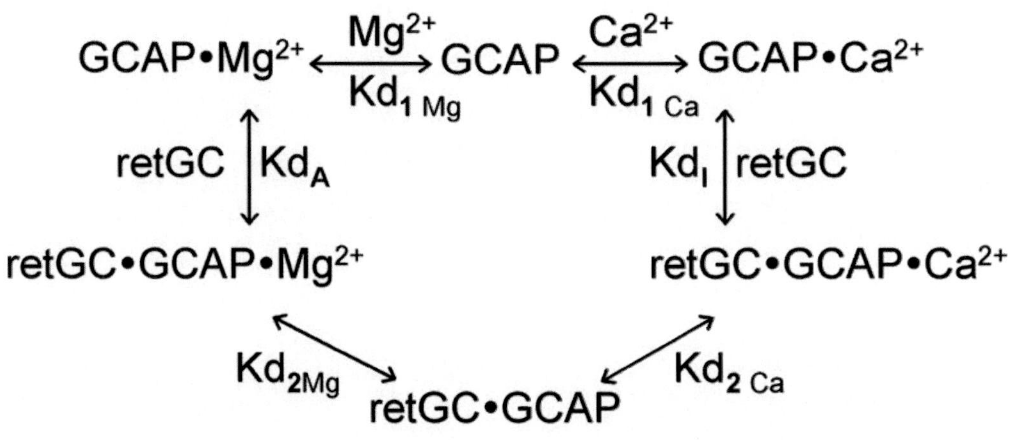

Figure 5. Factors that can affect Ca^{2+}-sensitivity of retGC regulation by GCAPs. See text for the detail.

3. Congenital Dominant Retinal Dystrophy Linked to Mutations in GCAP1 and GCAP2

Like many other components of phototransduction, both retGC and GCAPs have been linked to retinal degenerative diseases caused by mutations in human genes coding for these proteins.

One group of mutations has been linked to retinal degenerations causing primarily death of cones, sometimes followed by degeneration of rods. These include congenital dominant retinal dystrophies that develop between early adolescence and adulthood and result in progressive loss of color perception followed by loss of daylight vision, and, in some cases, complete loss of both rod and cone functions [59-62]. These mutations include substitutions in the third and the fourth EF hand domains of GCAP1. Among the first identified mutations of this type were substitutions, Tyr99Cys (Y99C) in EF hand 3 and Glu155Gly (E155G) in EF hand 4 [59,60]. Biochemical analysis of recombinant mutant GCAP1 *in vitro* demonstrates a prominent shift in Ca^{2+}-sensitivity of retGC toward much higher than normal free Ca^{2+} concentrations [60,63,64].

The biochemical properties of retGC regulation by the Y99C GCAP1 [63] lead us to a hypothesis that free cGMP and free Ca^{2+} concentrations in photoreceptors should both be affected by the mutation and rise above their normal levels in the dark [22,63] (Figure 6).

Evidently, retGC in normal photoreceptors should be able to produce just enough cGMP to open a small fraction of cGMP channels, because the influx of Ca^{2+} into the outer segment through these channels shuts down further production of cGMP as soon as free Ca^{2+} levels exceed ~250 nM. However, GCAP1 with lower Ca^{2+}-sensitivity, caused by the Y99C or similar mutations, would prevent shifting the cyclase to its inhibited state at normal concentrations of free Ca^{2+}, hence, even in the dark, cGMP production driven by the mutant GCAP will continue until higher than normal number of the channels is open to allow Ca^{2+} accumulate in the cell at higher than normal levels until they become able to shut down further production of cGMP.

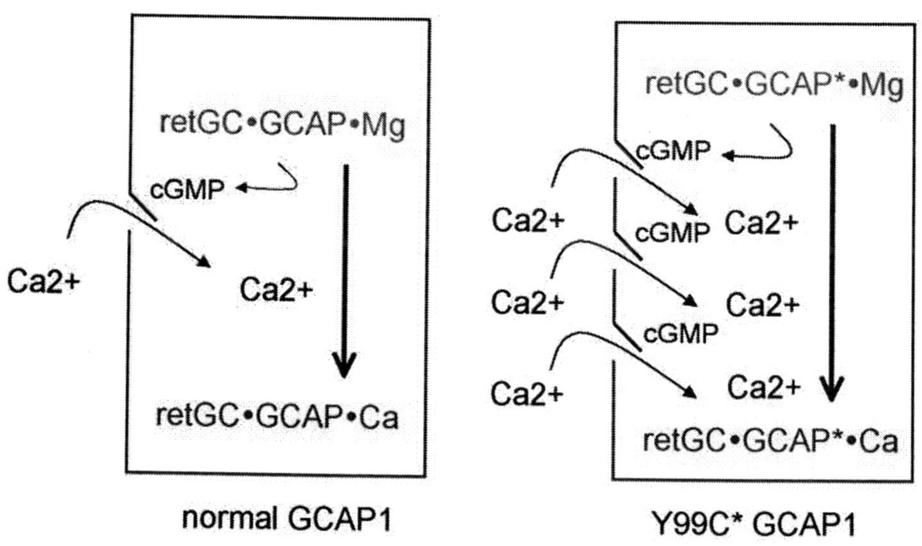

Figure 6. *Left.* Activation of retGC by normal GCAP1 stops when free Ca^{2+} concentration rises to its normal dark level, sufficient to substitute bound Mg^{2+} for Ca^{2+}, and shuts down further production of cGMP. Hence, only a small fraction of CNG channels, sufficient for making GCAP into an inhibitor of the cyclase, will be open in the dark to maintain the level of Ca^{2+}. *Right.* The retGC stimulating activity of Y99C GCAP1 is not suppressed at normal dark Ca^{2+} concentrations and continues to stimulate cGMP production, thus resulting in additional opening of the CNG channels and increase the Ca^{2+} influx into the outer segment, until free Ca^{2+} concentrations rise to the new high level, sufficient to convert the Y99C GCAP into its inhibitor form.

This hypothesis was recently verified using an *in vivo* model in which GCAP1, with or without the Y99C mutation, was expressed in photoreceptors of transgenic mice [65]. The mouse retinas expressing Y99C GCAP1 demonstrate the change in Ca^{2+}-sensitivity of their photoreceptor retGC activity, such that the EC50 for the Ca^{2+}-dependent inhibition of retGC becomes increased and the Hill coefficient decreased. Consequently, retGC dependence on Ca^{2+} in Y99C GCAP1 mouse retinas looks "shallow" and shifted above the physiological range of the free Ca^{2+} concentrations. The direct measurement of the free Ca^{2+} in mouse rods expressing Y99C GCAP1 using a fluorescent Ca^{2+} indicator dye demonstrates almost two-fold increase in resting free Ca^{2+} concentrations, well above the normal range [65]. Moreover, the rising phase of their single-photon responses becomes slower and reaches higher amplitude, consistently with the altered regulation of cGMP synthesis [65]. The Y99C GCAP1 expression in mouse rods *in vivo* does not render them non-responsive to light, but affected morphology and the size of their outer segments and gradually cause massive apoptotic death of photoreceptors (Figure 7).

Shortening of the outer segments observed at early stages in the L52H mouse rods expressing Y99C GCAP1 [65,66], may account for the comparatively minor change in their light sensitivity, despite the increased amplitudes of their single-photon responses [65]. Even with the amplitude of a single-photon event being increased nearly two-fold, the almost two-fold reduction in the size of the collecting area can apparently keep the photosensitivity of the affected rods at nearly normal range. Such behavior of the affected photoreceptors may account for the relatively late onset of the vision loss, unlike rapid inactivation of photoreceptor functions in severe forms of Leber congenital amaurosis that results in complete blindness at or soon after birth. Unlike human patients affected by the Y99C mutation, massive death of rods and cones occurred in mice, expressing the mutant GCAP1 under control of rhodopsin promotor [65,66]. This argues against a supposition, that rods may have special mechanisms helping them to resist the effects of the mutant GCAP1 [64]. More likely explanation for longer survival of rods in human patients is that the level of GCAP1 expression in human cones exceeds that in rods or that the levels of expression for the mutant versus normal GCAP1 allele are different in mouse and human rods, as well as in human rods versus cones. From using a mouse model for the *in vivo* study [65], it was also difficult to conclude whether or not the cones died directly of the Y99C GCAP1 expression or as a result of massive rod degeneration, because the level of transgenic GCAP1 expression in cones is impossible to quantify due to the low abundance of cones. Whatever the case may be, the Y99C mutation in GCAP1 found in human patients is clearly capable of creating *in vivo* conditions that trigger cell death.

Other recently reported mutations affecting EF hand 4 in GCAP1 found in patients with dominant cone-rod degenerations [61,62] demonstrate *in vitro* properties similar to those of Y99CGCAP1 and, most likely, act via similar mechanisms.

Recent report from Sato et al. [67] implicates GCAP2 in rare cases of retinitis pigmentosa-like retinal degeneration. Their analysis of several Asian families with a history of polymorphic retinal degenerations found a positive genetic link between a mutation, G157R, in EF-hand 4 of GCAP2 and the occurrence of retinal degeneration that affected primarily rods and, in one case, primarily cones. This mutation results in substitution of Gly in the 6th position of the Ca^{2+}-binding loop of the EF-4 and thus may potentially cause a change in Ca^{2+}-binding properties of the EF-hand. However, unlike mutations in GCAP1, the G157R substitution has not yet been shown to affect Ca^{2+}-sensitivity of GCAP2/retGC complex. Moreover, in the same family, one individual who inherited such mutation did not develop retinal degeneration. It therefore remains to be verified whether or not the mutation, G157R, in GCAP2 was indeed functionally responsible for the photoreceptor degeneration in human patients by affecting cGMP synthesis.

A B

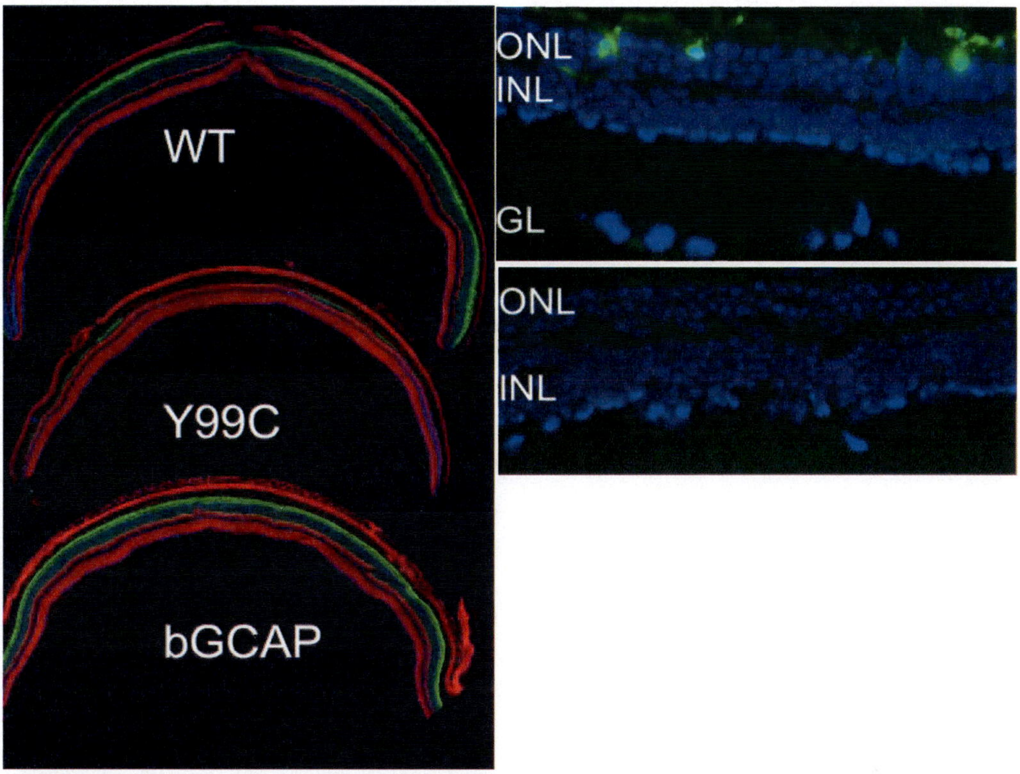

Figure 7. Retinal degeneration in two different lines of Y99C GCAP1-expressing mice. A, Retinal section from non-transgenic mice and mice expressing different amount of Y99C GCAP1 or wild type bovine GCAP1. The sections were probed with anti-α-transducin antibody (Santa Cruz Biotechnology) [53] to label rods (green fluorescence), anti- $G_{\alpha}0$ antibody, to label bipolar cells and plexiform layers (red), and the nuclei were counterstained with DAPI (blue). Red fluorescence in the inner retina was only observed in the presence of anti-$G_{\alpha}0$ antibody, red background in the sclera did not require primary antibodies (not shown). Each of the sections is a composite of several frames taken from the same retina. Obj. x 10. See reference [53] for other detail. B, Detection of the apoptotic DNA breakdown in photoreceptors expressing the Y99C GCAP1. The cryosections from 10-week old Y99C GCAP1 expressing retinas of fast degenerating mouse line L36 [53] were processed using a protocol for Promega DeadEnd Fluorimetric TUNEL System (Promega part # TB23), and the apoptotic nuclei incorporating fluorescein-12-dUTP by terminal deoxynucleotidyl transferase were visualized using fluorescent microscopy (green fluorescence, upper panel). The nuclei are counterstained using a Vector Laboratories DAPI/Vectashield mounting media (blue). The bottom panel: a control section incubated without terminal deoxynucleotidyl transferase. Obj. x 40. ONL-outer nuclear layer, INL-inner nuclear layer, GL ganglion cell layer.

4. Congenital Cone-Rod Dystrophy as a Result of Mutations in RetGC1

Mutations in retGC1 have been linked with several forms of blindness. One group of mutations that results in loss of the cyclase function associates with severe cases of Leber congenital amaurosis type A [68-71]. Most of these mutations are recessive, but some of them can reveal dominant-negative phenotype, possibly by suppressing the activity of retGC1 expressed from the non-mutant allele of the gene [70].

Another group of congenital dominant cone-rod degenerations, CORD6, results from gain-of-function mutations in retGC1 [72,73]. Multiple mutations in retGC1 dimerization domain have been found in patients from different families, affected by this disease, but the biochemical properties of the cyclase in those cases are primarily affected by substitutions of Arg838 by Cys, His or Ser [74-77]. These mutations shift Ca^{2+}-sensitivity of the recombinant retGC1 activated by *normal* GCAP1. Modeling the structural changes that may occur to the putative dimerization domain by using synthetic peptides indicates that the stability of a coiled-coil structure formed by the dimerization domain in the cyclase depends on R838. The extent to which different mutations at the position 838 stabilize coiled-coil complex of the alpha helixes apparently correlates with both the shift in Ca^{2+}-sensitivity of retGC1 *in vitro* and the severity of the clinical symptoms observed in patients. Among the mutations of this group, the substitution, R838S, produces the strongest effect on Ca^{2+}-sensitivity of retGC1 and associates with the most severe progression of retinal degeneration in patients [77].

The initial reports on biochemical properties of the mutant retGC1 indicated that the affinity of the cyclase for GCAP1 was increased by the R838 substitutions [78,79], and hence implied that the equilibrium between the free GCAP1 and GCAP1/retGC1 complex was likely altered by these mutations. It would be rather difficult to explain the shift in Ca^{2+}-sensitivity of retGC based on the assumption that the complex, retGC/GCAP1, is static. The subsequent detailed *in vitro* analysis of the GCAP1/R838S retGC1 interaction [80] revealed that (1) the affinity of the R838S retGC1 for *both* Mg^{2+}-bound and Ca^{2+}-bound GCAP1 increased compared to the wild type retGC1, (2) the cyclase affinity for the Mg^{2+}-bound (Ca^{2+}-free) GCAP1 increased more than the affinity for the Ca^{2+}-bound GCAP1 [80] (Figure 8).

If GCAP1 in retGC/GCAP1 complex is at equilibrium with the unbound GCAP1 and can be freely exchanged with the GCAP1 molecules outside the complex (with those that are either on the membrane or in solution), then the change in relative affinities for the Mg^{2+}-bound activator versus the Ca^{2+}-loaded inhibitor form of GCAP1 must result in altered Ca^{2+}-sensitivity of the cyclase regulation (Figure 5). Upon the increase in free Ca^{2+} concentration in the reaction mixture containing constant total concentration of GCAP and retGC, the ratio between the inhibitor and activator forms of GCAPs also increases; therefore the cyclase activity becomes suppressed. However, in case of the R838S retGC1, at the same free Ca^{2+} concentrations that are sufficient for inhibiting the normal cyclase, the Mg^{2+}-bound GCAP1 will more efficiently compete with the Ca^{2+}-bound inhibitor form. As a result, higher ratio of the Ca^{2+}-bound versus Mg^{2+}-bound GCAPs has to be created in order to inhibit the mutant retGC. Consequently, higher free Ca^{2+}-concentrations will be required to maintain the cyclase in complex with Ca^{2+}-bound GCAP1. There is also a second effect of the R838S mutation:

the mutant cyclase demonstrates residual activity even in the presence of fully Ca^{2+}-saturated GCAP1, although the absolute level of this activity is rather small compared to the maximal level of retGC stimulation by Mg^{2+}/GCAP1 [78-80]. Unlike the change in relative affinity for the Mg^{2+}-bound versus Ca^{2+}-bound GCAP1 that shifts Ca^{2+}-sensitivity of the cyclase, the residual activity of the R838S retGC1cannot be explained by its altered Ca^{2+}-sensitivity, and would more likely argue for the ability of even Ca^{2+}-bound GCAP1 to slightly enhance spontaneous dimerization of the R838S retGC1.

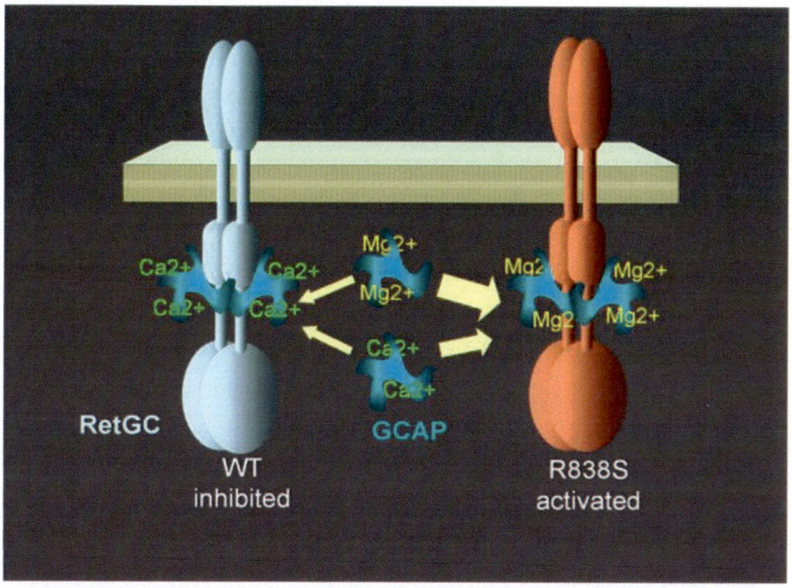

Figure 8. Relative affinity for the Mg^{2+}- versus Ca^{2+}-bound GCAP1 of the R838S retGC1 increases, therefore even at the concentration of Ca^{2+} that converts most of GCAP1 into Ca^{2+}-bound form, a significant portion of the R838S retGC1 can remain in complex with the Mg^{2+}-bound activator form of GCAP1 and will not be properly suppressed [70,80].

5. Conclusion

Two pathways in retGC regulation can potentially lead to its pathological, gain-of-function, conditions. The first pathway can reduce Ca^{2+}-binding activity of retGC Ca^{2+}-sensor proteins, GCAPs, thus rendering the cyclase active at high free Ca^{2+} levels, typical for normal dark-adapted photoreceptors (ca. 250 nM in mammals). This results in elevation of free cGMP in the outer segment and opening of a larger number of CNG channels, thus increasing the intracellular free Ca^{2+} levels. Both *in vitro* and *in vivo* models of human mutations in EF-hands 3 and 4 of GCAP1 are consistent with its affinity for calcium being reduced by these mutations, and a transgenic model demonstrates that the increase of free cGMP and/or free Ca^{2+} is a likely culprit for triggering apoptosis in photoreceptors.

Similar effect can be produced by mutations in retGC that change relative affinity of the cyclase for Mg^{2+}-bound (activator) versus Ca^{2+}-bound (inhibitor) form of GCAP1. This mechanism can potentially explain why even in the presence of normal GCAP1, retGC1 would require higher than normal free Ca^{2+} concentrations to suppress its GCAP-stimulated activity. While the *in vitro* modeling of the mutations in putative retGC1 dimerization domain is in a good agreement with this hypothesis, some serious questions remain about whether or not this takes place in live photoreceptors. First, the same mutation, R838S that decreases sensitivity of the Ca^{2+}-dependent inhibition of retGC1 by GCAP-1 produces the opposite effect in case of GCAP2. Therefore, in order for this mutation to alter retGC1 regulation *in vivo*, this isozyme would have to operate in a cellular compartment that is enriched by GCAP1 versus GCAP2. Second, the change in relative affinity of retGC1 in native photoreceptor membranes should be similar to that found in heterologous expression systems (such as HEK293 cells). Third, the complex between retGC and GCAPs has to be dynamic rather than static, in other words, even though the cyclase can be constantly bound to GCAPs, the Mg^{2+}-bound and the Ca^{2+}-bound GCAP molecules should be able to easily exchange within the retGC•GCAP1 complex. Whether or not this mechanism takes place *in vivo* remains a question that cannot be resolved without modeling the mutant retGC1 expression in live photoreceptor cell and requires developing new transgenic models for the future study.

Acknowledgments

This work was supported in parts by NIH grant EY11522, Formula 2004 grant from the Pennsylvania Department of Health and by Lions of Pennsylvania Sight Conservation Foundation. A.M.D. is the Martin and Florence Hafter Professor of Pharmacology.

References

[1] Pugh, E.N. Jr., Duda, T., Sitaramayya, A., and Sharma, R.K. 1997, *Biosci. Reports, 17*, 429.
[2] Arshavsky, V.Y., Lamb, T.D., and Pugh, E.N. Jr. 2002, *Annu Rev Physiol, 64*, 153.
[3] Pugh, E.N. Jr., Nikonov, S., and Lamb, T.D. 1999, *Curr. Opin. Neurobiol, 9*, 410.
[4] Makino, C.L., Wen, X.-H., and Lem, J. 2003, *Curr. Opin. Neurobiol, 13*, 404.
[5] Lem, J., and Makino, C.L. 1996, *Curr. Opin. Neurobiol., 6*, 453.
[6] Kilbride P., and Ebrey, T.G. 1979, *J. Gen. Physiol., 74*, 415.
[7] Gillesie, P.G., and Beavo, J.A. 1989, *Proc. Natl. Acad. Sci. USA, 86*, 4311.
[8] Govardovskii, V.I., and Berman, A.L. 1981, *Byophys. Struct. Mech., 7*, 125.
[9] Krispel, C.M., Chen, C.K., Simon, M.I., and Burns, M.E. 2003, *J. Gen. Physiol., 122*, 703.
[10] Koch, K.-W., and Stryer, L. 1988, *Nature, 334*, 64.
[11] Nakatani, K., and Yau, K.W. 1988, *Nature, 334*, 69.
[12] Koutalos, Y., Nakatani, K., Tamura., and T., Yau, K.-W. 1995, *J. Gen. Physiol., 106*, 863.
[13] Koutalos, Y., and Yau, K.-W. 1996, *Trends. Neurosci., 19*, 73.
[14] Fain, G.L., Lamb, T.D., Matthews, H.R., Murphy, R.L. 1989, *J. Physiol., 416*, 215

[15] Woodruff, M.L., Sampath, A.P., Matthews, H.R., Krasnoperova, N.V., Lem, J., and Fain G.L. 2002, *J. Physiol., 542*, 843.

[16] Woodruff, M.L., Wang, Z., Chung, H.Y., Redmond, T.M., Fain, G.L., and Lem, J. 2003, *Nat. Genet., 35,* 158.

[17] Gray-Keller, M.P., and Detwiler, P.B. 1994, *Neuron, 13*, 849.

[18] Lowe, D.G., Dizhoor, A.M., Liu, K., Gu, O., Laura, R., Lu, L., and Hurley, J.B. 1995, *Proc. Natl. Acad. Sci. USA, 92*, 5535.

[19] Yang, R.B., Foster, D.C., Garbers, D.L. and Fulle, H.J. 1995, *Proc. Natl. Acad. Sci. USA, 92*, 602.

[20] Goraczniak, R., Duda, T., and Sharma, R. K. 1997, *Biochem. Biophys. Res. Commun., 234*, 666.

[21] Dizhoor, A.M., Lowe, D.G., Olshevskaya, E.V., Laura, R.P., and Hurley, J. B. 1994, *Neuron, 12*, 1345

[22] Olshevskaya, E.V., Ermilov, A.N., and Dizhoor, A.M. 2002, *Molecular and Cellular Bioochemistry, 230*, 139.

[23] Yang, R.B., Robinson, S.W., Xiong, W.H, Yau, K.W., Birch, D.G., and Garbers, D.L. 1999, *J. Neurosci., 19*, 5889.

[24] Garbers, D. L. 1999, *Methods, 19*, 477.

[25] Garbers, D.L., and Lowe, D.G. 1994, *J. Biol. Chem., 269*, 30741.

[26] Dizhoor, A.M. 2000, *Cell. Signal., 12*, 711.

[27] Sharma, R.K. Mol. 2002, *Mol. Cell. Biochem., 230*, 3.

[28] Aparicio, J.G., and Applebury, M.L. 1996, *J. Biol. Chem., 271*, 27083. etc.

[29] Laura, R.P., Dizhoor, A.M., and Hurley, J.B. 1996, *J. Biol. Chem., 271*, 11646.

[30] Tucker, CL, Laura, RP, and Hurley, JB. 1997, *Biochemistry, 36*, 11995.

[31] Yamazaki, A., Yu, H., Yamazaki, M., Honkawa, H., Matsuura, I., Usukura, J., and Yamazaki, R. K. 2003, *J. Biol. Chem., 278*, 33150.

[32] Ramamurthy, V., Tucker, C., Wilkie, S.E., Daggett, V., Hunt, D.M., and Hurley, J.B. 2001, *J. Biol. Chem., 276*, 26218.

[33] Palczewski, K., Polans, A.S., Baehr, W., Ames, J.B. 2000, *Bioessays 22*, 337.

[34] Gorczyca, W.A., Gray-Keller, M.P., Detwiler, P.B., and Palczewski, K. 1994, *Proc. Natl. Acad. Sci. USA, 91*, 4014.

[35] Palczewski, K., Sokal, I., and Baehr, W. *Biochem. Biophys. Res. Commun.,* 2004, 322, 1123.

[36] Ermilov, A.N., Olshevskaya, E.V., and Dizhoor, A.M. 2001, *J. Biol. Chem., 276*, 48143.

[37] Sokal, I., Li, N., Klug, C. S., Filipek, S., Hubbell, W. L., Baehr, W., and Palczewski, K. 2001, *J. Biol. Chem., 276*, 43361.

[38] Ames, J.B., Dizhoor, A.M., Ikura, M., Palczewski, and K., Stryer, L. 1999, *J. Biol. Chem., 274*, 19329.

[39] Dizhoor, A.M., and Hurley, J.B. 1996, *J. Biol. Chem., 271*, 19346.

[40] Rudnicka-Nawrot, M., Surgucheva, I., Hulmes, J.D., Haeseleer, F., Sokal, I., Crabb, J.W., Baehr, W., and Palczewski, K. 1998, *Biochemistry, 37*, 248.

[41] Peshenko, I.V., and Dizhoor, A.M. 2004, *J. Biol. Chem., 279*, 16903.

[42] Chen C, Nakatani K, and Koutalos Y. 2003, *J. Physiol., 553*, 125.

[43] Yu, H., Olshevskaya, E., Duda, T., Seno, K., Hayashi, F., Sharma, R.K., Dizhoor, A.M., and Yamazaki, A. 1999, *J. Biol. Chem., 274*, 15547.
[44] Olshevskaya, E.V., Ermilov, A.N., and Dizhoor, A.M. 1999, J. Biol. Chem., 274, 25583.
[45] Duda, T., Venkataraman, V., Jankowska, A., Lange, C., Koch, K.W., and Sharma, R.K. 2000, *Biochemistry, 39*, 12522.
[46] Ramamurthy, V., Tucker, C., Wilkie, S. E., Daggett, V., Hunt, D. M., and Hurley, J. B. 2001, *J. Biol. Chem., 276*, 26218.
[47] Koch, K.W., Duda, and T., Sharma, R.K. 2002, *Mol. Cell. Biochem. 230*, 97.
[48] Lange, C., Duda, T., Beyermann, M., Sharma, R.K., and Koch, K.W. 1999, *FEBS Lett., 460*, 27.
[49] Sokal, I., Haeseleer, F., Arendt, A., Adman, E.T., Hargrave, P.A., and Palczewski, K. 1999, *Biochemistry, 38*, 1387.
[50] Krylov, D.M., and Hurley, J.B. 2001, *J. Biol. Chem., 276*, 30648.
[51] Duda, T., Fik-Rymarkiewicz, E., Venkataraman, V., Krishnan, R., Koch, K.W., and Sharma RK. 2005, *Biochemistry, 44*, 7336.
[52] Burns, M.E., Mendez, A., Chen, J., and Baylor, D.A. 2002, *Neuron, 36*, 81.
[53] Mendez, A., Burns, M.E., Sokal, I., Dizhoor, A.M., Baehr, W., Palczewski, K., Baylor, D.A., and Chen, J. 2001, *Proc. Natl. Acad. Sci USA, 98*, 9948.
[54] Howes, K.A., Pennesi, M.E., Sokal, I., Church-Kopish, J., Schmidt, B., Margolis, D., Frederick, J.M., Rieke, F., Palczewski, K., Wu, S.M., Detwiler, P.B., Baehr, W. 2002, *EMBO J., 21*, 1545.
[55] Hwang, J.Y., Lange, C., Helten, A., Hoppner-Heitmann, D., Duda, T., Sharma, R.K., and Koch, K.W. 2003, *Eur J Biochem, 270*, 3814.
[56] Dizhoor, A.M., Olshevskaya, E.V., Henzel, W.J., Wong, S.C., Stults, J.T., Ankoudinova, I., and Hurley, J.B. 1995, *J. Biol. Chem., 270*, 25200.
[57] Howes, K., Bronson, J. D., Dang, Y. L., Li, N., Zhang, K., Ruiz, C., Helekar, B., Lee, M., Subbaraya, I., Kolb, H., Chen, J., and Baehr, W. 1998, *Invest. Ophthalmol. Vis. Sci., 39*, 867.
[58] Kachi, S., Nishizawa, Y., Olshevskaya, E., Yamazaki, A., Miyake, Y., Wakabayashi, T., Dizhoor, A., and Usukura, J. 1999, *Exp. Eye. Res. 68*, 465.
[59] Payne, A.M., Downes, S.M., Bessant, D.A., Taylor, R., Holder G.E., Warren, M.J., Bird, A. C., and Bhattacharya, S.S. 1998, *Hum. Mol. Genet. 7*, 273.
[60] Wilkie, S.E., Newbold, R.J., Deery, E., Walker, C.E., Stinton, I., Ramamurthy, V., Hurley, J.B., Bhattacharya, S.S., Warren, M.J., and Hunt, D.M. 2000, *Hum. Mol. Genet., 9*, 3065.
[61] Nishiguchi, K.M., Sokal, I., Yang, L., Roychowdhury, N., Palczewski, K., Berson, E.L., Dryja, T.P., Baehr, W. 2004, *Invest. Ophthalmol. Vis. Sci. 45*, 3863.
[62] Sokal, I., Dupps, W.J., Grassi, M.A., Brown, J. Jr, Affatigato, L.M., Roychowdhury, N., Yang, L., Filipek, S., Palczewski, K., Stone, E.M., Baehr, W. 2005, *Invest. Ophthalmol. Vis. Sci. 46*, 1124-32.
[63] Dizhoor, A.M., Boikov, S.G. & Olshevskaya, E.V. 1998, *J. Biol. Chem., 273*, 17311.
[64] Sokal, I., Li, N., Surgucheva, I., Warren, M.J., Payne, A.M., Bhattacharya, S.S., Baehr, W., and Palczewski, K. 1998, *Mol. Cell, 2*, 129.

[65] Olshevskaya, E.V., Calvert P.D., Woodruff, M.L., Peshenko, I.V., Savchenko, A.B., Makino, C.L., Ho, Y-S., Fain, G.L., Dizhoor, A.M. 2004, *J. Neuroscience, 24*, 6078.

[66] Makino, C.L., Wen, X.-H., Michaud, N., Peshenko, I.V., Pawlyk, B., Brush, R.S., Solovyeva, M., Liu, X., Woodruff, M.L., Calvert, P.D., Savchenko, A.B., Anderson, R.E., Fain, G.L., Li, T., Sandberg, M.A., and Dizhoor, A.M. 2006 *Invest. Ophthalmol. Vis. Sci., in press*

[67] Sato, M., Nakazawa, M., Usui, T., Tanimoto, N., Abe, H., and Ohguro, H. 2005, *Graefes Arch. Clin. Exp. Ophthalmol., 243*, 235.

[68] Koenekoop, RK, Fishman, G.A., Iannaccone, A., Ezzeldin, H., Ciccarelli, M.L., Baldi, A., Sunness, J.S., Lotery, A.J., Jablonski, M.M., Pittler, S.J., and Maumenee, I. 2002, *Arch. Ophthalmol., 120,* 1325.

[69] Rozet, J.M., Perrault, I., Gerber S., Hanein, S., Barbet, F., Ducroq, D., Souied, E., Munnich, A., Kaplan, J. 2001, *Invest. Ophthalmol. Vis. Sci., 42*,1190.

[70] Tucker, C.L., Ramamurthy, V., Pina, A.L., Loyer, M., Dharmaraj, S., Li, Y., Maumenee, I.H., Hurley, J.B., and Koenekoop, R.K. 2004, *Mol. Vis., 10*, 297.

[71] Duda, T., Venkataraman, V., Goraczniak, R., Lange, C., Koch, K.W., Sharma, R.K. 1999, *Biochemistry, 38*, 509.

[72] Perrault, I., Rozet, J.M., Gerber, S., Kelsell, R. E., Souied, E., Cabot, A., Hunt, D.M., Munnich, A., and Kaplan, J. 1998, *Am. J. Hum. Genet., 63*, 651.

[73] Newbold, R.J., Deery, E.C., Payne, A.M., Wilkie, S.E., Hunt, D.M., and Warren, M.J. 2002, *Adv. Exp. Med. Biol., 514*, 411.

[74] Kelsell, R.E., Gregory-Evans, K., Payne, A.M., Perrault, I., Kaplan, J., Yang, R.B., Garbers, D.L., Bird, A.C., Moore, A.T., and Hunt, D. M. 1998, *Hum. Mol. Genet., 7*, 1179.

[75] Downes, S.M., Payne, A.M., Kelsell, R.E., Fitzke, F.W., Holder, G.E., Hunt, D.M., Moore, A.T., and Bird, A.C. 2001, *Arch. Ophthalmol., 119*, 1667.

[76] Gregory-Evans, K., Kelsell, R.E., Gregory-Evans, C.Y., Downes, S.M., Fitzke, F.W., Holder, G.E., Simunovic, M., Mollon, J.D., Taylor, R., Hunt, D.M., Bird, A.C., and Moore, A.T. 2000, *Ophthalmology, 107*, 55.

[77] Wilkie, S.E., Newbold, R.J., Deery, E., Walker, C.E., Stinton, I., Ramamurthy, V., Hurley, J.B., Bhattacharya, S.S., Warren, M.J., and Hunt, D.M. 2000, *Hum. Mol. Genet. 9*, 3065.

[78] Tucker, C.L., Woodcock, S.C., Kelsell, R.E., Ramamurthy, V., Hunt, D.M., and Hurley, J.B. 1999, *Proc. Natl. Acad. Sci. USA, 96*, 9039.

[79] Ramamurthy, V., Tucker, C., Wilkie, S.E., Daggett, V., Hunt, D.M., and Hurley, J.B. 2001, *J. Biol. Chem., 276*, 26218.

[80] Peshenko, I.V., Moyiseev, G.P, Olshevskaya, E.V., and Dizhoor, A.M. 2004, *Biochemistry*, 13796.

In: Neuronal Calcium Sensor Proteins
Editors: Pavel Philippov and Karl Koch, pp. 221-228
ISBN 1-59454-978-8

Chapter XII

GUANYLYL CYCLASE-ACTIVATING PROTEINS (GCAPS)

Wojciech A. Gorczyca*
Laboratory of Signaling Proteins, L. Hirszfeld Institute of Immunology and Experimental Therapy, Polish Academy of Sciences, Wroclaw, Poland.

Abstract

Cyclic GMP is synthesized in outer segments of photoreceptor cells by the retina-specific guanylyl cyclase (retGC) in response to lowered concentration of Ca^{2+}. This Ca^{2+}-dependent regulation of retGC activity is mediated by Ca^{2+}-binding proteins of the neuronal calcium sensors (NCS) family termed guanylyl cyclase-activating proteins (GCAPs). They activate retGC at low and inhibit it at a high cytoplasmic concentration of Ca^{2+} $[Ca^{2+}]_i$. Three different GCAPs (GCAP1, GCAP2 and GCAP3) are identified to date in the retina of mammals. Despite of common features with other NCS proteins, GCAPs reveal also properties which are characteristic exclusively to them.

Keywords: calcium-binding proteins, guanylyl cyclase, retina, signal transduction

Abbreviations

$[Ca^{2+}]$	concentration of free calcium ions
$[Ca^{2+}]_i$	cytoplasmic $[Ca^{2+}]$
EF	structural motif (EF-hand) with Ca^{2+}-binding site in calcium-binding proteins

* Correspondence concerning this article should be addressed to W.A Gorczyca; Laboratory of Signaling Proteins, L. Hirszfeld Institute of Immunology and Experimental Therapy, ul. R. Weigla 12, 53-114 Wroclaw, Poland. tel.: +4871 370 99 88; Fax: +4871 337 13 82; E-mail: gorczyca@immuno.iitd.pan.wroc.pl.

GCAP	guanylyl cyclase-activating protein
GCIP	guanylyl cyclase-inhibiting protein
NCS	neuronal calcium sensor
retGC	retina-specific guanylyl cyclase
ROS	rod outer segment
VILIP	visinin-like protein

1. Introduction

Cyclic GMP (cGMP) and calcium ions are involved together in many physiological processes in which their intracellular concentrations are interdependently regulated by feedback controlled mechanisms. In retinal photoreceptor cells, cGMP interacts directly with plasma membrane cationic channels and keeps them open. Light triggers the cascade of biochemical events, termed phototransduction, which leads to hydrolysis of cGMP by photoreceptor-specific phosphodiesterase, closure of the channels and decrease of $[Ca^{2+}]_i$. Lowered $[Ca^{2+}]_i$ is the signal for resynthesis of cGMP by retGC and contributes to the recovery of the dark state by photoreceptor cells [1]. The sensitivity of retGC to Ca^{2+} has been known for a long time [2,3]. It was observed that the enzyme activity in outer segments of rod photoreceptor cells (ROS) was low at high (above 500 nM) and markedly increased at low (below 100 nM) $[Ca^{2+}]_i$. The sensitivity of retGC to calcium was shown to disappear after washing the ROS membranes with low salt buffer and could be restored when the ROS extract was added back suggesting an existence of a soluble mediator of the Ca^{2+} effect. That postulate of Koch and Stryer [4] stimulated an extensive effort of several groups in the search of Ca^{2+}-dependent regulator(s) of retGC. Few years later Gorczyca [5] in the laboratory of Kris Palczewski and independently Dizhoor [6] in the laboratory of Jim Hurley succeeded in an isolation from the bovine retina of two distinct proteins which at $[Ca^{2+}]$ below 100 nM were able to activate retGC in washed ROS membranes. First of them was purified from ROS [5] and the other one from the whole retinas [6]. Both proteins shared 42% identity at the amino acid level and due to their ability to restore the Ca^{2+} sensitivity of retGC were named guanylyl cyclase-activating proteins 1 and 2 (GCAP1 and GCAP2), respectively [7,8]. The third guanylyl cyclase-activating protein (GCAP3) was cloned from the human retina [9]. Recently, several further new members of the GCAP subfamily were identified in lower vertebrates [10].

2. Properties of GCAPs

First studies on GCAP1 and GCAP2 were carried out using native proteins purified from the bovine retina. They were found to be soluble, highly acidic and hydrophobic Ca^{2+}-binding proteins which similarly to calmodulin exhibit different electrophoretic mobilities in Ca^{2+}-loaded and Ca^{2+}-free forms [5-7,11]. An apparent molecular mass of GCAP1 is 20 kDa in the presence of Ca^{2+} and is shifted to ~25 kDa in the presence of EGTA, while GCAP2 appears as a band of 18-19 kDa in the presence of Ca^{2+} and 24 kDa when Ca^{2+} is chelated [7].

More detailed studies were possible when recombinant, functionally active GCAPs (mainly GCAP1 and GCAP2) were expressed in bacteria [12-14] and in eucaryotic cells [7,9,13]. Mammalian GCAPs consist of 200-209 amino acids with calculated molecular mass of 23-24 kDa [7-9,11]. Individual GCAPs are highly conserved and GCAP1 as well as GCAP2 show about 70% sequence identity among vertebrates and even more than 90% identity among mammals [15]. Similarity between different GCAPs is lower. Human GCAP1, GCAP2, and GCAP3 share about 40% identity with the highest homology (57%) between GCAP1 and GCAP3 [9,15].

GCAPs regulate in a Ca^{2+}-dependent manner the activity of two known isoforms of retGC (retGC1 and retGC2) [6,7,16]. They stimulate activity of retGCs at low [Ca^{2+}] and inhibit it at high [Ca^{2+}]. Although all three GCAPs were shown to regulate both isoforms of retGC, GCAP1 activated recombinant bovine retGC1 about two times more efficiently than retGC2 [17] and did not activate human retGC2 [9], suggesting its higher specificity to retGC1. Interestingly, in the frog retina a GCAP-like protein was identified, which did not activate retGC at low but inhibited the enzyme activity at high [Ca^{2+}] and from that reason was named guanylyl cyclase-inhibiting protein (GCIP) [18]. All GCAPs contain four EF-hand motifs but only three of them (EF2-EF4) are able to bind Ca^{2+} and in GCIP EF4 is also disabled for Ca^{2+} binding [7,8,11,19]. As was shown for native and recombinant GCAPs, they bind Ca^{2+} cooperatively with Hill coefficients equal to 2-2.5 and 1.7-2.1 in the case of GCAP1 and GCAP2, respectively [5,6,8,12,20]. Calcium binding induces conformational rearrangements in GCAPs structure as proved by means of the fluorescence spectroscopy [11,21-25], NMR [26], circular dichroism [26], chemical modifications [23,27], site directed mutagenesis [22,28,29], and limited proteolysis [25,26,29]. Mutagenesis in Ca^{2+}-binding regions of GCAP1 and GCAP2 showed that Ca^{2+}-binding properties of GCAPs are critical in regulation of target cyclase(s) [21,25,28-31]. The retGC sensitivity to Ca^{2+} in the presence of GCAP1 differs from that measured in the presence of GCAP2 suggesting that if present together both GCAPs extend the dynamic range of the enzyme response to Ca^{2+} [15,32,33]. It was recently shown that Mg^{2+} affecting Ca^{2+}-sensitivity of GCAPs may influence their ability to regulate retGC [34].

GCAP1 and GCAP2 are acylated at N-terminal glycine by fatty acids [11,13]. GCAP3 was found to be myristoylated when expressed in insect cells [9]. The role of this modification of GCAPs is not clear yet but it was shown to change Ca^{2+}-binding characteristics of bacteria expressed GCAP1 and to influence its ability to activate retGC [12]. Also, expressed in insect cells GCAP1 mutants that lacked myristoylation, activated retGC much weaker than the wild-type protein [21,33,35]. On the other hand, N-terminal acylation by myristic acid did not influence the functional properties of GCAP2 [13,33,35].

3. Expression of GCAPs

The main site of GCAPs expression are the photoreceptor cells of the retina as demonstrated by Northern blotting, hybridization *in situ* [9,11,36], and immunohistochemical techniques [7-9,12,14,37,38]. With some differences among species, GCAP1 and GCAP2 have been shown to be expressed in rods and cones, while GCAP3 exclusively in cones (summarized in refs. 15,19). Besides the retina, expressed sequence tags for human GCAP1 and GCAP3 were cloned from testes and brain [19]. By means of biochemical and immunohistochemical methods, GCAP1 was detected also in the bovine pineal gland [39] and a GCAP-like protein was found in the rat pineal gland [40]. Moreover, expression of genes encoding GCAP1 and GCAP2 was reported for the chicken pineal gland [41]. Also, retGC was detected in pineal gland indicating that synthesis of cGMP may be regulated in this organ in a Ca^{2+}-dependent manner but the physiological significance of this mechanism is unclear [39,41].

The genes encoding GCAPs reveal similar intron/exon organization and contain 4 exons and 3 introns [19,42]. The human genes encoding GCAP1 and GCAP2 are located in a tail-to-tail array on the chromosome 6 (p21.1) [43], while the gene for GCAP3 is located on chromosome 3 (q13.1) [9]. It indicates that genes for different GCAPs appeared as a result of ancestral gene duplication and translocation. Mutations in the gene encoding GCAP1 (*GUCA1A*) have been found to be associated with autosomal dominant cone dystrophy [44,45] and with autosomal dominant cone-rod dystrophy [46]. Although earlier studies did not find relations between mutations in the gene encoding GCAP2 (*GUCA1B*) and retinal diseases [47], a very recent report indicates that mutation at codon 157 in *GUCA1B* might be a cause of autosomal dominant retinal dystrophy [48].

4. GCAPs as Neuronal Calcium Sensors

GCAPs are expressed in neuronal cells, mainly cones and rods. Sequence homology, similar molecular mass, consensus sequence for N-terminal acylation, the presence of three functionally active EF-hand motifs and residual fourth EF1 motif, located at N-terminus, indicate that GCAPs constitute a subfamily of Ca^{2+}-binding proteins, which originates from a common calmodulin-like ancestor [19,42]. Therefore, together with frequenins, visinin-like proteins (VILIPs), recoverins, and KChIPs they belong to the family of neuronal calcium sensors, NCS [49]. The three-dimensional structure of Ca^{2+}-bound GCAP2 was determined using NMR [20] and revealed its similarity to that of recoverin [50], neurocalcin [51], and frequenin [52]. They are compact proteins consisting of two globular domains connected by a short flexible linker. The N-terminal domain is formed by interacting EF1 and EF2 and the C-terminal half contains interacting EF3 and EF4 [20,50-52]. In spite of similarities, also differences between GCAPs and other NCS proteins have been noticed. Such N-terminally myristoylated NCS proteins as recoverin, neurocalcin, hippocalcin, VILIP1, and VILIP3 bind to biological membranes and to artificial hydrophobic surfaces mainly in the presence of Ca^{2+} using a mechanism termed Ca^{2+}-myristoyl switch [53]. In a Ca^{2+}-free protein, the myristoyl moiety is buried in a hydrophobic pocket and Ca^{2+} binding causes its extrusion. GCAPs do not 'sense calcium' in such a way since they have been shown not to use a Ca^{2+}-myristoyl switch [13,35] and their interactions with ROS membranes and hydrophobic surfaces were either independent of Ca^{2+} or occurred even weaker in the presence of Ca^{2+} [7,13,24,35]. Employing surface plasmon resonance technique Hwang and Koch showed that myristoylated GCAP2 interacted with lipid membranes better in the presence of EGTA than in the presence of Ca^{2+}, while the opposite relationship was detected in the case of myristoylated recoverin [35]. Experiments with ANS (1-anilinonaphthalene-8-sulfonate) used as a fluorescent probe of protein hydrophobicity supported these observations showing that native GCAPs only slightly change their hydrophobic properties upon Ca^{2+} binding [24] and rather in a reverse direction than it occurred in recoverin [24,54]. Therefore, the Ca^{2+}-dependent hydrophobic properties of GCAPs have been suggested to be a prerequisite for mechanisms by which they regulate retGCs [24]. The retGCs in the absence of GCAPs have only basal activity, while in the presence of GCAPs they are regulated in a Ca^{2+}-dependent manner [28,29]. Since hydrophobicity of GCAPs is high at low calcium and only slightly changes upon Ca^{2+} binding, it favors formation of the Ca^{2+}-independent complex GCAPs-retGC in which GCAPs changing their conformations in response to Ca^{2+} ('sensing calcium') serve as switches between active and inactive states of the target enzymes. Detailed mechanisms of retGC regulation are yet unknown but GCAPs have been shown to be complexed with target enzymes independently of Ca^{2+} [7,28,29,55].

5. Conclusion

More than ten years have passed since the discovery of the first GCAPs [5,6]. During that time new GCAPs have been identified, their structure and biochemical properties determined, and the role in regulation of retGC demonstrated. Using animal models a physiological role of GCAPs in phototransduction have been also studied (summarized in ref. 19). Another important source of information on GCAPs function was a correlation of GCAPs dysfunction with mutations in their genes which had led to retinal diseases [30,31,44]. Taken together, all studies and findings described in this chapter show that Ca^{2+}-induced conformational changes of GCAP structure are critical for their regulation of target retGC and therefore GCAPs are neuronal calcium sensors. However, detailed mechanisms that underlie this regulation still await an explanation. Based on the available data, two models have been proposed for mechanisms by which GCAP1 and GCAP2 may regulate their target enzymes. One of them assumes that GCAPs, being associated with retGC independently of [Ca^{2+}], in a Ca^{2+}-free form dimerize and thus change the conformation of catalytic region of adjacent cyclase stimulating cGMP synthesis [55]. The second model also assumes that GCAPs remain associated with retGC independently of [Ca^{2+}] but activation or inhibition of the enzyme activity is a result of Ca^{2+}-dependent conformational rearrangement around the central flexible helix in GCAP structure [22]. In determination of the exact molecular mechanisms underlying the retGC regulation by individual GCAPs, the knowledge of the structure of engaged proteins will be needed. However, the three dimensional structures of both Ca^{2+}-free and Ca^{2+}-loaded forms of myristoylated GCAP1 and GCAP2 are yet unsolved.

References

[1] Nakatani, K., Chen, C., Yau, K.W., and Koutalos, Y. 2002, *Adv. Exp. Med. Biol., 514*, 1.

[2] Lolley, R.N., and Racz, E. 1982, *Vision Res., 22*, 1481.

[3] Pepe, I.M., Panfioli, I., and Cugnoli, C. 1986, *FEBS Lett., 14*, 203, 73.

[4] Koch, K.W., and Stryer, L. 1988, *Nature, 334*, 64.

[5] Gorczyca, W.A., Gray-Keller, M.P., Detwiler, P.B., and Palczewski, K. 1994, *Proc. Natl. Acad. Sci. U.S.A., 91*, 4014.

[6] Dizhoor, A.M., Lowe, D.G., Olshevskaya, E.V., Laura, R.P., and Hurley, J.B. 1994, *Neuron, 12*, 1345.

[7] Gorczyca, W.A., Polans, A.S., Surgucheva, I.G., Subbaraya, I., Baehr, W., and Palczewski, K. 1995, *J. Biol. Chem., 270*, 22029.

[8] Dizhoor, A.M., Olshevskaya, E.V., Henzel, W.J., Wong, S.C., Stults, J.T., Ankoudinova, I., and Hurley, J.B. 1995, *J. Biol. Chem., 270*, 25200.

[9] Haeseleer, F., Sokal, I., Li, N., Pettenati, M., Rao, N., Bronson, D., Wechter, R., Baehr, W., and Palczewski, K. 1999, *J. Biol. Chem., 274*, 6526.

[10] Imanishi, Y., Yang, L., Sokal, I., Filipek, S., Palczewski, K., and Baehr, W. 2004, *J. Mol. Evol., 59*, 204.

[11] Palczewski, K., Subbaraya, I., Gorczyca, W.A., Helekar, B.S., Ruiz, C.C., Ohguro, H., Huang, J., Zhao, J., Crabb, J.W., Johnson, R.S., Walsh, K.A., Gray-Keller, M.P., Detwiler, P.B., and Baehr, W. 1994, *Neuron, 13*, 395.

[12] Frins, S., Bönigk, W., Müller, F., Kellner, R., and Koch, K.W. 1996, *J. Biol. Chem., 271*, 8022.

[13] Olshevskaya, E.V., Hughes, R.E., Hurley, J.B., and Dizhoor, A.M. 1997, *J. Biol. Chem., 272*, 14327.

[14] Otto-Bruc, A., Fariss, R.N., Haeseleer, F., Huang, J., Buczylko, J., Surgucheva, I., Baehr, W., Milam, A.H., and Palczewski, K. 1997, *Proc. Natl. Acad. Sci. U.S.A., 94*, 4727.

[15] Gorczyca, W.A., and Sokal, I. 2002, *Adv. Exp. Med. Biol., 514*, 319.

[16] Lowe, D.G., Dizhoor, A.M., Liu, K., Gu, Q., Spencer, M., Laura, R., Lu. L., and Hurley, J.B. 1995, *Proc. Natl. Acad. Sci. U.S.A., 92*, 5535.

[17] Krylov, D.M., Niemi, G.A., Dizhoor, A.M., and Hurley, J.E. 1999, *J. Biol. Chem., 274*, 10833.

[18] Li, N., Fariss, R.N., Zhang, K., Otto-Bruc, A., Haeseleer, F., Bronson, D., Qin, N., Yamazaki, A., Subbaraya, I., Milam, A.H., Palczewski, K., and Baehr, W. 1998, *Eur. J. Biochem., 252*, 591.

[19] Palczewski, K., Sokal, I., and Baehr, W. 2004, Biochem. Biophys. Res. Commun., 322, 1123.

[20] Ames, J.B., Dizhoor, A.M., Ikura, M., Palczewski, K., and Stryer, L. 1999, *J. Biol. Chem., 274*, 19329.

[21] Otto-Bruc, A., Buczylko, J., Surgucheva, I., Subbaraya, I., Rudnicka-Nawrot, M., Crabb, J.W., Arendt, A., Hargrave, P.A., Baehr, W., and Palczewski, K. 1997, *Biochemistry, 36*, 4295.

[22] Sokal, I., Otto-Bruc, A.E., Surgucheva, I., Verlinde, C.L., Wang, C.K., Baehr, W., and Palczewski, K. 1999, *J. Biol. Chem., 274*, 19829.

[23] Sokal, I., Li, N., Klug, C.S., Filipek, S., Hubbel, W.L., Baehr, W., and Palczewski, K. 2001, *J. Biol. Chem., 46*, 43361.

[24] Gorczyca, W.A., Kobiałka, M., Kuropatwa, M., and Kurowska, E. 2003, *Acta Biochim. Pol., 50*, 367.

[25] Hwang, J.Y., Schlesinger, R., and Koch, K.W. 2004, *Eur. J. Biochem., 271*, 3785.

[26] Hughes, R.E., Brzovic, P.S., Dizhoor, A.M., Klevit, R.E., and Hurley, J.B. 1998, *Protein Sci., 7*, 2675.

[27] Hwang, J.Y., Schlesinger, R., and Koch, K.W. 2001, *FEBS Lett., 508*, 355.

[28] Dizhoor, A.M., and Hurley, J.B. 1996, *J. Biol. Chem., 271*, 19346.

[29] Rudnicka-Nawrot, M., Surgucheva, I., Hulmes, J.D., Haeseleer, F., Sokal, I., Crabb, J.W., Baehr, W., and Palczewski, K. 1998, *Biochemistry, 37*, 248.

[30] Sokal, I., Li, Surgucheva, I., Warren, M.J., Payne, A.M., Bhattacharya, S.S., Baehr, W., and Palczewski, K. 1998, *Mol. Cell, 2*, 129.

[31] Dizhoor, A.M., Boikov, S.G., and Olshevskaya, E.V. 1998, *J. Biol. Chem., 273*, 17311.

[32] Peshenko, I.V., Olshevskaya, E.V., and Dizhoor, A.M. 2004, *J. Biol. Chem., 279*, 50342.

[33] Hwang, J.Y., Lange, C., Helten, A., Hoppner-Heitmann, D., Duda, T., Sharma, R.K., and Koch, K.W. 2003, *Eur. J. Biochem., 270*, 3814.

[34] Peshenko, I.V., and Dizhoor, A.M. 2004, *J. Biol. Chem., 279*, 16903.

[35] Hwang, J.Y., and Koch, K.W. 2002, *Biochemistry, 41*, 13021.

[36] Subbaraya, I., Ruiz, C.C., Helekar, B.S., Zhao, X., Gorczyca, W.A., Pettenati, M.J., Rao, P.N., Palczewski, K., and Baehr, W. 1994, *J. Biol. Chem., 269*, 31080.

[37] Cuenca, N., Lopez, S., Howes, K., and Kolb, H. 1998, *Invest. Ophthalmol. Vis. Sci., 39*, 1243.

[38] Kachi, S., Nishizawa, Y., Olshevskaya, E., Yamazaki, A., Miyake, Y., Wakabayashi, T., Dizhoor, A., and Usukura, J. 1999, *Exp. Eye Res., 68*, 465.

[39] Venkataraman, V., Nagele, R., Duda, T., and Sharma, R.K. 2000, Biochemistry, 39, 6042.

[40] Dejda, A., Matczak, I., and Gorczyca, W.A. 2002, *Acta Biochim. Polon., 49*, 899.

[41] Semple-Rowland, S.L., Larkin, P., Bronson, J.D., Nykamp, K., Streit, W.J., and Baehr, W. 1999, *Mol. Vis., 5*, 14.

[42] Palczewski, K., Polans, A.S., Baehr, W., and Ames, J.B. 2000, *Bioessays, 22*, 337.

[43] Surguchov, A., Bronson, J.D., Banerjee, P., Knowles, J.A., Ruiz, C., Subbaraya, I., Palczewski, K., and Baehr, W. 1997, *Genomics, 39*, 312.

[44] Payne, A.M., Downes, S.M., Bessant, D.A., Taylor, R., Holder, G.E., Warren, M.J., Bird, A.C., and Bhattacharya, S.S. 1998, *Hum. Mol. Genet., 7*, 273.

[45] Nishiguchi, K.M., Sokal, I., Yang, L., Roychowdhury, N., Palczewski, K., Berson, E.L., Dryja, T.P., and Baehr, W. 2004, *Invest. Ophthalmol. Vis. Sci., 45*, 3863.

[46] Sokal, I., Dupps, W.J., Grassi, M.A., Brown, J. Jr., Affatigato, L.M., Roychowdhury, N., Yang, L., Filipek, S., Palczewski, K., Stone, E.M., and Baehr, W. 2005, *Invest. Ophthalmol. Vis. Sci., 46*, 1124.

[47] Payne, A.M., Downes, S.M., Bessant, D.A., Plant, C., Moore, T., Bird, A.C., and Bhattacharya, S.S. 1999, *J. Med. Genet., 36*, 691.

[48] Sato, M., Nakazawa, M., Usui, T., Tanimoto, N., Abe, H., and Ohguro, H. 2005, *Graefes Arch. Clin. Exp. Ophthalmol., 243*, 235.

[49] Burgoyne, R.D. 2004, *Biochim. Biophys. Acta, 1742*, 59.

[50] Flaherty, K.M., Zozulya, S., Stryer, L., and McKay, D.B. 1993, *Cell, 75*, 709.

[51] Vijay-Kumar, S., and Kumar, V.D. 1999, *Nat. Struct. Biol., 6*, 80.

[52] Bourne, Y., Dannenberg, J., Pollmann, V., Marchot, P., and Pongs, O. 2001, *J. Biol. Chem., 276*, 11949.

[53] Ames, J.B., Ishima, R., Tanaka, T., Gordon, J.I., Stryer, L., and Ikura, M. 1997, *Nature, 389*, 198.

[54] Hughes, R.E., Brzovic, P.S., Klevit, R.E., and Hurley, J.B. 1995, *Biochemistry, 34*, 11410.

[55] Olshevskaya, E.V., Ermilov, A.N., and Dizhoor, A.M. 1999, *J. Biol. Chem., 274*, 25583.

In: Neuronal Calcium Sensor Proteins
Editors: Pavel Philippov and Karl Koch, pp. 229-241
ISBN 1-59454-978-8

Chapter XIII

DIFFERENT BIOCHEMICAL PROPERTIES OF GUANYLATE CYCLASE-ACTIVATING PROTEINS ALLOW FINE TUNING OF PHOTOTRANSDUCTION

Karl-Wilhelm Koch*

AG Biochemistry, Institute of Biology and Environmental Sciences, Faculty V, Carl von Ossietzky University Oldenburg, D-26111 Oldenburg, Germany.

Abstract

Guanylate cyclase-activating proteins (GCAPs) regulate membrane bound guanylate cyclases in the vertebrate retina in a calcium-dependent fashion. These calcium-sensitive steps constitute negative feedback loops to adjust light sensitivity and adaptation in photoreceptor cells. Among the best studied GCAP isoforms are GCAP1 and GCAP2. Both calcium sensor proteins show remarkable differences with respect to calcium-sensitivity, catalytic efficiency and target recognition. These differences allow a fine tuned regulation of the target guanylate cyclase. This review gives an overview on recent findings that deal with differences of GCAP1 and GCAP2 which in turn lead to a finely tuned regulation of guanylate cyclase. Further it discusses whether different calcium sensitivities in GCAPs determine certain physiological phenomena as manifested in the characteristics of the photoresponse of transgenic mice.

Keywords: phototransduction, adaptation, guanylate cyclase, cGMP

---

* Correspondence concerning this article should be addressed to Karl-Wilhelm Koch, AG Biochemistry, Institute of Biology and Environmental Sciences, Faculty V, Carl von Ossietzky University Oldenburg, D-26111 Oldenburg, Germany. Tel: +49-441-798-3640; Fax: +49-441-798-193640; E-mail: karl.w.koch@uni-oldenburg.de.

Abbreviations

GCAP	guanylate cyclase-activating protein
CNG channel	cyclic nucleotide-gated channel
cGMP	guanosine 3′,5′-cyclic monophosphate
ROS	rod outer segments
ROS-GC	rod outer segment guanylate cyclase
VILIP	visinin like protein

1. Introduction

Sensory perception requires constancy of reliable signal detection in the presence of a changing background. These adaptational phenomena are characteristic properties of the visual, auditory, olfactory and other sensory systems and large efforts are currently undertaken to understand sensory adaptation at the cellular and molecular level [1-6]. Although the visual system is among the best studied and understood, the interdependencies of the molecular components in rod and cone cells of the vertebrate retina are complex and still await full description and comprehension. Key feature of light adapation in rods and cones is the change in cytoplasmic [Ca^{2+}] triggered by illumination. Cytoplasmic [Ca^{2+}] in rods and cones is balanced by an influx via the cyclic nucleotide-gated (CNG) channel and by the efflux through the continous operation of a Na^+/Ca^{2+}; K^+-exchanger. These two transport routes, Ca^{2+}-influx and Ca^{2+}-extrusion, balance the intracellular free [Ca^{2+}] at a resting level of 500-600 nM. However, this value depends on the species that is studied, since it might be as high as almost 800 nM in salamander rods [7] and as low as 250 nM in mice [8].

Light absorption by rhodopsin triggers the visual transduction cascade: a heterotrimeric G-protein activates a photoreceptor specific phosphodiesterase that cleaves the intracelluar transmitter cyclic GMP (cGMP) with a high catalytic rate. Cyclic GMP binds to CNG-channels in the plasma membrane keepping them open in the dark. CNG-channels close after hydrolysis of cGMP, which prevents Ca^{2+} from entering the cell. Since Ca^{2+} is further extruded by the exchanger after a light pulse, the cytoplasmic [Ca^{2+}] will decrease to a level below 100 nM [1,2]. The change in intracellular [Ca^{2+}] is a trigger for several Ca^{2+}-binding proteins to modulate the function of their target proteins [9, 10]. These include recoverin that inhibits the activity of rhodopsin kinase at high [Ca^{2+}] [11-13] and calmodulin that controls the affnity of cGMP for the CNG-channel [14-16]. Another key event in this Ca^{2+}-mediated feedback system is the control of a membrane bound guanylate cyclase (ROS-GC) by guanylate cyclase-activating proteins (GCAPs) [10,17,18]. ROS-GC activity increases when intracellular [Ca^{2+}] drops. This mechanism replenishes the exhausted cGMP pool and works in concert with several inactivation steps to restore the dark state of the cell. The light sensitivity of a photoreceptor cell (light adaptation) depends critically on the precise operation of the inactivation steps and on the control of several Ca^{2+}-dependent processes [19,20].

2. Expression of GCAP Forms

The GCAP subfamily of NCS proteins comprises now at least 9 different homologues (GCAP1-8 and GCIP, guanylate cyclase-inhibitory protein) and several orthologues in fish, frog, mice and men [21]. Prominent isoforms in mammals are GCAP1 and GCAP2 [22-25]; however there is a larger variety of GCAP forms in fish. Up to 6 GCAP forms were found in zebrafish (*Danio rerio*) and up to 8 isoforms in pufferfish (*Fugu rubripes*) [26]. Extensive biochemical characterization is limited to mammalian GCAP1 and GCAP2 so far. Immunocytochemical studies revealed that GCAP-1 and GCAP-2 are expressed in photoreceptor cells; however it is unclear whether the expression of each GCAP form is restricted to either rods or cones. Use of different species in immunocytochemical studies made it difficult to assign a rod or cone specific form. For example, antibodies directed against GCAP1 stained bovine outer and inner segments of both rods and cones and staining of cone synaptic pedicles could be clearly defined [23,25]. Another study used human, monkey and bovine retina and described significantly less labelling of GCAP1 in rod outer segments (ROS), but strong labelling of cone outer segments in these species [27]. An antibody directed against GCAP2 labelled outer and inner segments of bovine rods and cones [24]. This labelling pattern was not observed in another study that used monkey retinas where outer segments of rods and cones and rod inner segments were weakly labelled by an anti-GCAP2 antibody [28]. A comparative study on several mammalian retinas showed a species dependent labelling pattern for GCAP1 and GCAP2 [29]. Furthermore, for some species as human and zebrafish (but not mice) GCAP3 appeared as a cone specific isoform of GCAP [30]. Finally the subcellular distribution of GCAP1 or GCAP2 within a rod or cone cell has not been satisfactorily determined.

Recent evidence suggests that at least in bovine ROS are almost equal amounts of both GCAP forms present (approximately 3 μM each) [31]. At a first glance it seems to be unnecessary for a cell to harbour two proteins of nearly the same properties. However, several detailed investigations of the biochemical properties of GCAP1 and GCAP2 have revealed significant differences between both isoforms. In addition, investigation of transgenic mice lacking GCAP1, GCAP2 or both provided essential information to further uncover the physiological meaning of GCAP redundancy in outer segments. The present article will give first an overview on the different biochemical properties of GCAP1 and GCAP2 and will then discuss these findings in the light of photoresponse characteristics of transgenic mice.

3. Targets of GCAPs

Targets of GCAP1-3 in mammalian photoreceptor cells are two membrane bound guanylate cyclases, ROS-GC1 and ROS-GC2 (alternatively dubbed GC-E and GC-F or retGC1 and retGC-2). These cyclases are mainly expressed in outer segments of rods and cones and in other cell layers of the retina [18]. Expression of ROS-GC1 has also been observed in other neurosensory tissue [32, 3]. Several target sites of GCAP1 and GCAP2 in ROS-GC1 have been identified by several groups [34-36] and a discussion of these contradictory results has been presented elsewhere [18]. Very recent findings have clearly shown that the binding sites of GCAP1 and GCAP2 in ROS-GC1 reside on different domains (Figure 1) [37], GCAP1 binds and regulates ROS-GC1 via the juxtamembrane domain, whereas GCAP2 interacts with a site on the catalytic domain, that partially overlaps with an additional binding site for the ubiquitous Ca^{2+}-binding protein S100β. However, regulation of ROS-GC1 by S100β is not taking place in the outer segment. Instead, it is restricted to the photoreceptor synaptic region [38].

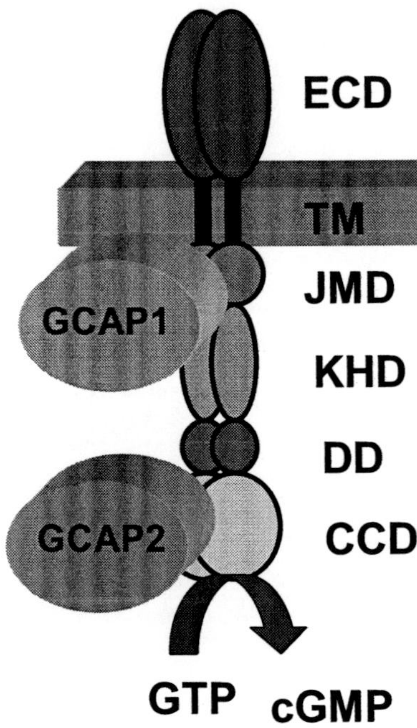

Figure 1. Model of ROS-GC1 containing two different target sites for GCAP1 and GCAP2. ROS-GC1 like other membrane bound guanylate cyclases consists of several domains: ECD, extracellular domain (intradiskal in the case of rod outer segment guanylate cyclases); TM, transmembrane domain; JMD, juxtamembrane domain (specific for GCs, not present in peptide-receptor GCs); KHD, kinase homology domain; DD, dimerization domain; CCD, cyclase catalytic domain. Membrane bound cyclases form dimers in order to catalyze the synthesis of cGMP from GTP. It is unclear, whether a complex of GCAP1, GCAP2 and ROS-GC1 as illustrated in the figure is the functional unit or whether mixed populations of ROS-GC1 with only one GCAP form are the physiological active unit in photoreceptor cells.

This current picture of GCAP modulation of ROS-GC1 activity comprising two nonoverlapping binding sites on different domains creates several questions. First, do both GCAPs bind with equal affinities to ROS-GC1? Do GCAP1 and GCAP2 bind under *in vivo* conditions to one ROS-GC1 molecule or does exist a mixed population of ROS-GC1 in complex with either GCAP1 or GCAP2?

The apparent affinity of GCAP1 and GCAP2 for native ROS-GC1 in bovine ROS membranes is almost equal with EC_{50} values of 0.13 μM and 0.29 μM, respectively [39]. When ROS-GC1 is heterologously expressed in COS cells the affinities are slightly lower (EC_{50} values of 0.8-1 μM and 6-8 μM have been reported for GCAP1 and GCAP2, respectively) [40, 41], but are still within one order of magnitude. Therefore it seems that different physiological roles of GCAP1 and GCAP2 are not determined by their affinities for the targets. Thus, the working unit of a heterogeneous complex such as in Figure 1 seems reasonable, but it needs further investigation to prove it.

4. Different Biochemical Properties of GCAP1 and GCAP2

GCAPs are heterogeneously acylated and harbor mostly a myristoyl group at the N-terminus [21-26]. Presence of a myristoyl group in GCAPs would suggest that GCAPs undergo a so-called Ca^{2+}-myristoyl switch like some NCS proteins as for example recoverin, neurocalcin δ, hippocalcin, VILIP1 and VILIP3 [42-47]. This switch is defined by two states of the protein. In its Ca^{2+}-bound form, the myristoyl group is exposed and anchors the protein to the membrane and in its Ca^{2+}-free form the myristoyl group is buried in a hydrophobic pocket. The reverse case, the myristoyl group is buried inside the protein in the Ca^{2+}-bound form and is exposed in the Ca^{2+}-free form, is also conceivable. Instead, GCAP1 and GCAP2 do not exhibit such a switch [48, 49]. The myristoyl group appears to be exposed in the Ca^{2+}-bound and Ca^{2+}-free form and therefore does not cause any Ca^{2+}-dependent transition of GCAPs from the cytoplasmic space to the membrane surface. Instead, myristoylation controls the properties of GCAPs in different ways as specified below.

O'Callaghan and Burgoyne have recently identified an amino acid sequence motif in the N-terminus of NCS-1, which might prevent the operation of a classical Ca^{2+}-myristoyl switch [50]. The corresponding motif EELTRK probably forces the N-terminal helix of NCS-1 in a rigid conformation when hydrogen bonds between the acid and basic amino acids are formed. Although an identical motif does not exist in GCAPs, one can find similar cluster of charged amino acids. For example, a similar interaction was proposed to occur in GCAP2 between glutamate residues at positions 26 and 33 and lysine residues at positions 29 and 30 [50]. The corresponding cluster in GCAP1 is found in the residues E16, E27, K22 and K23.

4.1. Shift in Ca^{2+}-Sensitivity

One of the most striking differences between GCAP1 and GCAP2 is the different Ca^{2+}-sensitivity with which they regulate guanylate cyclase activities. Myristoylated GCAP1 activates ROS-GC at higher free [Ca^{2+}] than myristoylated GCAP2; the IC_{50} values are 700 nM for GCAP1 and 100 nM for GCAP2 [31]. However, lack of acylation can shift the dose-response curve of GCAP1 to even higher [Ca^{2+}], whereas lack of acylation in GCAP2 does not cause such a shift [49]. A typical example of the Ca^{2+}-dependent dose-response curves for GCAP1 and GCAP2 is shown in Figure 2. In this particular case, the IC_{50} value was 644 nM Ca^{2+} for GCAP1 and 159 nM Ca^{2+} for GCAP2. It would broadly cover the range of drop in cytoplasmic [Ca^{2+}] during a light response. Such a response characteristic of two operating Ca^{2+}-sensors would enable the photoreceptor cell to react to incremental changes in [Ca^{2+}] with slightly different synthesis rates of cGMP.

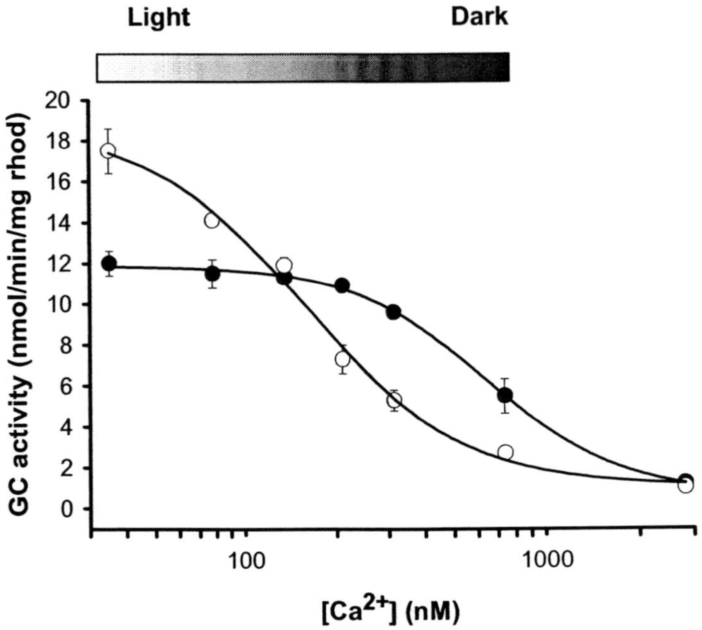

Figure 2. Membrane bound guanylate cyclase (ROS-GC1) activity in washed bovine ROS membranes as a function of [Ca^{2+}]. ROS were washed in hypotonic buffer to remove soluble proteins including endogenous GCAPs [31]. Membranes containing ROS-GC1 were reconstituted with recombinant forms of myristoylated GCAP1 (•) or myristoylated GCAP2 (○). ROS-GC1 activity was determined as a function of free [Ca^{2+}]. [Ca^{2+}] was adjusted with EGTA as a buffer. Final concentration of each GCAP was 1 μM. In order to ensure a sufficient high degree of myristoylation of GCAP1 during heterologous expression in E.*coli*, the mutant D^6S-GCAP1 was used. Halfmaximal activation expressed as EC_{50} value was at 644 nM Ca^{2+} for GCAP1 and 159 nM $Ca2^+$ for GCAP2. The horizontal bar in the figure illustrates the change of cytoplasmic [Ca^{2+}] during illumination in most vertebrate photoreceptor cells.

Another important factor that controls the Ca^{2+}-sensitivity of ROS-GC1 regulation by GCAPs is the free [Mg^{2+}]. Lowering [Mg^{2+}] lowers the concentration of Ca^{2+} at which the activation of ROS-GC1 is halfmaximal probably due to Mg^{2+}-binding properties of GCAPs [51,52]. Both activation curves in Figure 2 would therefore be shifted to lower [Ca^{2+}], but the relative difference in the IC_{50} values would remain.

The operation of two GCAP modes is consistent with distinct binding and regulatory sites for GCAP1 and GCAP2 in ROS-GC1 [34,37]. If the two modes operate in the same cell, many questions arise. Do these modes operate at the same time? Is there a trigger to switch from one mode to the other? And if so, what is the molecular identity of this trigger?

4.2. Catalytic Efficiency

The term k_{cat}/K_m is defined as catalytic efficiency. Both GCAPs increase k_{cat}/K_m of ROS-GC1, but this parameter depends on the presence of the myristoyl group on GCAP1 [31]. Catalytic efficiency of ROS-GC1 increases by a factor of 25, when GCAP1 is myristoylated, but only by a factor of 3.4 when it is nonmyristoylated. On the other hand, GCAP2 controls catalytic efficiency of ROS-GC1 almost independently of its myristoylation state.

4.3. Monomer-Dimer Equilibrium

GCAP2 dimerizes when Ca^{2+} is complexed by EGTA. It was therefore postulated that the dimeric form of GCAP2 activates ROS-GC1 and that the inhibitor-activator transition of GCAP2 coincides with a shift of the monomer-dimer equilibrium to the dimeric form [53]. In contrast, GCAP1 is only active as a monomer [54,55] and dimerization occurs as an irregular event due to point mutations in critical regions of the protein or due to denaturation and partial misfolding. One such critical region is EF-hand 1 in GCAP1. It is nonfunctional like in all other NCS proteins and serves an important role in target activation in both GCAP1 and GCAP2 [55, 56].

4.4. Phosphorylation

Recent evidence showed that GCAP-2 undergoes phosphorylation at its C-terminus by a retinal cyclic nucleotide-dependent protein kinase [57], but GCAP1 was not phosphorylated under the same conditions. Phosphorylation of GCAP2 has no significant effect on the regulatory properties of GCAP-2, so the physiological significance of these observations remains unclear and needs further investigation.

5. Ca^{2+}-Binding to GCAP1 and GCAP2

GCAPs harbour 4 EF-hand Ca^{2+}-binding motifs, of which 3 EF-hands – EF2, EF3 and EF4 – are capable of binding of Ca^{2+}. Thus, GCAPs share the domain structure with other homologues and orthologues of the NCS-protein family. The first EF-hand does not contain in the loop region the amino acid sequence to ensure binding of Ca^{2+} with high and moderate affinity. Instead, it serves as necessary constituent for targeting and regulating guanylate cyclases (see above). The Ca^{2+}-sensitivities of guanylate cyclase regulation suggests that binding of Ca^{2+} to EF-hands 2-3 must occur with moderate to high affinity with K_D values in the submicromolar range. Therefore, these EF-hands in GCAPs must sense changes of $[Ca^{2+}]$ in the submicromolar range. Binding and dissociation of Ca^{2+} then triggers Ca^{2+}-induced conformational changes in the GCAP molecule. Since GCAP1 and GCAP2 form a complex with their target guanylate cyclases at high and low $[Ca^{2+}]$ [34-37], it is presumed that the conformational changes in GCAP are transferred to the cyclase. By which mechanisms the Ca^{2+}-induced conformational changes in GCAP1 and GCAP2 trigger the activation and inhibition of guanylate cyclase is unclear at the moment.

5.1. GCAP1

Several lines of evidence indicate that the 3 Ca^{2+}-binding EF-hands in GCAP-1 and GCAP-2 fulfill different tasks. For example, only EF-hand 3 and 4 contribute to the Ca^{2+}-dependent activation of GCAP-1, since mutants with disabled Ca^{2+}-binding in either EF-hand 3 or 4 exhibits a shift in the Ca^{2+}-sensitivity and no complete inhibition at high $[Ca^{2+}]$, whereas mutants with a disabled EF-hand 2 show no shift in their Ca^{2+}-sensitivity [58]. When all 3 EF-hands were disabled, the resulting GCAP1 mutant is constitutively active.

Sensing of Ca^{2+} by GCAPs depends on the spatio-temporal distribution profile of Ca^{2+} in the outer segment [59, 60]. Decline of Ca^{2+} is delayed by radial and longitudinal diffusion of Ca^{2+}. This effect is particularly obvious in large amphibian ROS, it becomes less important in smaller sized mammalian rods [7,59,60]. The same consideration is true for the reverse step, the filling of EF-hands in GCAPs after reopening of CNG-channels and re-entry of Ca^{2+}. The association rate of Ca^{2+} with GCAP1 is close to the diffusion limit with a value of $k_1 > 2\times10^8$ $M^{-1}s^{-1}$ [61]. Thus, this rate is more affected by radial and longitudinal diffusion of Ca^{2+} in amphibian than in mammalian rods. Global dissociaton of Ca^{2+} from GCAP-1 appears to be rapid, but there exist no solid data about the affinities for Ca^{2+} of each single EF-hand. A first study in this direction has employed the accessibility of endogenous cysteine residues to monitor Ca^{2+}-induced conformational changes [62]. This approach allowed determining an apparent K_D of 2.9 μM for EF-hand 3. This rather low affinity for Ca^{2+} would result into a fast off-rate of 5.8×10^{-3} s^{-1} (taking into account the association rate described above). In addition, GCAP-1 must harbour at least one binding site for Ca^{2+} of very high affinity: if purified GCAP1 samples were passed over a Ca^{2+}-chelating (Chelex) column in order to remove any residual Ca^{2+}, only 2 Ca^{2+} can be bound per GCAP1 molecule in a subsequently performed $^{45}Ca^{2+}$-binding experiment. Denaturing Chelex treated-GCAP1 then releases the tighly bound Ca^{2+} [55]. This high affinity Ca^{2+}-binding site is probably located in EF-hand 4, since disabling of EF-hand 2 has no impact on the Ca^{2+}-sensitivity of GCAP1 [58]. One can conclude from these data, that rapid release of Ca^{2+} from EF-hand 3 triggers the conformational change in GCAP1, which subsequently leads to activation of ROS-GC1. The activating conformation of GCAP1 probably induces a reorientation of the dimerization domain in ROS-GC1 and thereby facilitates the formation of a better dimer interface. A correct dimer interface is essential for catalytic activity [41,63,64]. GCAPs influence the dimerization of ROS-GC1 in wildtype ROS-GC1 and in cone-rod dystrophy related mutants of ROS-GC1 that contain point mutations in the dimerization domain [41,64,65]. Finally, the transition state of cGMP synthesis is probably stabilized by the activating conformation of GCAP1 (and also GCAP2) [66].

5.2. GCAP2

GCAP2 also has 3 Ca^{2+}-binding EF-hands (the 2nd, 3rd and 4th) and undergoes a Ca^{2+}-induced conformational change that is not significantly different from the one observed in GCAP1. Binding of Ca^{2+} to GCAP2 occurs with an apparent K_D of 300 nM [67], but no specific assignment for the affinity of each EF-hand has been made so far. Imparting EF-hands in GCAP-2 showed the involvement of each EF-hand in the Ca^{2+}-sensitive regulation of guanlyate cyclase, but EF-hands 2 and 4 exert more influence on the Ca^{2+}-sensitivity than EF-hand 3 [68].

6. Physiological Role of GCAP1 and GCAP2 as Examined in Transgenic Mice

Disruption of GCAP-1 and GCAP-2 expression in transgenic mice confirmed previous conclusions about GCAPs′ central role in regulation of cGMP metabolismus in rods and cones. Recordings from transgenic animals demonstrated that the Ca^{2+}-dependent regulation of guanylate cyclase is an important determinant to shape the photoresponse of rods and cones [69-72]. Moreover, these experiments also showed that the contribution of GCAP-1 action to the response characteristics of a photoresponse differs from that of GCAP-2. Expression of GCAP-1 on a GCAP-1/GCAP-2 knockout background could rescue the normal wild type flash response [70], whereas the expression of GCAP-2 in GCAP null mice has led to a phenotype that is clearly different from the wildtype [69]. After reaching the maximum response amplitude it lacks the fast recovery that is typically seen in responses of wildtype mice. The first observation indicates that GCAP-1 would normally activate cyclase very early after a single flash. When GCAP1 is not expressed, acceleration of cGMP synthesis seems to be delayed. This situation is realized when the free [Ca^{2+}] in the cytoplasm has fallen to intermediate values that would normally trigger GCAP1 and in turn activate guanylate cyclase. Furthermore, in some cases the flash response ended in a prominent undershoot. The observed undershoot indicates a delayed activation of cyclase by GCAP-2. Together these observations are consistent with the results by Hwang et al. [31,49], that GCAP-2 is activated at lower free [Ca^{2+}] than GCAP-1. The operating ranges of free [Ca^{2+}] for GCAP-2 is narrower and is reached later after the onset of flash illumination. Expression of GCAP-1 can fully reverse the defects caused by the lack of GCAP-1 and GCAP-2. GCAP-1 activates ROS-GC1 over a larger range of free [Ca^{2+}] and it activates ROS-GC1 at higher free [Ca^{2+}] than GCAP-2. Therefore, it will operate earlier in the timeframe of a single-flash response than GCAP-2 does, which will compensate the lack of GCAP-2.

7. Conclusion

GCAP1 and GCAP2 share about 40 % identical amino acids and the arrangement of their main building blocks (EF-hands) is very similar. The three-dimensional structure of several NCS proteins including nonmyristoylated GCAP2 has been elucidated showing that the overall folding is also similar [67,73-76]. Therefore it has been suggested that GCAP1 and GCAP2 are almost identical in their main folding pattern. Although this is a reasonable assumption and might well account for the overall fold, it is nevertheless astonishing how diverse these proteins are with respect to their biochemical properties. It will be very interesting to learn which structural differences between GCAP1 and GCAP2 account for the biochemical differences summarized in this contribution. GCAPs seemed to be designed to respond to subtle differences in cytoplasmic [Ca^{2+}] and by this means provide a regulatory strategy for fine-tuning of phototransduction. The large variety of GCAPs found in teleost fish points to an even more complex network of Ca^{2+}-dependent regulatory features in rod and cone cells. Some GCAP forms might target to other proteins than guanylate cyclases, however all GCAPs share the EF-hand domain structure and very likely have a prominent function in neurosensory Ca^{2+}-signalling. How they perform the integration of different cellular Ca^{2+}-signals will be fascinating to unravel in future investigations.

Acknowledgements

Work in the laboratory of the author was funded by grants from the Deutsche Forschungsgemeinschaft (DFG).

References

[1] Pugh Jr, E.N., and, Lamb, T.D. 2000, *Handbook of Biological Physics*, D.G. Stavenga, W.J. DeGrip and E.N. Pugh Jr (Ed.), Elsevier Science B.V., 183.

[2] Burns, M.E., and Baylor, D.A. 2001, *Annu. Rev. Neurosci., 24*, 779.

[3] Matthews, H.R., and Reisert, J. 2003, *Curr. Opin. Neurobiol., 13*, 469.

[4] Lindemann, B. 2001, *Nature, 413*, 219.

[5] Fettiplace, R., and Ricci, A.J. 2003, *Curr. Opin. Neurobiol., 13*, 446.

[6] Jordt, S.-E., McKemy, D.D., and Julius, D. 2003, *Curr. Opin. Neurobiol., 13*, 487.

[7] Nakatani, K., Chen, C., Yau, K.-W., and Koutalos, Y., 2002, *Adv. Exp. Med. Biol., 514*, 1

[8] Woodruff, M. L., Sampath, A.P., Matthews, H.R., Krasnoperova, N.V., Lem, J., and Fain, G.L. 2002, *J. Physiol., 542*, 843.

[9] Koch, K.-W. 1995, *Cell Calcium, 18*, 314.

[10] Palczewski, K., Polans, A.S., Baehr, W., and Ames, J.B. 2000, *BioEssays, 22*, 337.

[11] Kawamura, S., Hisatomi, O., Kayada, S., Tokunaga, F., and Kuo, C.-H. 1993, *J. Biol. Chem., 268*, 14579.

[12] Gorodovikova, E.N., Senin, I.I., and Philippov, P.P. 1994, *FEBS Lett., 353*, 171.

[13] Klencin, V.A., Calvert, P.D., and Bownds, M.D. 1995, *J. Biol. Chem., 270*, 16147.

[14] Hsu, Y.-T., and Molday, R.S. 1993, *Nature, 361*, 76.

[15] Weitz, D., Zoche, M., Müller, F., Beyermann, M., Körschen, H.G., Kaupp, U.B., and Koch, K.-W. 1998, *EMBO J., 17*, 2273.

[16] Grunwald, M.E., Yu, W.-P., Yu, H.-H., and Yau, K.-W. 1998, J. Biol. Chem., 273, 9148.

[17] Dizhoor, A.M. 2000, *Cell. Signalling, 12*, 711-719.

[18] Koch, K.-W., Duda, T., and Sharma, R.K. 2002, *Mol. Cell. Biochem., 230*, 97.

[19] Pugh Jr, E.N., Nikonov, S., and Lamb, T.D. 1999, *Curr. Opin. Neurobiol., 9*, 410.

[20] Fain, G.L., Matthews, H.R., Cornwall, M.C., and Koutalos, Y. 2001, *Physiol. Rev., 81*, 117.

[21] Palczewski, K., Sokal, I., and Baehr, W. 2004, *Biochem. Biophys. Res. Comm., 322*, 1123.

[22] Palczewski, K., Subbaraya, I., Gorczyca, W.A., Helekar, B.S., Ruiz, C.C., Ohguro, H., Huang, J., Zhao, X., Crabb, J.W., Johnson, R.S., Walsh, K.A., Gray-Keller, M.P., Detwiler, P.B., and Baehr, W. 1994, *Neuron, 13*, 395.

[23] Frins, S., Bönigk, W., Müller, F., Kellner, R., and Koch, K.-W. 1996, *J. Biol. Chem., 271*, 8022.

[24] Dizhoor, A.M., Olshevskaya, E.V., Henzel, W.J., Wong, S.C., Stults, J.T., Ankoudinova, I., and Hurley, J.B. 1995, *J. Biol. Chem., 270*, 25200.

[25] Gorczyca, W.A., Polans, A.S., Surgucheva, I.G., Subbaraya, I., Baehr, W., and Palczewski, K. 1995, *J. Biol. Chem., 270*, 22029.

[26] Imanishi, Y., Yang, L., Sokal, I., Filipek, S., Palczewski, K., and Baehr, W. 2004, J. *Mol. Evol., 59*, 204.

[27] Kachi, S., Nishizawa, Y., Olshevskaya, E., Yamazaki, A., Miyake, Y., Wakabayashi, T., Dizhoor, A., and Usukura, J. 1999, *Exp. Eye Res., 68*, 465.

[28] Otto-Bruc, A., Fariss, R.N., Haeseleer, F., Huang, J., Buczylko, J., Surgucheva, I., Baehr, W., Milam, A.H., and Palczewski, K. 1997, *Proc. Natl. Acad. Sci., 94*, 4727.

[29] Cuenca, N., Lopez, S., Howes, K., and Kolb, H. 1998, *IOVS, 39*, 1243.

[30] Imanishi, Y., Li, N., Sokal, I., Sowa, M.E., Lichtarge, O., Wensel, T.G., Saperstein, D.A., Baehr, W., and Palczewski, K. 2002, *Eur. J. Neurosci., 15*, 63.

[31] Hwang, J.-Y., Lange, C., Helten, A., Höppner-Heitmann, D., Duda, T., Sharma, R.K., and Koch, K.-W. 2003, *Eur. J. Biochem., 270*, 3814.

[32] Duda, T., Venkataraman, V., Kishnan, A., Nagele, R.G., and Sharma, R.K. 2001, *Biochemistry, 40*, 4654.

[33] Duda, T., and Sharma, R.K. 2004, *FEBS Lett., 577*, 393.

[34] Lange. C., Duda, T., Beyermann, M., Sharma, R.K., and Koch, K.-W. 1999, *FEBS Lett., 460*, 27.

[35] Sokal, I., Haeseleer, F., Arendt, A., Adman, E.T., Hargrave, P.A., and Palczewski, K. 1999, *Biochemistry, 38*, 1387.

[36] Krylov, D.M., and Hurley, J.B. 2001, *J. Biol. Chem., 276*, 30648.

[37] Duda, T., Fik-Rymarkiewicz, E., Venkataraman, V., Krishnan, R., Koch, K.-W., and Sharma, R.K. 2005, *Biochemistry* - in press.

[38] Duda, T., Koch, K.-W., Venkataraman, V., Lange, C., Beyermann, M., and Sharma, R.K., *EMBO J., 21*, 2547.

[39] Hwang, J.-Y., and Koch, K.-W. 2002, *Biochim. Biophys. Acta, 1600*, 111.

[40] Krishnan, A., Goraczniak, R.M., Duda, T., and Sharma, R.K. 1998, *Mol. Cell. Biochem., 178*, 251.

[41] Duda, T., Venkataraman, V., Jankowska, A., Lange, C., Koch, K.-W., and Sharma, R.K. 2000, *Biochemistry, 39*, 12522.

[42] Zozulya, S., and Stryer, L. 1992, *Biochemistry, 89*, 11569.

[43] Kobayashi, M., Takamatsu, K., Saitoh, S., and Noguchi, T. 1993, *J. Biol. Chem., 268*, 18898.

[44] Ladant, D. 1995, *J. Biol. Chem., 270*, 3179.
[45] O'Callaghan, D.W., Ivings, L., Weiss, J.L., Ashby, M.C., Tepikin, A.V., and Burgoyne, R.D. 2002, *J. Biol. Chem., 277*, 14227.
[46] Spilker, C., Richter, K., Smalla, K.-H., Manahan-Vaughan, D., Gundelfinger, E.D., and Braunewell, K.-H. 2000, *Neuroscience, 96*, 121.
[47] Spilker, C., Dresbach, T., and Braunewell, K.-H. 2002, *J. Neurosci., 22*, 7331.
[48] Olshevskaya, E.V., Hughes, R.E., Hurley, J.B., and Dizhoor, A.M. 1997, *J. Biol. Chem., 272*, 14327.
[49] Hwang, J.-Y., and Koch, K.-W. 2002, *Biochemistry, 41*, 13021.
[50] O'Callaghan, D.W., and Burgoyne, R.D. 2004, *J. Biol. Chem., 279*, 14347.
[51] Peshenko, I.V., and Dizhoor, A.M. 2004, *J. Biol. Chem., 279*, 16903.
[52] Peshenko, I.V., Moiseyev, G.P., Olshevskaya, E.V., and Dizhoor, A.M. 2004, *Biochemistry, 43*, 13796.
[53] Olshevskaya, E.V., Ermilov, A.N., and Dizhoor, A.M. 1999, *J. Biol. Chem., 274*, 25583.
[54] Sokal, I., Li, N., Klug, C.S., Filipek, S.B., Hubbell, W.L., Baehr, W., and Palczewski, K. 2001, *J. Biol. Chem., 276*, 43361.
[55] Hwang, J.Y., Schlesinger, R., and Koch, K.-W. 2004, *Eur. J. Biochem., 271*, 3785.
[56] Ermilov, A.N., Olshevskaya, E.V., and Dizhoor, A.M. 2001, *J. Biol. Chem., 276*, 48143.
[57] Peshenko, I.V., Olshevskaya, E.V., and Dizhoor, A.M. 2004, *J. Biol. Chem., 279*, 50342.
[58] Rudnicka-Nawrot, M., Surgucheva, I., Hulmes, J.D., Haeseleer, F., Sokal, I., Crabb, J.W., Baehr, W., and Palczewski, K. 1998, *Biochemistry, 37*, 248.
[59] Gray-Keller, M., Denk, W., Shraiman, B., and Detwiler, P.B. 1999, J. Physiol., 519, 679.
[60] Nakatani, K., Chen, C., and Koutalos, Y. 2002, *Biophys. J., 82*, 728.
[61] Sokal, I., Otto-Bruc, A.E., Surgucheva, I., Verlinde, C.L.M.J., Wang, C.-K., Baehr, W., and Palczewski, K. 1999, *J. Biol. Chem., 274*, 19829.
[62] Hwang, J.-Y., Schlesinger, R., and Koch, K.-W. 2001, *FEBS Lett., 508*, 355.
[63] Liu, Y., Ruoho, A.E., Rao, V.D., and Hurley, J.H. 1997, *Biochemistry, 94*, 13414.
[64] Ramamurthy, V., Tucker, C., Wilkies, S.E., Daggett, V., Hunt, D.M., and Hurley, J.B. 2001, *J. Biol. Chem., 276*, 26218.
[65] Yu, H., Olshevskaya, E., Duda, T., Seno, K., Hayashi, F., Sharma, R.K., Dizhoor, A.M., and Yamazaki, A. 1999, *J. Biol. Chem., 274*, 15547.
[66] Koch, K.-W. 2002, *Adv. Exp. Med. Biol., 514*, 349.
[67] Ames, J.B., Dizhoor, A.M., Ikura, M., Palczewski, K., and Stryer, L. 1999, *J. Biol. Chem., 274*, 19329.
[68] Dizhoor, A.M., and Hurley, J.B. 1996, *J. Biol. Chem., 271*, 19346.
[69] Mendez, A., Burns, M.E., Sokal, I., Dizhoor, A.M., Baehr, W., Palczewsi, K., Baylor, D.A., and Chen, J. 2001, *PNAS, 98*, 9948.
[70] Howes, K.A., Pennesi, M.E., Sokal, I., Church-Kopish, J., Schmidt, B., Margolis, D., Frederick, J.M., Rieke, F., Palczewski, K., Wu, S.M., Detwiler, P.B., and Baehr, W. 2002, *EMBO J., 21*, 1545.

[71] Burns, M.E., Mendez, A., Chen, J., Baylor, D.A. 2002, *Neuron, 36*, 81.

[72] Pennesi, M.E., Howes, K.A., Baehr, W., and Wu, S.M. 2003, *PNAS, 100*, 6783.

[73] Ames, J.B., Ishima, R., Tanaka, T., Gordon, J.I., Stryer, L., and Ikura, M. 1997, *Nature, 389*, 198.

[74] Vijay-Kumar, S., and Kumar, V.D. 1999, *Nat. Struct. Biol., 6*, 80.

[75] Ames, J.B., Hendricks, K.B., Strahl, T., Huttner, I.G., Hamasaki, N., and Thorner, J. 2000, *Biochemistry, 39*, 12149.

[76] Bourne, Y., Dannenberg, J., Pollmann, V., Marchot, P., and Pongs, O. 2001, *J. Biol. Chem., 276*, 11949.

PART V: KChIPs

In: Neuronal Calcium Sensor Proteins
Editors: Pavel Philippov and Karl Koch, pp. 245-270
ISBN 1-59454-978-8

Chapter XIV

DREAM/KCHIP3/CALSENILIN: A MULTI-FUNCTIONAL PLAYER IN HEALTH AND DISEASE

***Hai-Ying M. Cheng*[1] *and Josef M. Penninger*[2,*]**
[1]Department of Neuroscience, The Ohio State University,
333 W. 10th Avenue, Columbus, Ohio 43210, USA;
[2]IMBA, Institute for Molecular Biotechnology of the Austrian Academy of Sciences,
Dr. Bohr Gasse 3-5, A-1030, Vienna, Austria.

Abstract

Between 1998 and 2000, the calcium-binding protein referred to as DREAM (downstream regulatory element antagonistic modulator), calsenilin or KChIP3 was identified in three independent studies, and attributed with three different cellular functions. As DREAM, it functions as a transcriptional repressor for the prodynorphin gene, amongst others. As calsenilin, it binds to the Alzheimer's disease-associated proteins known as presenilins. As KChIP3, it binds to Kv4 potassium channels and modulates their expression and function. This review will discuss the findings that have implicated DREAM/calsenilin/KChIP3 (hereafter referred simply as DREAM) in its various roles, in particular, key biochemical and genetic evidence upon which our current understanding of its mechanism of action is based. The latter half of the review focuses on the impact of DREAM function at the organismal level, including its involvement in pain modulation. Other biological events in which DREAM is likely to be implicated will also be discussed.

Keywords: DREAM, calsenilin, KChIP3, pain, addiction, calcium, gene expression, potassium channel, Alzheimer's disease

---

* Correspondence concerning this article should be addressed to (e-mail: hymcheng@yahoo.ca and josef.penninger@oeaw.ac.at).

Abbreviations

AA-NAT	arylalkylamine N-acetyltransferase
Aβ	amyloid β peptide
AD	Alzheimer's disease
AL	ad libitum
APP	amyloid precursor protein
cAMP	adenosine 3',5'-cyclic monophosphate
CBP	CREB-binding protein
CRE	cAMP response element
CREB	cAMP responsive element binding protein
CREM	cAMP responsive element modulator
DRE	downstream regulatory element
DREAM	downstream regulatory element antagonistic modulator
dsDNA	double-stranded DNA
Dyn	dynorphin
Fra-2	fos-related antigen-2
ICER	inducible cAMP early repressor
IFD	intermittent fasting diet
IL	interleukin
I_{SA}	subthreshold-activating transient K^+ current
I_{to}	transient outward current
KChIP	Kv channel interacting protein
LCD	leucine-charged residue-rich domain
LTP	long-term potentiation
MAPK	mitogen-activated protein kinase
ncAP-1	non-canonical activator protein-1 binding site
NCS	neuronal calcium sensor
PI3K	phosphatidylinositol-3 kinase
PKA	cAMP-dependent protein kinase A
PS	presenilin
RT-PCR	reverse transcription-polymerase chain reaction
Tg	thyroglobulin
THC	tetrahydrocannabinol
TSH	thyroid-stimulating hormone
TTF	thyroid transcription factor
UTR	untranslated region

1. Introduction

Calcium (Ca^{2+}) is one of the smallest but arguably most versatile second messengers in biological systems. An increase in intracellular Ca^{2+} concentration can induce changes in gene expression, cell proliferation, apoptosis, secretion, fertilization, muscular contraction and neuronal activities [1]. These Ca^{2+}-dependent events are mediated by a variety of Ca^{2+}-binding proteins, with members of the EF-hand-containing neuronal calcium sensor (NCS) protein superfamily garnering a great deal of scientific attention in the last decade [2].

Since its discovery in 1998 as a presenilin-binding protein and a putative player in Alzheimer's disease, DREAM, a novel NCS protein, has been attributed additional biological roles, including transcriptional repression of the prodynorphin gene and modulation of Kv4 potassium (K^+) channel function [3-5]. This review will provide a comprehensive examination of DREAM in each of these capacities. The review is organized in the following manner: 1) synopsis of the seminal publications leading to the cloning of DREAM; 2) cellular and anatomical distribution; 3) structure of the gene and protein; 4) *in vitro* effects of DREAM; 5) regulation of DREAM gene expression and/or protein function; 6) DREAM function as determined in animal models; and 7) prospective roles of DREAM in health and disease.

2. Identification of DREAM/Calsenilin/kchip3: A Retrospective

2.1. Calsenilin-A New Player in Alzheimer's Disease

In 1998, Buxbaum et al. [3] reported the discovery of a novel binding partner for the Alzheimer's disease (AD)-related proteins, presenilins (PS) 1 and 2. In brief, AD is characterized by a progressive decline in cognitive function, amyloid plaque aggregates in the brain, and neurodegeneration. Early-onset AD is an inherited disorder, with frequent mutations in the genes for amyloid precursor protein (APP), PS1 and PS2 [6-8]. Mutations in these three genes all lead to increased production of aggregate-prone amyloid Aβ42 peptides, which are proteolytic derivatives of APP. Presenilins have been strongly suggested to act as the γ-secretase that cleaves APP at the critical position to release Aβ42 [9-11]. Moreover, multiple lines of evidence indicate that presenilins may promote programmed cell death or apoptosis, and, in particular, that the C-terminal domain of PS mediates this pro-apoptotic effect. Based on the rationale that the C-terminus of PS2 plays an important role in apoptosis, Buxbaum et al. [3] used a yeast two-hybrid approach to search for protein factors that interacted physically with this domain, and in so doing cloned calsenilin (Figure 1).

The open reading frame of calsenilin encodes four EF-hand motifs and shows significant homology to the NCS family of proteins. Although the EF-hand motif(s) could coordinate with Ca^{2+} *in vitro* and the mobility of calsenilin could be shifted by Ca^{2+}, indicating a Ca^{2+}-dependent conformational change, in mammalian overexpression systems the calsenilin-presenilin interaction could be detected in the absence of an overt calcium signal [3]. Interestingly, mirroring the consequences of caspase-3 (a pro-apoptotic protease) activation, overexpression of calsenilin resulted in alternative proteolytic processing of PS2 to yield a smaller C-terminal cleavage fragment [3], the significance of which remained unknown. However, since altered Ca^{2+} homeostasis may underlie PS-mediated apoptosis, Buxbaum et al. [3] suggested that calsenilin, as both a Ca^{2+}- and a presenilin-binding protein, might be an important mediator between aberrant Ca^{2+} signaling and AD pathogenesis.

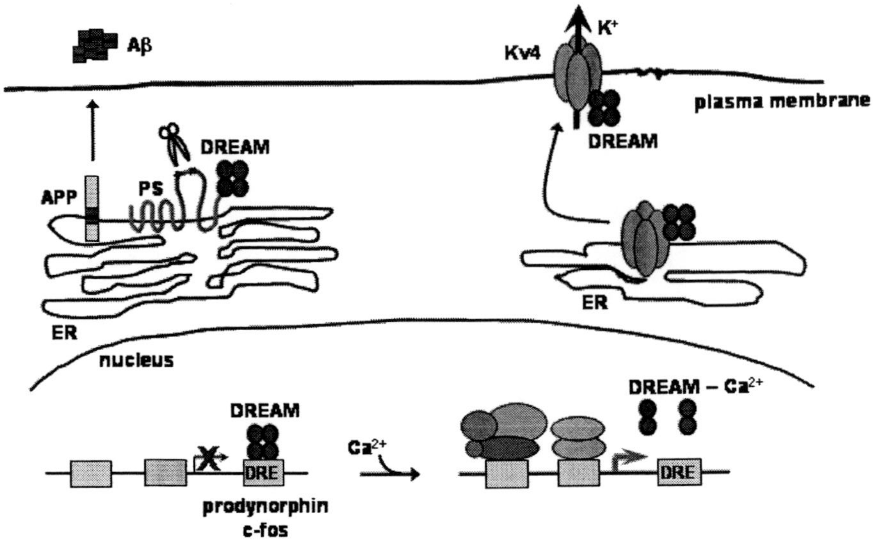

Figure 1. The multiple cellular roles of DREAM. DREAM has been demonstrated to interact physically with presenilins (PS), Kv4 potassium channels and the downstream regulatory elements (DRE) of the prodynorphin and c-fos genes. The interaction of DREAM with PS influences the proteolysis of PS as well as the production of amyloid peptides (Aβ) from amyloid precursor protein (APP). The interaction with Kv4 results in direct modulation of K+ channel activity and possibly trafficking of Kv4 subunits to the plasma membrane. In the nucleus, DREAM binds to DRE under low calcium concentration and represses gene transcription. Increased calcium levels abrogate the DREAM-DRE association, leading to transcriptional derepression. ER, endoplasmic reticulum.

2.2. DREAM-a Calcium-Dependent Transcriptional Repressor

In an independent study, Carrion et al. [4] set out to identify DNA-binding proteins that may regulate prodynorphin gene transcription. Prodynorphin is one of three genes encoding for the classical endogenous opioid peptides (the other two being proenkephalin and proopiomelanocortin). Within the central nervous system, prodynorphin gene expression is altered under various contexts, including exposure to addictive drugs, chronic pain conditions following tissue or nerve damage, specific forms of seizures, and processes of learning and memory, prompting the suggestion that prodynorphin contributes to the neuroadaptive response in each of these aforementioned conditions [12-14]. Studies addressing the mechanisms by which transcription of the prodynorphin gene is regulated have revealed enormous complexity (Figure 2). At least five elements in the rat prodynorphin gene promoter—cAMP response elements or CREs (DynCRE-1, -2 and -3) in the distal region, a proximal non-canonical activator protein-1 (ncAP-1)-binding site and a CRE-like element (DynCRE-4) situated past the transcriptional start site—contribute differentially to its induction in a tissue- and stimulus-specific manner [12,15,16]. A sixth regulatory element downstream of the +1 transcription start site of the human prodynorphin gene was identified by Carrion et al. [17], and was shown to repress transcription under basal conditions. In addition, cAMP-dependent protein kinase A (PKA) activation mediated transcriptional derepression at this downstream regulatory element (DRE) site. Importantly, dsDNA corresponding to the DRE sequence could bind to a 110-kDa nuclear protein complex expressed in human neuroblastoma cell lines, indicating the existence of a *trans*-acting protein factor(s) that could mediate transcriptional repression at the DRE site [17]. Subsequent screening of a human caudate expression library using DRE dsDNA led to the identification of DREAM, or downstream regulatory element antagonistic modulator (Figure 1) [4].

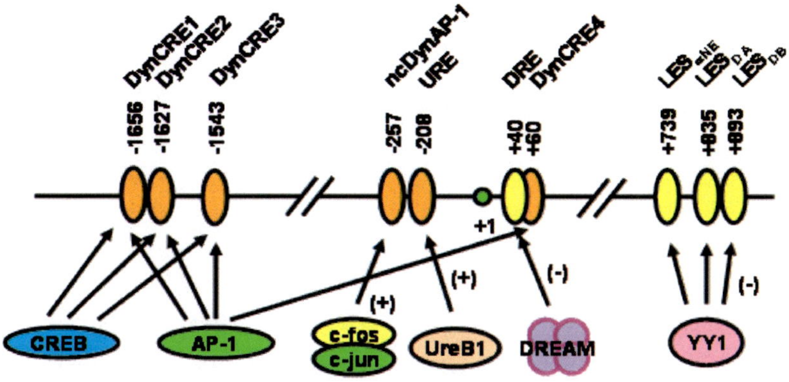

Figure 2. Transcriptional regulatory elements in the human and rat prodynorphin gene. The distal promoter of the rat prodynorphin gene contains three regulatory sequences which function as cyclic AMP-responsive element (CRE)/activator protein 1 (AP-1) mixed elements, and denoted DynCRE-1, -2 and –3. A noncanonical AP-1 (ncAP-1) element resides in the proximal rat prodynorphin gene promoter, and an asymmetric CRE-like element (DynCRE-4) in exon 1. DynCRE-1 to –4 have been shown to bind CRE-binding protein (CREB) and AP-1 transcriptional complexes, while the ncAP-1 element can associate with c-fos/c-jun heterocomplexes. An upstream regulatory element (URE) in the rat prodynorphin gene, and intragenic Leu-enkephalin encoding sequences (LES) of α-neoendorphin ($LES_{\alpha NE}$), dynorphin A (LES_{DA}) and dynorphin B (LES_{DB}) in the human counterpart, have been reported to associate with URE binding protein 1 (UreB1) and the transcription factor YY1, respectively. A downstream regulatory element (DRE) was discovered in the human prodynorphin gene which regulated both basal and protein kinase A (PKA)-induced transcription. The DRE antagonistic modulator (DREAM) was later found to bind specifically to the DRE site and to function as a transcriptional repressor of the prodynorphin gene under basal conditions. Numbers above the orange and yellow ovals indicate the nucleotide positions of the regulatory elements relative to the +1 transcription start site in the rat and human prodynorphin genes, respectively. (+) indicates positive effect on transcription, and (-) indicates a negative effect.

The original published amino acid sequences of DREAM and calsenilin were nearly identical, with the exception of an extended and divergent N-terminus in the case of DREAM [3,4]. DREAM was shown to bind as a tetramer to DRE sequences of the *prodynorphin* and *c-fos* genes with high affinity and specificity, with increasing calcium concentration abrogating this DREAM-DRE interaction [4]. Overexpression of DREAM inhibited basal transcription from a DRE-containing promoter in a calcium-reversible fashion. Calcium-dependent DNA binding and transcriptional repression were mediated by direct binding of calcium to EF-hands II, III and/or IV of DREAM, indicating that DREAM might be the first transcription factor to be directly regulated by calcium [4].

2.3. KChIPs-Novel Modulators of Kv4 Potassium Channels

Finally, a third study undertaken to identify binding partners for Kv4 potassium channels unearthed a family of proteins named KChIPs (Kv Channel Interacting Protein) [5]. A-type potassium (K+) currents (referred to as I_{SA} in neurons and I_{to} in cardiomyocytes), which are mediated by Kv4 channels, have distinguishing features, including the fact that they are gated by voltage, are activated at membrane potentials below the threshold for sodium spike generation, and are inactivated rapidly by depolarization [18]. Because of these attributes, A-type currents can regulate the timing and frequency of action potential discharges, and therefore govern the firing pattern and excitability of neurons and cardiomyocytes. A-type currents may contribute to such physiological processes as synaptic integration, vasodilation and relaxation, cardiac function under normal and diseased conditions, and learning and memory, amongst others [19]. However, heterologous expression of Kv4 channels do not recapitulate all of the kinetic properties of native A-type currents, suggesting that auxiliary factor(s) may work in concert with Kv4 channel subunits *in vivo* to generate A-type currents. Based on this speculation, An et al. [5] identified two structurally related proteins, KChIP1 and KChIP2, in a yeast two-hybrid screen in which the N-terminus of rat Kv4.3 was used as the bait. KChIP1 and 2 shared significant sequence similarity with DREAM, which was renamed KChIP3 by the authors of this study (Figure 1).

The *in vitro* study by An et al. [5] suggested a role for KChIPs in protein trafficking and modulation of K+ channel activity. While singly expressed Kv4.2 and KChIP1 were found in the ER/Golgi and diffusely in the cytoplasm, respectively, co-expression resulted in redistribution of both proteins to the plasma membrane [5]. Secondly, expression of Kv4.2 in combination with either KChIP1, 2 or 3, in Chinese hamster ovary cells resulted in the generation of A-type currents which more faithfully resembled properties of native I_{SA}/I_{to} (e.g., a shift in the mid-point of voltage activation to more hyperpolarized potentials, and an increase in the rate of recovery from inactivation), as well as increased A-current density by ten-fold [5]. Thus, it was suggested that KChIPs may play a role in promoting the stable expression of Kv4 at the plasma membrane, in addition to a direct modulatory effect on Kv4 activity [5].

3. DREAM/Calsenilin/KChIP3: Cellular and Anatomical Distribution

DREAM protein is expressed in nuclear as well as membrane-associated compartments of the cell [3-5, and H-YMC and JMP unpublished observations]. Using Northern blot analyses, DREAM mRNA has been detected in multiple tissues, including robust expression in the brain, spinal cord, reproductive tissues, thymus, thyroid, pituitary gland and bone marrow [20-22, and H-YMC and JMP unpublished observations]. However, RT-PCR profiling indicated a low-level expression pattern in nearly all tissues examined [H-YMC and JMP unpublished observations]. Within the central nervous system, DREAM immunoreactivity is found in the cerebellum, hippocampus, retrosplenial granular cortex and multiple sensory processing regions, including the olfactory bulb, optic tract, superior colliculus, and several thalamic relay centers such as the anterior dorsal, medial geniculate, dorsolateral geniculate, ventral posteromedial and ventral posterolateral nuclei [23]. *In situ* hybridization and real-time single-cell RT-PCR have indicated the presence of DREAM transcripts in the dorsal and ventral horns of the spinal cord, dopaminergic neurons of the substantia nigra, caudate putamen, globus pallidus and certain brainstem nuclei [20,22,24].

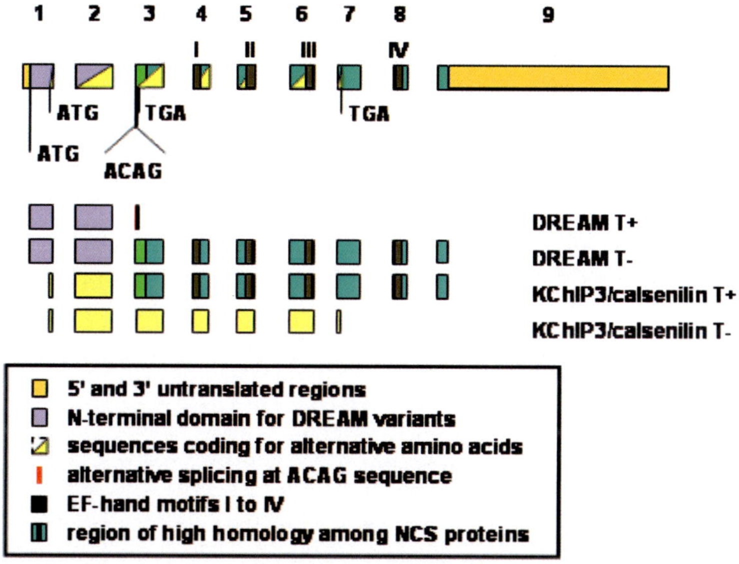

Figure 3. Potential protein isoforms encoded by the mouse DREAM/calsenilin/KChIP3 gene. The mouse DREAM/calsenilin/KChIP3 gene is comprised of nine exons. There are two potential ATG translation start codons which are out-of-frame relative to each other. Alternative splicing at the junction of exons 2 and 3 can result in the inclusion or exclusion of an ACAG sequence. EF-hand motifs I, II, III and IV are encoded by exons 4, 5, 6 and 8, respectively. If the first ATG were used and the ACAG sequence excluded, the resulting translation product would be an NCS-like protein with four EF-hands and a unique N-terminal domain (DREAM T-). Inclusion of the ACAG tetramer is predicted to give rise to an N-terminal truncation variant termed DREAM T+. A protein (KChIP3/calsenilin T+) which is identical to DREAM T- but bearing a different N-terminal domain would arise if translation began at the second ATG and the ACAG tetramer were present in the transcript. Use of the second ATG and the absence of the ACAG sequence would result in a protein (KChIP3/calsenilin T-) which is identical to KChIP3/calsenilin T+ in the N-terminus but carries a somewhat shorter C-terminal domain (due to introduction of an in-frame stop codon in exon 7) with no EF-hands or homology to any other known protein. The same color on different putative translation products schematically denotes regions where the amino acid sequence is identical. Exonic regions which may encode for different amino acids are indicated with an embedded triangle. (Reprinted with permission from Hai-Ying M Cheng, Josef M Penninger, When the DREAM is gone:from basic science to future prospectives in pain management and beyond, *Expert Opinion on Therapeutic Targets (*2003) 7(2), 249-263).

4. DREAM/calsenilin/KChIP3: Structure-Function Analysis of the Gene and Protein

Mouse DREAM, calsenilin and KChIP3 are encoded by a single gene comprised of nine exons on chromosome 2 [20] (Figure 3). Exon 1 contains two potential ATG translation start codons that are out-of-frame relative to each other but deemed to be equally well-suited for translation initiation [20]. In addition, alternative splicing at the junction of exons 2 and 3 results in two different pools of cDNAs that differ by the presence (T+ variant) or absence (T- variant) of an ACAG sequence immediately preceding the region of similarity with other NCS proteins [20]. Based on the potential for alternative splicing and alternative translation initiation sites, a maximum of four protein isoforms are predicted. Two of these, DREAM T- and calsenilin/KChIP3 T+, resemble some other NCS-like proteins in that they have four C-terminal EF-hand motifs but possess divergent N-terminal domains. Inclusion of the ACAG sequence in the DREAM T+ variant introduces an in-frame stop codon in exon 3 and gives rise to a protein containing only the N-terminal domain of DREAM T- (i.e., the EF-hand motifs are absent). The KChIP3/calsenilin T- variant is identical to KChIP3/calsenilin T+ in the N-terminus but carries a shorter C-terminal domain with no EF-hands or homology to any other known protein. Although the existence *in vivo* of all four 'hypothetical' protein isoforms of DREAM/calsenilin/KChIP3 remains to be proven, at the very minimum, alternative splicing is a shared feature of the KChIP family, and splicing variants of KChIP1 and KChIP2 which differ in the N-terminal region but retain the EF-hand motifs have been identified [25,26]. The possible significance of these alternative isoforms will be discussed in greater detail in Section 5, but may include functional diversification of DREAM due to targeting of distinct isoforms to certain subcellular compartments and/or differential regulation via isoform-specific post-translational modification.

5. *In Vitro* Effects of DREAM/Calsenilin/KChIP3

5.1. DREAM in Processes of Apoptosis and Amyloidogenesis

A number of *in vitro* studies have addressed the role of DREAM in regulating transcription, apoptosis and other presenilin-dependent processes, and potassium channel function.

In terms of apoptosis, overexpression of DREAM induced apoptosis spontaneously in HeLa cells [27] or enhanced apoptosis induced by serum withdrawal or thapsigargin (a potent inhibitor of ER/Golgi Ca^{2+} pumps which raises intracellular [Ca^{2+}]) in neuroglioma cells [28]. Conversely, antisense knockdown of DREAM expression suppressed apoptosis in Jurkat cells induced by Ca^{2+} ionophores or thapsigargin [27], while in cultured cortical and hippocampal neurons, blockade of DREAM expression protected against Aβ42-induced neurotoxicity [29]. Apoptosis induced by DREAM overexpression was potentiated by co-expression of PS2 [27], suggesting that presenilin and DREAM cooperate functionally to promote apoptosis. In two studies, Jo et al. [27,30] showed that DREAM-induced apoptosis was also potentiated by increased Ca^{2+} levels and AD-associated mutant forms of APP and PS, and was attenuated by γ-secretase inhibitors. In contrast, a role for DREAM in inhibiting apoptosis via a separate mechanism (i.e., transcriptional derepression of the pro-apoptotic gene, *hrk*) was demonstrated by Sanz et al. [31,32]. Although the exact mechanism of DREAM-induced cell death remains unknown, overexpression of DREAM increased the proteolytic activity of calpain and caspase following the induction of apoptosis [28]. DREAM itself was shown to be an *in vitro* substrate for caspase-3-mediated cleavage [33]. The cumulative data suggest a role of DREAM in regulating apoptotic processes.

Furthermore, overexpression of DREAM may contribute to amyloid peptide accumulation. Aβ42 production that is induced by overexpression of wild-type APP, AD-associated APP mutants or wild-type PS2 is further augmented by co-expression of DREAM [27]. Although endogenous DREAM protein is not associated with the γ-secretase complex [34], a recent study suggested that overexpression of DREAM/calsenilin/KChIP3 enhanced γ-secretase activity in cells and that purified DREAM protein enhanced Aβ production in an *in vitro* cell-free system [35].

5.2. DREAM as a Transcriptional Regulator: Effects of Calcium

The role of DREAM as a DNA-binding protein and transcriptional regulator has been explored in a number of studies (Figure 4). Tetrameric and, to a lesser extent, dimeric forms of DREAM have the ability to bind to DNA *in vitro* [4]. Tetramerization of DREAM occurs when the EF-hand motifs are not bound by Ca^{2+} (apo-state), whereas Ca^{2+}-bound DREAM exists preferentially as a dimer [36]. Thermodynamic studies reinforced the observation that the affinity for DRE-binding is much greater for apo-DREAM (tetramer) than for Ca^{2+}-bound DREAM (dimer) [36].

5.3. Transcriptional Regulation by the DREAM-CREM Interaction

Besides Ca^{2+} other signals have been shown to mediate DREAM-dependent derepression. Activation of PKA results in both prodynorphin gene induction and loss of DREAM binding to the DRE (Figure 4) [17,37]. Subsequently, it was discovered that DREAM could physically associate with cAMP responsive element modulator α (αCREM), a protein mediator of cAMP signals (Figure 4) [38]. DREAM-dependent derepression at DRE sites could be achieved by overexpressing αCREM or εCREM, but not cAMP responsive element binding protein (CREB) or other isoforms of the CREM gene (βCREM, ICER-I) [38]. The functional interaction between αCREM and DREAM was shown to depend on protein interaction motifs known as leucine-charged residue-rich domains (LCD) present in both molecules, as well as PKA-dependent phosphorylation of αCREM [38]. Moreover, using LCD mutants of DREAM which were unable to interact with αCREM, it was shown that that Ca^{2+}- versus αCREM-mediated derepression at DRE sites could occur independently [38].

5.4. PI3K-Dependent Regulation of DREAM

The phosphatidylinositol-3 kinase (PI3K) pathway has also been implicated in the regulation of DREAM (Figure 4) [31]. In hematopoietic cells, the proapoptotic gene, *hrk*, which is constitutively repressed under normal growth conditions, is rapidly induced upon withdrawal of the cytokine, interleukin (IL)-3 [39]. A functional DRE site was identified in the *hrk* gene, and IL-3 withdrawal (or increased Ca^{2+} concentration) abrogated the binding of DREAM to DRE-*hrk* [31]. Blocking PI3K activity also abrogated this interaction, suggesting that an active or ongoing signal, leading to PI3K-dependent phosphorylation of DREAM, is required to maintain DREAM-dependent repression of the *hrk* gene in this system.

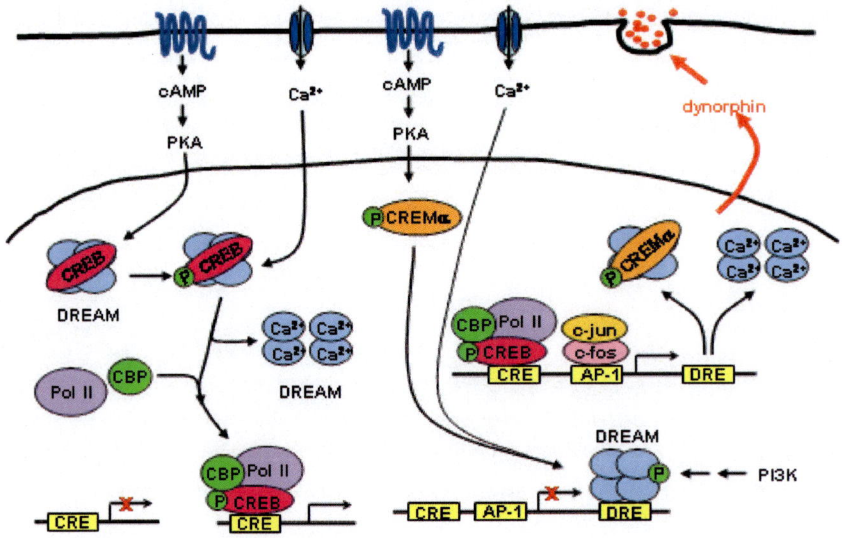

Figure 4. Mechanisms regulating transcriptional repression by DREAM at DRE and CRE sites. Under low calcium concentrations, DREAM can block the expression of cAMP-responsive genes *via* binding with CREB. The DREAM-CREB physical interaction prevents the binding of activated (phospho-) CREB both to CRE sites as well as to the transcriptional coactivator, CBP. Increasing calcium concentration abrogates the binding of DREAM to CREB, the latter now free to recruit CBP to CRE sites and activate transcription. In the case of DRE-dependent repression, low calcium levels result in direct occupancy of the DRE site by DREAM. Binding of Ca^{2+} or phosphorylated CREMα to DREAM relieves this repression, and allows transcription to occur following transactivating events at upstream regulatory sites. For example, induction of prodynorphin expression requires DREAM-dependent derepression at the DRE site as well as transactivation at CRE and/or AP-1 sites. In the case of the *hrk* gene, repression at its DRE site requires PI3K-dependent phosphorylation of DREAM. (Reprinted with permission from Hai-Ying M Cheng, Josef M Penninger, When the DREAM is gone:from basic science to future prospectives in pain management and beyond, *Expert Opinion on Therapeutic Targets (*2003) 7(2), 249-263).

5.5. DREAM and the Pineal Gland: Implications for Circadian Rhythms

Besides *hrk*, *prodynorphin* and *c-fos*, DRE sites were also identified within the genes for ICER, Fos-related antigen-2 (Fra-2), and arylalkylamine N-acetyltransferase (AA-NAT), all three of which contribute to the day-night rhythm of melatonin synthesis in the pineal gland [40]. DREAM was demonstrated to bind to the DRE sites of ICER, Fra-2 and AA-NAT, and repressed both basal and inducible transcription from reporters driven by these promoters in which the DRE sites were present [40]. DREAM-DRE binding itself exhibited circadian rhythmicity in the rat pineal gland, peaking in the day, a time when ICER, Fra-2 and AA-NAT mRNA levels are at their lowest [40]. The rhythmic DRE-binding activity may be regulated at the level of clock-controlled post-translational modification(s) of DREAM that may affect tetramerization and/or DREAM protein stability [40]. Regardless of the mechanisms underlying rhythmic DREAM activity in the pineal gland, these data suggest a role of DREAM in regulating the expression of genes involved in circadian rhythms.

5.6. DREAM-Dependent Transcriptional Regulation in the Thyroid Gland

Another tissue in which DREAM may function in its capacity as a transcriptional repressor is the thyroid gland. The thyroid-specific transcription factors Pax8 and Thyroid Transcription Factor-2 (TTF-2), which are critical for thyroid cell growth and differentiation, contain DRE sites in their 5'-untranslated regions (5'-UTR) [41]. Recombinant DREAM protein was able to bind to the DREs of Pax8 and TTF-2 *in vitro*, and blockade of intracellular Ca^{2+} release by thapsigargin in a DREAM-expressing rat thyroid cell line enhanced the DREAM-DRE association [41]. The thapsigargin-dependent enhancement of DREAM-DRE binding correlated with a decrease in Pax8 and TTF-2 expression in these cells [41]. Overexpression of an EF-hand mutant form of DREAM, which is unable to bind to Ca^{2+}, reduced proliferation of thyroid cells by altering cell cycle progression [41].

Another thyroid-specific gene, thyroglobulin (*Tg*), carries a DRE consensus sequence 70 base pairs upstream from the TATA box, in a region that overlaps with the binding sites for Pax-8 and TTF-1, which, along with TTF-2, are the main regulators of *Tg* transcription [42]. The DRE sequence of thyroglobulin bound to DREAM *in vitro* and functioned as a transcriptional silencer in reporter assays where DREAM was co-expressed [42]. In addition to direct DNA binding to *Tg*DRE, DREAM could physically interact with the N-terminus of TTF-1 protein, which contains the main transactivating domain, and block its activity [42]. Overexpression of DREAM in *Tg*-expressing thyroid cell lines decreased *Tg* mRNA and Tg protein levels [42]. Furthermore, DREAM-dependent repression of *Tg* gene expression was abrogated by activating the thyroid-stimulating hormone (TSH)/cAMP pathway or increasing intracellular Ca^{2+} levels [42]. Together, the data suggest that DREAM is able to block TTF-1-mediated transactivation of *Tg* gene expression in a cAMP- and Ca^{2+}-dependent manner.

5.7. The Effect of DREAM on CREB/CRE-Dependent Transcription

In addition to DRE-mediated repression, DREAM may affect CRE-dependent transcription via a direct physical interaction with CREB *in vivo* (Figure 4) [43]. This physical association blocked the binding of CREB to both CRE sites and to the transcriptional coactivator, CREB-binding protein (CBP) [43]. The DREAM-CREB interaction was not dictated by the activation (i.e., phosphorylation) status of CREB but, instead, was sensitive to increased concentration of Ca^{2+}, which bound to the EF-hand domains of DREAM and abrogated the binding to CREB [43]. The authors postulated a direct competition between DREAM and CBP for binding to CREB, all three of which carry LCD motifs [43]. DREAM-dependent derepression at CRE sites appeared to be a common event downstream of various stimulations, including K^+-induced depolarization, phorbol ester-dependent activation of protein kinase C, and EGF- and NGF-dependent MAPK activation [43]. Antisense knockdown of endogenous DREAM confirmed its role in basal transcriptional repression at both CRE and DRE sites [43]. The study by Ledo et al. [43] suggested that in addition to activity-dependent phosphorylation of CREB, transcriptional activation at CRE sites depends on an intermediate regulatory step which releases CREB from DREAM binding and repression. It further implied that DREAM's ability to modulate transcription is dependent on a variety of stimuli related to neuronal excitation, and that there are potentially many other downstream target genes that are transcriptionally regulated by DREAM either in a CRE- or DRE-dependent manner.

5.8. The Role of DREAM in Potassium Channel Modulation, Protein Trafficking and Subunit Assembly

The original observation by An et al. [5] indicated a role for KChIPs, including DREAM (KChIP3), in direct modulation of the kinetic properties of A-type currents, via a physical association with the N-termini of Kv4 subunits. Heterologous expression of Kv4 genes alone produces K^+ currents that differ from native A-type currents, in that half-activation and half-inactivation voltage values are less hyperpolarized, the kinetics of inactivation are much slower, and kinetic properties are discrete rather than fall along a broad continuum [44,45]. Co-expression of DREAM or KChIP1 or 2 restored most of the native features of A-type currents that were otherwise absent when Kv4 was singly expressed, with the exception that the rate of inactivation showed a further reduction [5,46]. The precise effects on the kinetic properties of A-type currents appear to depend on which KChIP member, and which particular splice variant of that KChIP, is expressed [47,48]. Secondly, since KChIP co-expression resulted in a 10-fold increase in A-current density, an additional role in protein trafficking was proposed [5]. Extending on the observations of An et al. [5], Shibata et al. [49] demonstrated that overexpression of Kv4.2 alone in COS-1 cells resulted in distribution of Kv4.2 in a perinuclear fashion typical for proteins which are retained in the endoplasmic reticulum. Co-expression of DREAM (or KChIP1 or 2) caused redistribution of Kv4.2 protein to the cell surface, with residual intracellular Kv4.2 being localized to the Golgi rather than the endoplasmic reticulum [49]. The mechanism by which DREAM or KChIPs affect protein trafficking is unclear, but it has been proposed that the N-terminal cytoplasmic domain of Kv4 mediates retention of the protein in the endoplasmic reticulum, and that KChIPs may serve to mask this domain, facilitating channel trafficking to the plasma membrane [25]. Palmitoylation of DREAM and other KChIPs might also contribute to trafficking of Kv4 channels to the plasma membrane, since mutation of these cysteines reduced plasma membrane localization of KChIPs and the associated Kv4 as well as KChIP-mediated enhancement of A-type current density [50].

A recent study indicated that DREAM may have an additional role in promoting subunit assembly of Kv4 channel subunits [51]. Mutants of Kv4.2 incapable of binding an intersubunit zinc ion (Zn^{2+}) did not form functional tetramers at the plasma membrane, but remained in the endoplasmic reticulum [51]. DREAM co-expression rescued this defect, promoting tetramerization of mutant Kv4.2 and the subsequent expression of A-type currents, albeit with lower current density and altered kinetic properties compared with wild-type Kv4.2 plus DREAM [51].

6. Regulatory Mechanisms Underlying DREAM Gene Expression and Protein Function

6.1. The Effects of Calcium

The original studies indicated the critical involvement of Ca^{2+} in regulating DREAM function [3-5]. The affinity of DREAM for DNA-binding is greatly diminished when the EF-hand domains (most likely EF-hands III and/or IV) are complexed with Ca^{2+} [4,36,52]. Furthermore, Ca^{2+} binding results in a conformational change in the secondary and/or tertiary structure of DREAM [36,53]. However, recent data indicate that magnesium (Mg^{2+}) may be constitutively bound to EF-hand II under physiological concentrations, and that bound Mg^{2+} is essential for sequence-specific binding of DREAM to DRE sites [52].

In addition to altering DNA binding, Ca^{2+} may alter the subcellular distribution of DREAM. In human neuroglioma cells overexpressing DREAM, treatment with caffeine or calcium ionophores, both of which elevate intracellular Ca^{2+} concentration, resulted in enhanced translocation of DREAM to the nucleus [54]. This effect was not mediated by direct binding of Ca^{2+} to DREAM, since EF-hand mutants of DREAM retained the ability to translocate to the nucleus in response to increased calcium levels [54]. Despite the apparently contradictory observations that increased calcium concentration serves both to recruit a transcriptional repressor into the nucleus as well as to abrogate its DNA binding capacity, the authors suggest that the calcium signal functions to effect transcriptional derepression, at the same time ensuring that once intracellular calcium levels return to baseline there will be sufficient quantities of DREAM in the nucleus to provide adequate repression of target genes [54].

6.2. The Effects of Palmitoylation

Subcellular distribution of DREAM, and, consequently, DREAM's capacity to function in different biological contexts (as a transcriptional repressor, presenilin-binding protein or K^+ channel modulator), are governed by different post-translational modifications. Palmitoylation of cysteine residues in the N-termini of DREAM, as well as other KChIP splicing variants, may play a key role in their trafficking to the plasma membrane. When co-expressed with Kv4.3 channels in HEK293 cells, DREAM protein was detected at the plasma membrane as well as in other intracellular compartments [50]. Mutations of cysteines 47 and 48 in the N-terminus of DREAM greatly reduced plasma membrane localization [50]. Furthermore, the palmitoylation-deficient mutant form of DREAM showed reduced ability to enhance A-type current density compared with wild-type DREAM, suggesting that palmitoylation promotes DREAM/KChIP-mediated trafficking of Kv4 channels to the plasma membrane [50]. Interestingly, DREAM is a substrate for caspase-3-mediated cleavage, which separates the conserved EF-hand-containing C-terminus from the N-terminal domain in which the palmitoylated cysteines reside [55]. Therefore, caspase-3-dependent proteolysis may be another mechanism that regulates subcellular localization of DREAM.

6.3. Phosphorylation and Dephosphorylation of DREAM

Phosphorylation and dephosphorylation of DREAM have been demonstrated in a number of studies. Choi et al. [55] detected phospho-DREAM under basal conditions in stably transfected neuroglioma cells. Increased calcium levels promoted phosphorylation of DREAM in a manner that was independent of direct Ca^{2+}-binding to the EF-hands of DREAM [55]. *In vitro* kinase assays indicated that casein kinase I can phosphorylate DREAM [55]. Phospho-DREAM is a substrate for dephosphorylation by protein phosphatase 1 and/or protein phosphatase 2A [55]. Interestingly, caspase-3-dependent cleavage of DREAM was inhibited by phosphorylation at serine 63 of DREAM [55]. In a study of hematopoietic progenitor cells, Sanz et al. [31] demonstrated that DREAM was phosphorylated following IL-3 stimulation via a PI3K-dependent but Akt-independent pathway. The authors suggested that PI3K-dependent phosphorylation of DREAM would increase its DNA binding affinity for the DRE sequence of the *hrk* gene [31]. Notably, it has been suggested by Campos et al. [56] that forskolin, which increases cAMP levels, promotes prodynorphin gene transcription in a kinase-dependent manner, whereas caffeine, which increases Ca^{2+} levels, does so in a kinase-independent fashion [56]. Whether or not kinase activity (in the case of forskolin stimulation) results in direct phosphorylation of DREAM was not resolved in this study [56], but at the very least, it suggested that DREAM-dependent derepression can occur by distinct mechanisms.

6.4. Protein-Protein Interactions

Direct interaction with other proteins has been demonstrated to modulate DREAM's ability to regulate transcription. As mentioned previously, DREAM can physically associate with certain CREM isoforms (α and ε) via the LCD domains present in both molecules [38]. The DREAM-CREM interaction led to derepression at DRE sites [38]. On the other hand, a physical association between DREAM and CREB, again via LCD motifs, interfered with the ability of CREB to bind to CRE sites as well as to CBP, resulting in a reduction in CRE-mediated transcription [43].

6.5. Regulation of DREAM Gene and Protein Expression

Regulation of DREAM gene expression has also been investigated in a number of studies. Hong et al. [57] found that DREAM protein expression was decreased in the cortex and the CA3 region of the hippocampus in mice 24 hours following kainic acid-induced seizure and status epilepticus. A similar reduction in DREAM expression was seen in a cell culture model of seizure-like activity using hippocampal neurons, as well as in the hippocampus of epilepsy patients [57]. In an independent study, Matsu-Ura et al. [58] demonstrated that seizures induced by pentylenetetrazol (an antagonist of GABA receptors expressed by inhibitory neurons) or kainic acid in mice produced a modest increase in DREAM mRNA levels in the hippocampus 7-8 hours following injection and returned to baseline after 2 days. The authors of this study [58] correlated the intensity of the seizures, and presumably the degree of neuronal activity, with the expression of DREAM: in mice which showed stronger convulsions, induction of DREAM mRNA was both stronger and faster. Pentylenetetrazol-induced DREAM expression was dependent on N-methyl-D-aspartate (NMDA) receptor activity [58]. Other regions of the mouse brain, including the cerebral cortex, amygdala and piriform cortex, also showed enhanced DREAM mRNA levels following pentylenetetrazol injections [58]. The discrepancy in the findings by Hong et al. [57] and Matsu-Ura [58] is unclear, as is the functional significance of changes in DREAM expression following seizures. Hong et al. [57] proposed that downregulation of DREAM may have a role in cell death associated with status epilepticus, while Matsu-Ura et al. [58] suggested that induction of DREAM may contribute to feedback repression of late responsive genes after seizure stimulation or, more generally, to neuroplasticity-related mechanisms.

DREAM protein expression has been found to be upregulated in the cortex of Alzheimer's disease patients [29,59]. While DREAM immunoreactivity was predominantly neuronal in normal brains, Jin et al. [59] detected expression of DREAM in reactive astrocytes of AD brains. Furthermore, they found that the caspase-derived proteolytic fragment of DREAM was detected only in the cytosol of AD brains, prompting the suggestion that DREAM may play a role in apoptosis and in Alzheimer's disease pathology [59]. Secondarily, Jo et al. [29] demonstrated that DREAM protein levels were also elevated in the neocortex and hippocampus of aged transgenic mice expressing an AD-related mutant form of APP, and that exposure of cultured cortical and hippocampal neurons to Aβ42 peptides increased DREAM mRNA and protein expression *in vitro*.

6.6. Splicing Variants of DREAM

Some studies have indicated a differential expression of individual splicing variants of DREAM in certain cell types. Liss et al. [24] demonstrated that the long splice variant of DREAM (termed KChIP3.1 in their study) was exclusively expressed in the dopaminergic neurons of the substantia nigra. Using single-cell RT-PCR, it was shown that the expression level of KChIP3.1 was linearly correlated with the A-type current density in these neurons [24]. Within the rat pituitary, a long (ACAG+) and a short (ACAG-) variant were identified in all cell types examined, with the long variant comprising two-thirds of the DREAM expression in mammotropes, somatotropes, mammosomatotropes, and corticotropes, and only less than one-half of the total expression in thyrotropes and gonadotropes [21]. The function of DREAM in the pituitary, and particularly of the different splicing variants, remains to be clarified.

7. The *In Vivo* Roles of DREAM/Calsenilin/ KChIP3: Insights Obtained from Knock-Out Mice

7.1. Overview of the Phenotype of DREAM Knock-Out Mice

In vitro studies suggested a plethora of biological functions for DREAM/calsenilin/KChIP3. In order to investigate the essential function(s) of DREAM in a physiological setting, we generated dream knock-out (*dream*$^{-/-}$) mice and asked the question, "What are the biological consequences resulting from the lack of DREAM expression?" Given the pleiotropic functions of DREAM *in vitro*, *dream*$^{-/-}$ mice were initially screened for any structural, functional and/or behavioural deficits using a battery of assays [22]. In general, *dream*$^{-/-}$ mice were phenotypically uncompromised compared with wild-type littermate controls. General health, longevity and fertility of *dream*$^{-/-}$ animals appeared normal. Although DREAM was shown to bind to the DRE site of the *c-fos* gene (which has a pivotal role in immune function and bone development) [4,60-63], *dream*$^{-/-}$ mice exhibited no deficit in terms of the development and function of the immune system. This included: cellularity of the thymus, spleen, lymph nodes and bone marrow; relative abundance of subpopulations of hemapoietic cells in each of these compartments; *in vitro* proliferation of T and B cells; thymocyte development; thymocyte apoptosis; and calcium flux of T and B cells (H-YMC and JMP, unpublished observations). Moreover, a comprehensive histological examination indicated that the absence of DREAM had no overt effect on the structural integrity of any tissue, including the brain and heart, where presenilins and Kv4 channels, respectively, have prominent functions (H-YMC and JMP, unpublished observations) [27,28,64,65]. There was no evidence for the existence of supernumerary neurons in *dream*$^{-/-}$ brains, as predicted from the pro-apoptotic function of DREAM (H-YMC and JMP, unpublished observations) [27]. Gross malformation of, and disorganization of neurons within, *dream*$^{-/-}$ brains, which would otherwise suggest problems related to neuronal excitability, were also not observed (H-YMC and JMP, unpublished observations). Heart morphology and heart-to-body mass ratios were likewise phenotypically normal in *dream*$^{-/-}$ mice (H-YMC and JMP, unpublished observations).

Initial behaviour phenotyping of *dream*$^{-/-}$ mice indicated the lack of a deficit in many of the assays. These included: locomotor responses under non-stressful (locomotor activity box) and stressful (open-field test) conditions; the degree of anxiety in the open-field test and light/dark activity box; motor coordination and balance in three independent tasks (beam walking, balance task, rotarod); fine motor control in tasks of skilled reaching of the forelimbs; performance in the standard Morris watermaze task to assess acquisition of a spatial task; and passive avoidance to footshock (H-YMC and JMP, unpublished observations). However, a more in-depth analysis revealed that *dream*$^{-/-}$ mice suffered from a mild impairment in limb posturing in a grid-walking task, as well as a strong deficit in the reversal learning paradigm of the Morris watermaze, which requires the subject to learn a new spatial task (platform location) on subsequent days of training (H-YMC and JMP, unpublished observations). The overall conclusion from this set of observations is that DREAM does not play an *essential* and *generalized* role in locomotion, motor coordination, anxiety, and spatial and non-spatial learning and memory, although specific aspects of spatial learning and motor control may be influenced by DREAM expression.

7.2. DREAM as a Modulator of Pain Processing

The most striking phenotype displayed by *dream*$^{-/-}$ mice was a marked reduction in pain behaviour across a panel of different pain tests (see Table 1 for a definition of terms related to pain processes) [22]. *Dream*$^{-/-}$ mice exhibited a delayed response to noxious heat applied to the tail (in the tail-flick test), as well as an increased pain threshold to mechanical pressure applied to the paw. However, normal sensory perception of innocuous tactile stimuli remained intact in these animals, indicating that the deficits in pain response are restricted to nociception and arguing against a global impairment of sensory processing.

Pain behaviour provoked by tonic rather than acute nociceptive stimulation was examined using the formalin model. Subcutaneous injection of formalin into the paw produces characteristic licking and flinching behaviour over a protracted time, which can be divided into three phases: an early phase (0-5 min) resulting from the direct activation of peripheral afferent nerves/fibres; a quiescent phase known as the 'interphase' (5-10 min); and a late phase (10-60 min) that is believed to arise from central sensitization as well as ongoing peripheral afferent activity [66,67]. *Dream*$^{-/-}$ mice exhibited a reduction in the magnitude of the pain response in both early and late phases of the formalin test compared with wild-type controls [22].

To determine whether the pain phenotype in *dream*$^{-/-}$ mice was specific to noxious stimulation of somatic tissues only or was, in fact, a more generalized phenomenon, we examined the behaviour of these animals in response to chemical irritation of the viscera [22]. Intraperitoneal injection of magnesium sulfate ($MgSO_4$) or acetic acid evoked less pain behaviour (abdominal writhes) in *dream*$^{-/-}$ mice [22]. Given the link between DREAM and prodynorphin, it is interesting to note that visceral pain is enhanced in κ-opioid receptor 1 knockout mice, which lack the receptor selective for dynorphin peptides, implying that this endogenous antinociceptive system is activated following chemical insult to the viscera [68]. Furthermore, the utility of κ-agonists against visceral pain has been widely described [69 70].

Inflammatory pain of a neurogenic and non-neurogenic origin was assessed using the capsaicin and carrageenan models, respectively. Injection of capsaicin or carrageenan into the paw produces mechanical hypersensitivity (allodynia), but capsaicin additionally elicits an acute licking response similar to the acute phase of the formalin test. Also, capsaicin-induced allodynia is short-lasting compared with that induced by carrageenan (<1 hr versus < 4 days, respectively). In both tests, *dream*$^{-/-}$ mice displayed a blunted allodynic response, and capsaicin-evoked licking behaviour was also reduced [22]. DREAM may function like a rheostat rather than an on-off switch in allowing a certain level of pain transmission: in the carrageenan model, only the magnitude of the allodynia, but not the kinetics of recovery, was attenuated in *dream*$^{-/-}$ mice [22]. Furthermore, the blunted pain response in *dream*$^{-/-}$ mice was not attributed to a primary deficit in the inflammatory response, as the extent of paw swelling induced by either agent was indistinguishable between wild-type and *dream*$^{-/-}$ mice [22].

Finally, we assessed chronic neuropathic pain in *dream*$^{-/-}$ mice following peripheral nerve injury. Neuropathic pain is a manifestation of a dysfunctional nervous system, usually arising from lesions to the peripheral or central nervous system. Symptoms of neuropathic pain include spontaneous pain (pain in the absence of sensory stimuli) and tactile alloydnia. Treatment of neuropathic pain poses a major clinical challenge, with controversy surrounding whether it is resistant to opioid analgesics such as morphine, or whether it simply responds to higher opioid doses compared with other forms of pain [71-75]. Human peripheral nerve injury can be mimicked in rodents by chronic constriction of the sciatic nerve, producing a state of long-lasting and severe tactile allodynia [76]. Relative to wild-type controls, *dream*$^{-/-}$ mice maintained a marked attenuation in tactile allodynia up to 3 weeks following nerve cuffing [22]. Furthermore, spontaneous behaviour, which was observed in nerve-constricted wild-type mice, was not evident in *dream*$^{-/-}$ mice [22].

Table 1. Definition of 'Pain' Terms.

allodynia	Pain in response to a stimulus which does not normally provoke pain
analgesia	Absence of pain in response to a stimulus which would normally provoke pain
hypoalgesia	Diminished pain in response to a pain-provoking stimulus
hyperalgesia	Increased (exaggerated) pain in response to a stimulus which is normally painful
neuropathic pain	Pain initiated or caused by a primary lesion or dysfunction in the nervous system
nociception	The detection of tissue-damaging (noxious) stimuli; the sensory component of pain
noxious stimulus	A stimulus which is damaging to normal tissues
pain	An unpleasant sensory and emotional experience associated with actual or potential tissue damage, or described in terms of such damage

As defined by the International Association for the Study of Pain (http://www.iasp-pain.org/terms-p.html)

Our study [22] indicated that the absence of DREAM suppressed pain behaviour irrespective of the modality (thermal, mechanical or chemical) of the noxious stimulus, the temporal nature (acute, tonic or chronic) of the pain state, or the tissue type (somatic or visceral) affected (see Table 2 for a summary of pain behaviour of *dream*$^{-/-}$ mice and Figure 5 for a proposed model of DREAM's involvement in pain modulation). The analgesia displayed by *dream*$^{-/-}$ mice in the thermal and mechanical pain tests, as well as the carrageenan model, could be blocked by the administration of naloxone (a *pan*-opioid receptor antagonist) or nor-BNI (a κ-opioid receptor-selective antagonist) [22]. Thus, the hypoalgesic phenotype arising from DREAM ablation is likely due to tonic activation of κ-opioid receptors, or enhanced κ-opioid receptor activation following somatosensory stimulation. Consistent with the pharmacological data, we found that levels of prodynorphin mRNA and dynorphin peptides were elevated in the spinal cords of *dream*$^{-/-}$ mice compared with wild-type animals [22].

Table 2. Summary of Pain Behaviours in *dream*$^{-/-}$ Mice.

	Response	sensitive to		
		naloxone	nor-BNI	MK-801
Tail Flick	7.0 ± 0.6 (4.5 ± 0.3)	yes (no)	yes (no)	no (no)
Von Frey (tactile sensitivity)	42 ± 5 (36 ± 3)	n/d	n/d	n/d
Paw Pressure (noxious)	239 ± 51 (109 ± 32)	yes (no)	yes (no)	n/d
Formalin				
• acute (0-5 min)	58 ± 9 (106 ± 11)	n/d	yes (no)	n/d
• tonic (10-50 min)	141 ± 25 (440 ± 81)	n/d	yes (yes)	n/d
Visceral Pain				
• $MgSO_4$	1.3 ± 0.6 (8.1 ± 0.8)	n/d	n/d	n/d
• acetic acid	15 ± 4 (38 ± 5)	n/d	n/d	n/d
Capsaicin				
• licking (0-5 min)	33 ± 3 (48 ± 7)	n/d	n/d	n/d
• von Frey	13 ± 2 (6 ± 1)	n/d	n/d	n/d
Carrageenan				
• von Frey	21 ± 4 (12 ± 2)	n/d	yes (no)	n/d
Neuropathic Pain				
• von Frey	28 ± 6 (7 ± 1)	yes (no)	yes (no)	no (yes)

Responses of wild-type animals are given in parentheses ( ).
Tail flick, formalin and capsaicin (licking) data are represented in seconds.
Von Frey and paw pressure data are represented in g/mm^2.
Visceral pain data are represented as number of writhes within 5 min ($MgSO_4$) or 20 min (acetic acid) post-injection.
Von Frey sensitivity determined at 20 min (capsaicin) or 4 hr (carrageenan) post-injection.
Neuropathic model: Mosconi and Kruger. Von Frey sensitivity determined at 18 days post-cuff implantation.
Naloxone (1 mg/kg, i.p.); nor-BNI (2 mg/kg, s.c.); MK-801 (0.05 mg/kg, s.c.).
n/d = not done.

In the neuropathic model, naloxone or nor-BNI administered 3 weeks following nerve injury had only a partial effect in *dream*$^{-/-}$ mice in terms of enhancing tactile allodynia [22], indicating that the attenuated allodynic behaviour of *dream*$^{-/-}$ mice is not fully dependent on ongoing opioid receptor activation. It is possible that other non-opioid mechanisms are contributing to the lessened degree of allodynia in *dream*$^{-/-}$ mice at later stages of peripheral neuropathy. Alternatively, activation of opioid receptors prior to and in the early stages following initial nerve injury may potentially reduce subsequent neuroplastic changes (e.g., central sensitization), either by reducing the afferent nociceptive drive via a presynaptic mechanism and thereby inhibiting excitatory neurotransmitter release, or directly depressing the excitability of central (spinal) nociceptive neurons via a post-synaptic mechanism. Therefore, acute pharmacological blockade of κ-opioid receptors would not be expected to restore the pain behaviour of *dream*$^{-/-}$ mice to wild-type levels at late time-points post-nerve injury. Consistent with the argument that central sensitization is dampened in the absence of DREAM, our study demonstrated that an MK-801 (an NMDA receptor antagonist)-sensitive component of the tactile allodynia, normally found in nerve-injured animals, was absent in *dream*$^{-/-}$ mice, implying that NMDA receptor activation, or its behavioural consequences, is suppressed in these animals [22]. Since dynorphins and other κ-agonists have been demonstrated to inhibit neurotransmission in spinal cord neurons [77,78], the lack of NMDA receptor involvement in the hypoallodynic response of neuropathic *dream*$^{-/-}$ mice suggests to us that there may be insufficient activity of spinal neurons in these animals to engage NMDA receptors.

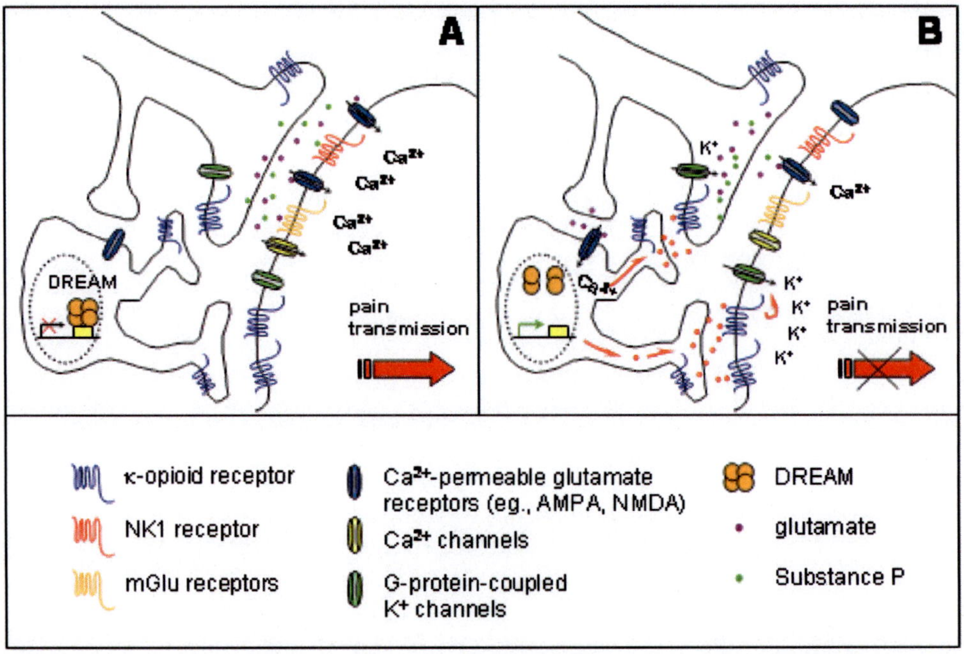

Figure 5. A proposed model for DREAM's potential to gate pain transmission. A) A noxious stimulus from the periphery elicits release of excitatory neurotransmitters, principally glutamate and, if the intensity is sufficiently high, substance P, from the central terminals of C-fibers. Under basal conditions or at the time of noxious stimulation, DREAM is repressing prodynorphin expression. Pain transmission is therefore permitted. B) Noxious input may concomitantly activate DREAM-expressing cells, presumably inhibitory interneurons. Calcium influx leads to DREAM-dependent derepression of the prodynorphin gene. Production and secretion of dynorphin peptides lead to the activation of κ-opioid receptors (κOR), which are coupled to potassium channels. If the κORs are located post-synaptically on second-order dorsal horn neurons, their activation would hyperpolarize these neurons, rendering them less excitable to further neurotransmitter stimulation. If the κORs are located pre-synaptically on the central terminals of afferent fibers, their activation may inhibit the release of excitatory neurotransmitters, preventing subsequent noxious input from being transmitted to the dorsal horn.

The role of DREAM in pain modulation has been suggested in two additional studies. Lilliehook et al. [79] independently generated calsenilin-deficient mice, and assessed their pain response to acute thermal noxious stimuli using the tail-flick test. Similar to our results [22], *calsenilin*$^{-/-}$ mice demonstrated a four-fold increase in response latency to noxious heating of the tail [79]. The mechanistic basis underlying the reduced pain response was not determined in this study [79]. However, in the electric foot shock assay, a lower current intensity was sufficient to elicit stereotypical responses (i.e., vocalization and running) in *calsenilin*$^{-/-}$ mice compared with wild-type controls [79]. The discrepancy between the behaviour of *calsenilin*$^{-/-}$ mice in these two tests was not addressed. In contrast, de los Santos-Arteaga et al. [80] were investigating the effect of diet on pain modulation. Compared with mice that were fed *ad libitum* (AL), mice that were on an intermittent fasting diet (IFD) displayed reduced pain responses to thermal heating of the paws (hot-plate test) and intraperitoneal injections of acetic acid [80]. The analgesia observed in IFD mice was reversed by naloxone or nor-BNI, indicating a κ-mediated effect [80]. Importantly, prodynorphin mRNA levels were increased in the spinal cords of IFD mice [80]. Furthermore, nuclear protein extracts prepared from the spinal cords of IFD mice showed less DREAM-DNA binding activity as determined in electrophoretic mobility shift assays using DRE probes [80]. Since c-Fos expression (a marker of neuronal activity) following noxious stimulation was reduced in IFD mice compared with AL mice, it was suggested that the increased κ-tone in IFD mice serves to dampen nociceptive transmission [80]. From the studies by Cheng et al. [22], Lilliehook et al. [79] and de los Santos-Arteaga [80], it is clear that DREAM plays an important physiological role in the modulation of pain processes.

7.3. A Role for DREAM in Amyloidogenesis and Hippocampal LTP

The study by Lilliehook et al. [79] implicated DREAM/calsenilin in presenilin- and Kv4-dependent processes. Levels of Aβ40 (the non-amyloidogenic form) and Aβ42 were decreased in the cortex and cerebellum of *calsenilin*$^{-/-}$ mice [79]. Long-term potentiation (LTP) in the dentate gyrus of the hippocampus of *calsenilin*$^{-/-}$ was enhanced relative to wild-type controls [79]. Enhanced LTP in *calsenilin*$^{-/-}$ mice may be due to a reduction in A-type current, since application of phrixotoxin, a spider venom that specifically blocks Kv4 channels, facilitated the induction of LTP in wild-type but not *calsenilin*$^{-/-}$ animals [79].

7.4. DREAM in Drug Addiction: Modulating the Effects of Cannabinoids

In a subsequent study, we investigated the role of DREAM in processes of drug addiction [81]. Drugs of abuse, such as cocaine, morphine (opioid), and marijuana (cannabinoid), can have rewarding (positive motivational) and/or aversive (negative motivational) effects, as a result of their action on neural circuits which mediate motivation. In addition, chronic drug exposure can result in physical dependence, which becomes evident in the form of physical (somatic) symptoms of withdrawal during periods when drug use is ceased and the neuroadaptive changes brought about by chronic drug exposure are unmasked.

The conditioned place paradigm—which uses preference or aversion for an environment previously associated with drug exposure as an index of drug reward or aversion, respectively—was employed to determine the role of DREAM in the positive and negative motivational effects of morphine, cocaine, and the cannabinoid, delta(9)-tetrahydrocannabinol (THC). Neither the rewarding effects of morphine, nor the aversive effects of naloxone-induced withdrawal in morphine-dependent subjects, were impaired in *dream*$^{-/-}$ mice [81]. Physical dependence to morphine was also not altered in these animals [81]. Moreover, cocaine reward and aversion were unaffected by the absence of DREAM as determined using the conditioned place preference and conditioned taste aversion paradigms, respectively [81]. However, the sensitivity of *dream*$^{-/-}$ mice to the aversive effects of THC was greatly potentiated: a low dose of THC, which produced place neutrality in wild-type mice, elicited place aversion in *dream*$^{-/-}$animals [81]. Given the fact that the phenotype of 'ongoing' analgesia in *dream*$^{-/-}$ mice was due to an enhanced action of the dynorphin/κ - opioid receptor system, it is interesting to note that administration of κ-agonists by themselves is aversive [82]. Therefore, an altered κ-opioid system may underlie the hypersensitivity of *dream*$^{-/-}$ mice to the aversive effects of THC. Indeed, pretreatment of *dream*$^{-/-}$ mice with nor-BNI prior to THC conditioning reduced the aversive effects of low-dose THC in these animals [81], supporting the argument that the potentiation in THC aversion is a consequence of enhanced κ-opioid receptor activation. Furthermore, the involvement of DREAM in drug-induced aversion is specific for THC, since the aversive properties of such compounds as lithium chloride and naloxone were unaffected in *dream*$^{-/-}$ mice [81].

8. Concluding Remarks: Prospective Roles of Dream in Health and Disease

The cumulative data obtained thus far from animal models indicate that DREAM is critically involved in pain modulation and motivational processes engaged by cannabinoids, as a result of its role as a transcriptional repressor for the prodynorphin gene and, hence, a regulator of the κ-opioid system [22,81]. Additionally, DREAM (calsenilin) is required to maintain a balanced production of amyloid peptides and to regulate hippocampal LTP, likely through an effect on presenilin-dependent proteolytic cleavage and A-type current expression [79]. The impact of DREAM on the prodynorphin/κ-opioid system, as well as on presenilin and K^+ channel function, suggests that the biological consequences of loss of DREAM expression may be more far-reaching than initially envisioned. A more thorough investigation into the role of DREAM in cognition is warranted, given that all three biological systems in which DREAM has been implicated have been shown to contribute to processes of learning and memory. The observation that DREAM expression is altered by seizure-inducing compounds raises the issue of the role of DREAM in epilepsy. Furthermore, the expression of DREAM in non-neural tissues such as the pituitary, thyroid, reproductive organs and immune cells promises to extend our understanding of the biological significance of DREAM beyond the nervous system.

References

[1] Berridge, M.J., Lipp, P., and Bootman, M.D. 2000, *Nat. Rev. Mol. Cell. Biol., 1*, 11.

[2] Burgoyne, R.D., and Weiss, J.L. 2001, *Biochem. J., 353*, 1.

[3] Buxbaum, J.D., Choi, E.K., Luo, Y., Lilliehook, C., Crowley, A.C., Merriam, D.E., and Wasco, W. 1998, *Nat. Med., 4*, 1177.

[4] Carrion, A.M., Link, W.A., Ledo, F., Mellstrom, B., and Naranjo, J.R. 1999, *Nature, 398*, 80.

[5] An, W.F., Bowlby, M.R., Betty, M., Cao, J., Ling, H.P., Mendoza, G., Hinson, J.W., Mattsson, K.I., Strassle, B.W., Trimmer, J.S., and Rhodes, K.J. 2000, *Nature, 403*, 553.

[6] Goate, A., Chartier-Harlin, M.C., Mullan, M., Brown, J., Crawford, F., Fidani, L., Giuffra, L., Haynes, A., Irving, N., James, L., et al. 1991, *Nature, 349*, 704.

[7] Sherrington, R., Rogaev, E.I., Liang, Y., Rogaeva, E.A., Levesque, G., Ikeda, M., Chi, H., Lin, C., Li, G., Holman, K., et al. 1995, *Nature, 375*, 754.

[8] Levy-Lahad, E., Wasco, W., Poorkaj, P., Romano, D.M., Oshima, J., Pettingell, W.H., Yu, C.E., Jondro, P.D., Schmidt, S.D., Wang, K., et al. 1995, *Science, 269*, 973.

[9] Herreman, A., Serneels, L., Annaert, W., Collen, D., Schoonjans, L., and De Strooper, B. 2000, *Nat. Cell. Biol., 2*, 461.

[10] Li, Y.M., Xu, M., Lai, M.T., Huang, Q., Castro, J.L., DiMuzio-Mower, J., Harrison, T., Lellis, C., Nadin, A., Neduvelil, J.G., Register, R.B., Sardana, M.K., Shearman, M.S., Smith, A.L., Shi, X.P., Yin, K.C., Shafer, J.A., and Gardell, S.J. 2000, *Nature, 405*, 689.

[11] Zhang, Z., Nadeau, P., Song, W., Donoviel, D., Yuan, M., Bernstein, A., and Yankner, B.A. 2000, *Nat. Cell. Biol., 2*, 463.

[12] Naranjo, J.R., Mellstrom, B., Achaval, M., and Sassone-Corsi, P. 1991, *Neuron, 6*, 607.

[13] Cole, R.L., Konradi, C., Douglass, J., and Hyman, S.E. 1995, *Neuron, 14*, 813.

[14] Romualdi, P., Bregola, G., Donatini, A., Capobianco, A., and Simonato, M. 1999, *J. Mol. Neurosci., 13*, 69.

[15] Douglass, J., McKinzie, A.A., and Pollock, K.M. 1994, *Mol. Endocrinol., 8*, 333.

[16] McMurray, C.T., Devi, L., Calavetta, L., and Douglass, J.O. 1989, *Endocrinology, 124*, 49.

[17] Carrion, A.M., Mellstrom, B., and Naranjo, J.R. 1998, *Mol. Cell. Biol., 18*, 6921.

[18] Connor, J.A., and Stevens, C.F. 1971, *J. Physiol., 213*, 21.

[19] Shieh, C.C., Coghlan, M., Sullivan, J.P., and Gopalakrishnan, M. 2000, *Pharmacol. Rev., 52*, 557.

[20] Spreafico, F., Barski, J.J., Farina, C., and Meyer, M. 2001, *Mol. Cell. Neurosci., 17*, 1.

[21] Leclerc, G.M., Leclerc, G.J., Shorte, S.L., Stephen Frawley, L., and Boockfor, F.R. 2002, *Gen. Comp. Endocrinol., 129*, 45.

[22] Cheng, H.Y., Pitcher, G.M., Laviolette, S.R., Whishaw, I.Q., Tong, K.I., Kockeritz, L.K., Wada, T., Joza, N.A., Crackower, M., Goncalves, J., Sarosi, I., Woodgett, J.R., Oliveira-dos-Santos, A.J., Ikura, M., van der Kooy, D., Salter, M.W., and Penninger, J.M. 2002, *Cell, 108*, 31.

[23] Hammond, P.I., Craig, T.A., Kumar, R., and Brimijoin, S. 2003, *Brain Res. Mol. Brain Res., 111*, 104.

[24] Liss, B., Franz, O., Sewing, S., Bruns, R., Neuhoff, H., and Roeper, J. 2001, *EMBO J., 20*, 5715.

[25] Bahring, R., Dannenberg, J., Peters, H.C., Leicher, T., Pongs, O., and Isbrandt, D. 2001, *J. Biol. Chem., 276*, 23888.

[26] Van Hoorick, D., Raes, A., Keysers, W., Mayeur, E., and Snyders, D.J. 2003, *Mol. Cell. Neurosci., 24*, 357.

[27] Jo, D.G., Kim, M.J., Choi, Y.H., Kim, I.K., Song, Y.H., Woo, H.N., Chung, C.W., and Jung, Y.K. 2001, *FASEB J., 15*, 589.

[28] Lilliehook, C., Chan, S., Choi, E.K., Zaidi, N.F., Wasco, W., Mattson, M.P., and Buxbaum, J.D. 2002, *Mol. Cell. Neurosci., 19*, 552.

[29] Jo, D.G., Lee, J.Y., Hong, Y.M., Song, S., Mook-Jung, I., Koh, J.Y., and Jung, Y.K. 2004, *J. Neurochem., 88*, 604.

[30] Jo, D.G., Chang, J.W., Hong, H.S., Mook-Jung, I., and Jung, Y.K. 2003, *Biochem. Biophys. Res. Commun., 305*, 62.

[31] Sanz, C., Mellstrom, B., Link, W.A., Naranjo, J.R., and Fernandez-Luna, J.L. 2001, *EMBO J., 20*, 2286.

[32] Sanz, C., Horita, M., and Fernandez-Luna, J.L. 2001, *Haematologica, 87*, 903.

[33] Choi, E.K., Zaidi, N.F., Miller, J.S., Crowley, A.C., Merriam, D.E., Lilliehook, C., Buxbaum, J.D., and Wasco, W. 2001, *J. Biol. Chem., 276*, 19197.

[34] Esler, W.P., Kimberly, W.T., Ostaszewski, B.L., Ye, W., Diehl, T.S., Selkoe, D.J., and Wolfe, M.S. 2002, *Proc. Natl. Acad. Sci. USA, 99*, 2720.

[35] Jo, D.G., Jang, J., Kim, B.J., Lundkvist, J., and Jung, Y.K. 2005, Neurosci. Lett., 378, 59.

[36] Osawa, M., Tong, K.I., Lilliehook, C., Wasco, W., Buxbaum, J.D., Cheng, H.Y., Penninger, J.M., Ikura, M., and Ames, J.B. 2001, *J. Biol. Chem., 276*, 41005.

[37] Messersmith, D.J., Gu, J., Dubner, R., Douglass, J., and Iadarola, M.J. 1994, *Mol. Cell. Neurosci., 5*, 238.

[38] Ledo, F., Carrion, A.M., Link, W.A., Mellstrom, B., and Naranjo, J.R. 2000, *Mol. Cell. Biol., 20*, 9120.

[39] Sanz, C., Benito, A., Inohara, N., Ekhterae, D., Nunez, G., and Fernandez-Luna, J.L. 2000, *Blood, 95*, 2742.

[40] Link, W.A., Ledo, F., Torres, B., Palczewska, M., Madsen, T.M., Savignac, M., Albar, J.P., Mellstrom, B., and Naranjo, J.R. 2004, *J. Neurosci., 24*, 5346.

[41] D'Andrea, B., Di Palma, T., Mascia, A., Motti, M.L., Viglietto, G., Nitsch, L., and Zannini, M. 2005, *Exp. Cell. Res., 305*, 166.

[42] Rivas, M., Mellstrom, B., Naranjo, J.R., and Santisteban, P. 2004, *J. Biol. Chem., 279*, 33114.

[43] Ledo, F., Kremer, L., Mellstrom, B., and Naranjo, J.R. 2002, *EMBO J., 21*, 4583.

[44] Serodio, P., Kentros, C., and Rudy, B. 1994, *J. Neurophysiol., 72*, 1516.

[45] Shibata, R., Nakahira, K., Shibasaki, K., Wakazono, Y., Imoto, K., and Ikenaka, K. 2000, *J. Neurosci., 20*, 4145.

[46] Nadal, M.S., Amarillo, Y., Vega-Saenz de Miera, E., and Rudy, B. 2001, *J. Physiol., 537*, 801.

[47] Decher, N., Uyguner, O., Scherer, C.R., Karaman, B., Yuksel-Apak, M., Busch, A.E., Steinmeyer, K., and Wollnik, B. 2001, *Cardiovasc. Res., 52*, 255.

[48] Patel, S.P., Campbell, D.L., Morales, M.J., and Strauss, H.C. 2002, *J. Physiol., 539*, 649.

[49] Shibata, R., Misonou, H., Campomanes, C.R., Anderson, A.E., Schrader, L.A., Doliveira, L.C., Carroll, K.I., Sweatt, J.D., Rhodes, K.J., and Trimmer, J.S. 2003, *J. Biol. Chem., 278*, 36445.

[50] Takimoto, K., Yang, E.K., and Conforti, L. 2002, *J. Biol. Chem., 277*, 26904.

[51] Kunjilwar, K., Strang, C., DeRubeis, D., and Pfaffinger, P.J. 2004, *J. Biol. Chem., 279*, 54542.

[52] Osawa, M., Dave, A., Tong, K.I., Valiveti, A., Ikura, M., and Ames, J.B. 2005, *J. Biol. Chem., 280*, 18008.

[53] Craig, T.A., Benson, L.M., Venyaminov, S.Y., Klimtchuk, E.S., Bajzer, Z., Prendergast, F.G., Naylor, S., and Kumar, R. 2002, *J. Biol. Chem., 277*, 10955.

[54] Zaidi, N.F., Thomson, E.E., Choi, E.K., Buxbaum, J.D., and Wasco, W. 2004, *J. Neurochem., 89*, 593.

[55] Choi, E.K., Miller, J.S., Zaidi, N.F., Salih, E., Buxbaum, J.D., and Wasco, W. 2003, *Mol. Cell. Neurosci., 23*, 495.

[56] Campos, D., Jimenez-Diaz, L., Naranjo, J.R., and Carrion, A.M. 2003, *Mol. Cell. Neurosci., 22*, 135.

[57] 57. Hong, Y.M., Jo, D.G., Lee, M.C., Kim, S.Y., and Jung, Y.K. 2003, *Neurosci. Lett., 340*, 33. 1.

[58] Matsu-Ura, T., Konishi, Y., Aoki, T., Naranjo, J.R., Mikoshiba, K., and Tamura, T.A. 2002, *Brain Res. Mol. Brain Res., 109*, 198.

[59] Jin, J.K., Choi, J.K., Wasco, W., Buxbaum, J.D., Kozlowski, P.B., Carp, R.I., Kim, Y.S., and Choi, E.K. 2005, *Neuroreport, 16*, 451.

[60] Ruther, U., Garber, C., Komitowski, D., Muller, R., and Wagner, E.F. 1987, *Nature, 325*, 412.

[61] Ruther, U., Muller, W., Sumida, T., Tokuhisa, T., Rajewsky, K., and Wagner, E.F. 1988, *Cell, 53*, 847.

[62] Grigoriadis, A.E., Wang, Z.Q., Cecchini, M.G., Hofstetter, W., Felix, R., Fleisch, H.A., and Wagner, E.F. 1994, *Science, 266*, 443.

[63] Okada, S., Wang, Z.Q., Grigoriadis, A.E., Wagner, E.F., and von Ruden, T. 1994, *Mol. Cell. Biol., 14*, 382.

[64] Castro, P.A., Cooper, E.C., Lowenstein, D.H., and Baraban, S.C. 2001, *J. Neurosci., 21*, 6626.

[65] Wickenden, A.D., Lee, P., Sah, R., Huang, Q., Fishman, G.I., and Backx, P.H. 1999, *Circ. Res., 85*, 1067.

[66] Coderre, T.J., Vaccarino, A.L., and Melzack, R. 1990, *Brain Res., 535*, 155.

[67] Abbadie, C., Taylor, B.K., Peterson, M.A., and Basbaum, A.I. 1997, *Pain, 69*, 101.

[68] Simonin, F., Valverde, O., Smadja, C., Slowe, S., Kitchen, I., Dierich, A., Le Meur, M., Roques, B.P., Maldonado, R., and Kieffer, B.L. 1998, *EMBO J., 17*, 886.

[69] Burton, M.B., and Gebhart, G.F. 1998, *J. Pharmacol. Exp. Ther., 285*, 707.

[70] Sandner-Kiesling, A., Pan, H.L., Chen, S.R., James, R.L., DeHaven-Hudkins, D.L., Dewan, D.M., and Eisenbach, J.C. 2002, *Pain, 96*, 13.

[71] Arner, S., and Meyerson, B.A. 1988, *Pain, 33*, 11.

[72] Kupers, R.C., Konings, H., Adriaensen, H., and Gybels, J.M. 1991, *Pain, 47*, 5.

[73] Jadad, A.R., Carroll, D., Glynn, C.J., Moore, R.A., and McQuay, H.J. 1992, *Lancet, 339*, 1367.

[74] Portenoy, R.K., Foley, K.M., and Inturrisi, C.E. 1990, *Pain, 43*, 273.

[75] Benedetti, F., Vighetti, S., Amanzio, M., Casadio, C., Oliaro, A., Bergamasco, B., and Maggi, G. 1998, *Pain, 74*, 205.

[76] Mosconi, T., and Kruger, L. 1996, *Pain, 64*, 37.

[77] Randic, M., Cheng, G., and Kojic, L. 1995, *J. Neurosci., 15*, 6809.

[78] Ogura, M., and Kita, H. 2000, *J. Neurophysiol., 83*, 3366.

[79] Lilliehook, C., Bozdagi, O., Yao, J., Gomez-Ramirez, M., Zaidi, N.F., Wasco, W., Gandy, S., Santucci, A.C., Haroutunian, V., Huntley, G.W., and Buxbaum, J.D. 2003, *J. Neurosci., 23*, 9097.

[80] de los Santos-Arteaga, M., Sierra-Dominguez, S.A., Fontanella, G.H., Delgado-Garcia, J.M., and Carrion, A.M. 2003, *J. Neurosci., 23*, 11120.

[81] Cheng, H.Y., Laviolette, S.R., van der Kooy, D., and Penninger, J.M. 2004, *Eur. J. Neurosci., 19*, 3033.

[82] Bechara, A., and van der Kooy, D. 1987, *Pharmacol. Biochem. Behav., 28*, 227.

In: Neuronal Calcium Sensor Proteins
Editors: Pavel Philippov and Karl Koch, pp. 271-284
ISBN 1-59454-978-8

Chapter XV

A Role for Calsenilin in Multiple Aspects of Neuronal Function

Wilma Wasco[1,*] and Joseph D. Buxbaum[2]
[1]Cellular & Molecular Mechanisms of Neurodegeneration Laboratory, MassGeneral Institute for Neurodegenerative Disease, Genetics and Aging Research Unit, Harvard Medical School, Massachusetts General Hospital, 114 16th Street, Charlestown, MA 02129-9142, USA;
[2]Laboratory of Molecular Neuropsychiatry, Departments of Psychiatry, Neuroscience, and Geriatrics and Adult Development, Mount Sinai School of Medicine, One Gustave L. Levy Place, Box 1668, New York, NY 10029, USA.

Abstract

The protein variously called calsenilin, downstream regulatory element antagonist modulator (DREAM), or potassium-channel interacting protein-3 (KChIP3) has been serially discovered three times, accounting for the different names currently in use. Calsenilin is a member of a small group of neuronal calcium sensors, the calsenilin-like proteins (CLPs), that also include KChIP1, KChIP2, and calsenilin-like protein (CALP)/KChIP4. These proteins, which are highly conserved across most of the protein but have divergent amino-termini, have three, perhaps independent, functions. CLPs can bind and modulate some of the effects of the Alzheimer disease-related protein presenilin, and presenilin can modulate some of the effects of CLPs. CLPs binds specific DNA elements and regulate gene expression. CLPs binds Kv4 channels and modulate potassium conductance, playing a role in long-term potentiation (LTP) as well as in other important plastic pathways. In cell and animal studies, the CLP family has been directly implicated in Alzheimer's Aβ protein formation, in nociception, and, because of its effects on potassium conductance, in neuronal excitability, in LTP, and in neuronal

* Correspondence concerning this article should be addressed to Tel: 617-726-8307; Fax: 617-724-1823; E-mail: wasco@helix.mgh.harvard.edu. or Tel: 212-659-8862; Fax: 212-828-4221; E-mail: Joseph.Buxbaum @mssm.edu.

pacemaker activity. Further functional dissection of this family of proteins will provide insight into numerous aspects of neuronal function and will illuminate the role of the CLPs in disease.

Keywords: KChIP, DREAM, Alzheimer's disease, transcription, Kv4 channel

1. Introduction

Calsenilin, also known as DREAM, or KChIP3, has been independently identified three times, and consequently, the protein is currently referred to by three different names. It was first identified as a calcium binding protein during a two hybrid screen for proteins that interact with the cytoplasmic, carboxyl terminal domain of presenilin 2 [1]. In that first description, where the protein was termed calsenilin, it was shown to interact with both presenilin 1 and presenilin 2 and to have the ability to regulate the levels of a presenilin fragment. Subsequently, in a search that was designed to identify proteins that bind to a downstream regulatory element (DRE) located in the promoter of the dynorphin gene, Carrion et al. [2] identified calsenilin (there called DRE antagonist modulator or DREAM) and demonstrated that it interacts with the DRE and represses the expression of dynorphin. Finally, An et al. [3] used the carboxyl-terminal of the alpha channel of Kv4 A-type potassium channels to identify Kv4-interacting proteins and identified two proteins they called potassium-channel interacting protein-1 (KChIP-1) and 2 (KChIP-2). Using the sequence from these two genes to search the public cDNA database, An et al [3] also identified calsenilin as an A-type potassium channel interactor and called it KChIP3. More recently, Morohashi et al. [4], also using the carboxyl-terminal of presenilin for two-hybrid screening, identified a protein they termed calsenilin-like protein or CALP/KChIP4 and demonstrated that it interacts with the presenilins as well as the A-type potassium channels. Thus, to date, four immediate members of the calsenilin family – KChIP1, KChIP2, calsenilin/DREAM/KChIP3, and CALP/KChIP4 – have been identified (Figure 1). Each of these proteins is a calcium-binding protein and each is structurally related to the extended family of recoverin family of calcium binding proteins that includes visinin, frequenin and neuronal calcium sensor-1, as well as the guanalyl cyclase activating proteins. Unlike the other members of the superfamily, however, calsenilin and its homologues have an N-terminal domain that is not present in the other proteins (Figure 2). The members of the superfamily, as well as the calsenilin subfamily, are evolutionarily conserved and calsenilin homologues are encoded in the Drosophila and Dictyostelium genome (WW, unpublished observations).

2. Calsenilin is a Member of a Conserved Protein Family

Like all of the members of the recoverin superfamily, each of the proteins in the calsenilin subfamily has 4 conserved EF hand motifs, although it appears that only the EF 2,

3 and 4 motifs actually have the ability to bind calcium [5] and that EF-2 constitutively binds magnesium [6]. In fact, the interaction of calsenilin with magnesium appears to function to stabilize the protein as a monomer, while the binding of calcium results in dimer formation [6]. The binding of calcium has also been shown to reduce the interaction of calsenilin with DNA targets [2,6]. Taken together, these findings indicate that the monomeric forms of calsenilin have the ability to interact with DNA and repress the transcription of target genes and that the interaction of calcium with EF-3 and EF-4 modulates this interaction. The EF-hands are located in the C-terminal domain of the protein, and it is this C-terminal domain that is highly conserved, not only between each of the members of the calsenilin subfamily, but between this subfamily and the other members of the recoverin superfamily (see Figures 1 and 2). Notably, KChIP1, KChIP2, calseniln/Dream/KChIP3 and CALP/KChIP4 all have an N-terminal domain that is not present in the other members of the family, nor is it conserved within the calsenilin subfamily. Taken together these observations suggest that the N-terminal domain may provide functional specificity to the members of the calsenilin subfamily in general as well as a specific function to each member of the subfamily. This hypothesis is supported by the finding that there is a caspase cleavage site located at the junction of the N and C-terminal domains and that cleavage at this site by caspase-3 is regulated by the phosphorylation of calsenilin at a serine residue located within the caspase site (see Figure 1) [7]. Whether this regulated cleavage is simply part of programmed cell death pathways or a developmental process, or whether it is associated with a non-apoptotic caspase event will require further investigation. It is intriguing to speculate that cleavage at the juncture of these two independent domains has a functional or regulatory role, especially since the N-terminal domain has recently been shown to interact with a transcriptional co-repressor (see below and [8]) and there is increasing evidence that caspase activation does not always lead to cell death [5].

3. Calsenilin is a Neuronal Protein

Northern blot analysis indicates that calsenilin mRNA is expressed primarily in the brain [1] as are the mRNAs for KChIP1, KChIP2 and CALP/KChIP4 (although KChIP2 expression is much higher in the heart than in the brain). Immunohistochemical analysis has been used to demonstrate that while calsenilin is expressed in regions throughout the brain, the levels are highest in the cortex and the hippocampus [9]. This study also showed that calsenilin co-localized with the somatodendritic marker microtubule associated protein-2, primarily in the granular cell layer of the cerebellum. In primary cultured neurons, calsenilin immunoreactivity was observed in cell bodies as well as in neuronal processes. Co-localization experiments using specific axonal and dendritic markers indicate that these processes were mainly axonal in nature, although a smaller subset of dendrites also appears to contain calsenilin. In a recent study. immunoblot analysis and *in situ* hybridization were combined and used to assess the localization of each to the family members in the brain [10]. KChIP1 was found in a subpopulation of neurons that were widely distributed in brain and enriched in Purkinje cells of the cerebellum and in the reticular thalamic and medial habenular nuclei. KChIP2 levels were high in layer IV of the cerebral cortex, striatum and hippocampus, but low in cerebellum. Calsenilin/DREAMKChIP3 transcripts were detected primarily in layer V and deep layer VI of the cerebral cortex, the hippocampus, and the entire cerebellum, and CALP/KChIP4 was highly expressed in the neurons of layers II-IV of cortex and in hippocampus, thalamus and the Purkinje cell layer of the cerebellum.

```
                                          N-terminal domain

KChIP1        ---------------MGAVMGTFS-SLQTKQRRPSKD------------------------KIEDELEMTMVHHRPEGLE
KChIP2        MRGQGRKESLSDSRDLDGSYDQLTGHPPGPTKKALKQRFLKLLPCCGPQALPSVSE----NSVDDEFELSTVCHRPEGLE
CALP/KChIP4   -MNVRRVESISAQLEEASSTGGFL-YAQNSIKRNIKERLMKLLPCSAAKTSSPAIQ----NSVEDELEMATVRHRPEALE
Calsenilin    MQPAKEVTKASDGSLLGDLGHTPLSKKEGIKWQRPRLSRQALMRCCLVKWILSSTAPQGSDSSDSELELSTVRHQPEGLD
                                                 :   :                 . :.*:*:: * *:**.*

                            EF-1                                      EF-2
KChIP1        QLEAQTNFTKRELQVLYRGFKNECPSGVVNEDTFKQIYAQFFPHGDASTYAHYLFNAFDTTQTGSVKFEDFVTALSILLR
KChIP2        QLQEQTKFTRKELQVLYRGFKNECPSGIVNEENFKQIYSQFFPQGDSSTYATFLFNAFDTNHDGSVSFEDFVAGLSVILR
CALP/KChIP4   LLEAQSKFTKKELQILYRGFKNECPSGVVNEETFKEIYSQFFPQGDSTTYAHFLFNAFDTDHNGAVSFEDFIKGLSILLR
Calseniln    QLQAQTKFTKKELQSLYRGFKNECPTGLVDEDTFKLIYAQFFPQGDATTYAHFLFNAFDADGNGAIHFEDFVVGLSILLR
               *: *::**::*** **********:*:*:*:.** **:****:**::*** :******:   *:: ****: .**::**

                        EF-3                                           EF-4
KChIP1        GTVHEKLRWTFNLYDINKDGYIVKEEMMDIVKAIYDMMGKYTYPVLKEDTPRQHVDVFFQKDKNMDGIVTLDEFLESCQ
KChIP2        GTVDDRLNWAFNLYDLNKDGCITKEEMLDIMKSIYDMMGKYTYPALREEAPREHVESFFQKMDRNKDGVVTIEEFIESCQ
CALP/KChIP4   GTVQEKLNWAFNLYDINKDGYITKEEMLDIMKAIYDMMGKCTYPVLKEDAPRQHVETFFQKMDKNKDGVVTIDEFIESCQ
Calseniln    GTVHEKLKWAFNLYDINKDGYITKEEMLAIMKSIYDMMGRHTYPILREDAPAEHVERFFEKMDRNQDGVVTIEEFLEACQ
              ***.::*.*:*****:**** *.****: *:*:******: *** *:*::* :**: **:***:* **:**::**:*:**

KChIP1        EDDNIMRSLQLFQNVM
KChIP2        KDENIMRSMQLFDNVI
CALP/KChIP4   KDENIMRSMQLFENVI
Calsenilin    KDENIMSSMQLFENVI
              :*:*** *:***:**:
```

Figure 1. Comparison of amino acid sequences of the proteins in the calsenilin subfamily of neuronal calcium sensor proteins. An alignment of the amino acid sequences of calsenilin, KChIP1, KChIP2 and KChIP4/CALP is shown. Amino acids that are identical in all 4 proteins are denoted by a (*) symbol, highly similar amino acids are denoted by a (:) symbol, and similar amino acids are denoted by a (.) symbol. The four EF hands are shaded and the location of the variable N-terminal domain is indicated. Note that this domain is not conserved among the proteins. The caspase cleavage site (DSSD) in calsenilin in shown in bold, and the serine 63, which is can be phosphorylated is underlined. Phosphorylation at serine 63 inhibits caspase cleavage.

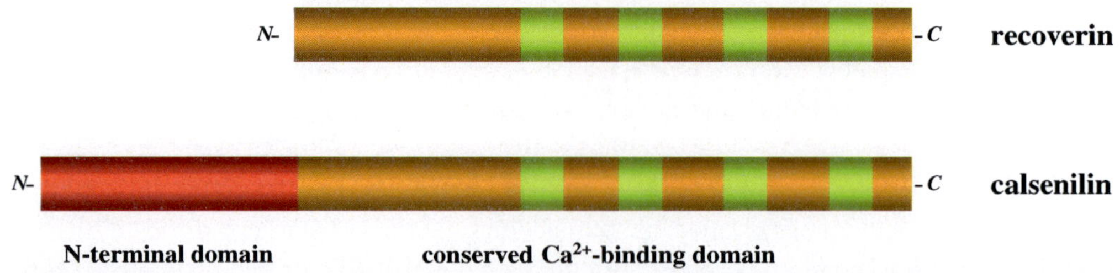

Figure 2. Calsenilin is a calcium binding protein that is a member of the recoverin superfamily. The calcium binding domain of recoverin, which contains 4 EF hand motifs, is conserved in calsenilin. Calsenilin has an extended N-terminal that is not present in recoverin, nor is it present in any of the other members of the recoverin superfamily.

4. Calsenilin is a Cytoplasmic Protein that is Recruited to Intracellular Membranes

In the first description of calsenilin, it was shown that when the protein was overexpressed in cultured cells it had a diffuse cytoplasmic location, however when it was co-expressed with presenilin 2 the two proteins co-localized to the membranes of the ER and Golgi. Similarly, when CALP was identified, it was shown that CALP had the ability to colocalize with the presenilins when the both of the proteins were both overexpressed [4]. Subsequently it was demonstrated that endogenous calsenilin can be co-immunoprecipitated with endogenous presenilin 1 or presenilin 2 [9]. These observations suggest that calsenilin and presenilin form a complex *in vivo* and that the presenilins, which are located in the plasma membrane as well as in the membranes of the ER and Golgi, have the ability to recruit calsenilin and CALP to the membrane. This hypothesis was supported by the finding that although calsenilin is not predicted to be a membrane protein and does not contain a membrane spanning domain, the majority of calsenilin is very tightly associated with membrane fractions isolated from brain extracts [9]. Interestingly this non-covalent association is strong enough to survive treatment with high salt, as well as high and low pH, suggesting that the association is unusually strong. This study also found that calcium did not have a significant effect on the levels of calsenilin associated with the membrane, suggesting that the association of calsenilin with the membrane is not regulated by calcium binding to calsenilin. Whether this membrane association is a result of calsenilin interactions with the presenilins, the Kv4 A-type potassium channel, other as of yet unidentified proteins, or a combination of the above remains unclear. This membrane association is reminiscent of the myristoyl-dependent association of the other proteins of the recoverin superfamily that have a myristoylation site located at the amino terminal end of the protein. When the levels of intracellular calcium are low, the myristoyl group of recoverin and other related proteins is held in a hydrophobic pocket of the proteins. As local calcium levels increase, the interaction of calcium with the EF hands of recoverin causes a conformational change that alters that hydrophobic pocket and results in the unleashing of the myristoyl group, which is now free to interact with intracellular membranes. This "switch" is believed to provide a mechanism for the proteins to be cytosolic at low or resting concentrations of calcium and to interact with intracellular membranes as local levels of calcium increase. Unlike these proteins, the members of the calsenilin subfamily do not contain a myristoylation site and do not appear to be myristoylated (JDB, unpublished observations). One current hypothesis is that the interaction with membrane associated proteins serves as a similar "switch" for the specific subcellular localization of calsenilin/DREAM/KChIP3 and the other members of the calsenilin subfamily. It is also possible that palmitoylation is responsible for the association of calsenilin with the membrane fraction. Takimoto et al [11] found that mutating potential palmitoylation cysteines residues in calsenilin as well those as in KChIP2, reduces their plasma membrane localization.

5. Calsenilin Translocates from the Cytoplasm to the Nucleus

Although calsenilin is principally a cytoplasmic protein that does not contain a classic nuclear localization signal [9], it is clear that it must translocate from the cytoplasm to the nucleus to function as a transcriptional repressor. In a study designed to determine the cellular events that are associated with the translocation of calsenilin from the cytoplasm to the nucleus, Zaidi and coworkers found that the nuclear translocation of calsenilin was enhanced following serum deprivation [12]. A similar effect was observed when cells were treated with pharmacological agents that directly increased the levels of intracellular calcium (caffeine and the calcium ionophore A23187), as does serum deprivation. This study also showed that EF-hand mutant forms of calsenilin that were incapable of binding calcium retained the ability to translocate to the nucleus. Taken together, these findings indicate that the level of intracellular calcium can modulate the translocation of calsenilin from the cytoplasm to the nucleus and that this process does not involve the direct binding of calcium to calsenilin. Notably, data from Choi and colleagues [7] demonstrate that calsenilin can be phosphorylated by casein kinase I and that calcium plays a role in this process, perhaps by altering the activity of the casein kinase I (see below). It has been proposed that the calcium modulated increase in nuclear translocation of calsenilin is associated with its phosphorylation state [12]. Such a mechanism has been reported for other proteins; for instance the phosphorylation of the insulin like growth factor binding protein by deoxyribonucleic acid dependent protein kinase, which enhances the nuclear import of this protein [13].

6. Calsenilin is a Phosphoprotein

Choi and coworkers have demonstrated that calsenilin is a phosphoprotein that can be constitutively phosphorylated *in vivo* [7]. In this study it was shown that treatment of cultured cells with thapsigargin or calyculin A results in higher levels of phosphorylated calsenilin, suggesting that the process can be regulated by levels of intracellular calcium. *In vitro* phosphorylation assays utilizing a series of purified protein kinases and protein kinase inhibitors indicated that calsenilin could be phosphorylated by casein kinase 1. In addition, treatment with inhibitors of serine/threonine protein phosphatase 1 (PP1) and 2A (PP2A), results in augmented levels of phosphorylated calsenilin both *in vitro* and *in vivo,* while inhibitors of PP2B (cyclosporine A or FK506) do not, which suggests that phosphorylated calsenilin is a substrate for PP1and/or PP2A. Site directed mutagenesis was used to identify a major phosphorylation site at serine residue 63 and minor sites at serine residues 50, 62 and 65, all of which are located in the N-terminal domain of calsenilin. Finally, this study showed that the phosphorylation of calsenilin at Ser 63 inhibits its cleavage by caspase-3. Because the N-terminal domain of calsenilin is not conserved in either the other KChIP/CALP proteins or in the NCS family overall, it was hypothesized that the phosphorylation of calsenilin may regulate a functional role that is unique to this member of the superfamily.

7. Calsenilin Interacts with CTBP2, a Transcriptional Co-Repressor

Perhaps the most notable characteristic of the calsenilin subfamily is an N-terminal extension that is not present in other members of the superfamily (see Figures 1 and 2). Interestingly, although all four of the proteins (KChIP1, KChIP2, calsenilin/DREAM/KChIP3, CALP/KChIP4) have this novel domain, it varies in length, shows extensive splice variation, and is not well conserved, suggesting that it may provide specificity of function to the individual members of this subfamily. In an effort to learn more about the role of this domain in calsenilin function, the N-terminal of calsenilin was used in a two-hybrid screen that identified C-terminal Binding Protein 2 (CtBP2) [14] as a calsenilin N-terminal interacting protein [8]. CtBP2 and its homologue CtBP1 have been demonstrated to act as co-repressors for a growing list of transcriptional repressors [14] and are thought to act by recruiting histone deacetylase to the promoter. It is possible that the interaction of CtBP2 with calsenilin modulates the ability of calsenilin to repress the transcription of target genes containing a DRE (see below). It is not yet known whether the N-terminal domains of the other proteins have the ability to interact with CtBP1/2 or other transcriptional corepressors.

8. Evidence for Calsenilin's Role in Presenilin Function

When calsenilin was originally identified [1] it was observed that when it was overexpressed in cultured cells, increased levels of a presenilin fragment were detected. This finding, which provided the first evidence of a potentially functional interaction between calsenilin and the presenilins, was later extended by studies showing that calsenilin preferentially interacted with an apoptosis-related presenilin fragment [15]. Subsequently, to assess the role of calsenilin in presenilin mediated modulation of intracellular calcium concentrations, Leissring and his colleagues [16] injected cDNA encoding wild-type or familial Alzheimer's disease associated mutant presenilin in the presence or absence of calsenilin into *Xenopus* oocytes. These investigators had previously shown that overexpression of presenilin in oocytes increased inositol-tris-phophate (IP_3)-mediated increases in intracellular calcium, and that wild-type presenilin augments IP_3-mediated calcium releases, with mutant presenilin further enhancing this effect. They postulated a role for presenilin in endoplasmic reticulum (ER) function. When calsenilin was added to the system, it was shown that calsenilin was able to reverse the effect of presenilin on IP_3-mediated calcium release. To address the possibility that this result was due to a buffering effect of calsenilin, as it is a calcium binding protein, Leissring et al. then showed that when calsenilin was introduced without the addition of presenilin, there was no effect on IP_3-mediated calcium release. This study strongly supported a functional interaction between calsenilin and presenilin and placed calsenilin and presenilin in the pathway of ER calcium signaling. These studies were followed up by an analysis of mammalian cells that had been stably transfected with calsenilin under the control of the tetracycline inducible system [17] which showed, as previously postulated [1], that calsenilin was involved in the process of apoptosis. Induction of calsenilin expression increased apoptosis that was produced either by serum withdrawal or by thapsigargin, an inhibitor of SERCA, the sarcoplasmic and endoplasmic reticulum ATPase. Given the potential role for calsenilin in ER calcium signaling, these investigators then looked at ER function in cells in the absence or presence of calsenilin and found that calsenilin overexpression dramatically increased levels of calcium in the calcium stores. In the presence of an apoptotic stimulus, this calcium was released and potentially contributed to apoptosis, likely through activation of caspase and calpain. In a further study on the role of calsenilin in cell death, Jo et al. [18] demonstrated that overexpression of calsenilin increased cell death and this was further augmented by coexpression of wild type or mutant presenilin. In aggregate, these studies demonstrated that calsenilin and presenilin can interact in a way that modulates calcium stores and also contributes to apoptosis.

Additional findings suggesting a functional relationship between calsenilin and presenilin come from studies demonstrating a role for calsenilin in Alzheimer's beta-amyloid (Aβ) formation. In a first study [19], transfecting cells with increasing amounts of calsenilin was found to lead to increased formation of Aβ42. In addition, the co-transfection of calsenilin with presenilin 2 increased Aβ formation over the levels seen with transfection of presenilin- 2 alone. Additional evidence for a role for calsenilin in Aβ formation came from studies in calsenilin knockout mice [20]. In mice with a disruption of the calsenilin gene, Aβ40 and Aβ42 levels were significantly decreased in the cerebellum, an area with high calsenilin expression. In the cortex, with more restricted calsenilin expression, a significant decrease in Aβ40 levels and a trend towards decreased Aβ42 levels were observed. Recently, Jo et al [21] provided direct evidence for a role for calsenilin in the modulation of presenilin associated γ-secretase activity that is associated with the generation of Aß. These investigators found that the overexpression of calsenilin enhanced γ-secretase activity in cultured cells, that the effect was blocked by a γ-secretase inhibitor, and that the addition of calsenilin to isolated membrane fractions resulted in increased Aß production in a cell free system. These data suggest that calsenilin may be one of the regulatory factors for γ-secretase and demonstrate a role for calsenilin in Aβ formation that directly involves presenilin, supporting a functional interaction between calsenilin and presenilin.

Two interesting additional studies support a role for calsenilin in presenilin function, and perhaps even Alzheimer's pathology. Jo et al. [22] demonstrated that incubating cultured cells with Aβ leads to a profound increase in calsenilin expression. Furthermore, these researchers showed that the toxic effects of Aβ on cultured neurons can be ameliorated by the use of antisense calsenilin construct. The implication is that cell death induced by multiple causes, including Aβ, can involve calsenilin and calsenilin-induced apoptosis. In the same study, it was shown that there was an up-regulation of calsenilin in Alzheimer's disease and, by immunocytochemistry, that there was calsenilin associated with Alzheimer's lesions. Additionally, a report from Jin et al. [23] also shows that there are increased levels of calsenilin mRNA and protein expression in neurons as well as in reactive astrocytes in the Alzheimer's disease brain.

9. Evidence for a Role for Calsenilin in Modulating Gene Expression

The first report implicating calsenilin in regulation of transcription demonstrated that calsenilin was able to bind a probe including the DRE. This binding and the resulting electrophoretic mobility shift could not be competed by AP-1, CRE or Oct-1 oligonucleotides [2]. Using a transcription reporter assay, these researchers went on to show that binding of calsenilin to the DRE sequence functioned as a repressor. Binding of calsenilin to the DRE was inhibited in the presence of calcium [2, 24]. These studies were then followed up with *in vivo* studies making use of calsenilin knockout mice [25]. In the spinal cord of these animals, there was a robust induction of dynorphin expression, with no change in the levels of enkephalin. The increased expression of dynorphin found in the absence of calsenilin expression is consistent with the observation that calsenilin may act as a repressor of dynorphin expression.

With the increase in dynorphin expression, it became immediately interesting as to whether there are alterations in pain sensitivity in these animals. Remarkably, these animals proved less sensitive to quite a few different modalities of pain measurement [25]. This included increased latency in a tail-flick assay, increased latency for paw withdrawal, decreased responses to injection of noxious substances, and in the case of formalin injection, where there are both acute and longer term pain responses, both the short term and the longer term responses (the latter to do with inflammation) were associated with significantly lower sensitivity in the knockout animals. Most remarkably, neuropathic pain was profoundly decreased in the animals for weeks after the implantation of a cuff. By all these measures, calsenilin is critically involved in pain modulation.

One issue that came up with these studies, was the discovery that while these animals are in fact less sensitive to certain measures of pain, they appear to be more sensitive to shock. Thus in a parallel study with a calsenilin knockout [20] it was observed that knockout animals are actually more sensitive to shock while being less sensitive in the tail flick assay. It was generally assumed that shock sensitivity and tail flick latency are both equivalent measures of pain sensitivity, but the fact that these animals are more sensitive in one case and less in the other suggests that these paradigms are not equivalent. Another interesting point is that some of the areas with highest dynorphin expression, including for example the basal ganglia layer V of the cortex and the habenula, do not correspond to areas of high calsenilin expression. Similarly, areas of high calsenilin expression, such as the piriform cortex and layer VIb of the cortex, do not show significant dynorphin expression. The suggestion here would be that calsenilin might have additional roles in these other areas and that dynorphin expression may be regulated by different mechanisms in different brain areas.

Since the original studies showing that calsenilin could modulate dynorphin expression, calsenilin has also been shown to repress the expression of other DRE containing genes such as c-fos [2], hrk [26], aryalkylamine N-acetyl transferase, the inducible cAMP repressor (ICER) and Fos related antigen (fra-2) [27]. Calsenilin has also demonstrated to bind to DREs in the promoter of the transcription factors Pax8 and TTF-2/Foxe1, suggesting that calsenilin can modulate genes associated with thyroid gene expression [28]. Finally, a recent report demonstrates that calsenilin can also act as a transcriptional activator [29]. In this study, calsenilin was shown to have the ability to stimulate basal and ligand-dependent activation of promoters containing vitamin D and retinoic acid response elements when the elements were located upstream, but not downstream, of the transcription initiation site. This interaction was shown *in vitro* as well as in chromatin immunoprecipitation assays. These findings are intriguing in that they indicate that calsenilin can function as an activator of transcription as well as a repressor, and suggest that calsenilin has the potential to modulate genes containing binding sites for nuclear receptors.

10. Evidence for a Role for Calsenilin in Modulating Potassium Channel Function

It has long been known that expressing the cDNA for A-type channels in cultured cells result in channel properties that differ from those seen *in vivo*. This formed the basis of the hypothesis that there is an obligate beta subunit for these channels that had yet to be identified. Several groups have now identified potential beta subunits. In one case, researchers identified proteins called potassium-channel interacting proteins (KChIPs) that interact with the Kv4.2, Kv4.3 and Kv4.4 channels [3, 4]. After identifying the KChIP proteins, these authors then demonstrated that the KChIPs are co-localized with the A-type potassium channels in tissue, as well as in cultured cells. Subsequently, it was shown that co-expressing KChIP1, KChIP2, calsenilin/DREAM/KChIP3, or CALP/KChIP4, with Kv4.2, Kv4.3 or Kv4.4 channels had a profound effect on the electrophysiology of the resultant channels. The current observed when the Kv4 channels were co-transfected with calsenilin or a calsenilin homolog was dramatically higher then in the absence of these proteins. Furthermore, there was increased conductance, and there was a more rapid recovery from an inactivation of the currents. Mutating the calcium binding motif of CALP led to a loss of the effect of CALP on the current [4], but deleting the end terminal of CALP did not affect current density.

With the development of knockout animals of calsenilin, the effects of calsenilin on channel function *in vivo* could be studied. Making use of the connection between the perforant path and the dentate gyrus, Lilliehook et al. [20] examined long-term potentiation (LTP) in wild-type and calsenilin knockout mice. This pathway was chosen because of the high levels of expression of calsenilin in the dentate gyrus. These studies demonstrated an increase in LTP in the knockout animals, which was associated with decreased current density for potassium channels. Furthermore, the increased LTP could be mimicked by the addition of phrixotoxin-2 in wild-type animals. (Phrixotoxin-2 blocks the A-type channels). The data are therefore consistent with decreased A-type currents in the knockout animals leading to increased LTP, which can be mimicked in the wild-type animals by blocking these same channels.

In another *in vivo* study, the related protein KChIP2 was looked at functionally in the heart. KChIP2 is expressed in the heart, and Kuo et al. [30], looked at the effects of disrupting the KChIP2 gene on potassium channels in the heart. They observed that ECG traces from these animals showed prolonged elevation of the ST segment and, in fact, a complete loss of the I_{TO} current. Animals with a disruption of the KChIP2 gene became susceptible to ventricular tachycardia. To confirm the loss of the I_{TO} current, Kuo et al. looked at myocites in the knockout animals and show a complete lack of the transient I_{TO} current.

11. A Potential Role for Calsenilin in Dopaminergic Signaling

One remarkable study has shown that calsenilin is expressed in the substantia nigra in dopaminergic neurons [31]. In fact, the pacemaker frequency of the individual dopaminergic neurons in the substantia nigra appears to be directly related to the levels of Kv4.3, which are in turn directly related to the levels of calsenilin in individual cells. With increased levels of calsenilin in single cells, there is increased A-type current density. Because the levels of the A-type channels in the substantia nigra modulates the pacemaker activity (such that with decreased A-current the pacemakers fire more rapidly) it appears that calsenilin can directly effect pacemaker tuning in these very important cells. It would be worth further looking at the role for calsenilin in these cells because of their potential relevance to Parkinson's disease, drug abuse, and perhaps the action of antipsychotics.

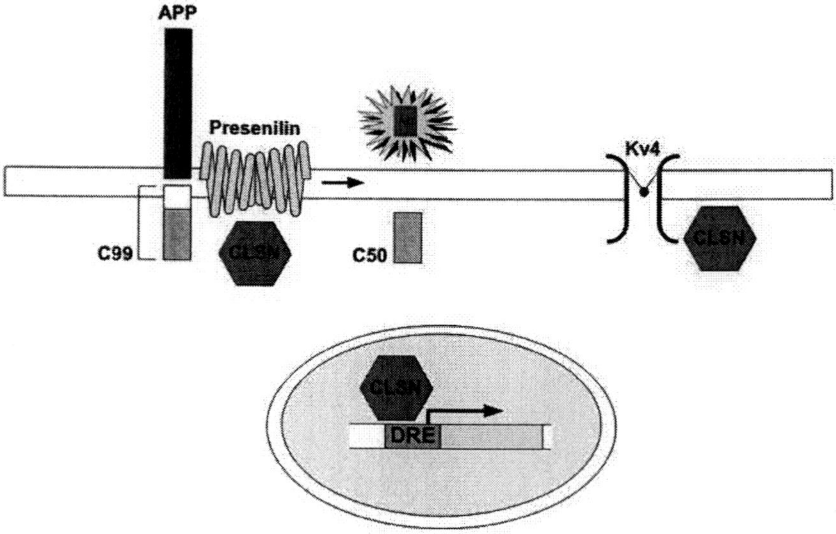

Figure 3. The three functions of calsenilin. Calsenilin (CLSN) interacts with presenilin and modulate its activities. Here, the cleavage of the C99 fragment of the Alzheimer amyloid protein precursor (APP) is shown to involve presenilin and to be modulated by calsenilin. This cleavage results in the liberation of Aß and the formation of the C50 cytoplasmic fragment involved in nuclear signaling. In addition, calsenilin can bind the dynorphin response element (DRE) in the nucleus inhibiting expression of dynorphin. Finally, calsenilin can bind potassium channels of the Kv4 type and modulate the levels of these channels on the cell surface and their functioning.

Conclusion

Calsenilin is a neuronal protein that is a member of a subfamily of the recoverin super family of neuronal calcium binding proteins. Calsenilin appears to have three potentially independent roles in the cell (Figure 3). In the cytoplasm, it binds presenilin and modulates some of the effects of presenilin, and interestingly, presenilin appears to have the ability to modulate some of the effects of calsenilin. Also in the cytoplasm, calsenilin binds to Kv4 channels and modulates potassium conductance, playing a role in LTP as well as potentially other important plastic pathways. The evidence for a role for calsenilin in dopaminergic neurons in the substantia nigra is very interesting, and requires further study. Finally, in the nucleus, calsenilin binds the downstream regulatory element in the dynorphin response element and modulates dynorphin expression, hence regulating nociception. It also binds to the regulatory elements of a number of other genes, and although as with dynorphin, to date the majority of the interactions characterized repress the transcription of target genes, recent data suggests that calsenilin also has the ability act as a transcriptional activator.

Acknowledgements

These studies were supported by National Institute on Aging, National Institute of Health Grants AG005138 (JDB), AG021792 (JDB), and AG16361 (WW). The authors thank Christopher Plescia for preparing the Figure 3.

References

[1] Buxbaum, J.D., Choi, E.K., Luo, Y., Lilliehook, C., Crowley, A.C., Merriam, D.E., and Wasco, W. 1998, *Nat. Med., 4*, 1177.

[2] Carrion, A.M., Link, W.A, Ledo, W.A., Mellstrom, F.B., and Naranjo, J.R. 1999, *Nature, 398*, 80.

[3] An, W.F., Bowlby, M.R., Betty, M., Cao, J., Ling, H.P., Mendoza, G., Hinson, J.W., Mattsson, K.L., Strassle, B.W., Trimmer, J.S., and Rhodes, K.J. 2000, *Nature, 403*, 553.

[4] Morohashi, Y., Hatano, N., Ohya, S., Takikawa, R., Watabiki, T., Takasugi, N., Imaizumi, Y., Tomita, T., and Iwatsubo, T. 2002, *J. Biol. Chem., 277*, 14965.

[5] Abraham, M.C., and Shaham, S. 2004, *Trends Cell Biol., 14*, 184.

[6] Osawa, M., Tong, K.I., Lilliehook, C., Wasco, W., Buxbaum, J.D., Cheng, H.Y., Penninger, J.M., Ikura, M., and Ames, J.B. 2001, *J. Biol. Chem., 276*, 41005.

[7] Choi, E.K., Miller, J.S., Zaidi, N.F., Salih, E., Buxbaum, J.D., and Wasco, W. 2003, *Mol. Cell Neurosci., 23*, 495.

[8] Zaidi, N.F., Kuplast, K.G., Buxbaum, J.D., and Wasco, W. 2004, *Society for Neuroscience Abstract, 90*, 2.

[9] Zaidi, N.F., Berezovska, O., Choi, E.K., Chan, H., Lilliehook, C., Hyman, B.T., and Wasco, W. 2002, *Neuroscience, 114*, 247.

[10] Xiong, H., Kovacs, I., and Zhang, Z. 2004, *Brain Res., Mol. Brain Res., 128*, 103.

[11] Takimoto, K., Yang, E.K., and Conforti, L. 2002, *J. Biol. Chem., 277*, 26904.

[12] Zaidi, N.F., Thomson, E.E., Choi, E.K., Buxbaum, J.D., and Wasco, W. 2004, *J. Neurochem., 89*, 593.

[13] Schedlich, L.J., Nilsen, T., John, A.P., Jans, D.A., and Baxter, R.C. 2004, *Endocrinology, 144*, 1984.

[14] Hildebrand, J.D., and Soriano, P. 2002, *Mol. Cell Biol., 15*, 5296.

[15] Choi, E.K., Zaidi, N.F., Miller, J.S., Crowley, A.C., Merriam, D.E., and Lilliehook, C. 2001, *J. Biol. Chem., 276*, 19197.

[16] Leissring, M.A., Yamasaki, T.R., Wasco, W., Buxbaum, J.D., Parker, I., and LaFerla, F.M. 2000, *Proc. Natl. Acad. Sci., USA, 97.*

[17] Lilliehook, C., Chan, S., Choi, E.K., Zaidi, N.F., Wasco, W., Mattson, M.P., and Buxbaum, J.D. 2002, *Mol. Cell Neurosci., 552.*

[18] Jo, D.G., Chang, W.J., Hong, H.S., Mook-Jung, I., and Jung, Y.K. 2003, *Biochem. Biophys. Res. Commun., 305*, 62.

[19] Jo, D.G., Kim, M.J., Choi, Y.H., Kim, I.K., Song, Y.H., Woo, H.N., Chung, C.W., and Jung, Y.K. 2001, *FASEB J., 589.*

[20] Lilliehook, C., Bozdagi, O., Yao, J., Gomez-Ramirez, M., Zaidi, N.F., Wasco, W., Gandy, S., Santucci, A.C., Haroutunian, V., Huntley, G.W., and Buxbaum, J.D. 2003, J. *Neurosci., 23*, 9097.

[21] Jo, D.G., Jang, J., Kim, B.J., Lundkvist, J., and Jung, Y.K. 2005, *Neurosci. Lett., 378*, 59.

[22] Jo, D.G., Lee, J.Y., Hong, Y.M., Song, S., Mook-Jung, I., Koh, J.Y., and Jung, Y.K. 2004, *J. Neurochem., 88*, 1570.

[23] Jin, J.K., Choi, J.K., Wasco, W., Buxbaum, J.D., Kozlowski, P.B., Carp, R.I., Kim, Y.S., and Choi, E.K. 2005, *Neuroreport, 16*, 451.

[24] Osawa, M., Tong, K.I., Lilliehook, C., Wasco, W., Buxbaum, J.D., Cheng, H.Y., Penninger, J.M., Ikura, M., and Ames, J.B. 2001, *J. Biol. Chem., 276*, 41005.

[25] Cheng, H.Y., Pitcher, G.M., Laviolette, S.R., Whishaw, I.Q., Tong, K.I., Kockeritz, L.K., Wada, T., Joza, N.A., Crackower, M., Goncalves, J., Sarosi, I., Woodgett, J.R., Oliveira-dos-Santos, A.J., Ikura, M., van der Kooy, D., Salter, M.W., Penninger, J.M., 2002, *Cell, 108*, 31.

[26] Sanz, C., Mellstrom, B., Link, W.A, Naranjo, J.R., and Fernandez-Luna, J.L. 2001, *EMBO J., 20*, 2286.

[27] Link, W.A., Ledo, F., Torres, B., Palczewska, M., Madsen, T.M., Savignac, M., Albar, J.P., Mellstrom, B., and Naranjo, J.R. 2004, *J. Neurosci., 24*, 5346.

[28] D'Andrea, B., Di Palma, T., Mascia, A., Motti, M.L., Viglietto, G., Nitsch, L., and Zannini, M. 2005, *Experimental Cell Research, 30*, 166.

[29] Scsucova, S., Palacios, D., Savignac, M., Mellstrom, B., Naranjo, J.R., and Aranda, A. 2005, *Nucleic Acids Res., 33*, 2269.

[30] Kuo, H.C., Cheng, C.F., Clark, R.B., Lin, J.J., Lin, J.L., Hoshijima, M., Nguyen-Tran, V.T., Gu, Y., Ikeda, Y., Chu, P.H., Ross, J., Giles, W.R., and Chien, K.R. 2001, *Cell, 107*, 801.

[31] Liss, B., Franz, O., Sewing, S., Bruns, R., Neuhoff, H., and Roeper, J. 2001, *EMBO J., 20*, 5715.

In: Neuronal Calcium Sensor Proteins
Editors: Pavel Philippov and Karl Koch, pp. 285-297
ISBN 1-59454-978-8

Chapter XVI

KChIPs as Integral K^+ Channel Subunits

Kenneth J. Rhodes ***and Robert H. Scannevin***

CNS Research Team, Johnson & Johnson Pharmaceutical Research & Development, L.L.C., Spring House, PA, USA.

Abstract

Rapidly inactivating, or A-type, voltage-gated K^+ (Kv) channels are important regulators of excitability in neurons, cardiac myocytes and many other excitable cells. Several years ago we identified a family of calcium binding proteins, termed KChIPs (for Kv Channel Interacting Proteins), which are now widely accepted to be integral subunits of A-type K^+ channels. KChIPs modulate the surface density, solubility, voltage-dependent activation and inactivation kinetics of expressed Kv4 subunits and native A-type currents. Remarkably, they do not appear to confer Ca^{2+} sensitivity on their targets. Although roles for KChIP family members as transcriptional regulators and modulators of presenilin function have been reported (KChIP3 = DREAM = Calsenilin; see related chapters in this volume), the structural and functional relationships underlying their role as Kv channel accessory subunits has been most extensively studied. This chapter will review recent structural data describing the molecular interaction between KChIPs and the Kv4 channel amino-terminus. We will also review anatomical data revealing the sites of interaction between individual KChIP isoforms and Kv4 subunits in mammalian brain. Together, these data provide considerable information about the role of KChIPs as modulators of cell excitability.

Keywords: Shal, EF-hand, memory, plasticity, hippocampus

* Correspondence concerning this article should be addressed to Kenneth J. Rhodes, Johnson & Johnson Pharmaceutical Research and Development, 1000 Route 202 Raritan, NJ 08869. E-mail: krhodes2@prdus.jnj.com.

Abbreviations

K^+	potassium
Kv channel	voltage-gated K^+ channel
KChIP	Kv Channel Interacting Protein
DREAM	downstream regulatory element antagonistic modulator
YTH	yeast two-hybrid
GC	granule cell
NCS	neuronal calcium sensor

1. Introduction

In many excitable cells rapidly inactivating, or A-type, voltage-gated potassium (Kv) currents shape responses to plasma membrane depolarization [1-3]. For example, in mammalian neurons these channels regulate the amplitude and duration of dendritic action potentials [4], modulate excitability in response to soluble and membrane associated second messengers [5-7] and contribute to the types of membrane responses that support long-term changes in synaptic excitability (e.g., long-term potentiation [8]). On the basis of immunohistochemical analyses (reviewed in [9], see also [10]), pharmacological sensitivity [7,11] and biophysical properties [12-14] it is now well established that the somatodendritic and cardiac A-type Kv currents are formed by pore-forming α-subunits from the Kv4, or *Shal*-related gene family [15]. However, it is also clear that the A-type currents formed by expression of individual Kv4 α-subunits in heterologous cells are different from those recorded from native cells, suggesting that in the native A-type channels, other factors co-associate with the pore-forming Kv4 α-subunit and alter its biophysical properties. In support of this, co-expression of Kv4 α-subunits together with low molecular weight fractions of brain mRNA was shown to generate A-type currents with a more "native" phenotype [12,16].

In light of the importance of Kv4 channels in neuronal and cardiac excitability, their potential as targets for therapeutic drugs and our long-standing interest in modulatory Kv channel subunits, several years ago we sought to identify accessory subunits for the Kv4 channel family. As the well-characterized "β-subunits" that regulate the properties of Kv1 channels tightly associate with the cytoplasmic Kv1 amino-terminus, we reasoned that the cytoplasmic N-terminus of Kv4 channels might also contain a docking site for accessory subunits and used amino acids 1-180 from the cytoplasmic N-terminus of the human Kv4.3 subunit as "bait" in a yeast two-hybrid (YTH) screen to "fish" for proteins that interact with Kv4 channels. Recognizing that there is a very high level of Kv4.3 mRNA expression in the rat midbrain [14] and assuming a higher probability of finding accessory subunits in a library enriched for Kv4 α-subunit expression, we constructed a rat midbrain YTH library for our screen to identify Kv4 interactors (see [17] for details). Among the hits from this screen, we independently isolated several cDNAs predicted to encode two highly-related proteins. Analysis of the cDNA and predicted polypeptide sequences indicated that these cDNAs encoded two members of a new family of EF-hand containing, putative calcium binding proteins. We named the new proteins KChIPs 1 and 2. Subsequent bioinformatics searching and sequence analyses resulted in the identification of two additional KChIP family members, which we called KChIPs 3 and 4.

All KChIP family members were shown to bind $^{45}Ca^{2+}$ and based on Northern and *in situ* hybridization analyses to be highly expressed in the central nervous system [17]. Importantly, KChIP2 is also highly expressed in mammalian heart [17-19]. All KChIPs interact with Kv4 family Kv channels but do not appear to interact with α-subunits from other Kv channel families [17]. Remarkably, co-expression of KChIP1, 2, 3 or 4 with Kv4 α-subunits in heterologous cells dramatically increases the density, slows the inactivation kinetics, shifts the membrane voltage for Kv4 channel activation to more negative membrane potentials and speeds the rate of channel recovery from inactivation [17,20-23]. Although there are subtle differences in the modulation of Kv4 currents by individual KChIPs, as well as subtle differences in modulation as a result of alternative splicing within the N-termini of individual KChIP isoforms, all KChIP family members described to date modulate Kv4 currents in the manner described above. KChIP activity is contained within the highly conserved C-terminal "core" domain of KChIPs [17] – expression of KChIP constructs that contain only this core domain recapitulates all of the modulatory activity of the full-length polypeptides [17]. Overall, the effects of KChIPs enable Kv4 currents to exert greater control over membrane excitability (greater current density, activation at more negative potentials, slower inactivation, faster recovery from inactivation), and make the resulting currents more closely resemble native A-type currents recorded from some cell types.

Importantly, KChIPs have been shown to be components of Kv channel complexes in native tissues. In our initial report [17], we described generation of antibodies to the Kv4 and KChIP polypeptides and demonstrated that in cells transfected with Kv4 and KChIP cDNAs and in detergent extracts of rodent brain membranes, KChIP polypeptides co-associate with Kv4 channel subunits. Thus, the initial observation of a KChIP:Kv4 protein-protein interaction in yeast translated faithfully to heterologous expression systems where KChIPs directly modulate Kv4 currents and also to native tissues where KChIPs have been shown to be integral subunits of Kv4 channel complexes. A comprehensive discussion of the effects of KChIPs on the biophysical properties of Kv4 channels is well beyond the scope of this review (but see [17,24], others for primary data). Rather, we focus here on recent anatomical and structure-function analyses of the co-localization, co-association and protein-protein interaction between KChIPs and Kv4 channel subunits.

2. Localization of KChIPs in the Brain

In our original paper describing the KChIP:Kv4 interaction, we used subunit-specific antibodies to demonstrate an association and overlapping distribution of Kv4 α-subunits and KChIPs in rodent brain [17]. More recently, we extended those early findings by generating monoclonal antibodies that specifically recognize each KChIP and Kv4 polypeptide [10]. We used those reagents to map the distribution of each KChIP isoform in rat brain using light microscopic immunohistochemistry and for reciprocal co-immunoprecipitation experiments to demonstrate a tight association between KChIP and Kv4 polypeptides in membranes isolated from rat hippocampus [10]. Moreover, we used those reagents to demonstrate, at the light microscopic and ultrastructural levels, that there is an overlapping plasma membrane distribution of Kv4 subunits and KChIPs in many brain regions [25].

Generating specific antibodies for each KChIP isoform was challenging: the KChIP family members share ~70% amino acid identity throughout the C-terminal "core" region [17]. In our strategy we used as an immunogen the entire coding sequence of each individual KChIP, expressed as a fusion with glutathione-S-transferase. Through brute-force screening of hundreds of monoclonal antibody tissue culture supernatants we were able to identify anti-KChIP mAbs that were specific for their respective antigen. Surprisingly, the epitope for each of these specific antibodies was mapped to the conserved "core" region and therefore the antibodies to individual KChIP isoforms are likely to recognize all of the N-terminal splice variants related to the corresponding KChIP family member [10]. Because of this, the staining patterns we reported are likely to reflect the distribution of all possible N-terminal splice forms for each individual KChIP family member. This point must be considered in interpreting the immunohistochemical staining patterns, as the contribution of KChIP expression to the phenotype of Kv4-mediated currents may depend on the specific splice variant that is expressed [21,23,26,27].

Although it is beyond the scope of this review to recapitulate all of the features of the immunohistochemical staining patterns for each KChIP, some general themes emerged from our immunohistochemical analyses [28-30]. In general, we observed that the staining patterns for KChIPs 2, 3 and 4 have considerable overlap with Kv4.2 in cortical and hippocampal CA1 pyramidal cells, in granule cells of the hippocampal dentate gyrus, in the granule cells of the cerebellar cortex and in the output neurons and large cholinergic interneurons of the striatum. In contrast, the staining pattern for KChIP1 overlaps with Kv4.3, most notably in inhibitory interneurons throughout the cortex and hippocampus as well as in striatal interneurons. Co-immunoprecipitation analysis in detergent solubilized hippocampal membranes revealed that all four KChIPs co-associate with Kv4 α-subunits in hippocampal membranes, confirming that the KChIPs are tightly associated with Kv4 α-subunits in native brain A-type channel complexes [10].

Somatodendritic A-currents have been most extensively studied by patch-clamp in hippocampal pyramidal cells and in the large layer V pyramidal neurons in motor and somatosensory cortex (e.g., [2,4,6,11,31-35]). In CA1 pyramidal cells, A-type currents play a critical role in synaptic integration and plasticity by controlling sub-threshold excitation and the amplitude of back-propagating action potentials [4,31]. In both hippocampal and neocortical pyramidal cells, we observed a strong overlap in the distribution of immunoreactivity for KChIPs 2, 3 and 4 isoforms and for Kv4.2 subunits. The tight overlap in KChIP and Kv4 immunostaining was most apparent in sections processed using multi-label immunofluorescence methods and visualized by confocal microscopy [10]. Taken together with co-immunoprecipitation analyses, the immunohistochemical data provide very strong evidence to indicate that in the dendrites of hippocampal and neocortical pyramidal cells, and in interneurons in many brain regions, KChIPs associate with native Kv4 channels.

Another brain region in which there is a striking overlap in the distribution of Kv4 and KChIP polypeptides is in the granule cells (GC) of the cerebellar cortex. In these cells, KChIP and Kv4 immunoreactivity is concentrated in the somata and dendrites at specialized synapses known as glomeruli. Using double-label immunofluorescence staining, we have shown that Kv4.2 and Kv4.3 α-subunits are expressed at high levels and are colocalized in the GC glomeruli. Previous studies had shown that Kv4.2 [14,36,37] and Kv4.3 [14,36] mRNAs and Kv4.2 [17,37-39] and Kv4.3 [40] immunoreactivity are concentrated in GC somata and glomeruli. Together with this concentration of immunoreactivity for Kv4 subunits, there is robust staining for KChIP1, KChIP3 and KChIP4 (but not KChIP2) in GC somata and glomeruli, suggesting that these subunits have the opportunity to interact with Kv4 α-subunits present at these sites. Moreover, as these KChIPs are cytoplasmic subunits that exhibit a diffuse cytoplasmic localization when expressed in heterologous cells, the strong membrane-associated staining on GC somata, and the trafficking to dendritic glomeruli, may reflect an association with the transmembrane Kv4 α-subunits. As native A-type channels in GCs are modulated by second messengers including arachidonic acid, and the effects of arachidonic acid on Kv4 channels require KChIP coexpression [7], the overlapping staining and functional properties the A-type currents in granule cells strongly implies the presence of KChIP:Kv4 channel complexes.

Although in all cell types where we observed KChIP and Kv4 immunoreactivity we saw strong staining at or near the plasma membrane, in some cells we also observed KChIP (and to a lesser extent Kv4) immunoreactivity that was diffusely localized within the neuronal cytoplasm. The cytoplasmic KChIP immunoreactivity might represent newly synthesized KChIP protein not associated with Kv4 subunits or associated with newly synthesized Kv4 channel complexes in the endoplasmic reticulum or Golgi apparatus or in transit to the plasma membrane. It is also possible that this cytoplasmic KChIP immunoreactivity is associated with an intracellular protein, e.g., presenilin [41,42]. The protein we identified as KChIP3 [17] was identified independently as DREAM, a Ca^{2+}-dependent transcriptional repressor [43] and as calsenilin, a presenilin interacting protein [41]. Recently, Zaidi et al. [42] described the distribution of calsenilin immunoreactivity in mouse brain, using a rabbit polyclonal antibody raised against a full-length calsenilin/KChIP3 fusion protein, which is likely to cross react with other KChIP family members. Interestingly, neither the Zaidi et al. [42] polyclonal antibody nor our KChIP3-specific mAb exhibit staining in the nuclei of neurons or glial cells, which might be expected for a Ca^{2+}-dependent regulator of transcription [43,44]. The remarkable correspondence in the patterns of immunoreactivity for Kv4.2 and KChIP3 [10,42] suggests that the preferred binding partner for KChIP3 in neurons is probably Kv4.2, raising questions as to whether KChIP3 in native cells has a role in gene transcription.

3. Regulation of Kv4 Channels by KChIP is Dependent upon the Kv4 N-Terminus

As described above, the initial identification of KChIPs was uncovered through its interaction with the Kv4 cytoplasmic N-terminus [17]. In an effort to identify specific domains or motifs within the Kv4 N-terminus that were responsible for the interaction with Kv4 subunits, we used a multidisciplinary approach that combined the techniques of electrophysiology, cell biochemistry, x-ray crystallography and yeast genetics. To assess KChIP-dependent effects on Kv4 channel function, *Xenopus* oocytes were co-injected with cRNAs encoding various forms of Kv4 channels and KChIP1. To explore the physical interaction between Kv4 N-terminus and KChIP1 more directly, we used the YTH system together with biochemical analyses of bacterially expressed recombinant proteins. Using a deletion mutagenesis strategy, we found that large deletions of the N-terminus of Kv4 completely disrupted the expression of currents [45]. It is possible that these larger deletions of the first 70 or 50 amino acids in Kv4 channels prevented the channels from forming functional tetramers, resulting in a lack of detectable current. Nevertheless, we could express these larger mutants in yeast and bacteria and assess the effects of the amino acid deletions on the interaction with KChIP. However, we observed that these large deletions within the Kv4 N-terminus completely disrupted the KChIP:Kv4 interaction [45]. In targeting smaller domains in the N-terminus, we found that limiting the deletions to the first 21 amino acids resulted in the expression of functional channels. These mutant Kv4 channels exhibited kinetic properties that were substantially similar to wild-type channels [45]. Interestingly, KChIP1 did not modulate the density, inactivation, or recovery from the inactivation of Kv4 channels in which residues 2-21 or even 2-11 were deleted from the Kv4 N-terminus [45]. Moreover, the Kv4Δ2-11 and Kv4Δ2-21 mutants did not interact with KChIP1 in the YTH system, suggesting there was no physical interaction between these mutants and KChIP1 [45]. Surprisingly, a smaller deletion mutant of the Kv4 N-terminus (Kv4Δ2-6) was modulated by KChIP1 in a manner comparable to wild-type channels, and as assessed by YTH, this Kv4 mutant did indeed interact with KChIP1 [17,45]. Together, these data suggested that residues 7-11 within the Kv4 N-terminus played an important role in the interaction and we sought to confirm this by generating a mutant that lacked these amino acids. KChIP1 did not significantly modulate the biophysical properties of the Kv4Δ7-11 mutant, and the interaction with KChIP1 in the YTH system was also greatly reduced [45]. Identification of residues within the Kv4 7-11 region that were essential for the interaction proved to be more difficult, as in our hands individual point mutations within the 7-11 domain had no effect on modulation of Kv4 currents or on the protein-protein interaction [45]. Even when all residues within this region were mutated simultaneously to generate radical changes in amino acid charge or side-chain (e.g., AWLPF to WGEFA), modulation of the expressed currents by KChIP1 was only slightly diminished and the mutants retained the interaction with KChIP1 in the YTH system [45]. However, it is important to point out that other studies have shown that mutating only two residues within this domain (W8 and F11) can disrupt the interaction with KChIP1, suggesting these residues may have side-chain interactions critical for Kv4-KChIP binding [46]. Taken together, these results all indicate that the proximal N-terminus of Kv4 channels is critical for KChIP1 association and for modulation of the kinetic properties of expressed Kv4 channels.

Based on results obtained using bacterially expressed recombinant proteins as well as studies using full-length channel subunits transiently expressed in heterologous cells, we have suggested that the proximal N-terminus of Kv4 channels plays a role in Kv4 solubility and surface trafficking [22,45]. Overall, the proximal N-terminus of Kv4 channels is remarkably aliphatic, as 16 of the first 23 amino acids (70%) are hydrophobic. Expression of full-length Kv4 subunits in heterologous cells indicates that in the absence of KChIP co-expression, Kv4 channels containing an intact N-terminus are extensively aggregated. In contrast, a Kv4 N-terminal deletion mutant that lacks the first 10 amino acids is soluble [45].

These results are consistent with those of earlier studies that reported a redistribution of Kv4 from perinuclear regions to the cell surface in response to KChIP co-expression [20,22]. Other studies have demonstrated a role for the Kv4 N-terminus in channel aggregation and endoplasmic reticulum retention, as deleting portions of the hydrophobic N-terminus results in increased surface trafficking and increased current density even in the absence of KChIP [20,22].

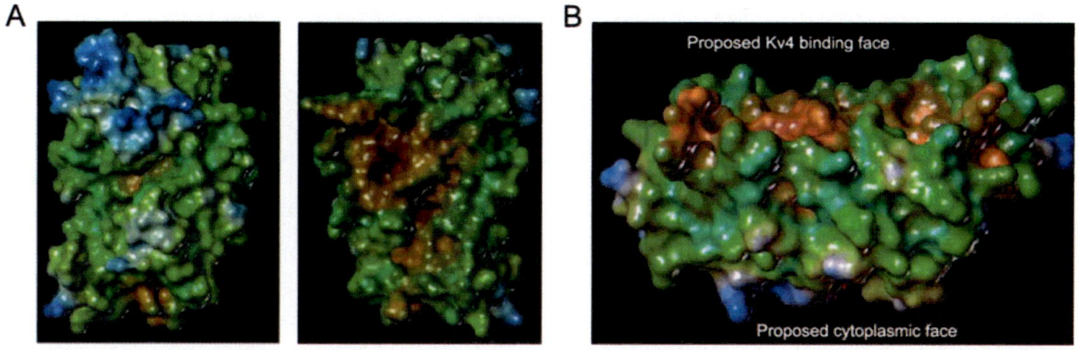

Figure 1. Crystal structure of KChIP1. (a) 180° rotated views of the KChIP1 structure. Structure has been mapped with a calculated MOLCAD surface. The surfaces are color-coded by lipophilicity, where blue is highly hydrophilic and brown is highly lipophilic. Note the deep hydrophobic pocket indicated by the brown colored surface in the right hand panel. This region is proposed to serve as the acceptor domain for interaction with the hydrophobic Kv4 N-terminus. (b) Side-on view of the KChIP1 structure rotated into an orientation such that the lipophilic region, as denoted by the orange and brown color, is facing upwards. The hydrophilic regions, as indicated by the green and blue colors, are orientated outwards relative to the lipophillic domain. *Taken from Scannevin et al. 2004, Neuron, 41, 587.*

The recent description of the KChIP1 structure revealed the presence of a deep hydrophobic pocket formed by portions of the first two EF-hands (Figure 1) [45,46]. Data from a chimeric molecule in which the N-terminal 30 amino acids of Kv4.2 were fused to the C-terminus of KChIP1 suggests that critical residues (W8, F11) in the proximal Kv4 N-terminus make contacts with residues contained within this hydrophobic pocket on KChIP1 [46]. Taken together with our data, we have hypothesized that this pocket in KChIP1 accommodates the hydrophobic N-terminus of Kv4 α-subunits, and this interaction stabilizes the Kv4:KChIP1 complex. The "masking" of the hydrophobic Kv4 N-terminus may thereby facilitate channel trafficking by preventing aggregation and endoplasmic reticulum retention of the nascent Kv4 channel polypeptide. This would also have the effect of promoting solubility and increased surface trafficking, leading to the observed increase in the density of macroscopic Kv4 currents. Interestingly, the hydrophobic pocket described for KChIP1 has also been observed in several other neuronal calcium sensor (NCS) proteins such as frequenin/NCS-1, recoverin and neurocalcin [47-49]. We have reported that in the YTH system, the Kv4 N-terminus did not interact with hippocalcin, a related NCS molecule [17]. Other reports have indicated that Kv4 channels do associate with frequenin/NCS-1 in brain [50] and heart [51], but this interaction with Kv4 channels may have decreased affinity as compared to KChIPs [52]. Interestingly, co-transfection of Kv4 channels and frequenin/NCS-1 in heterologous cells lead to dramatic increases in current density, but no modulation of channel kinetics was observed [51]. This suggests the intriguing possibility that the association of Kv4 N-termini with the KChIP hydrophobic pocket is only responsible for the chaperone-like effects of KChIP1 and that these effects on trafficking can be dissociated from their effects on the biophysical properties of Kv4 channels.

4. A Loop Structure in Kv4 is Critical for KChIP Modulation and Interaction

Earlier work suggested that Kv4 channels may contain more than one binding site for KChIPs, and these sites may have distinct functional roles [24,51]. With this possibility in mind, we extended our analysis of Kv4 N-terminal deletions to identify other domains that might be critical for the interaction between Kv4 subunits and KChIP1. As we observed for deletions within the proximal N-terminus of Kv4 subunits, many of these other Kv4 mutants either failed to form functional channels upon expression in Xenopus oocytes or had drastically different properties as compared to wild-type Kv4 channels [17,45]. Some of these mutants also failed to associate with KChIP1 in the YTH system, perhaps as a result of vastly altered folding [45]. After our preliminary analysis of a panel of Kv4 internal deletions, our efforts focused on a particular region of the Kv4 N-terminus, residues 71-90. Deletion of these amino acids in Kv4.2 yielded functional channels with kinetic properties similar to wild-type Kv4.2 [45]. Remarkably, the kinetic properties of the Kv4Δ71-90 mutant were not modulated upon co-expression of KChIP1 [45]. Assessing the physical interaction in yeast or using recombinant proteins confirmed that Kv4 mutants that were lacking residues 71-90, or even smaller 10 amino acid segments within this region, did not interact with KChIP1 [45]. The affects of mutations within the 71-90 domain of Kv4.2 on the KChIP1 interaction was further explored using single amino acid mutations. Of several mutations tested, only one (Kv4.2F74R) showed reduced modulation by KChIP1 [45]. This was not the case, however in the YTH system. Several mutations within the 71-90 region, including F74R, were sufficient to disrupt the interaction with KChIP1 [45]. Although is still unclear why some mutations that disrupt the protein-protein interaction in yeast had no effect on the functional consequences of interaction, these data suggest that within the Kv4 71-90 domain there are amino acid side-chain contacts that form an important component of the binding site for KChIP1.

Given the specificity of this 71-90 domain in mediating the interaction between Kv4 and KChIP, we sough to determine if this region would be sufficient to confer KChIP1 binding upon a Kv channel subunit with which KChIP1 does not normally interact. Using a chimera mutagenesis strategy, we changed individual residues in the Kv1.2 N-terminus into the corresponding residues from Kv4.2. A chimera was constructed in which amino acids 5-9 in Kv1.2 were changed to the amino acids encoding residues 7-11 in Kv4.2. This chimera, along with wild-type Kv1.2 N-terminus did not interact with KChIP in the YTH system [45], suggesting that amino acids 7-11 in Kv4.2 were not sufficient to confer KChIP1 binding. In a second chimera, we changed amino acids 68-86 in Kv1.2 to the correlate amino acids encoding residues 71-90 in Kv4.2. Interestingly, this chimera showed a robust interaction with KChIP1 in the YTH system [45], indicating that Kv4 residues 71-90 are sufficient to confer an interaction with KChIP1. To determine the interaction specificity of this chimera, we introduced the E79R point mutation in the region of the chimera corresponding to residues 71-90 from Kv4.2. When compared to the unaltered chimera, this point mutation had an attenuated interaction with KChIP in the YTH system [45]. Although not completely disrupted, the greatly diminished interaction suggests that the binding between KChIP1 and the Kv1/Kv4 71-90 chimera is specific, and that individual residues within this region of Kv4 channels form side chain contacts with KChIP1.

A

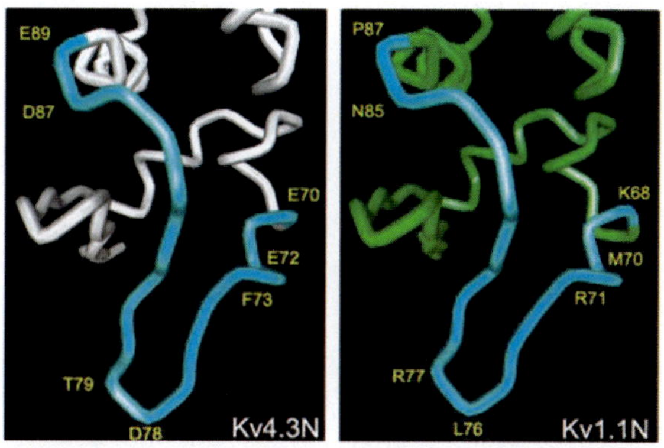

B

```
Kv1.1  (68)KRMRYFDPLRNEYFFDRNRP
Kv1.2  (64)KRMRYFDPLRNEYFFDRNRP

Kv2.1  (67)QVCDDYSLEDNEYFFDRHPG

Kv3.1  (43)HSHFDYDPRADEFFFDRHPG

Kv4.1  (70)EKEFFYDAESGEYFFDRDPD
Kv4.2  (71)ERDFFYHPETQQYFFDRDPD
Kv4.3  (70)EKEFFFNEDTKEYFFDRDPE
           * **       **
```

Figure 2. Crystal structure of the Kv4.3 T1 domain. (a) Views of the extended loop domains of both Kv4.3N (left panel) and Kv1.1N (right panel). In each panel, the loop region thought to be important for in auxiliary subunit binding is highlighted in cyan. Flanking residues are denoted in yellow, as are residues that were conserved in Kv4 channels, yet divergent with most other Kv α-subunits. Note the gross structural similarity of these two domains. The Kv4-conserved residues may indicate points in the Kv4.3 loop that are specific for interaction with KChIPs. (b) Optimal sequence alignment of rodent Kv family loop domains. Residues in red denote residues that are conserved in five or more sequences. Residues in cyan indicate amino acids that are conserved in Kv4 subfamily members but are divergent in most other Kv α-subunits. Asterisks denote residues that when mutated in Kv4.2N disrupted the interaction with KChIP1 in YTH. *Taken from Scannevin et al. 2004, Neuron, 41, 587.*

In our recently solved structure of the Kv4.3 N-terminus, one striking feature is a loop that extends from the surface of the molecule [45]. This loop structure contains residues 71-90 of the Kv4 N-terminus. In comparison to the published structure of the N-terminus from the related channel Kv1.1 [53], we noticed that the 71-90 loop in Kv4.3 exhibits a high degree of structural similarity to a loop structure within the N-terminus of Kv1.1 that was found to be critical for docking the β-subunit. (Figure 2a) [53]. This structural homology suggests the intriguing possibility that the Kv4 loop serves as a docking site for KChIP, similar to the function of the corresponding region in Kv1.1. Within the 71-90 region of Kv4 channels there are several residues that are conserved across the Kv4 channel family but different from the corresponding residues in most other Kv channel families (Figure 2b). Given the specificity of Kv4 channels for interacting with KChIPs [17,45], it is possible that together with other domains the amino acids within this region of the N-terminus determine the specificity of auxiliary subunit binding. In Kv4 channels, this region thus serves as a second interaction site, and although difficult to directly test, we speculate that this domain confers the ability of KChIPs to modulate Kv4 currents. This complements the role of the

Kv4 proximal N-terminal interaction, which serves a chaperone-like function to promote channel solubility and surface trafficking and subsequently increase in current density. While additional studies have provided an exciting "macroscopic" glimpse of the Kv4-KChIP complex [54], future work aimed at determining a higher resolution structure will undoubtedly shed more light onto the nature of this interaction.

5. Unanswered Questions and Summary

From many perspectives, the role of KChIPs as modulators of Kv4-family Kv channels is well described. The KChIPs exert many effects on the biophysical properties of Kv channels and in addition serve as integral subunits of somatodendritic and cardiac A-type channels where they regulate the solubility and surface trafficking of the channel's pore-forming α-subunits. Although the precise role of alternative splicing within the N-terminus of individual KChIP family members remains to be elucidated, the effects on Kv channel function mediated by the KChIP "core" domain will likely dominate more subtle contributions of the variable N-termini. Clearly, the major unanswered questions in KChIP biology revolve around the role of this family of proteins in diverse aspects of cellular physiology, including Kv channel modulation, presenilin biology and transcriptional control, and the role of the Ca2+-binding EF-hand domains in KChIP function. For example, although all KChIPs bind Ca2+ ions and the coordination of Ca2+ ions within EF-hands 3 and 4 of KChIP1 is similar to that observed in other members of the NCS protein family, KChIPs do not appear to confer Ca2+ sensitivity upon Kv channels, and therefore the role of Ca^{2+} in KChIP function remains to be elucidated. In addition, the structural basis for KChIP interaction with a diverse set of binding partners, ranging from Kv4 channels, presenilins and regulatory sequences of DNA, remains to be determined. Although it is conceivable that KChIP3/DREAM/calsenilin is the only truly multifunctional member of the KChIP family, this does not appear to be likely and in fact our own unpublished studies indicate that all of the known KChIP family members interact with presenilins, at least in heterologous expression systems (L. Doliveira and K. Rhodes, unpublished observations). From a structural perspective, it is difficult to reconcile how KChIPs bind to these structurally unrelated proteins and this raises important questions about the nature of the binding interactions between KChIPs, specifically KChIP3, and other molecules, e.g. presenilins [41,42] and the DNA sequence of the DREAM regulatory element [43,55]. Clearly, future studies aimed at revealing the structural basis of these interactions with KChIP3 will aid in elucidating the role of this molecule in such a diverse array of biological processes.

References

[1] Johnston, D., Hoffman, D.A., and Poolos, N.P. 2000, *Epilepsia, 41*, 1072.

[2] Storm, J.F. 2000, *J. Physiol., 525 Pt 3*, 565.

[3] Johnston, D., Christie, B.R., Frick, A., Gray, R., Hoffman, D.A., Schexnayder, L.K., Watanabe, S., and Yuan, L.L. 2003, *Philos Trans. R. Soc. Lond. B. Biol. Sci., 358*, 667.

[4] Hoffman, D.A., Magee, J.C., Colbert, C.M., and Johnston, D. 1997, *Nature, 387*, 869.

[5] Nakamura, T.Y., Coetzee, W.A., Vega-Saenz De Miera, E., Artman, M., and Rudy, B. 1997, *Am. J. Physiol., 273*, H1775.

[6] Hoffman, D.A. and Johnston, D. 1998, *J. Neurosci., 18*, 3521.

[7] Holmqvist, M.H., Cao, J., Knoppers, M.H., Jurman, M.E., Distefano, P.S., Rhodes, K.J., Xie, Y., and An, W.F. 2001, *J. Neurosci., 21*, 4154.

[8] Frick, A., Magee, J., and Johnston, D. 2004, *Nat. Neurosci., 7*, 126.

[9] Trimmer, J.S. and Rhodes, K.J. 2004, *Annu. Rev. Physiol., 66*, 477.

[10] Rhodes, K.J., Carroll, K.I., Sung, M.A., Doliveira, L.C., Monaghan, M.M., Burke, S.L., Strassle, B.W., Buchwalder, L., Menegola, M., Cao, J., An, W.F., and Trimmer, J.S. 2004, *J. Neurosci., 24*, 7903.

[11] Wu, R.L. and Barish, M.E. 1992, *J. Neurosci., 12*, 2235.

[12] Serodio, P., Kentros, C., and Rudy, B. 1994, *J. Neurophysiol., 72*, 1516.

[13] Serodio, P., Vega-Saenz de Miera, E., and Rudy, B. 1996, *J. Neurophysiol., 75*, 2174.

[14] Serodio, P. and Rudy, B. 1998, *J. Neurophysiol., 79*, 1081.

[15] Gutman, G.A., Chandy, K.G., Adelman, J.P., Aiyar, J., Bayliss, D.A., Clapham, D.E., Covarriubias, M., Desir, G.V., Furuichi, K., Ganetzky, B., Garcia, M.L., Grissmer, S., Jan, L.Y., Karschin, A., Kim, D., Kuperschmidt, S., Kurachi, Y., Lazdunski, M., Lesage, F., Lester, H.A., McKinnon, D., Nichols, C.G., O'Kelly, I., Robbins, J., Robertson, G.A., Rudy, B., Sanguinetti, M., Seino, S., Stuehmer, W., Tamkun, M.M., Vandenberg, C.A., Wei, A., Wulff, H., and Wymore, R.S. 2003, *Pharmacol. Rev., 55*, 583.

[16] Rudy, B., Hoger, J.H., Lester, H.A., and Davidson, N. 1988, *Neuron, 1*, 649.

[17] An, W.F., Bowlby, M.R., Betty, M., Cao, J., Ling, H.P., Mendoza, G., Hinson, J.W., Mattsson, K.I., Strassle, B.W., Trimmer, J.S., and Rhodes, K.J. 2000, *Nature, 403*, 553.

[18] Rosati, B., Grau, F., Rodriguez, S., Li, H., Nerbonne, J.M., and McKinnon, D. 2003, *J. Physiol., 548*, 815.

[19] Rosati, B., Pan, Z., Lypen, S., Wang, H.S., Cohen, I., Dixon, J.E., and McKinnon, D. 2001, *J. Physiol., 533*, 119.

[20] Bahring, R., Dannenberg, J., Peters, H.C., Leicher, T., Pongs, O., and Isbrandt, D. 2001, *J. Biol. Chem., 276*, 23888.

[21] Holmqvist, M.H., Cao, J., Hernandez-Pineda, R., Jacobson, M.D., Carroll, K.I., Sung, M.A., Betty, M., Ge, P., Gilbride, K.J., Brown, M.E., Jurman, M.E., Lawson, D., Silos-Santiago, I., Xie, Y., Covarrubias, M., Rhodes, K.J., Distefano, P.S., and An, W.F. 2002, *Proc. Natl. Acad. Sci. USA, 99*, 1035.

[22] Shibata, R., Misonou, H., Campomanes, C.R., Anderson, A.E., Schrader, L.A., Doliveira, L.C., Carroll, K.I., Sweatt, J.D., Rhodes, K.J., and Trimmer, J.S. 2003, *J. Biol. Chem., 278*, 36445.

[23] Patel, S.P., Parai, R., and Campbell, D.L. 2004, *J. Physiol., 557*, 19.

[24] Beck, E.J., Bowlby, M., An, W.F., Rhodes, K.J., and Covarrubias, M. 2002, *J. Physiol., 538*, 691.

[25] Strassle, B.W., Menegola, M., Rhodes, K.J., and Trimmer, J.S. 2005, *J. Comp. Neurol., 484*, 144.

[26] Boland, L.M., Jiang, M., Lee, S.Y., Fahrenkrug, S.C., Harnett, M.T., and O'Grady, S.M. 2003, *Am. J. Physiol. Cell Physiol., 285*, C161.

[27] Decher, N., Barth, A.S., Gonzalez, T., Steinmeyer, K., and Sanguinetti, M.C. 2004, *J. Physiol., 557*, 761.

[28] Rhodes, K.J., Keilbaugh, S.A., Barrezueta, N.X., Lopez, K.L., and Trimmer, J.S. 1995, *J. Neurosci., 15*, 5360.

[29] Rhodes, K.J., Monaghan, M.M., Barrezueta, N.X., Nawoschik, S., Bekele-Arcuri, Z., Matos, M.F., Nakahira, K., Schechter, L.E., and Trimmer, J.S. 1996, *J. Neurosci., 16*, 4846.

[30] Rhodes, K.J., Strassle, B.W., Monaghan, M.M., Bekele-Arcuri, Z., Matos, M.F., and Trimmer, J.S. 1997, *J. Neurosci., 17*, 8246.

[31] Johnston, D., Hoffman, D.A., Magee, J.C., Poolos, N.P., Watanabe, S., Colbert, C.M., and Migliore, M. 2000, *J. Physiol., 525 Pt 1*, 75.

[32] Golding, N.L., Mickus, T., and Spruston, N. 1999, *J. Neurosci., 19*, 8789.

[33] Bekkers, J.M. 2000, *J. Physiol., 525 Pt 3*, 593.

[34] Bekkers, J.M. 2000, *J. Physiol., 525 Pt 3*, 611.

[35] Korngreen, A. and Sakmann, B. 2000, *J. Physiol., 525 Pt 3*, 621.

[36] Tsaur, M.L., Chou, C.C., Shih, Y.H., and Wang, H.L. 1997, *FEBS Lett., 400*, 215.

[37] Sheng, M., Tsaur, M.L., Jan, Y.N., and Jan, L.Y. 1992, *Neuron, 9*, 271.

[38] Shibata, R., Nakahira, K., Shibasaki, K., Wakazono, Y., Imoto, K., and Ikenaka, K.. 2000, *J. Neurosci., 20*, 4145.

[39] Tsaur, M.L., Wu, Y.L., Huang, F.L., and Shih, Y.H. 2001, *Chin. J. Physiol., 44*, 133.

[40] Hsu, Y.H., Huang, H.Y., and Tsaur, M.L. 2003, *Eur. J. Neurosci., 18*, 601.

[41] Buxbaum, J.D., Choi, E.K., Luo, Y., Lilliehook, C., Crowley, A.C., Merriam, D.E., and Wasco, W. 1998, *Nat. Med., 4*, 1177.

[42] Zaidi, N.F., Berezovska, O., Choi, E.K., Miller, J.S., Chan, H., Lilliehook, C., Hyman, B.T., Buxbaum, J.D., and Wasco, W. 2002, *Neuroscience, 114*, 247.

[43] Carrion, A.M., Link, W.A., Ledo, F., Mellstrom, B., and Naranjo, J.R. 1999, *Nature, 398*, 80.

[44] Cheng, H.Y., Pitcher, G.M., Laviolette, S.R., Whishaw, I.Q., Tong, K.I., Kockeritz, L.K., Wada, T., Joza, N.A., Crackower, M., Goncalves, J., Sarosi, I., Woodgett, J.R., Oliveira-dos-Santos, A.J., Ikura, M., van der Kooy, D., Salter, M.W., and Penninger, J.M. 2002, *Cell, 108*, 31.

[45] Scannevin, R.H., Wang, K., Jow, F., Megules, J., Kopsco, D.C., Edris, W., Carroll, K.C., Lu, Q., Xu, W., Xu, Z., Katz, A.H., Olland, S., Lin, L., Taylor, M., Stahl, M., Malakian, K., Somers, W., Mosyak, L., Bowlby, M.R., Chanda, P., and Rhodes, K.J. 2004, *Neuron, 41*, 587.

[46] Zhou, W., Qian, Y., Kunjilwar, K., Pfaffinger, P.J., and Choe, S. 2004, *Neuron, 41*, 573.

[47] Flaherty, K.M., Zozulya, S., Stryer, L., and McKay, D.B. 1993, *Cell, 75*, 709.

[48] Vijay-Kumar, S. and Kumar, V.D. 1999, *Nat. Struct. Biol., 6*, 80.

[49] Bourne, Y., Dannenberg, J., Pollmann, V., Marchot, P., and Pongs, O. 2001, *J. Biol. Chem., 276*, 11949.

[50] Nakamura, T.Y., Pountney, D.J., Ozaita, A., Nandi, S., Ueda, S., Rudy, B., and Coetzee, W.A. 2001, *Proc. Natl. Acad. Sci. USA, 98*, 12808.

[51] Guo, W., Malin, S.A., Johns, D.C., Jeromin, A., and Nerbonne, J.M. 2002, *J. Biol. Chem., 277*, 26436.

[52] Ren, X., Shand, S.H., and Takimoto, K. 2003, *J. Biol. Chem., 278*, 43564.

[53] Gulbis, J.M., Zhou, M., Mann, S., and MacKinnon, R. 2000, *Science, 289*, 123.

[54] Kim, L.A., Furst, J., Gutierrez, D., Butler, M.H., Xu, S., Goldstein, S.A., and Grigorieff, N. 2004, *Neuron, 41*, 513.

[55] Osawa, M., Tong, K.I., Lilliehook, C., Wasco, W., Buxbaum, J.D., Cheng, H.Y., Penninger, J.M., Ikura, M., and Ames, J.B. 2001, *J. Biol. Chem., 276*, 41005.

PART VI: CaBP

In: Neuronal Calcium Sensor Proteins
Editors: Pavel Philippov and Karl Koch, pp. 301-313
ISBN 1-59454-978-8

Chapter XVII

Caldendrin – A Neuronal Calcium Binding Protein Involved in Synapto-Dendritic Ca^{2+} Signaling

Michael R. Kreutz**, *Constanze I. Seidenbecher* and *Eckart D. Gundelfinger
AG Molecular Mechanisms of Plasticity, Department of Neurochemistry and Molecular Biology, Leibniz Institute for Neurobiology, 39118 Magdeburg, Germany.

Abstract

During recent years significant progress has been made in our understanding how neuronal calcium sensor (NCS) proteins transform discrete Ca^{2+} signals into long-lasting changes in neuronal excitability, intracellular signaling cascades and gene expression. Members of the caldendrin/CaBP family of calmodulin (CaM) domain-containing proteins, which are the closest relatives of CaM in the central nervous system, may play a major role in these processes. Despite its close similarity to CaM, caldendrin has a specific set of binding partners. This in conjunction with its selective cellular and sub-cellular localization and its unique biochemical characteristics renders the protein a synapto-dendritic Ca^{2+} sensor most likely involved in synaptic plasticity processes. In this article, we will review the current knowledge about the cellular functions of the caldendrin/CaBP gene family in brain and retina with special reference to a putative synapto-dendritic signaling function.

Keywords: Caldendrin, EF-hands, neuronal calcium sensor, synapse, post-synaptic density, Ca^{2+} channels, cytoskeleton.

* Correspondence concerning this article should be addressed to Michael R. Kreutz AG Molecular Mechanisms of Plasticity, Department of Neurochemistry and Molecular Biology, Leibniz Institute for Neurobiology, Brenneckestr. 6, 39118 Magdeburg, Germany. E-mail: michael.kreutz@ifn-magdeburg.de.

Abbreviations

CaBP	calcium-binding protein
CaM	calmodulin
$Ca_v1.2$	L-type voltage-dependent Ca^{2+} channel
$Ca_v2.1$	P/Q-type voltage-dependent Ca^{2+}-channel
CNS	central nervous system
HCA	hydrophobic cluster analysis
$InsP_3R$	inositol trisphosphate receptor
LC3	microtubule-associated protein light chain 3
LTP	long-term potentiation
NCS	neuronal calcium sensor
PSD	post-synaptic density
TRPC5	transient receptor potential canonical type 5 Ca^{2+}-channel

1. Introduction

Ca^{2+} is a crucial signaling ion for a variety of processes in neurons ranging from neurotransmitter release, nuclear gene transcription to synaptic plasticity. Another important aspect of neuronal function is the Ca^{2+}-dependent signaling from dendrites and synapses to the nucleus, a process that is thought to couple synaptic activity to nuclear gene transcription and in turn long-term changes in neuronal characteristics. A number of signaling events have been proposed during recent years that might account for the underlying molecular mechanisms. Coupling of neurotransmitter receptors to signal-transducing intracellular cascades is realized by the postsynaptic density (PSD), a highly specialized cytomatrix structure, which functions as a molecular machine in tuning excitatory signals [1,2]. The major calcium-sensor protein involved in triggering synapto-dendritic Ca^{2+}-transients to downstream effectors is thought to be calmodulin (CaM). However, with caldendrin a novel calcium-binding protein was identified, which has a C-terminal domain very similar to CaM but is specifically expressed in neurons. Caldendrin is a cytoskeleton-associated neuronal Ca^{2+} sensor (NCS) protein, which is highly enriched in the PSD [3,4]. It harbors a number of features that make it a particularly good candidate for a signaling function in the synapse as well as to effectors downstream of the synapse that might couple synaptic activity to the control of gene expression.

2. The Molecular Structure of Caldendrin

Caldendrin is the first described member of a novel family of Ca^{2+}-binding proteins (CaBPs; [3,5,6]) that comprises thus far seven members (Figure 1). In contrast to the other members of the family, caldendrin displays a unique primary structure with a bipartite organization of the molecule into a very basic N-terminus (P_I 11.5) and a rather acidic C-terminus (P_I 4.6) resulting in an overall P_I of 7.1. The N-terminal half shows no homology to other polypeptide sequences in public databases and is predicted to start with a short hydrophobic helix comprising residues with rather small side chains (HCA in Figure 1 and PHD prediction; [7]). The caldendrin gene is alternatively spliced and gives rise to two shorter isoforms (termed caldendrin-S1 and -S2) with an unrelated N-terminus (Figure 1; [5,8]). Thus, although the isoform caldendrin-S1 harbors an N-terminal helical region, this helix has an amphipathic character and is primarily composed of amino acid residues with large side chains. Furthermore, both shorter isoforms as well as most other family members are myristoylated at their N-termini (Figure 1).

The entire C-terminal half of caldendrin and the other family members is highly homologous to CaM, the prototypical member of the family of EF-hand calcium-binding proteins. In fact, apart from the CaM-like protein, caldendrins are the closest relatives of CaM identified to date. The comparison between CaM and caldendrins/CaBPs, however, also shows that evolutionary modifications of one or more of the EF hands presumably disturb EF-hand Ca^{2+}-binding, but keep the overall structural topology of the molecules conserved. Troponin C and CaBP3, for instance, have no functional EF-hands 1 and 2 [5], whereas caldendrin and all other CaBPs lack Ca^{2+}-binding to EF-hand 2 [3,5,9]. Finally, the most distantly related member of the family, calneuron has no functional EF-hands 3 and 4 but exhibits two presumably functional EF-hands 1 and 2 [10].

Based on the solved crystal structures for CaM and troponin C with and without bound peptide ligands or calcium ions we have modeled the structure of the CaM-like domain in caldendrin/CaBP (Figure 1D). The topology can be described as a dumb bell formed by two "halves" or "lobes" connected by a flexible helical linker. That dumb-bell shape is stabilized by the 4 EF hand helix-loop-helix motifs (Figure 1). Although the structure of the C-terminal half of caldendrin as determined by modeling perfectly fits the CaM structure, a different surface of the molecules has to be considered because most conserved structure-determining residues are part of the interior hydrophobic cavity in both molecules while the outer surface is covered by residues that clearly differ between CaM and caldendrin. Furthermore, caldendrin as well as most of the CaBPs harbor a longer hinge region between EF-hands 2 and 3 than CaM possibly enhancing the flexibility of the two halves [5]. As caldendrin is not present in the *Drosophila* and *C. elegans* genomes, while the CaM gene is found in all known eukaryotic organisms [11,12], the caldendrin-specific structural features may have evolved after duplication of the CaM gene and fusion with other exons during vertebrate development. Interestingly, the point mutations in EF-hand 2 of caldendrin/CaBP1, which leave the overall structure intact but render this EF-hand cryptic and non-functional [9], are highly conserved in different species during vertebrate evolution with almost identical solvent exposed amino acid residues at corresponding positions. This observation suggests that the singularity of the caldendrin surface is intrinsic and independent from insertions or deletions and we, therefore, hypothesize that this might be due to adaptations of its surface to a specific localization and function in neurons of higher vertebrates (see also below).

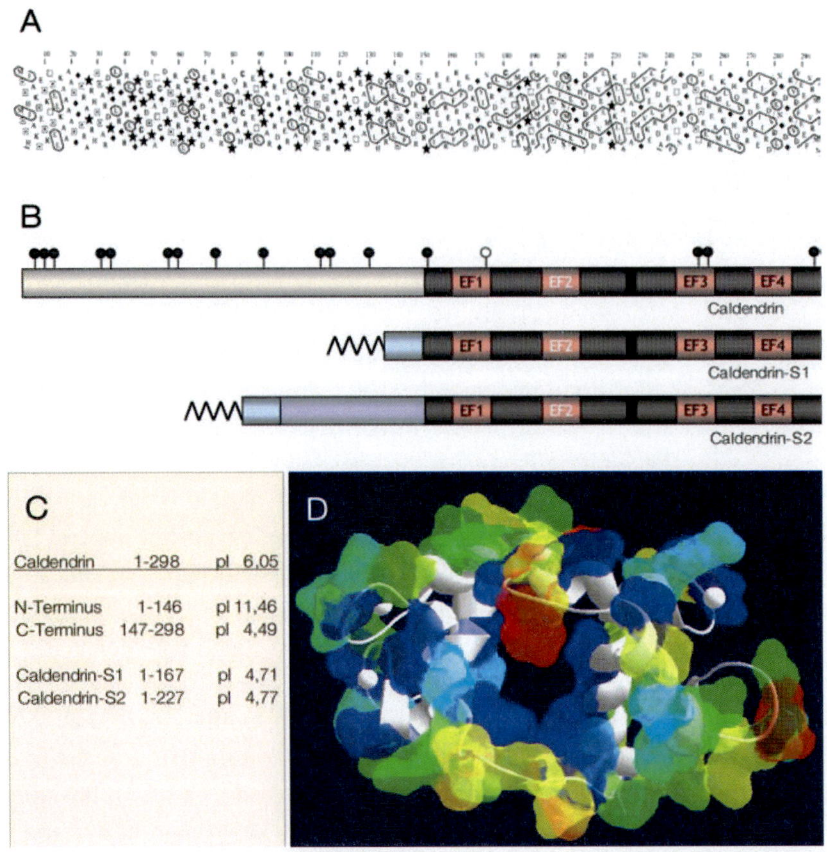

Figure 1. The molecular structure of caldendrin: isoforms, domains and functional sites. A. Hydrophobic cluster analysis (HCA) of caldendrin obtained with the drawhca program (http://bioserv.rpbs.jussieu.fr). As depicted by the graphic output, the protein consists of two entirely different structural parts. The C-terminal half displays highly ordered hydrophobic clusters as structural determinants (framed amino acids in single letter code). The N-terminal half is almost devoid of hydrophobic clusters but rich in structure-loosening proline residues (asterisks). Numbering indicates the amino acid positions. B. Schematic drawing of the primary structures of caldendrin and its shorter splice isoforms -S1 and -S2. Dots indicate putative Ser/Thr (filled dots) or Tyr (open circle) phosphorylation sites. Myristoylation sites are depicted as zic-zac lines. EF hand domains are indicated. The black bar marks the extended hinge region between EF hands 2 and 3. C. Calculated isoelectric point values for full-length caldendrin, its halfs and the shorter splice variants. Note, the two shorter isoforms have a significantly more acidic pI than the long isoform. D. Model structure of the caldendrin C-terminus based on coordinates of CaM (pdb: 1CKK). The protein backbone is drawn as white ribbon and calcium ions as space-filling objects. The surface is generated for solvent-exposed amino acids only. The surface similarity is colour-coded from blue (identical residues) to red (no homology). The red surface in front of the peptide-binding site represents a tyrosine, which replaces a glutamic acid. The proximal surface of caldendrin is different, while the binding pocket is highly similar to CaM.

3. Caldendrin is a Neuron-Specific Protein

A number of studies have dealt with the localization of caldendrin and other family members in brain and retina. Immunolocalization studies in rat and human central nervous system (CNS) revealed that caldendrin expression is clearly restricted to neurons and is absent from glial cells [4]. Caldendrin immunoreactivity is prominently present in the somato-dendritic compartment of neurons but could not be detected in axons. At the subcellular level, caldendrin seems to be a component of a discrete subset of mature excitatory synapses [3,4]. Moreover, *in situ* hybridization experiments and immunostaining have shown that caldendrin transcripts and protein are preferentially found in CNS regions with a laminar cytoarchitecture, i.e. cerebral and cerebellar cortices, hippocampus, olfactory bulb and the retina. Caldendrin is localized mainly to principal neurons in brain and is prominently present in pyramidal cells [3,4,13]. Within the cerebral cortex and the hippocampus, the expression is highly heterogeneous, with highest levels of caldendrin in layer III of the piriform and layers II/III of the somatosensory cortex, as well as the CA3 region and the dentate gyrus of the hippocampus. This difference in expression levels may reflect different requirements for dendritic Ca^{2+} signaling in these neurons [4].

Interestingly, the caldendrin gene is alternatively spliced (see above). The shorter variants lacking the N-terminal part of caldendrin are only detectable in the retina and to a minor degree in the cerebellum [4,5]. Most importantly, with the notable exception of calneuron, all other family members are expressed only at a very low level in brain and are only abundant in retina [4,5]. Thus, caldendrin and calneuron can be considered as the only family members with a more general role in Ca^{2+} signaling in the brain. Furthermore, the developmental expression in brain shows that only minor amounts of caldendrin mRNA and the protein are present in the early postnatal brain [3,4] as compared to adult levels. Caldendrin transcript levels increase during the second postnatal week coinciding with the period of dendrite maturation and synaptogenesis. This indicates a role for caldendrin in the maturing brain but not during early development.

Several studies have dealt with the distribution of caldendrin immunoreactivity in the rodent retina [4,14,15]. All these reports document a predominant localization of the protein to the inner retina. Müller glial cells, the main retinal glia cells, were found to be devoid of caldendrin. These data are consistent with the exclusively neuronal expression pattern in the brain. Moreover, caldendrin antibodies prominently stain the inner plexiform but not the outer plexiform layer where synaptic contacts between presynaptic photoreceptor cells and postsynaptic bipolar and horizontal cells reside. CaBP5, a cognate member of the caldendrin family, is detectable in rod and cone bipolar cell perikarya as well as in fine processes including dendrites in the outer plexiform layer and axons ramifying in the inner plexiform layer [5]. Despite the close relation to caldendrin, the distributions of both molecules differ significantly [14]. Furthermore, caldendrin immunoreactivity does not seem to be congruent with that of CaBP4 [16]. Thus, although the C-terminus of the different family members is highly similar, the differential distribution in retina suggests that they might have cell-specific functions and do not merely substitute for each other.

A recent study has demonstrated the localization of caldendrin in the postsynaptic compartment of ribbon synapses between photoreceptors and horizontal cells in the fish retina [17]. Comparison of light- and dark-adapted retinae revealed a light-dependent shift of membrane-associated caldendrin from the terminal dendrites into spinules formed by these dendrites. This finding indicates that caldendrin might be involved in the regulation of spinule dynamics and supports the view of an important role of the protein in the regulation of synaptic plasticity (see below).

4. Caldendrin is a Component of the PSD

Most calcium-binding proteins are either soluble cytoplasmic proteins or membrane-associated via N-terminal myristoylation. Upon subcellular fractionation of brain proteins, caldendrin partitions into both, the soluble as well as the detergent-resistant particulate fraction [3]. Further biochemical analysis reveals a tight association of the insoluble caldendrin pool with the neuronal cytoskeleton including the subsynaptic cytomatrix of the PSD. Since most NCS proteins are fairly soluble molecules, this issue is of particular interest because it may shed some light on the challenging question how components of the specialized neuronal cytoskeleton contribute to the transduction of segregated Ca^{2+} signals.

Caldendrin immunoreactivity migrates in SDS-PAGE as a doublet of two protein bands (33 and 36 kDa), both of which derive from the same mature transcript [3]. The discrete nature of these two bands is not completely solved, but the 36-kDa isoform can only be released by very harsh treatments, like 8M urea, suggesting rather tight cytoskeletal association. Moreover, purification of the highly specialized cytomatrix of the postsynapse, the PSD, showed a significant enrichment of caldendrin immunoreactivity in this subcellular compartment. Importantly, further extraction of peripheral protein components from the PSD even lead to a further enrichment of caldendrin suggesting that the protein is a core component of this cytoskeletal meshwork (see also below) and that especially the 36-kDa isoform seems to be particularly abundant in this synaptic fraction [3]. The highly basic N-terminus of caldendrin is thought to mediate the tight association of the protein with the subsynaptic cytoskeleton. The smaller caldendrin splice variants S1 and S2 appear not to be enriched in synapse fractions from the retina [12]. Finally, similar fractionation experiments were performed for the neuronal calcium sensor-1 (NCS-1; [18]). These showed that NCS-1 equally partitions between the cytosolic and detergent-soluble membrane fractions but is not detectable in the detergent-insoluble cytoskeleton fraction, where caldendrin is enriched. This finding further supports the hypothesis that caldendrin's association with the cytoskeleton is a rather unique feature for calcium-binding proteins.

At present it is not completely understood how the different subcellular localization of caldendrin into an insoluble cytoskeletal and a soluble cytoplasmic pool is mediated. Our preliminary data suggest that phosphorylation of caldendrin may play an important role in this process. Caldendrin harbors 14 putative Ser/Thr phosphorylation sites (Figure 1B), and *in situ* phosphorylation experiments revealed that caldendrin is a kinase substrate. On two-dimensional protein gels it became obvious that the 33 and 36 kDa isoforms differ in their P_I: whereas the 36-kDa isoform displays a P_I of approx. 6.8-6.9, the 33-kDa variant separates into a chain of spots within the range of 5.9-6.8 [3]. Moreover, a functional casein kinase II site has been identified in the C-terminus of caldendrin [19]. In conclusion, it can be hypothesized that phosphorylation of caldendrin by protein kinase C, casein kinase II or other kinases might release the protein from the cytoskeleton, whereas its dephosphorylation via phosphatases might have the opposite effect. Important in this context is the observation that the association of caldendrin with the PSD appears to be stimulus-dependent. In the rat, kainate-induced epileptic seizures, a model of human temporal lobe epilepsy, induce a massive recruitment of the protein to the PSD [20]. Similar observations were made in other *in vivo* models of synaptic plasticity like long-term potentiation (LTP; Balschun, Smalla, Kreutz, unpublished observations) and *in vitro* after glutamate stimulation of hippocampal primary neuronal cultures (Kreutz & Kreutz, unpublished observations). We therefore assume that the equilibrium between phosphorylation and de-phosphorylation might account for caldendrin's subcellular distribution and its activity-dependent translocation. The above described light-dependent relocalization of caldendrin in horizontal cells of the fish retina might reflect such a translocation process [17].

In summary, caldendrin in contrast to most other calcium-binding proteins is tightly associated with the cortical cytoskeleton, enriched in the PSD fraction [3], associates late during development with the synaptic cytomatrix in the brain [4] and does so in a stimulus-dependent manner [20]. Moreover, both its two-domain structure and the differential association with subcellular compartments suggest that caldendrin serves functions different from those of other known neuronal Ca^{2+} sensors and Ca^{2+} buffers.

5. Caldendrin Binding Partners and Dendritic Ca^{2+} Signaling

Concerning the function of caldendrin two hypotheses that are not mutually exclusive might be considered. (1) Caldendrin might compete locally with the ubiquitous calcium sensor CaM (or other NCS proteins) for binding partners and thus be involved in fine-tuning of their effects. (2) Caldendrin might interact with a specific group of targets and modulate their functions. Several lines of evidence support the notion that due to its unique properties caldendrin might not merely substitute for CaM in specific functions. This is of vital importance since it has to be assumed that in neurons CaM expression levels are much higher than those of caldendrin. Therefore, it is very unlikely that caldendrin can efficiently compete with CaM for binding to a target interaction if the binding mode and binding affinities are comparable and protein levels in a neuronal subcompartment are similar. To date, several potential binding partners for caldendrin and the other family members have been identified. The emerging picture from these data is rather complex but some general conclusions can be drawn from the published studies. Caldendrin has a number of specific binding partners like the microtubule-associated protein light chain 3 (LC3) [21], and two recently identified molecules from our lab that we termed Esau and Jacob, which are not bound by CaM (Figure 2; unpublished data). In particular, the interaction with Jacob is interesting, as this molecule seems to be involved in activity-regulated synapto-nuclear signaling processes within neurons (Dieterich et al., submitted). Vice versa caldendrin/CaBP1 does not bind to some prominent interaction partners of CaM in neurons such as the phosphatase calcineurin [22] or the neuronal nitric oxide synthase [21]. Similarly, caldendrin is only a very weak activator for the well-established CaM-interacting protein Ca^{2+}/calmodulin-dependent kinase II [5]. On the other hand, also a number of interaction partners were identified that are bound by both caldendrin and CaM. In these cases, however, caldendrin binds either with substantially different Ca^{2+} requirements and/or its binding can not be competed by CaM, indirectly suggesting the existence of non-overlapping binding regions (see also below).

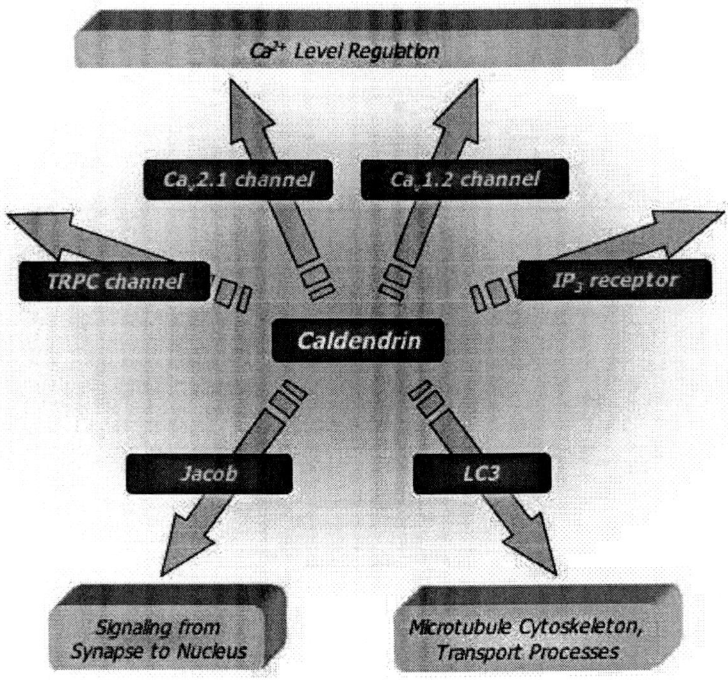

Figure 2. Caldendrin as a multifunctional regulator of neuronal Ca^{2+} signals. The cartoon illustrates protein interactions with different cation channels at the plasma membrane ($Ca_V2.1$, $Ca_V1.2$, TRPC5) and at the ER ($InsP_3R$), and with Jacob, a putative downstream effector. The interaction with LC3 is thought to be implicated in neuronal transport processes to and from the synapse (For details see text).

In more detail, for caldendrin-S1/CaBP1 a CaM- and Ca^{2+}-independent binding has been reported to P/Q-type ($Ca_V2.1$) Ca^{2+}-channels [23]. This splice isoform of caldendrin positively shifts the voltage dependence of $Ca_V2.1$ activation, accelerates inactivation, and prevents paired-pulse facilitation in a Ca^{2+}-independent manner [23-25]. Myristoyl-deficient mutants of this splice isoform behaved largely like CaM and did not show Ca^{2+}-independent paired-pulse facilitation [24]. This raises the interesting possibility that N-terminal myristoylation might not only provide a membrane anchor but will also alter the functional consequences of a target interaction. Accordingly, caldendrin, which cannot be N-terminally myristoylated, may have different effects than the shorter splice isoforms. At present it is unclear whether this interaction is physiologically relevant, since $Ca_V2.1$ channels are considered to be largely presynaptic while caldendrin localization seems to be confined to postsynaptic sites [2]. Recently, an example of postsynaptic and dendritic localization of $Ca_V2.1$ immunoreactivity has been identified for cerebellar Purkinje cells [26]. However, these cells do not express caldendrin and a common appeareance of the two proteins in the same subcellular compartment still remains to be demonstrated. In addition, the shorter splice isoforms that can be N-terminally myristoylated are virtually only present in retina [12] and in neurohypophyseal axons [8]. Therefore, the physiological significance of the findings by Catteral and colleagues in brain remains to be established [23-25].

A further interaction between caldendrin/CaBP1 and a Ca^{2+} channel was reported for the alpha1-subunit of L-type voltage-dependent $Ca_V1.2$ Ca^{2+} channels [27]. This interaction is Ca^{2+}-dependent and it can be competed by CaM. Importantly, co-immunoprecipitation of caldendrin/CaBP1 and $Ca_V1.2$ Ca^{2+} channels could be shown from synaptic protein preparations, which in conjunction with the reported postsynaptic localization of $Ca_V1.2$ Ca^{2+} channels [27] suggests that the interaction might take place at the postsynaptic membrane. The functional consequence of caldendrin binding seems to be a facilitation of Ca^{2+} currents whereas CaM binding leads in contrast to an inactivation of Ca^{2+} influx [27]. The regulation of $Ca_V1.2$ Ca^{2+} channels seems to be, however, more complex. Caldendrin/CaBP1 and CaM bind to an IQ-domain in the cytoplasmic C-terminal domain of the alpha1-subunit of $Ca_V1.2$ Ca^{2+} channels, but caldendrin/CaBP1 also binds to a second more N-terminal region in a Ca^{2+}-independent manner [28]. CaM binding to this region is in contrast Ca^{2+}-dependent. Deletion of this domain abolished the effects of caldendrin/CaBP1 on $Ca_V1.2$ Ca^{2+} currents, but not the Ca^{2+}-dependent inactivation mediated by CaM [28]. Accordingly, the functionally distinct effects of caldendrin/CaBP1 and CaM seem to be controlled by binding to two discrete domains with different Ca^{2+}-requirements, allowing a high degree of fine-tuning of synapto-dendritic Ca^{2+}-responses.

Interestingly, also two caldendrin binding sites were identified in the microtuble-binding protein LC3 (see below), from which only one exhibited a strict Ca^{2+}-dependency for the interaction to take place [21]. In this case, however, CaM does not bind. More recently, an interaction of caldendrin/CaBP1 with transient receptor potential canonical type 5 (TRPC5) Ca^{2+}-channels was reported [29]. The C-terminal part of caldendrin/CaBP1 but not CaM inhibits TRPC5 activity. Importantly, the caldendrin/CaBP1 binding site within the C-terminal region of TRPC5 is close to but not identical with the CaM-binding site [29]. TRPC channels have been localized to growth cones and will presumably play a role in neurite extension [30]. It is therefore conceivable that caldendrin/CaBP1 might regulate such processes. The experiments by Kinoshita-Kawada et al. [29], however, were performed in *Xenopus* oocytes, and further studies will be needed to establish the physiological relevance in neurons.

One of the most important interactions in functional terms was recently shown by a Ca^{2+}-, CaM- and ATP-independent interaction of caldendrin/CaBP1 with the inositol trisphosphate ($InsP_3$) receptor [19,31,32] (but see also [33] for conflicting results). Interestingly, the affinity of the caldendrin/CaBP1-$InsP_3R$ interaction is greater than that between CaM and $InsP_3Rs$. Caldendrin/CaBP1 does, in contrast to CaM, not bind to ryanodine receptors. Of particular interest was the observation that two binding sites for caldendrin/CaBP1 exist in $InsP_3Rs$. Both interactions are, in contrast to those with CaM, Ca^{2+}-independent. Interestingly, the interaction requires phosphorylation of caldendrin/CaBP1 via casein kinase II [19] whereas N-terminal myristoylation is not required. The functional consequence of caldendrin binding is a reduction of $InsP_3$-triggered intracellular Ca^{2+}-release [19,31]. This is most likely mediated by an inhibition of $InsP_3$ binding [31,32]. Taken together, these data suggest that caldendrin/CaBP1 might play a crucial role in dendritic Ca^{2+} signaling as a constitutive, i.e. Ca^{2+}-independent, inhibitor of $InsP_3Rs$.

Most of the hitherto characterized caldendrin/CaBP1 interactions with its binding partners have been established *in vitro* or in heterologous systems with the short isoforms caldendrin-S1 and -S2, which are in fact extremely rare as compared to the long isoform. Therefore it will be a challenging task to verify these observations with the abundant caldendrin isoform under physiological conditons.

6. Caldendrin/CaBP1: A Regulator of Neuronal Ca^{2+} Channels?

Interestingly, all proteins interacting with both CaM and caldendrin/CaBP1 constitute Ca^{2+} channels, leading to the notion that caldendrin/CaBPs are a family of calcium sensors that modulate the activity of neuronal Ca^{2+} channels in a CaM-independent manner. Although this idea is intriguing, it should be stated that these Ca^{2+} channels constitute established and functionally easily accessible CaM binding targets that were not identified by conventional protein-protein interaction screening methods. It is therefore possible that a different picture will emerge with the identification of additional binding partners by less biased methods that also fulfill the criteria of a specific binding mode that cannot be competed by CaM.

A further point that deserves mentioning is that caldendrin's interaction with the majority of its binding partners is Ca^{2+}-independent (i.e. $InsP_3Rs$, $Ca_V2.1$ and presumably $Ca_V1.2$ Ca^{2+} channels). This suggests that protein complexes exist at resting Ca^{2+} levels that are not directly regulated by Ca^{2+} transients and thereby neuronal activity. One might therefore speculate that the localization of caldendrin is an important factor to trigger its target interactions. Of interest in this regard is the observation that caldendrin translocates from the cytoplasm to the cortical cytoskeleton after sustained synaptic activation [20] (see also above). In a recent study, we could show that a significant pool of caldendrin is associated with the microtubule cytoskeleton and that this interaction might be mediated via binding to the small microtubule-associated protein LC3[21]. CaM does not bind to LC3. Two interaction sites exist for caldendrin, only one of which is Ca^{2+}-dependent [21]. LC3 shows sequence homology to the $GABA_A$ receptor-associated protein, which is involved in the trafficking of GABA receptors [34]. The interaction to LC3 is well suited for a cellular transport process and might be therefore important for caldendrin's localization in neurons.

7. A Model for Caldendrin's Function in Synapto-Dendritic Ca^{2+} Signaling

Based on the present knowledge we would like to provide an admittedly highly speculative model of caldendrin's function in synapto-dendritic compartments. The model is based on the assumption that synaptic activity controls caldendrin's subcellular localization. Increased synaptic activity will eventually lead to a de-phosphorylation of caldendrin possibly via an activity-dependent phosphatase like calcineurin and to a tighter association with the subsynaptic cytoskeleton. In addition, this translocation might require binding of caldendrin to LC3. At the synapse, caldendrin can bind to L-type voltage-dependent $Ca_V1.2$ Ca^{2+} channels or other targets (Figure 2). The interaction with L-type voltage-dependent $Ca_V1.2$ Ca^{2+} channels will probably lead to increased Ca^{2+} currents following synaptic activation and thereby indirectly via increased synaptic activity promote caldendrin's localization in the synapse. Low synaptic activity and, hence, low synapto-dendritic Ca^{2+} levels will instead favor caldendrin's binding to the $InsP_3$-R. This interaction will most likely take place at the smooth ER of the dendrite and to a much lesser degree in spine synapses.Thus, the association of caldendrin with the PSD and the cortical cytoskeleton of the cell will be critical for its functional role. The scenario predicts that caldendrin will – controlled via its activity-dependent localization to different subcellular mircodomains – trigger different target interactions and thereby amplify existing differences in synapto-dendritic Ca^{2+} signaling within the neuron. This regulatory function will affect a number of Ca^{2+}-controlled events and possibly also feed back to caldendrin's own Ca^{2+}-dependent target interactions.

Taken together, although we have only begun to appreciate the role of caldendrin/CaBP1 within the network of neuronal Ca^{2+} regulation, it appears that this protein may exert unique functions in principal neurons of stratified brain regions. Future studies including the analysis of mouse mutants for caldendrin are required to establish the protein's contribution to neuronal operation especially in higher order brain functions of vertebrates.

Acknowledgements

The authors gratefully acknowledge caldendrin modeling by Carsten Reissner. Work in the authors' laboratories is supported by grants from the BMBF (01GZ0307/ 01GA0505), the DFG (Kr1879/2-1, 2-2; Se952/2-1, 2-2; SFB 426), the Fonds der Chemischen Industrie, the Schram-Foundation, the DAAD and the Land Sachsen-Anhalt (FKZ: 3293A, 2508A and 3004A).

References

[1] Gundelfinger, E.D., and tom Dieck, S. 2000, *Naturwiss., 87*, 513.

[2] Kennedy, M.B. 2000, *Science, 290*, 750.

[3] Seidenbecher, C.I., Langnaese, K., Sanmarti-Vila, L., Boeckers, T.M., Smalla, K.H., Sabel, B.A., Garner, C.C., Gundelfinger, E.D., and Kreutz, M.R. 1998, *J. Biol. Chem., 273*, 21324.

[4] Laube, G., Seidenbecher, C.I., Richter, K., Dieterich, D.C., Hoffmann, B., Landwehr, M., Smalla, K.H., Winter, C., Böckers, T.M., Wolf, G., Gundelfinger, E.D., and Kreutz, M.R. 2002, *Mol. Cell. Neurosci., 19*, 459.

[5] Haeseleer, F, Sokal, I., Verlinde, L.M.J., Erdjumen-Bromage, H., Tempst, P., Pronin, A.N., Benovic, J.L., Fariss, R.N., and Palczewski, K.2000, *J. Biol. Chem., 275*, 1247.

[6] Burgoyne, R.D., O'Callaghan, D.W., Hasdemir, B., Haynes, L.P., and Tepikin, A.V. 2004, *Trends Neurosci., 27*, 203.

[7] Rost, B and Sander, C. 1993, *J. Mol. Biol., 232*, 584.

[8] Landwehr, M., Redecker, P., Dieterich, D.C., Richter, K., Böckers, T.M., Gundelfinger, E.D., and Kreutz, M.R. 2003, *FEBS Lett., 547*, 189.

[9] Wingard, J.N., Chan, J., Bosanac, I., Haeseleer, F., Palczweski, K., Ikura, M., and Ames, J.B. 2005. *J. Biol. Chem* (Epub ahead of print, Sep 7, 2005)

[10] Wu, Y.Q., Lin, X., Liu, C.M., Jamrich, M., and Shaffer, L.G. 2001, *Mol. Genet. Metab., 72*, 343.

[11] Rogers, M.S., and Strehler, E.E., 1996, Calmodulin. In: *Guidebook to the calcium-binding proteins*. Eds. Celio MR, Pauls TL, Schwaller B. Oxford Univ Press.

[12] Seidenbecher, C.I., Reissner, C., and Kreutz, M.R. 2002, *Adv. Exp. Med. Biol., 514*, 451.

[13] Bernstein, H.G., Seidenbecher, C.I., Smalla, K.H., Gundelfinger, E.D., Bogerts, B., and Kreutz M.R. 2003, *J Histochem Cytochem., 51*, 1109.

[14] Haverkamp, S., Ghosh, K.K., Hirano, A.A., and Wässle, H. 2003, *J. Comp. Neurol., 455*, 463.

[15] Menger, N., Seidenbecher, C.I., Gundelfinger, E.D., and Kreutz, M.R. 1999, *Cell Tissue Res., 298*, 21.

[16] Haeseleer, F., Imanishi, Y., Maeda, T., Possin, D.E., Maeda, A., Lee, A., Rieke, F., and Palczewski, K. 2004, *Nat. Neurosci., 7*, 1079.

[17] Schultz, K., Janssen-Bienhold, U., Gundelfinger, E.D., Kreutz, M.R., and Weiler, R., 2004 *J. Comp. Neurol., 479*, 84.

[18] Reynolds, A.J., Bartlett, S.E., and Morgans, C. 2001, *Neuroreport, 12*, 725.

[19] Kasri, N.N., Holmes, A.M., Bultynck, G., Parys, J.B., Bootman, M.D., Rietdorf, K., Missiaen, L., McDonald, F., De Smedt, H., Conway, S.J., Holmes, A.B., Berridge, M.J., and Roderick, H.L. 2004, *EMBO J., 23*, 312.

[20] Smalla ,K.H., Seidenbecher, C.I., Tischmeyer, W., Schicknick, H., Wyneken, U., Böckers, T.M., Gundelfinger, E.D., and Kreutz, M.R. 2003, *Brain Res. Mol. Brain Res., 116*, 159.

[21] Seidenbecher, C.I., Landwehr, M., Smalla, K.H., Kreutz, M., Dieterich, D.C., Zuschratter, W., Reissner, C., Hammarback, J.A., Böckers, T.M., Gundelfinger, E.D., and Kreutz, M.R. 2004, *J Mol Biol., 336*, 957.

[22] Sokal, I., Li, N., Verlinde, C.L., Haeseleer, F., Baehr, W., and Palczewski, K. 2000, *Biochim. Biophys. Acta, 1498*, 233.

[23] Lee, A., Westenbroek, R.E., Haeseleer, F., Palczewski, K., Scheuer, T., and Catterall, W.A. 2002, *Nat. Neurosci., 3*, 210.

[24] Few, A.P., Lautermilch, N.J., Westenbroek, R.E., Scheuer, T., and Catterall W.A. 2005, *J. Neurosci., 25*, 7071.

[25] Lautermilch, N.J., Few, A.P., Scheuer, T., and Catterall, W.A. 2005, *J. Neurosci., 25*, 7062.

[26] Kulik, A., Nakadate, K., Hagiwara, A., Fukazawa, Y., Lujan, R., Saito, H., Suzuki, N., Futatsugi, A., Mikoshiba, K., Frotscher, M., and Shigemoto, R. 2004, *Eur. J. Neurosci., 19*, 2169.

[27] Zhou, H., Kim, S.A., Kirk, E.A., Tippens, A.L., Sun, H., Haeseleer, F., and Lee, A. 2004, *J. Neurosci., 24*, 4698.

[28] Zhou, H., Yu, K., McCoy, K.L., and Lee, A. 2005, J Biol Chem., 280, 29612.

[29] Kinoshita-Kawada, M., Tang, J., Xiao, R., Kaneko, S., Foskett, J.K., and Zhu M.X. 2005, *Pflügers Arch., 450*, 345.

[30] Wang, G.X., and Poo, M.M. 2005, *Nature, 434*, 898.

[31] Haynes, L.P., Tepikin, A.V., and Burgoyne, R.D. 2004, *J. Biol. Chem., 279*, 547

[32] Bosanac, I., Yamazaki, H., Matsu-Ura, T., Michikawa, T., Mikoshiba, K., and Ikura, M. 2005, *Mol. Cell., 17*, 193.

[33] Yang, J., McBride, S., Mak, D.O., Vardi, N., Palczewski, K., Haeseleer, F., and Foskett, JK. 2002, *Proc Natl Acad Sci U S A., 99*, 7711.

[34] Kittler, J.T., Rostaing, P., Schiavo, G., Fritschy, J.M., Olsen, R., Triller, A., and Moss, S.J. 2001, *Mol Cell Neurosci., 18*, 13.

Part VII: Biophysical and Chemical Approaches to Understand NCS Protein Function

In: Neuronal Calcium Sensor Proteins
Editors: Pavel Philippov and Karl Koch, pp. 317-329
ISBN 1-59454-978-8

Chapter XVIII

MOLECULAR STRUCTURE AND TARGET RECOGNITION OF NEURONAL CALCIUM SENSOR PROTEINS

James B. Ames[1,*] and Mitsuhiko Ikura[2]

[1]Center for Advanced Research in Biotechnology, University of Maryland Biotechnology Institute, Rockville, MD 20850;

[2]Division of Molecular and Structural Biology, Ontario Cancer Institute and Department of Medical Biophysics, University of Toronto, 610 University Avenue, Toronto, Ontario, Canada M5G 2M9;

Abstract

Neuronal calcium sensor (NCS) proteins, a sub-branch of the EF-hand superfamily, are expressed predominantly in the brain and retina where they serve a variety of functions related to signal transduction. Retinal recoverin regulates rhodopsin kinase in retinal rod cells, neuronal frequenin and related proteins modulate synaptic activity and vesicle secretion, K^+ channel interacting proteins (KChIPs) regulate ion channel gating kinetics, and DREAM controls the transcription of various genes. Here we review the atomic-level structures of various NCS proteins, including recoverin, neurocalcin, frequenin, guanylate cyclase activating proteins (GCAPs), and KChIPs to understand structural determinants that specify their functional differences. The three-dimensional structures of many NCS proteins look remarkably similar and yet their functions are non-overlapping and highly diverse. The key to understanding their functional differences will ultimately involve deciphering the structural interaction with different target proteins. The recent molecular structure of KChIP1 in complex with an N-terminal fragment of A-type K^+ channel reveals the first atomic-resolution view of an NCS target complex. The

* Correspondence concerning this article should be addressed to Dr. James B. Ames, Center for Advanced Research in Biotechnology, 9600 Gudelsky Drive, Rockville, MD 20850. Phone: (301) 738-6120; Fax: (301) 738-6255; e-mail: james@carb.nist.gov.

structures of NCS proteins and their target interactions will be discussed and compared with previous structures of calmodulin and calcineurin in complex with their respective target proteins. An emerging theme for target recognition by NCS proteins is the hydrophobic association of a target helix with an exposed concave surface (or groove) formed by helices from N-terminal EF-hands.

Keywords: EF-hand, calcium, nuclear magnetic resonance, recoverin, frequenin, vision.

1. Introduction

Intracellular calcium (Ca^{2+}) regulates a variety of neuronal signal transduction processes in the brain and retina [1,2]. The effects of changes in neuronal Ca^{2+} are mediated primarily by an emerging class of neuronal calcium sensor (NCS) proteins [3-6] that belong to the EF-hand superfamily [7,8]. The human genome encodes 14 members of the NCS family [9]. The amino acid sequences of NCS proteins are highly conserved from yeast to humans (Fig. 1). Recoverin, the first NCS protein to be discovered, and the guanylate cyclase activating proteins (GCAPs) are expressed exclusively in the retina where they serve as Ca^{2+} sensors in vision [10-13]. Other NCS proteins are expressed in the brain and spinal cord such as neurocalcin [14], frequenin (NCS-1) [15,16], K^+ channel interacting proteins (KChIPs) [17], DREAM/calsenilin [18,19] and hippocalcin [20,21]. Frequenin is also expressed outside of the central nervous system [22] as well as in invertebrates including flies [15], worms [23] and yeast (Frq1) [24-26]. The common features of these proteins are an approximately 200-residue chain containing four EF-hand motifs, the sequence CPXG in the first EF-hand that markedly impairs its capacity to bind Ca^{2+}, and an amino-terminal myristoylation consensus sequence.

Table 1. Functions of NCS Proteins.

NCS protein	Function	Reference
Recoverin	Inhibit rhodopsin kinase in retinal rods.	[27,29.31]
GCAP-1	Activate retinal guanylate cyclases.	[12,35]
GCAP-2	Activate retinal guanylate cyclases.	[11]
GCIP	Inhibit guanylate cyclase in frog photreceptors.	[72]
KChIP1	Regulate K^+ channel gating kinetics in brain.	[17,73]
KChIP2	Regulate K^+ channel gating kinetics in cardiac cells.	[17,74]
Calsenilin/DREAM	Repress transcription of prodynorphin and c-fos genes.	[17-19]
Frequenin (Frq1)	Activate PI(4) kinase; regulate Ca^{2+} and K^+ channels	[15,24,41,42]
Neurocalcin δ	?	[14]
Hippocalcin	Activate phospholipase D; MAP kinase signaling.	[20,75,76]
VILIP-1	Activate guanylate cyclase; traffic nicotinic receptors.	[77,78]

```
                                                                                         1           10
KChIP1          ~~~~~~~~~~ ~~~~~~~~~~ ~~~~~~~~~~ ~~~~~~~~~~ ~~MGAVMGTF SSLQTKQRRP SKDKIEDELE MTMVCHRPEG
DREAM           MQPAKEVTKA SDGSLLGDLG HTPLSKKEGI KWQRPRLSRQ ALMRCCLVKW ILSSTAPQ.. GSDSSDSELE LSTVRHQPEG
Neurocal        ~~~~~~~~~~ ~~~~~~~~~~ ~~~~~~~~~~ ~~~~~~~~~~ ~~~~~~~~~~ ~~~~~~~~~~ ~~~~~~~MGK QNS.KLRPEV
Hippocal        ~~~~~~~~~~ ~~~~~~~~~~ ~~~~~~~~~~ ~~~~~~~~~~ ~~~~~~~~~~ ~~~~~~~~~~ ~~~~~~~MGK QNS.KLRPEM
Frequenin       ~~~~~~~~~~ ~~~~~~~~~~ ~~~~~~~~~~ ~~~~~~~~~~ ~~~~~~~~~~ ~~~~~~~~~~ ~~~~~~~MGK SNS.KLKPEV
Frq1            ~~~~~~~~~~ ~~~~~~~~~~ ~~~~~~~~~~ ~~~~~~~~~~ ~~~~~~~~~~ ~~~~~~~~~~ ~~~~~~~MGA KTS.KLSKDD
Recoverin       ~~~~~~~~~~ ~~~~~~~~~~ ~~~~~~~~~~ ~~~~~~~~~~ ~~~~~~~~~~ ~~~~~~~~~~ ~~~~~~~MGN SKSGALSKEI

                        20         30         40         50         60         70         80         90
KChIP1          LEQLEAQTNF TKRELQVLYR GFKNECPSGV VNEDTFKQIY AQFFPHGDAS TYAHYLFNAF DTTQTGSVKF EDFVTALSIL
DREAM           LDQLQAQTKF TKKELQSLYR GFKNECPTGL VDEDTFKLIY AQFFPQGDAT TYAHFLFNAF DADGNGAIHF EDFVVGLSIL
Neurocal        MQDLLESTDF TEHEIQEWYK GFLRDCPSGH LSMEEFKKIY GNFFPYGDAS KFAEHVFRTF DANGDGTIDF REFIIALSVT
Hippocal        LQDLRENTEF SELELQEWYK GFLKDCPTGI LNVDEFKKIY ANFFPYGDAS KFAEHVFRTF DTNSDGTIDF REFIIALSVT
Frequenin       VEELTRKTYF TEKEVQQWYK GFIKDCPSGQ LDAAGFQKIY KQFFPFGDPT KFATFVFNVF DENKDGRIEF SEFIQALSVT
Frq1            LTCLKQSTYF DRREIQQWHK GFLRDCPSGQ LAREDFVKIY KQFFPFGSPE DFANHLFTVF DKDNNGFIHF EEFITVLSTT
Recoverin       LEELQLNTKF TEEELSSWYQ SFLKECPSGR ITRQEFQTIY SKFFPEADPK AYAQHVFRSF DANSDGTLDF KEYVIALHMT

                       100        110        120        130        140        150        160        170
KChIP1          LRGTVHEKLR WTFNLYDINK DGYINKEEMM DIVKAIYDMM GKYTYPVLKE D..TPRQHVD VFFQKMDKNK DGIVTLDEFL
DREAM           LRGTVHEKLK WAFNLYDINK DGYITKEEML AIMKSIYDMM GRHTYPILRE D..APAEHVE RFFEKMDRNQ DGVVTIEEFL
Neurocal        SRGKLEQKLK WAFSMYDLDG NGYISKAEML EIVQAIYKMV S..SVMKMPE DESTPEKRTE KIFRQMDTNR DGKLSLEEFI
Hippocal        SRGRLEQKLM WAFSMYDLDG NGYISREEML EIVQAIYKMV S..SVMKMPE DESTPEKRTE KIFRQMDTNN DGKLSLEEFI
Frequenin       SRGTLDEKLR WAFKLYDLDN DGYITRNEML DIVDAIYQMV G..NTVELPE EENTPEKRVD RIFAMMDKNA DGKLTLQEFQ
Frq1            SRGTLEEKLS WAFELYDLNH DGYITFDEML TIVASVYKMM G..SMVTLNE DEATPEMRVK KIFKLMDKNE DGYITLDEFR
Recoverin       SAGKTNQKLE WAFSLYDVDG NGTISKNEVL EIVTAIFKMI SPEDTKHLPE DENTPEKRAE KIWGFFGKKD DDKLTEKEFI

                       180        190
KChIP1          ESCQEDDNIM RSLQLFQNVM  216
DREAM           EACQKDENIM SSMQLFENVI  256
Neurocal        RGAKSDPSIV RLLQCDPSSA GQF  193
Hippocal        RGAKSDPSIV RLLQCDPSSA SQF  193
Frequenin       EGSKADPSIV QALSLYDGLV  190
Frq1            EGSKVDPSII GALNLYDGLI  190
Recoverin       EGTLANKEIL RLIQFEPQKV KEKLKEKKL  202
```

Figure 1. Amino-acid sequence alignment of human KChIP1 (Swiss-Prot accession no. Q9NZI2), mouse DREAM (Q9QXT8), bovine neurocalcin δ (P61602), human hippocalcin (Q5U068), drosophila frequenin (P37236), S. cerevisae Frq1 (Q06389), and bovine recoverin (P21457). The 29-residue EF-hand motifs are highlighted in color: EF-1 (green), EF-2 (red), EF-3 (cyan), EF-4 (yellow).

The structurally similar NCS proteins have remarkably different physiologic functions (Table 1). Perhaps the best characterized NCS protein is recoverin that serves as a calcium sensor in retinal rod cells. Recoverin prolongs the lifetime of light-excited rhodopsin [27-29] by inhibiting rhodopsin kinase only at high Ca^{2+} levels [30-32]. Hence, recoverin makes receptor desensitization Ca^{2+}-dependent, and the resulting shortened lifetime of rhodopsin at low Ca^{2+} levels may promote visual recovery and contribute to the adaptation to background light. Recoverin was also identified as the antigen in cancer-associated retinopathy, an autoimmune disease of the retina caused by a primary tumor localized outside the nervous system [33,34]. (Recoverin as a paraneoplastic antigen is discussed in another chapter of this issue). Other NCS proteins in retinal rods include the GCAP proteins that activate retinal guanylate cyclase only at low Ca^{2+} levels and inhibit the cyclase at high Ca^{2+} [11,12,35]. GCAPs are important for regulating the recovery phase of visual excitation and particular mutants are linked to various forms of retinal degeneration [36,37]. Yeast and mammalian frequenins bind and activate a particular PtdIns 4-OH kinase isoform (Pik1 gene in yeast) [22,24,38] required for vesicular trafficking in the late secretory pathway [39,40]. Mammalian frequenin (NCS-1) also regulates voltage-gated Ca^{2+} and K^+ channels [41,42]. The KChIPs regulate the gating kinetics of voltage-gated, A-type K^+ channels [17]. The DREAM/calsenilin/KChIP3 protein binds to specific DNA sequences in the prodynorphin and c-fos genes [18,43] and serves as a calcium sensor and transcriptional repressor for pain modulation [44,45]. Hence, the functions of the NCS proteins appear to be quite diverse and non-overlapping.

Mass spectrometric analysis of retinal recoverin and some of the other NCS proteins revealed that they are myristoylated at the amino terminus [21,46,47]. Recoverin contains an N-terminal myristoyl (14:0) or related fatty acyl group (12:0, 14:1, 14:2). Retinal recoverin and myristoylated recombinant recoverin, but not unmyristoylated recoverin, bind to membranes in a Ca^{2+}-dependent manner. Likewise, bovine neurocalcin and hippocalcin contain an N-terminal myristoyl group and both exhibit Ca^{2+}-induced membrane binding. These findings led to the proposal that NCS proteins possess a Ca^{2+}-myristoyl switch (Fig. 2). The covalently attached fatty acid is highly sequestered in recoverin in the calcium-free state. The binding of calcium to recoverin leads to the extrusion of the fatty acid, making it available to interact with lipid bilayer membranes or other hydrophobic sites.

In this chapter we review the atomic resolution structures of NCS proteins. Ca^{2+}-induced conformational changes as well as structural interactions with membranes and/or downstream signaling proteins will be examined to understand mechanisms of target recognition.

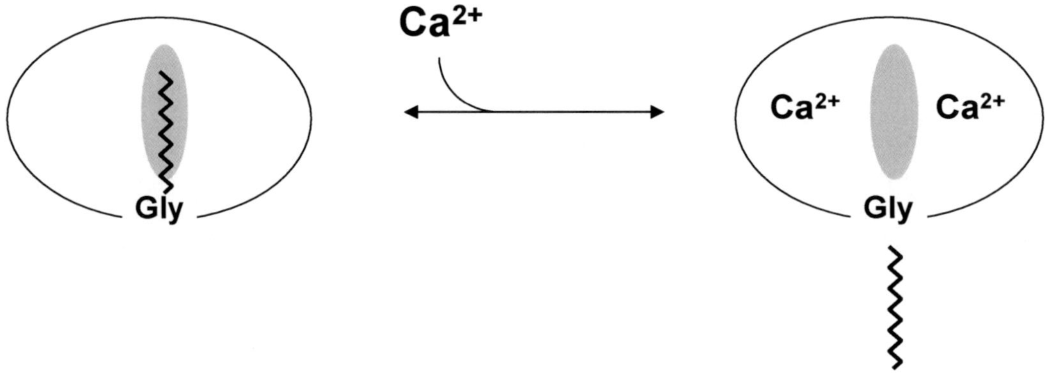

Figure 2. Schematic diagram of calcium-myristoyl switch in recoverin. The binding of two Ca^{2+} ions promotes the extrusion of the myristoyl group and exposure of other hydrophobic residues (marked by the shaded oval).

2. Three-Dimensional Structures of Recoverin

The x-ray crystal structure of recombinant unmyristoylated recoverin [48,49] showed it to contain a compact array of EF-hand motifs, in contrast to the dumbbell shape of calmodulin [50] and troponin C [51]. The four EF-hands are organized into two domains: The first EF-hand, EF-1 (residues 27-56, colored green in Figs. 1 and 3), interacts with EF-2 (residues 63-92, red) to form the N-terminal domain, and EF-3 (residues 101-130, cyan) and EF-4 (residues 148-177, yellow) form the C-terminal domain. The linker between the two domains is U-shaped rather than α-helical. Ca^{2+} is bound to EF-3 and Sm^{3+} (used to derive phases) is bound to EF-2. The other two EF hands possess novel features that prevented ion binding. EF-1 is disabled by a Cys-Pro sequence in the binding loop. EF-4 contains an internal salt bridge in the binding loop that competes with Ca^{2+} binding. Myristoylated recoverin, the physiologically active form, has thus far eluded crystallization.

The structures of myristoylated recoverin in solution with 0, 1 and 2 Ca^{2+} bound were determined by nuclear magnetic resonance (NMR) spectroscopy [52-54] (Fig. 3). In the Ca^{2+}-free state, the myristoyl group is sequestered in a deep hydrophobic cavity in the N-terminal domain. The cavity is formed by five α-helices. The two helices of EF-1 (residues 26-36 and 46-56), the exiting helix of EF-2 (residues 83-93), and entering helix of EF-3 (residues 100-109) lie perpendicular to the fatty acyl chain and form a box-like arrangement that surrounds the myristoyl group laterally. A long, amphipathic α-helix near the N-terminus (residues 4-16) packs closely against and runs antiparallel to the fatty acyl group, and serves as a lid on top of the four-helix box. The N-terminal residues Gly 2 and Asn 3 form a tight hairpin turn that connects the myristoyl group to the N-terminal helix. This turn positions the myristoyl group inside the hydrophobic cavity and gives the impression of a cocked trigger. The bond angle strain stored in the tight hairpin turn may help eject the myristoyl group from the pocket once Ca^{2+} binds to the protein.

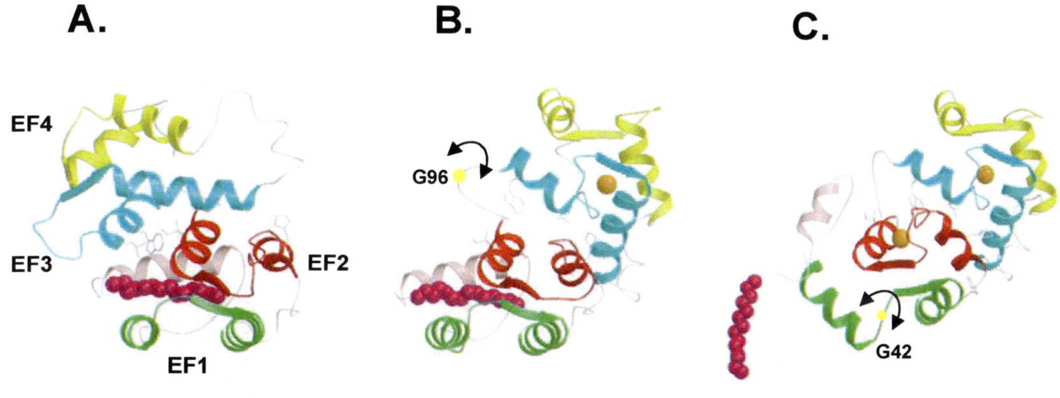

Figure 3. Three-dimensional structures of myristoylated recoverin with 0 Ca^{2+} bound (A), 1 Ca^{2+} bound (B) and 2 Ca^{2+} bound (C). The first step of the mechanism involves the binding of Ca^{2+} to EF-3 that causes minor structural changes within the EF-hand that sterically promote a 45° swiveling of the two domains, resulting in a partial unclamping of the myristoyl group and a dramatic rearrangement at the domain interface. The resulting altered interaction between EF-2 and EF-3 facilitates the binding of a second Ca^{2+} to the protein at EF-2 in the second step, which causes structural changes within the N-terminal domain that directly lead to the ejection of the fatty acyl group.

The structure of myristoylated recoverin with one Ca^{2+} bound at EF-3 (half saturated recoverin, Fig. 3B) [53] represents a hybrid structure of the Ca^{2+}-free and Ca^{2+}-saturated states. The structure of the N-terminal domain (residues 2-92, green and red in Fig. 3) of half saturated recoverin (Fig. 3B) resembles that of Ca^{2+}-free state (Fig. 3A) and is very different from that of the Ca^{2+}-saturated form (Fig. 3C). Conversely, the structure of the C-terminal domain (residues 102-202, cyan and yellow in Fig. 3) of half saturated recoverin more closely resembles that of the Ca^{2+}-saturated state. Most striking in the structure of half saturated recoverin is that the myristoyl group is flanked by a long N-terminal helix (residues 5-17) and is sequestered in a hydrophobic cavity containing many aromatic residues from EF-1 and EF-2 (F23, W31, Y53, F56, F83, and Y86). An important structural change induced by Ca^{2+}

binding at EF-3 is that the carbonyl end of the fatty acyl group in the half saturated species is displaced far away from hydrophobic residues of EF-3 (W104 and L108, Figs. 3A-B) and becomes somewhat solvent exposed. By contrast, the myristoyl group of Ca^{2+}-free recoverin is highly sequestered by residues of EF-3 [52].

The structure of myristoylated recoverin with two Ca^{2+} bound shows the amino-terminal myristoyl group to be extruded [54] (Figure 3C). The N-terminal eight residues are solvent exposed and highly flexible and thus serve as a mobile arm to position the myristoyl group outside the protein when Ca^{2+} is bound. The flexible arm is followed by a short α-helix (residues 9-17) that precedes the four EF-hand motifs, arranged in a tandem array as was seen in the x-ray structure. Calcium ions are bound to EF-2 and EF-3. EF-3 has the classic "open conformation" similar to the Ca^{2+} occupied EF-hands in calmodulin and troponin C. EF-2 is somewhat unusual and the helix-packing angle of Ca^{2+}-bound EF-2 (120°) in recoverin more closely resembles that of the Ca^{2+}-free EF-hands (in the "closed conformation") found in calmodulin and troponin C. The overall topology of Ca^{2+}-bound myristoylated recoverin is similar to the x-ray structure of unmyristoylated recoverin described above. The RMS deviation of the main chain atoms in the EF-hand motifs is 1.5 Å in comparing Ca^{2+}-bound myristoylated recoverin to unmyristoylated recoverin. Hence, in Ca^{2+}-saturated recoverin, the N-terminal myristoyl group is solvent exposed and does not influence the interior protein structure.

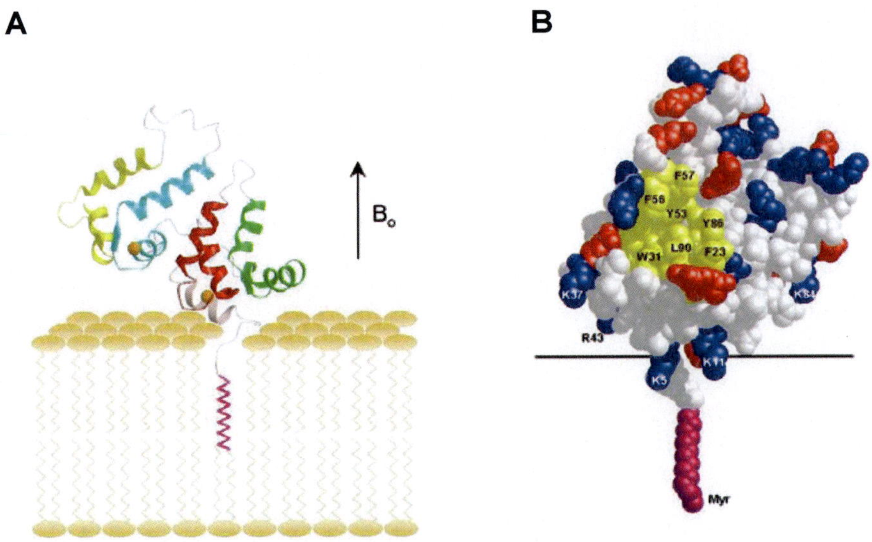

Figure 4. Main chain structure (A) and space-filling representation (B) of myristoylated recoverin bound to oriented lipid bilayers determined by solid-state NMR [55]. Hydrophobic residues are yellow, bound Ca^{2+} ions are orange, and charged residues are red and blue.

The Ca^{2+}-induced exposure of the myristoyl group (Figs. 2 and 3) enables recoverin to bind to membranes only at high Ca^{2+}. Recent solid-state NMR studies have determined the structure of Ca^{2+}-bound myristoylated recoverin bound to oriented lipid bilayer membranes (Fig. 4) [55]. Membrane-bound recoverin appears to retain approximately the same overall structure as it has in solution. The protein is positioned on the membrane surface such that its long molecular axis is oriented 45° with respect to the membrane normal. The N-terminal region of recoverin points toward the membrane surface, with close contacts formed by basic residues K5, K11, K22, K37, R43, and K84. This orientation of membrane-bound recoverin allows an exposed hydrophobic crevice (lined primarily by residues F23, W31, F35, I52, Y53, F56, Y86 and L90), near the membrane surface that may serve as a potential binding site for the target protein, rhodopsin kinase (Fig. 4B).

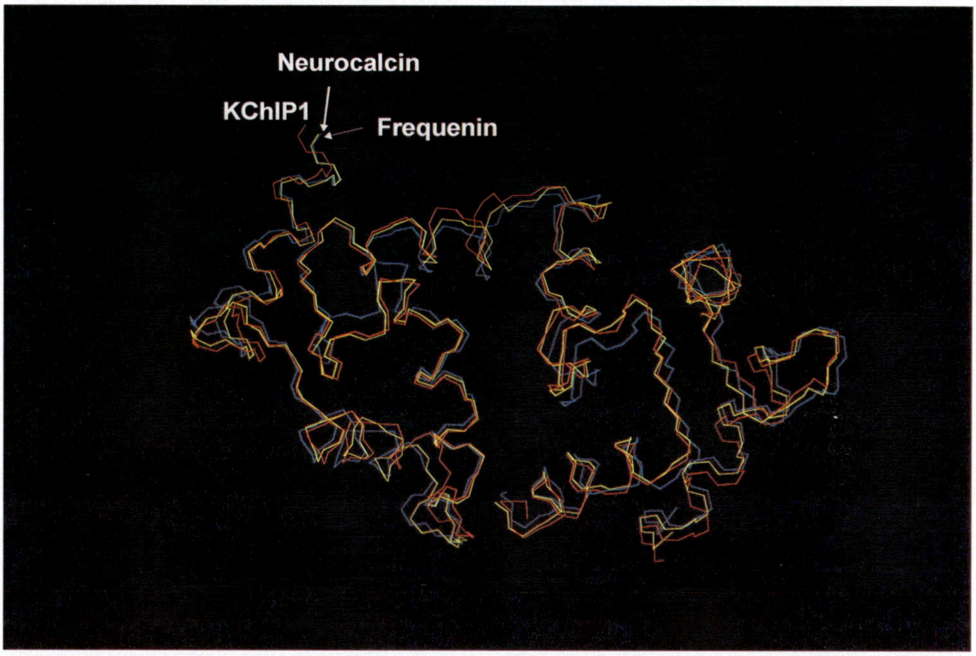

Figure 5. Overlay of the main chain structures of neurocalcin (yellow), human frequenin (cyan) and KChIP1 (red).

3. Structural Characteristics of NCS Proteins

Three-dimensional structures have been determined recently for Ca^{2+}-bound forms of neurocalcin [56], frequenin [57], KChIP1 [58,59], Frq1 [60] and GCAP2 [61], and their main chain structures are superimposed with respect to Ca^{2+}-bound recoverin (Fig. 5). The overall RMS deviation is 2.3 Å when comparing main chain atoms of EF-hand regions, suggesting that their main chain conformations are highly similar to one another, which is not too surprising given their sequence relatedness. However, if the main chain structures are so similar, then how can one explain their non-overlapping functional differences? One distinguishing structural property is the number and location of bound Ca^{2+}. Recoverin has Ca^{2+} bound at EF-2 and EF-3; KChIP1 has Ca^{2+} bound at EF-3 and EF-4; and frequenin, neurocalcin and GCAP2 have Ca^{2+} bound at EF-2, EF-3 and EF-4. Another important structural property is the distribution of charged and hydrophobic residues on the protein surface. Surface representations of hydrophobicity and charge density of the various NCS structures are shown in Fig. 6. All NCS structures exhibit a similar exposed hydrophobic surface located on the N-terminal half of the protein, formed primarily by residues in EF-1 and EF-2 (F35, W31, F56, F57, Y86 and L90 for recoverin in Fig. 6A). The exposed hydrophobic residues in this region are highly conserved (labeled and colored yellow in Fig. 6) and correspond to residues of recoverin that interact with the myristoyl group in the Ca^{2+}-free state (see Fig. 3A). A similar hydrophobic patch is also seen in membrane-bound recoverin (see Fig. 4B). These exposed residues in the hydrophobic patch have been implicated in target recognition from mutagenesis studies [62-65], and these residues very likely form intermolecular contacts with target proteins as has been demonstrated in the recent crystal structure of KChIP1 (see below).

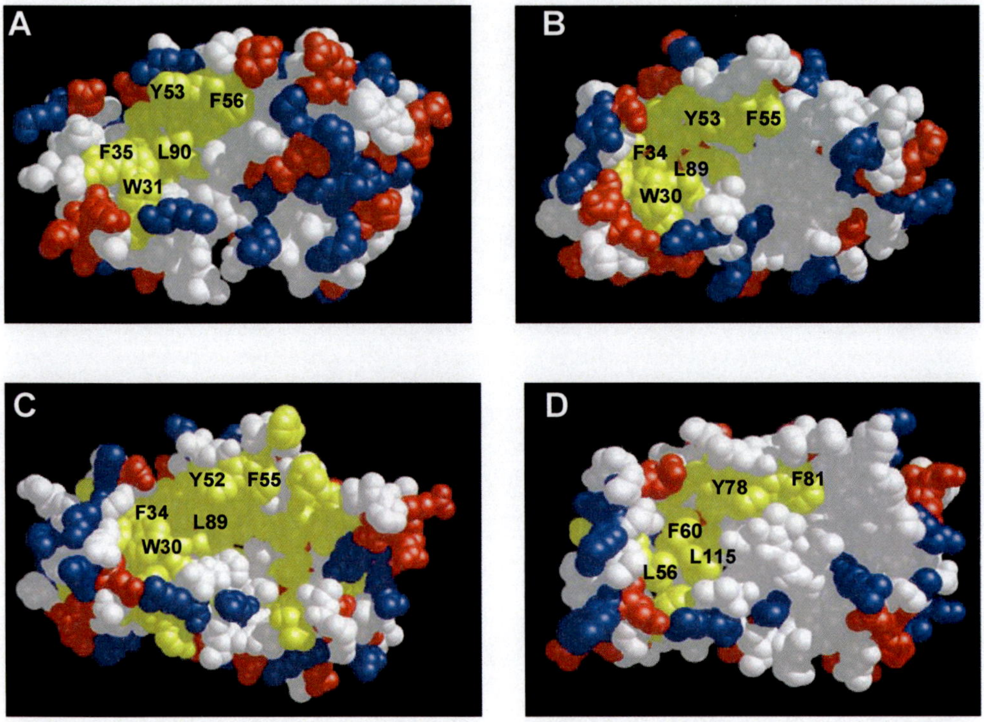

Figure 6. Space-filling representations of the Ca^{2+}-bound structures of recoverin (A), neurocalcin (B), frequenin (C) and KChIP1 (D). Exposed hydrophobic residues are yellow, neutral residues are white and charge residues are red and blue.

The distribution of charged (red and blue) and hydrophobic (yellow) residues on the surface of the C-terminal half of the NCS proteins is highly variable (Fig. 6). Frequenin exhibits exposed hydrophobic residues in the C-terminal domain that fuse together with the exposed hydrophobic crevice in the N-terminal domain, forming one continuous and elongated patch (Fig. 6C). By contrast, recoverin (Fig. 6A) has mostly charged residues on the surface of the C-terminal half, whereas neurocalcin (Fig. 6B) and KChIP1 (Fig. 6D) have mostly neutral residues shown in white. The different patterns of charge distribution on the C-terminal surface of NCS proteins might be important for conferring target specificity.

Ca^{2+} sensitive dimerization of NCS proteins is another structural characteristic that could influence target recognition. Neurocalcin [56], recoverin [48] and KChIP1 [58] exist as dimers in their x-ray crystal structures. Hydrodynamic studies have confirmed that neurocalcin and DREAM form dimers in solution at high Ca^{2+} and are monomeric in the Ca^{2+}-free state [66,67]. By contrast, GCAP2 forms a dimer only in the Ca^{2+}-free state and is monomeric at high Ca^{2+} [67]. Ca^{2+} sensitive dimerization of GCAP-2 has been demonstrated to control its ability to activate retinal guanylate cyclase. Ca^{2+}-sensitive protein oligomerization is also important physiologically for DREAM: the Ca^{2+}-free DREAM protein serves as a transcriptional repressor by binding to DNA response elements as a protein tetramer [18]. Ca^{2+}-induced dimerization of DREAM appears to disrupt DNA binding and may activate transcription of prodynorphin and c-fos genes [18]. In a related fashion, Ca^{2+}-bound KChIP1 forms a dimer in solution and in complex with an N-terminal fragment of the Kv4.2 K^+ channel [58]. By contrast, the full-length Kv4.2 channel tetramer binds to KChIP1 with a 4:4 stoichiometry [68], suggesting that KChIP1 dimers may assemble as a protein tetramer to recognize the channel. Such a protein tetramerization of KChIP1 may be Ca^{2+} sensitive like it is for DREAM. In short, the oligomerization properties of some NCS proteins appear to be Ca^{2+} sensitive, which may play a role in target recognition.

4. Structural Basis of Target Recognition

The N-terminal hydrophobic patch conserved in the structures of NCS proteins (see Fig. 6) has been implicated previously in target recognition [63-65]. The recent x-ray structure of KChIP1 in a target complex with the Kv4.2 channel definitively demonstrated that the N-terminal hydrophobic patch of KChIP1 is indeed located at the target interface [58]. Hydrophobic residues L56, I77, Y78, F81, F82, F111 and L115 of KChIP1 form close intermolecular contacts with A223, W224, F227, A228 and A231 of Kv4.2. The corresponding residues in GCAP2 and recoverin are also important for target recognition [64,65]. The hydrophobic interfacial residues of Kv4.2 are located on the same face of an amphipathic helix (magenta) that interacts with a concave surface (or groove) of KChIP1 formed by helices of EF-1 (green) and EF-2 (red) (Fig. 7A).

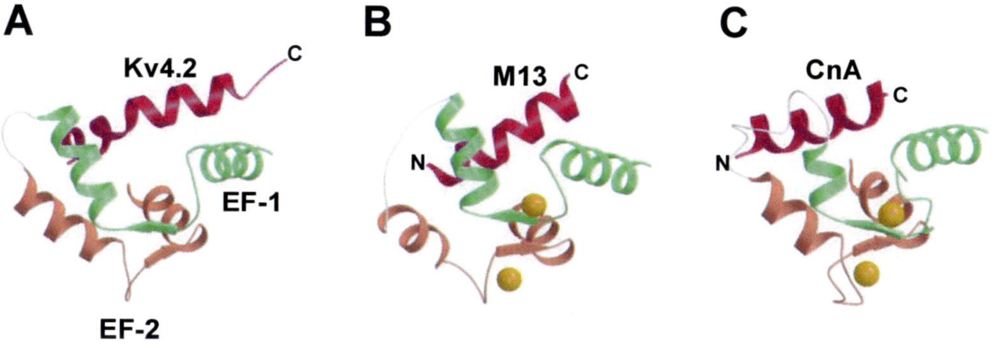

Figure 7. Ribbon diagrams illustrating intermolecular interactions for KChIP1 bound to Kv4.2 [58] (A), calmodulin bound to M13 peptide [69] (B), and calcineurin B bound to calcineurin A [71] (C). In each case, a target helix (magenta) is inserted in groove formed by the helices of EF-1 (green) and EF-2 (red). The intermolecular interactions are mostly hydrophobic as described in the text.

A similar type of hydrophobic target helix interaction was seen previously in the structures of calmodulin bound to the M13 segment of myosin light chain kinase (Fig. 7B) [69,70] and calcineurin B-subunit interacting with the catalytic A-subunit (Fig. 7C) [71]. This common structural framework appears to be a recurring theme for target recognition by EF-hand proteins generally. What distinguish the different target interactions are the precise contour properties of the target-binding site in each case formed by particular residues from EF-1 and EF-2. For example, in calmodulin, a number of methionine residues (M36, M51 and M72) help form deep holes beneath the surface of the binding site that are recognized by aromatic rings of Trp residues strategically positioned in the target helix. By contrast, KChIP1 contains a number of aromatic residues (F53, Y78, F81 and F111), forming an exposed flat surface that interacts with Ala methyl groups on the opposing face of the target helix. The hydrophobic residues of KChIP1 at the target interface are highly conserved in all NCS proteins, suggesting a common structural motif for target recognition. However, a few important interfacial residues of KChIP1 are not conserved (e.g. V55, L56, L118 and L119), suggesting that residues at these positions in other NCS proteins may be important for conferring target specificity. Indeed, these residues in yeast Frq1 correspond to Q29, W30, T92 and S93 that have been implicated to interact specifically with phosphatidylinositol 4-kinase isoform, Pik1 in our preliminary structural analysis (unpublished results, also see ref. [25]).

In summary, we reviewed the molecular structures of NCS proteins and examined structural determinants important for target recognition. The overall main chain structures and the presence of an exposed hydrophobic patch in the N-terminal domain are highly conserved in all NCS proteins. Differences in their surface charge density and protein dimerization properties may help to explain target specificity and functional diversity of NCS proteins. In the future, atomic resolution structures of additional NCS proteins bound to their respective target proteins are needed to improve our understanding of how this structurally conserved family of proteins can uniquely recognize their diverse biological targets.

Acknowledgements

This work was supported by grants to J.B.A from the NIH, and to M.I. from the Medical Research Council of Canada.

References

[1] Augustine, G.J., Santamaria, F., and Tanaka, K. 2003, *Neuron, 40*, 331.

[2] Berridge, M.J., Lipp, P., and Bootman, M.D. 2000, *Nat. Rev. Mol. Cell Biol., 1*, 11.

[3] Burgoyne, R.D., O'Callaghan, D.W., Hasdemir, B., Haynes, L.P., and Tepikin, A.V. 2004, *Trends Neurosci., 27*, 203.

[4] Ames, J.B., Tanaka, T., Stryer, L., and Ikura, M. 1996, *Curr. Opin. Struct. Biol., 6*, 432.

[5] Braunewell, K.H., and Gundelfinger, E.D. 1999, *Cell Tissue Res., 295*, 1.

[6] Burgoyne, R.D., and Weiss, J.L. 2001, *Biochem. J., 353*, 1.

[7] Ikura, M. 1996, *Trends Biochem. Sci., 21*, 14.
[8] Moncrief, N.D., Kretsinger, R.H., and Goodman, M. 1990, *J. Mol. Evol., 30*, 522.
[9] Weiss, J.L., and Burgoyne, R.D. 2002, In *Handbook of cell signaling*, ed. R Bradshaw, pp. 79, San Diego, Academic Press.
[10] Dizhoor, A.M., Ray, S., Kumar, S. Niemi, G., Spencer, M., Rrolley, D., Walsh, K.A., Philipov, P.P., Hurley, J.B. and Stryer, L. 1991, *Science 251*, 915.
[11] Dizhoor, A.M., Lowe, D.G., Olsevskaya, E.V., Laura, R.P., and Hurley, J.B. 1994, *Neuron, 12*, 1345.
[12] Palczewski, K., Subbaraya, I., Gorczyca, W.A., Helekar, B.S., Ruiz, C.C., et al. 1994, *Neuron, 13*, 395.
[13] Palczewski, K., Polans, A.S., Baehr, W., and Ames, J.B. 2000. *Bioessays, 22*, 337.
[14] Hidaka, H., and Okazaki, K. 1993, *Neuroscience Research, 16*, 73.
[15] Pongs, O., Lindemeier, J., Zhu, X.R., Theil, T., Engelkamp, D., Krah-Jentgens, Lambrecht, H.G., Kock, K.W., Schwerner, J., Rivosecchi, R., Mallart, A., Galceran, J., Canal, I., Barbas, J.A., and Ferrus, A. 1993, *Neuron, 11*, 15.
[16] McFerran, B.W., Graham, M.E., and Burgoyne, R.D. 1998, *J. Biol. Chem., 273*, 22768.
[17] An, W.F., Bowlby, M.R., Betty, M., Cao, J., Ling, H.P., et al. 2000, *Nature, 403*, 553.
[18] Carrion, A.M., Link, W.A., Ledo, F., Mellstrom, B., and Naranjo, J.R. 1999, *Nature, 398*, 80.
[19] Buxbaum, J.D., Choi, E.K., Luo, Y., Lilliehook, C., Crowley, A.C., et al. 1998, *Nat. Med., 4*, 1177.
[20] Kobayashi, M., Takamatsu, K., Saitoh, S., Miura, M., and Noguchi, T. 1992, *Biochemical and Biophysical Research Communications, 189*, 511.
[21] Kobayashi, M., Takamatsu, K., Saitoh, S., and Noguchi, T. 1993, *Journal of Biological Chemistry, 268*, 18898.
[22] Kapp, Y., Melnikov, S., Shefler, A., Jeromin, A., and Sagi, R. 2003, *J. Immunol., 171*, 5320.
[23] Gomez, M., De Castro, E., Guarin, E., Sasakura, H., Kuhara, A., et al. 2001, *Neuron, 30*, 241.
[24] Hendricks, K.B., Wang, B.Q., Schnieders, E.A., and Thorner, J. 1999, *Nature Cell Biol., 1*, 234.
[25] Huttner, I.G., Strahl, T., Osawa, M., King, D.S., Ames, J.B., and Thorner, J. 2003, *J. Biol. Chem., 278*, 4862.
[26] Hamasaki-Katagiri, N., Molchanova, T., Takeda, K., and Ames, J.B. 2004, *J. Biol. Chem., 279*, 12744.
[27] Kawamura, S. 1993, *Nature, 362*, 855.
[28] Erickson, M.A., Lagnado, L., Zozulya, S., Neubert, T.A., Stryer, L., and Baylor, D.A. 1998, *Proc. Natl. Acad. Sci. USA, 95*, 6474.
[29] Makino, C.L., Dodd, R.L., Chen, J., Burns, M.E., Roca, A., et al. 2004, *J. Gen. Physiol., 123*, 729.
[30] Calvert, P.D., Klenchin, V.A., and Bownds, M.D. 1995, *J. Biol. Chem., 270*, 24127.
[31] Klenchin, V.A., Calvert, P.D., and Bownds, M.D. 1995, *J. Biol. Chem., 270*, 16147.
[32] Chen, C.K., Inglese, J., Lefkowitz, R.J., and Hurley, J.B. 1995, *J. Biol. Chem., 270*, 18060.

[33] Polans, A.S., Buczylko, J., Crabb, J., and Palczewski, K. 1991, *J. Cell Biol., 112*, 981.
[34] Subramanian, L., and Polans, A.S. 2004, *Biochem. Biophys. Res. Commun., 322*, 1153.
[35] Palczewski, K., Sokal, I., and Baehr, W. 2004, *Biochem. Biophys. Res. Commun., 322*, 1123.
[36] Semple-Rowland, S.L., Gorczyca, W.A., Buczylko, J., Helekar, B.S., Ruiz, C.C., et al. 1996, *FEBS Lett., 385*, 47.
[37] Sokal, I., Li, N., Surgucheva, I., Warren, M.J., Payne, A.M., et al. 1998, Mol. Cell, 2, 129.
[38] Strahl, T., Grafelmann, B., Dannenberg, J., Thorner, J., and Pongs, O. 2003, *J. Biol. Chem., 278*, 49589.
[39] Hama, H., Schnieders, E.A., Thorner, J., Takemoto, J.Y., and DeWald, D.B. 1999, *J. Biol. Chem., 274*, 34294.
[40] Walch-Solimena, C., and Novick, P. 1999, *Nature Cell Biol., 1*, 523.
[41] Nakamura, T.Y., Pountney, D.J., Ozaita, A., Nandi, S., Ueda, S., et al. 2001, *Proc. Natl. Acad. Sci. USA, 98*, 12808.
[42] Weiss, J.L., Archer, D.A., and Burgoyne, R.D. 2000, *J. Biol. Chem., 275*, 40082.
[43] Carrion, A.M., Mellstrom, B., and Naranjo, J.R. 1998, *Mol. Cell Biol., 18*, 6921.
[44] Cheng, H.Y., Pitcher, G.M., Laviolette, S.R., Whishaw, I.Q., Tong, K.I., et al. 2002, *Cell, 108*, 31.
[45] Lilliehook, C., Bozdagi, O., Yao, J., Gomez-Ramirez, M., Zaidi, N.F., et al. 2003, *J. Neurosci., 23*, 9097.
[46] Dizhoorm A.M., Ericsson, L.H., Johnson, R.S., Kumar, S., Olshevskaya, E., et al. 1992, *J. Biol. Chem., 267*, 16033.
[47] Ladant, D. 1995, *J. Biol. Chem., 270*, 3179.
[48] Flaherty, K.M., Zozulya, S., Stryer, L., and McKay, D.B. 1993, *Cell, 75*, 709.
[49] Weiergraber, O.H., Senin, I.I., Philippov, P.P., Granzin, J, and Koch, K.W. 2003, J. *Biol. Chem., 278*, 22972.
[50] Babu, Y.S., Bugg, C.E., and Cook, W.J. 1988, *J. Mol. Biol., 204*, 191.
[51] Herzberg, O., and James, M.N. 1988, *J. Mol. Biol., 203*, 761.
[52] Tanaka, T., Ames, J.B., Harvey, T.S., Stryer, L., Ikura, M. 1995, *Nature, 376*, 444.
[53] Ames, J.B., Hamasaki, N., and Molchanova, T. 2002, *Biochemistry, 41*, 5776.
[54] Ames, J.B., Ishima, R., Tanaka, T., Gordon, J.I., Stryer, L., and Ikura, M. 1997, *Nature, 389*, 198.
[55] Valentine, K.G., Mesleh, M.F., Opella, S.J., Ikura, M., and Ames, J.B. 2003, *Biochemistry, 42*, 6333.
[56] Vijay-Kumar, S., and Kumar, V.D. 1999, *Nature Struct. Biol., 6*, 80.
[57] Bourne, Y., Dannenberg, J., Pollmann, V.V., Marchot, P., and Pongs, O. 2001, *J. Biol. Chem., 276*, 11949.
[58] Zhou, W., Qian, Y., Kunjilwar, K., Pfaffinger, P.J., and Choe, S. 2004, *Neuron, 41*, 573.
[59] Scannevin, R.H., Wang, K., Jow, F., Megules, J., Kopsco, D.C., et al. 2004, *Neuron, 41*, 587.
[60] Ames, J.B., Hendricks, K.B., Strahl, T., Huttner, I.G., Hamasaki, N., and Thorner, J. 2000, *Biochemistry, 39*, 12149.

[61] Ames, J.B., Dizhoor, A.M., Ikura, M., Palczewski, K., and Stryer, L. 1999, *J Biol Chem, 274*, 19329.

[62] Ermilov, A.N., Olshevskaya, E.V., and Dizhoor, A.M. 2001, *J. Biol. Chem., 276*, 48143.

[63] Krylov, D.M., Niemi, G.A., Dizhoor, and A.M., Hurley, J.B. 1999, *J. Biol. Chem., 274*, 10833.

[64] Tachibanaki, S., Nanda, K., Sasaki, K., Ozaki, K., and Kawamura, S. 2000, *J. Biol. Chem., 275*, 3313.

[65] Olshevskaya, E.V., Boikov, S., Ermilov, A., Krylov, D., Hurley, J.B., and Dizhoor, A.M. 1999, *J. Biol. Chem., 274*, 10823.

[66] Osawa, M., Dace, A., Tong, K.I., Valiveti, A., Ikura, M., and Ames, J.B. 2005, *J Biol Chem*, in press.

[67] Olshevskaya, E.V., Ermilov, A.N., and Dizhoor, A.M. 1999, *J. Biol. Chem., 274*, 25583.

[68] Kim, L.A., Furst, J., Butler, M.H., Xu, S., Grigorieff, N., and Goldstein, S.A. 2004, *J Biol Chem, 279*, 5549.

[69] Ikura, M., Clore, G.M,. Gronenborn, A.M., Zhu, G., Klee, C.B., and Bax, A. 1992, *Science, 256*, 632.

[70] Meador, W.E., Means, A.R., and Quiocho, F.A. 1992, *Science, 257*, 1251.

[71] Griffith, J.P., Kim, J.L., Kim, E.E., Sintchak, M.D., Thomson, J.A., et al. 1995, *Cell, 82*, 507.

[72] Li, N., Fariss, R.N., Zhang, K., Otto-Bruc, A., Haeseleer, F., et al. 1998, *European Journal of Biochemistry, 252*, 591.

[73] Holmqvist MH, Cao, J., Hernandez-Pineda, R., Jacobson, M.D., Carroll, K.I., et al. 2002, *Proc. Natl. Acad. Sci. USA, 99*, 1035.

[74] Kuo, H.C., Cheng, C.F., Clark, R.B., Lin, J.J., Lin, J.L., et al. 2001, *J. Biol. Chem., 278*, 36445.

[75] Hyun, J.K., Yon, C., Kim, Y.S., Noh, D.Y., Lee, K.H., and Han, J.S. 2000. *Mol. Cells, 10*, 669.

[76] Nagata, K., Puls, A., Futter, C., Aspenstrom, P., Schaefer, E., et al. 1998, *EMBO J., 17*, 149.

[77] Braunewell, K.H., Brackmann, M., Schaupp, M., Spilker, C., Anand, R., and Gundelfinger, E.D. 2001, *J. Neurochem., 78*, 1277.

[78] Lin, L., Jeanclos, E.M., Treuil, M., Braunewell, K.H., Gundelfinger, E.D., and Anand, R. 2002, *J. Biol. Chem., 277*, 41872.

In: Neuronal Calcium Sensor Proteins
Editors: Pavel Philippov and Karl Koch, pp. 331-349
ISBN 1-59454-978-8

Chapter XIX

The Use of Physical Methods in Studies of Neuronal Calcium Sensor Proteins

Eugene A. Permyakov*
Institute for Biological Instrumentation of the Russian Academy of Sciences, Pushchino, Moscow region, 142290 Russia.

Abstract

Physical methods, such as x-ray crystallography, luminescent spectroscopy, circular dichroism spectroscopy, nuclear magnetic resonance (NMR), scanning and titration calorimetry and others, are widely used in studies of calcium sensor proteins. They give valuable information about structural, physical and chemical properties of these proteins. The information provides a basis for understanding of physiological functions of calcium sensor proteins. Examples of physical studies of such calcium sensor proteins as calmodulin, recoverin, parvalbumin, S100 proteins and others demonstrate ability of the physical methods. (Surface plasmon resonance spectroscopy is considered in other Chapters of the present issue).

Keywords: calcium sensor proteins, physical methods, spectroscopy, crystallography, calorimetry, NMR, recoverin, parvalbumin, calmodulin

1. Introduction

Nearly all normal intracellular and probably most extracelluar processes involving calcium are based on its interactions with proteins. Calcium physiology is mostly calcium - protein interactions. It is well known that the free or weakly bound Ca^{2+} concentration in

* Correspondence concerning this article should be addressed to Eugene A. Permyakov, Institute for Biological Instrumentation of the Russian Academy of Sciences Pushchino, Moscow region, 142290 Russia.

extracellular fluids is about $2 \cdot 10^{-3}$ M, while the resting free Ca^{2+} concentration within cells is about 10^{-7} - 10^{-8} M or less. In response to an external stimulus, the low intracellular free Ca^{2+} concentration may increase about 10 to 100 times. Ca^{2+} entry can be activated either by voltage changes or by binding of some exogenous and endogenous transmitters. The changes in intracellular Ca^{2+} concentration may be uniform through the whole cell or may be located in certain areas. The changes can vary throughout stimulation resulting in various spatial and temporal signalling patterns. At the same time, each cell contains a set of calcium binding proteins for which the Ca^{2+} dissociation constants are within the micromolar and sub-micromolar range (10^{-6}-10^{-8} M). The binding of Ca^{2+} induces changes in conformation of these proteins, which in turn changes the character of their interactions with other cell systems and triggers various cell processes.

The intracellular calcium binding proteins are characterized by special properties and their study requires the use of special methods. This chapter describes the principle physical methods used to study calcium binding proteins. Some of the methods allow studying the binding of calcium and other metal cations to the proteins, others are used to study structure and properties of the proteins. Some of the methods are common for studies of any classes of proteins, others are specific and their use is connected with some specific properties of calcium binding proteins.

2. Atomic Absorption Spectroscopy

Atomic flame absorption spectroscopy is used to measure the total concentration of metals including total calcium in whatever liganded state by measuring the intensity of characteristic emission lines of these metals relative to the intensities of those lines in standard samples (reviewed in [1,2]). Concentration measurements are usually carried out using a working curve after calibrating the instrument with standards of known concentration.

Ions or atoms in a sample should be vaporized in a high-temperature source such as a flame or graphite furnace. The atoms are excited by ultraviolet or visible light and make transitions to higher electronic energy levels. The amount of absorbed light at a characteristic wavelength is a quantitative measure of the concentration of the element to be analyzed. The light source is usually a hollow cathode lamp of the element that is being measured. Lasers are also used in research instruments. The flame is arranged in such a way that it is laterally long and not deep and the height of the flame must also be controlled by the flow of the fuel mixture. For a low temperature flame, acetylene and air are used; a hotter flame can be produced using acetylene and pure oxygen. An even hotter flame can be produced using acetylene and pure nitrous oxide.

A disadvantage of the atomic absorption technique is the non-linearity of the calibration curves when absorbance become higher than 0.5. The relative standard deviations are between 0.3 and 1.0% for absorbances of 0.1 to 0.2. Detection limits for flame atomic absorption spectroscopy vary enormously: from 1 - 5 ppb (e.g. Ca, Cd, Cr, Cu) to more than 1000 ppb (e.g. P).

3. Radioactivity

^{45}Ca has a half life of 162.7 days. It emits β particles with the maximum energy of 0.252 MeV. β particles can be counted quantitatively by a liquid scintillation counter. The β particles from ^{45}Ca travel of about 60 cm in air. Like atomic flame absorption spectroscopy, it is well suited to determining total calcium or it can be used to measure free calcium in a solution separated from the protein solution by porous membrane in equilibrium dialysis procedure (reviewed in [3,4]). Frequently ^{45}Ca-labeled proteins can be measured semi-quantitatively in gel electrophoresis experiments by direct measurement of β particles emission by liquid scintillation counting or by film darkening.

4. Ion Selective Electrodes

An ion selective electrode measures the potential of a specific ion in solution, thus the pH electrode is an ion selective electrode for the hydrogen ion (reviewed in [2,5,6]). This potential is measured against a stable reference electrode of constant potential. The potential difference between the two electrodes depends upon the activity of the specific ion in solution. The activity is related to the concentration of that specific ion, thereby allowing the user to make an analytical measurement of that specific ion. Ion selective electrodes have been developed for a variety of different ions. Ion selective electrodes, long available for H^+, are available for K^+, Na^+, and Ca^{2+} ions as well. They can measure $[Ca^{2+}]$ from 10^{-3} to 5×10^{-7} M at 5 - 10% accuracy. Microelectrodes appropriate for intracellular use are also available. Unfortunately, they are easily contaminated, therefore they should be calibrated with each use. Despite these problems, they are extremely valuable because, in contrast to flame absorption spectroscopy and radioactivity measurements, they selectively measure free calcium as opposed to total calcium. With the exception of calcium specific dyes and fluors, these electrodes are the only technique that directly measures free calcium.

5. Calcium Buffers

Some small organic molecules bind calcium with moderate to high affinity and with reasonable selectivity relative to magnesium. Ethylene diamine tetra-acetic acid (EDTA) and ethylene glycol-bis(β-aminoethyl ether)-N,N,N',N'-tetraacetic acid (EGTA) are well characterized and do not interact with (most) proteins (reviewed in [2,7]). Frequently the concentration of free Ca^{2+} ion is cited not from direct measurement (by ion specific electrode) but calculated on the base of dissociation constants of Ca^{2+} complexes with EDTA/EGTA. The affinities of most calcium buffers are pH dependent. Hence, not only should the pH of the medium be measured but one should also be aware that the addition of the EDTA/EGTA itself can change the pH of the medium, thereby affecting the reaction in question. Magnesium will also be buffered and to some extent consumed by the EDTA/EGTA.

Programs that calculate the final pCa, pMg, and pH of EDTA/EGTA buffered solutions are available.

The use of calcium buffers can be complicated by direct interaction of the low molecular mass calcium chelators with proteins. For example, EGTA and EDTA in millimolar concentrations interact with calcium binding protein α-lactalbumin [8]. For this reason, the use of high concentrations of EGTA and EDTA in this case is restricted.

6. Calcium Specific Dyes

Since calcium is spectrally inert, it is reasonable to use organic ligands whose spectral properties are changed upon the coordination of calcium (reviewed in [9,10]). The selectivity of the dye with respect to calcium should be at least three decades relative to magnesium, sodium, and potassium – the ions most likely to compete and (perhaps) give a false signal – and the selectivity should be four or five decades to these cations for use *in vivo*. The second condition is that the signal from the dye should be strong enough, and different enough in the calci- *vs*. the apo-form of the protein, so that the dye can be used in small quantity and not function as a calcium buffer. Finally, if the dye is used *in vivo,* it should not be toxic.

The development of calcium dyes, such as FURA that can be injected into the cytosol, has revolutionized cellular calcium physiology. Normally the dye is derivatized as an ester. The molecule is uncharged and easily taken up by the cell whose esterases cleave off the alcohol moiety; the resulting charged molecule diffuses through the cell.

Single-wavelength calcium dyes (Table 1) show no change in position of absorption or emission maxima upon the binding of calcium. As a rule, the intensity of emission increases (with the exception of Fura-red) when they bind calcium. This means that they are dependent on the intracellular dye concentration, and other non calcium associated effects, for example changes in membrane thickness.

Table 1. Spectral properties of single-wavelength calcium dyes (data from Molecular Probes catolog).

		Absorption maxima, nm		Emission maxima, nm	
Indicator	**K_d, nM**	**Ca^{2+} free**	**Ca^{2+} bound**	**Ca^{2+} free**	**Ca^{2+} bound**
Quin-2	115	352	332	492	498
Fluo-3	400	503	506	526	526
Calcium Green-1	189	506	506	534	533
Calcium Green-2	574	506	506	531	531
Calcium Green-5N	3.3	506	506	531	531
Rhod-2	1.0	556	553	576	576
Calcium Orange	328	554	555	575	576
Calcium Crimson	205	588	588	611	611

Ratiometric calcium dyes (Table 2) show spectral peaks at different wavelengths in the bound and free states. This means that some parameters such as spectral maximum position or ratio of intensities at two fixed wavelengths are not dependent on the intracellular concentration of the dye. The most commonly used dyes need UV light for excitation, which can lead to problems, for example intrinsic fluorescence of cells or photolysis of photosensitive caged compounds.

Table 2. Spectral properties of ratiometric calcium dyes (data from Molecular Probes catalog).

		Absorption maxima, nm		Emission maxima, nm	
Indicator	**K_d, nM**	**Ca^{2+} free**	**Ca^{2+} bound**	**Ca^{2+} free**	**Ca^{2+} bound**
Fura-2	224	362	335	512	505
Fura Red	133	472	436	645	640
Indo-1	250	349	331	485	410

Aequorin is an EF-hand containing photoprotein originally isolated from luminescent jellyfish and other marine organisms (reviewed in [11,12]). The aequorin complex comprises a 22,000-dalton apo-aequorin protein, molecular oxygen, and the luminophore coelenterazine. When three Ca^{2+} ions bind to this complex, coelenterazine is oxidized to coelenteramide, with a concomitant release of carbon dioxide and blue light. The approximately third-power dependence of aequorin's bioluminescence on $[Ca^{2+}]$ gives it a broad detection range, allowing the measurement of Ca^{2+} concentrations from ~0.1 μM to >100 μM. Unlike fluorescent Ca^{2+} indicators, Ca^{2+}-bound aequorin can be detected without illuminating the sample, thereby eliminating interference from intrinsic cell fluorescence and allowing simultaneous labeling with caged probes. The injected aequorin is depleted over time. The genes encoding aequorin and the enzymes required to synthesize coelenterazine can be incorporated into the genome of the cell under study.

7. Isothermal Titration Calorimetry

Isothermal titration calorimetry (ITC) is a thermodynamic technique that allows the study of interactions of two molecules or ions (reviewed in [13-15]). When these molecules interact, heat is either generated or absorbed. By measuring these interaction heats, dissociation constants (K_d), binding stoichiometry (n), and thermodynamic parameters including enthalpy (ΔH) and entropy (ΔS) can be accurately determined. ITC is the only method for the direct determination of ΔH. Varying the temperature of the experiment allows the determination of the heat capacity change (ΔC_p) for the reaction. The titration experiment is performed at a constant temperature by titrating one binding partner (titrant) into a solution containing the other binding partner (titrand) in the sample cell of the calorimeter. After each addition of a small aliquot of titrant, the heat released or absorbed in the sample cell is measured with respect to a reference cell filled with buffer. The heat change is proportional to the electrical power required to maintain a constant small temperature difference between the sample cell and the reference cell, both of which are placed in an adiabatic jacket.

The experimental data can be plotted in the differential mode when the total heat accumulated up to injection *i* is normalized to the total ligand concentration at step *i* or against the ratio of the total ligand concentration at step *i* to the total receptor concentration. In the integral mode, the total cumulative heat is plotted against the total ligand concentration to yield a hyperbolic saturation curve. Deconvolution analysis of the binding curves yields accurate values of K_d (in the range of 10^{-2} to 10^{-8} M), n, ΔH, and ΔS. In addition, kinetic data (for slow reactions) may be obtained in a batch injection experiment. The deconvolution has a firm thermodynamic foundation and avoids additional assumptions as is necessary in the analysis of spectroscopic binding data.

For very tight binding reactions, optimal concentrations of the reagents are too low to get measurable heat changes. It is for this reason that even with the most sensitive instruments presently available, K_d lower than 10^{-9} M (ΔG about -50 kJ mol^{-1} at room temperature) cannot be very accurately measured. The method can be used to study interactions of proteins with calcium ions, nevertheless in most cases the lower limit for calcium dissociation constant, which can be measured by this method, usually does not exceed 10^{-5}-10^{-6} M.

Determination of binding parameters for metal ion binding to proteins usually requires preceding steps to remove protein-bound metal ions, which is often associated with decreased stability and inactivation of the protein. Nielsen et al. [16] presented two simple isothermal titration calorimetric procedures that eliminate separate metal ion removal steps and directly monitor the exchange of metal ions between buffer, protein, and chelator. The concept is to add either excess of chelator or metal ion to the protein under investigation and subsequently titrate with metal ion or chelator, respectively. It is thereby possible in the same experimental trial to obtain both chelator-metal ion and protein-metal ion binding parameters due to the different thermodynamic "fingerprints" of chelator and protein. Verifications of the models were carried out by titrations of mixtures of calcium chelators (BAPTA, HEDTA, and EGTA) and calcium ions and they were both able to account satisfactorily for the observed binding isotherms.

8. Differential Scanning Calorimetry

The previous techniques – flame spectrophotometry, radio isotopes, selective electrodes, calcium buffers, specific dyes and even ITC – are intended primarily to measure and/or regulate directly free calcium and/or total calcium concentration. The following techniques measure changes that occur in the protein that binds calcium.

Differential scanning calorimetry (DSC) is the most direct experimental technique to study energetics of thermally induced conformational transitions of biological macromolecules (reviewed in [17-20]). DSC measures heat capacity of a solution of a molecule under study as a function of temperature. Scanning calorimeters usually have twin cells and operate in a differential mode. The solution containing the solute is loaded in the sample cell and an equal volume of solvent is placed in the reference cell. The system is heated quasi-adiabatically at a constant rate and a certain amount of electrical power is required to compensate the temperature difference between the two cells. The power difference is a direct measure of the heat capacity difference between the solution and the solvent ΔC_p. When a protein solution is heated, partial specific heat capacity of the protein $C_p(T)$ follows a peak-shaped curve if thermal denaturation of the protein is cooperative and reversible. In the case of two interacting substances, statistical analysis of the differential scanning calorimetry traces can give parameters of the association reaction. Certainly, if a sharp difference is found between the apo- and the calci-form(s), that difference(s) can be exploited in the analysis to determine stoichiometry, affinity, etc. The DSC method is often used to study properties of calcium binding proteins. Figure 1 shows heat sorption curves for recoverin in Ca-loaded and Ca, Zn-loaded states measured on a DASM-4M differential adiabatic scanning microcalorimeter.

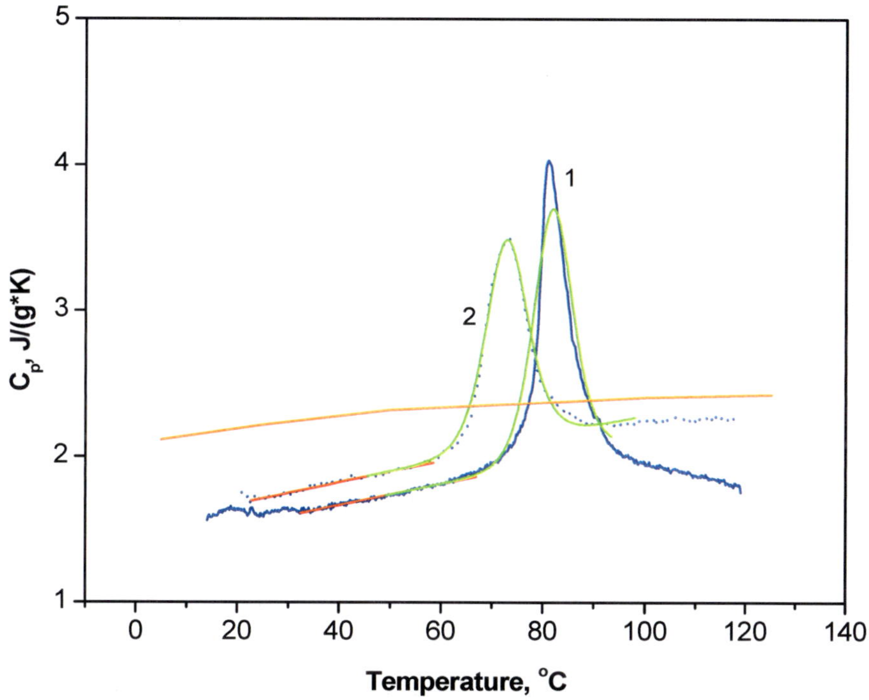

Figure 1. Heat sorption peaks for Ca-loaded (1) and Ca,Zn-loaded recoverin measured on DASM-4M scanning microcalorimeter.

9. Absorption Spectroscopy

9.1. UV- and IR-Spectroscopy

The polypeptide chain itself, free of ligands or post-translational modifications, has low absorbance in the visible region. At the same time, the side chains of Phe, Tyr, and especially Trp have characteristic absorbances in the ultraviolet region (210-300 nm). To good approximation the UV-absorption of a protein is the sum of absorptions of its constituent amino acids. Usually, there is little change in UV-absorption associated with changes in conformation of the protein, including those induced by calcium binding. As a rule, such changes can be detected only in differential mode of measurement (reviewed in [21]). The most pronounced calcium-induced changes in UV absorption of tryptophan was found for whiting and cod parvalbumin: the Ca^{2+}-loaded form of this protein has an unusually pronounced accessory band of $^1L_a \leftarrow A$ transition of tryptophan at 291 nm, which is reduced to a slight shoulder in the metal-free form [22].

Both sources and detector of infrared radiation have been improved dramatically recently. Certain infrared bands are exquisitely sensitive to local changes in conformation. The peptide group gives nine characteristic infrared bands named amide A, B, I, II ... VII (reviewed in [23–27]). The amide A band (about 3500 cm^{-1}) and amide B band (about 3100 cm^{-1}) originate from a Fermi resonance between the first overtone of amide II and the N-H stretching vibration. Amide I and amide II bands are two major bands of the protein infrared spectrum. The amide I band (between 1600 and 1700 cm^{-1}) is associated mainly with the C=O stretching vibration (70-85%) and is directly related to the backbone conformation. Amide II results from the N-H bending vibration (40-60%) and from the C-N stretching vibration (18-40%). This band is sensitive to changes in conformation of the protein backbone. Amide III and IV are very complex bands resulting from a mixture of several coordinate displacements. The out-of-plane motions are found in amide V, VI and VIII. The shape of the amide I band of globular proteins is characteristic of their secondary structures. It was shown, for example, that the infrared amide I' and III' regions for parvalbumin are very much affected by Ca^{2+} removal and are indicative of significant perturbation of secondary structure [28].

9.2. FTIR - Spectroscopy

The determination of secondary structures in proteins can be carried out by means of Fourier Transform Infrared Spectroscopy (FTIR) (reviewed in [25,29,30]). High potential of this method appeared due to the availability of high signal-to-noise ratio, digitized spectra obtained by the FTIR spectrometer and by the access to computers and software able to perform many operations on the spectra in a short time. The concept of Fourier self deconvolution is based on the assumption, that a spectrum of single bands (each narrow band is characteristic for a secondary structure) is broadened in the liquid or solid state. Therefore, the bands overlap and cannot be distinguished in the amide envelope. A curve fitting procedure can be applied to estimate quantitatively the area of each component representing a type of secondary structure.

The binding of metal cations can change protein secondary structure, which can be detected by FTIR spectroscopy (reviewed in [31]). To investigate the coordination structures of metal-bound Ca^{2+} binding sites in recoverin, FTIR studies were performed [32]. It is well known that the recoverin EF-hands, Ca^{2+}-binding regions each comprising 12 residues, are arranged to coordinate Ca^{2+} with seven oxygen ligands, two of which are provided by a conserved bidentate Glu at the 12th relative position in the EF-hand. FTIR analysis confirmed that Sr^{2+}, Ba^{2+}, Cd^{2+}, and Mn^{2+} were coordinated to COO^- of Glu by a bidentate state as well as Ca^{2+}, while coordination of COO^- with Mg^{2+} was a pseudo-bridging state with six-coordinate geometry.

10. Fluorescence Spectroscopy

The interaction of metal cations with proteins usually causes changes in conformation that can change the environments of some of the side chains of Phe, Tyr, and Trp, which often induces changes in their fluorescence quantum yield, position of fluorescence spectrum maximum, fluorescence spectrum shape, and/or fluorescence lifetime (reviewed in [33,34]). These changes can be used to determine the fraction of protein with bound cation(s) and sometimes the proximity of binding site(s) to the fluorescence side chain(s). However, characteristics of the fluorescence cannot yet be predicted from the structure of the protein, nor the structure from the fluoresecence. The absence of change in fluorescence does not prove an absence of cation-protein interaction.

Intrinsic fluorescence method was successfully used for studies of interactions of calcium binding proteins with metal cations.

Figure 2 shows experimental data on spectrofluorimetric Ca^{2+}- and EGTA-titration of wild type recoverin. The binding of calcium results in an increase of tryptophan fluorescence intensity at 370 nm. These data can be used for evaluation of calcium binding constants.

Intrinsic fluorescence allows measurements of kinetics of metal cation binding to calcium binding proteins.

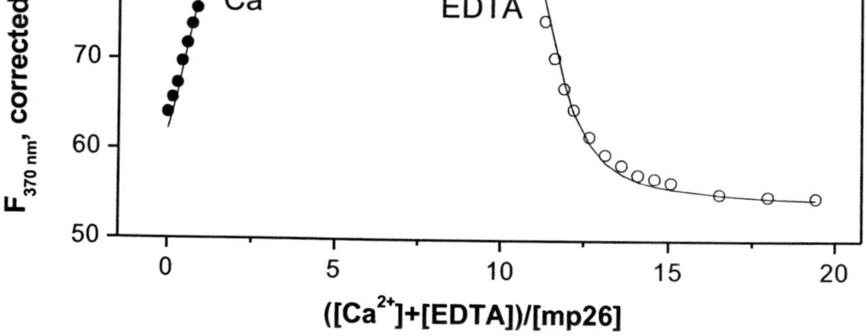

Figure 2. Spectrofluorimetric Ca^{2+}- and EDTA-titration of wild type recoverin monitored by tryptophan fluorescence intensity at 370 nm.

11. Circular Dichroic and Optical Rotatory Dispersion Spectroscropy

Asymmetric chromophores or symmetric chromophores in asymmetric environments interact differently with right- and left-circularly polarized light resulting in two related phenomena. Circular dichroism (CD) refers to the difference in absorption of right- and left-circularly polarized light at some wavelengths due to differences in extinction coefficients for the two polarized rays. The right and the left components of circularly polarized light travel through an optically active medium with different velocities due to the different indices of refraction for right- and left-circularly polarized light; this is called optical rotation or circular birefringence. The variation of optical rotation as a function of wavelength is called optical rotary dispersion (ORD).

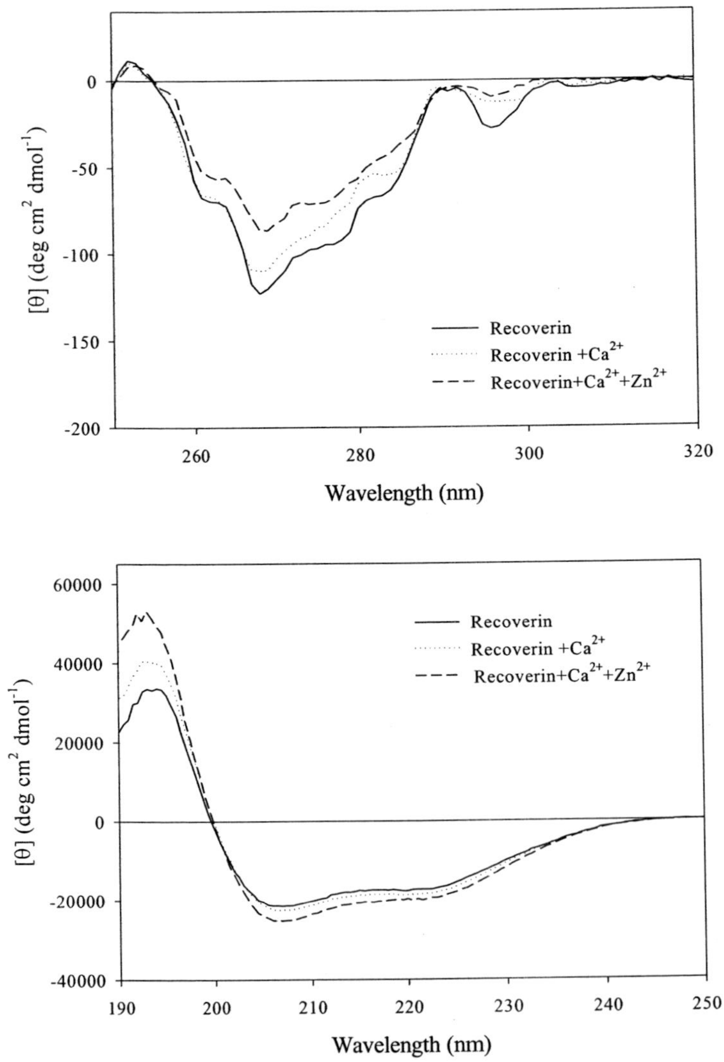

Figure 3. Circular dichroism spectra of various states of wild type recoverin.

Protein chromophores can be divided into three classes: the peptide bond, the amino acid side chaines, and prosthetic groups. The lowest energy transition in the peptide chromophore is an n → p* transition observed at 210-220 nm with very weak intensity. The n → p* transition appears in the α-helical structure as a small shoulder near 220 nm on the tail of a much stronger absorption band centered at 190 nm. This intense band, responsible for the majority of the peptide bond absorbance, is a p → p* transition.

CD is very sensitive to the secondary structure of polypeptides and proteins (reviewed in [35-37]). A CD spectrum between 180 and 260 nm can be analyzed to estimate the relative amounts of α-helix, β-strands, and turns. CD is absorptive and can occur only within a normal absorption band, and thus requires either an inherently asymmetric chromophore (uncommon) or a symmetric one in an asymmetric environment. The aromatic side chains of Phe, Tyr, and Trp also have absorption bands in the ultraviolet spectrum. CD is particularly powerful in monitoring conformational changes including those induced by binding of metal ions. In the region of 178 - 230 nm, one expects to observe effects of backbone conformational changes, while CD effects at longer wavelengths (>230 nm) should monitor aromatic chromophore contributions and should reflect more global, three-dimensional properties of the protein.

CD spectroscopy is very often used to monitor changes in the secondary and tertiary structure of calcium binding proteins induced by various reasons including amino acid substitutions. For example, Permyakov et al. [38] studied the effects of substitutions in the second, third and fourth EF-hand sites of recoverin on its Ca^{2+}-binding properties and some other characteristics, using intrinsic fluorescence, CD spectroscopy and differential scanning microcalorimetry. Figure 3 shows CD spectra of various states of wild type recoverin.

12. Mass Spectroscopy

Mass spectrometers consist of three main parts: the ionization source, analyzer, and detector. The sample under investigation has to be introduced into the ionization source of the instrument. Inside the ionization source the sample molecules are ionized and these ions go into the analyzer region of the mass spectrometer where they are separated according to their mass (m) -to-charge (z) ratios (m/z). The separated ions are detected and this signal sent to a data system where the m/z ratios are stored together with their relative intensities for presentation in the format of a m/z spectrum. The analyzer, detector, and ionization source are maintained under high vacuum to give the ions a chance of travelling from one end of the instrument to the other without any collisions with air molecules (reviewed in [4,39-42]).

The ionization methods used for the majority of biochemical analyses are Electrospray Ionisation (ESI) and Matrix Assisted Laser Desorption Ionisation (MALDI). ESI is one of the atmospheric pressure ionization (API) techniques and is well-suited to the analysis of polar molecules ranging from less than 100 Da to more than 1,000,000 Da including proteins. There are a number of mass analyzers currently available, the better known of which include quadrupoles, time-of-flight (TOF), magnetic sectors, and both Fourier transform and quadrupole ion traps. Using electrospray or nanospray ionization, a mass accuracy of within 0.01% of the molecular weight can be achieved.

Both (i) the range of molecular masses that can be resolved and (ii) the sensitivity and accuracy of their detections are quite remarkable. Often one needs to know whether or not a protein(s) binds calcium and, if so, the stoichiometry of the binding. In some cases, one can compare the mass of an intact protein, only partially purified, with its mass after treatment with a chelator, EGTA for example. Many metal ions bind strongly enough, so that they are not dissociated during ionization. Even a small fraction of the total mass with a shift in mass of 40, or 80, etc., is a strong evidence of a bound Ca^{2+} ion(s). Certainly such experiments should be controlled by testing well characterized calcium binding proteins under identical conditions.

13. Small Angle X-Ray Scattering

The small x-ray scattering (SAXS) of a protein that is not interacting with its neighbors in solution can be computed as the spherically averaged Fourier transform of that protein. From a measurement of that transform one cannot, even in principle, determine the detailed shape of the protein. However, if a protein is modeled as an ellipsoid of revolution, or a cylinder, or a dumbbell, one can compare the measured intensity with that calculated (reviewed in [43-45]). Although one treats the absolute determination of shape with caution, SAXS is exquisitely sensitive to changes in structure, such as that induced by binding calcium or a target protein.

14. Crystallography

Most of our understanding of protein structure comes from x-ray crystallography and from the nuclear magnetic resonance (NMR) method (see next Section). It is reassuring that results from crystallography and NMR are consistent with one another and with those from other techniques described above (reviewed in [46-49]). Nonetheless, it is often difficult to relate the structure of a protein to its thermodynamic, chemical, or functional properties.

Obtaining reasonable crystals has become the rate limiting step in many investigations (reviewed in [50]). However, not all created crystals are of equal quality. Usually, the content of solvent, or crystallization medium, varies from 80 to 40% of the crystal volume.

X-ray diffraction data are measured to a resolution, usually expressed in terms of the Bragg equation – $\sin \Theta = \lambda / 2 \times d$, in which $2 \times \Theta$ is the angle relative to the direct beam at which the reflection is recorded; λ is the wave length of x-ray used (usually between 1.0 and 2.0 Å), and d is the nominal spacing giving rise to that reflection, or constructive interference of diffracted intensity. The functional resolution in protein crystallography is better than d. The structure of the protein is refined by varying the dihedral angles, Φ and Ψ, of the main chain and of the side chain (χ's).

Proteins in crystals undergo several types of motion or disorder. The entire protein, as a rigid unit, may be in slightly different orientation, at any instant of time, in different subunits, hence smearing the computed electron density associated with each atom. Side chains, or loops of proteins, may have alternate orientations within one crystal. The cumulative effects of various oscillations and vibrations are approximated by a "temperature factor" (B). The higher the B factor, the greater the disorder. In a well ordered protein whose diffraction data were measured from crystals maintained at liquid nitrogen temperature, the B's for main chain atoms should be between 10 and 14, those for side chains should be between 12 and 18. Helices have lower B's than do coil regions. If a stretch of residues has main chain B's over 20, there is probably some disorder in that region. The B assigned to a single Ca^{2+} ion should be less than 15 for 2.0 Å data.

Crystal structure of many calcium binding proteins is well known now. Among them: calmodulin, troponin C, parvalbumin, calbindin, recoverin, sorcin, S100 proteins, annexins, synaptotagmin, and many others. Moreover, now we know three-dimensional structure of many mutant calcium binding proteins.

Just the x-ray crystallography allowed elucidation in calcium binding proteins of a characteristic conformation, consisting of a helix, calcium-binding loop, and second helix, referred to as the "EF hand" (first found and reviewed by Kretsinger [51]). EF-hands have been identified in numerous Ca^{2+}-binding proteins by similarity of amino acid sequence and confirmed in many crystal structures (reviewed in [52]). Advances in the three-dimensional structural analysis of various EF-hand proteins have led to new insights into the structure and functional relationship of this large family of Ca^{2+}-binding proteins (reviewed in [53]). Now the structure of one of the EF-hand calcium binding proteins, calmodulin, is known not only for its various states (see for example recent publications [54, 55]) but also for its complexes with various peptides [56,57].

15. Nuclear Magnetic Resonance

The phenomenon of magnetic resonance results from the interaction of the magnetic moment of an atomic nucleus (μ) with an external magnetic field. The nature of this magnetic moment is the quantum mechanical angular momentum (spin angular momentum) of all nuclei that are no (g,g) nuclei (even number of protons and neutrons). In an external magnetic field, the magnetic moment orients according to:

$$J_z = - m * h/(2\pi) \Rightarrow \mu_z = m * \gamma * h/(2\pi),$$

where the constant γ is characteristic for each isotope and is called the gyromagnetic ratio. The magnetic quantum number m can be an integer number between $-I$ and $+I$. Thus, the external field leads to a splitting of the energy levels. For spin ½ nuclei, two energy levels exist according to a parallel or antiparallel orientation of the magnetic moment with respect to the magnetic field. The energy of these levels is given by the classical formula for a magnetic dipole in a homogenous magnetic field of the strength B_0:

$$E = - \mu_z * B_0 = - m * \gamma * h/(2\pi) * B_0$$

The magnetic moment of each nucleus precesses around B_0. The frequency of this precession is the Larmor frequency (ω_0) which is equivalent to the resonance frequency of the nucleus and the energy difference between the two levels is:

$$\gamma * h/(2\pi) * B_0 = \Delta E = h * \nu = h/(2\pi) * \omega_0$$
$$\omega_0 = \gamma * B_0$$

The Larmor frequency depends on the gyromagnetic ratio and the strength of the magnetic field, i.e. it is different for each isotope. At a magnetic field of 18.7 T, the Larmor frequency of protons is 800 MHz. Transitions between different energy levels occur if the frequency of radiation is equivalent to the energy difference between the two levels (the NMR method was reviewed in numerous publications, for example, in [58-64]).

In Fourier transform-NMR, the resonances are not measured one after another but all nulei are excited at the same time by a radio frequency pulse. Normally, a radio emitter works at a fixed frequency ν_0. If the radiation is emitted as a very short pulse (some μs in NMR), the pulse frequency becomes 'uncertain'. A short radio frequency pulse contains many frequencies in a broad band around ν_0 and thus excites the resonances of all spins in a sample at the same time. The excited spins emit the absorbed radiation after the pulse. The emitted signal is a superposition of all excited frequencies and its evolution in time is recorded. The intensities of the several frequencies, which give the observed signal in their superposition, are calculated by a mathematical operation, the Fourier transformation, which translates the time data into the frequency domain.

The electrons in a molecule surround the nuclei and create a small magnetic field which shields the nuclei slightly from the external field. Therefore, the Larmor frequencies of different nuclei vary due to their different chemical environments. This phenomenon is called 'chemical shift' and it is one of the major parameters of NMR spectroscopy since it is reflected in the different positions of the signals of different nuclei in a NMR spectrum. Different chemical groups have different chemical shifts. This is one of the major information sources in NMR spectra.

The use of multidimensional spectra was the major leap in NMR spectroscopy apart from the introduction of Fourier transform NMR. The 2D experiment has an indirect evolution time t_1 and a mixing sequence. This scheme includes the following steps: preparation, evolution, mixing, and detection.

After preparation, the spins can preccess freely for a given time t_1. During this time the magnetization is 'labeled' with the chemical shift of the first nucleus. During the mixing time magnetization is then transferred from the first nucleus to a second one. Mixing sequences utilize two mechanisms for magnetization transfer: scalar coupling or dipolar interaction. Data are acquired at the end of the experiment (detection); during this time the magnetization is 'labeled' with the chemical shift of the second nucleus. The cross signals originate from nuclei that exchanged magnetization during the mixing time (frequencies of the first and second nucleus in each dimension, respectively). They indicate an interaction of these two nuclei. Therefore, the cross signals contain the really important information of 2D NMR spectra.

The Nuclear Overhauser Effect Correlation Spectroscopy (NOESY) experiment is crucial for the determination of protein structure. It uses the dipolar interaction of spins (the nuclear Overhauser effect, NOE) for correlation of protons. The intensity of the NOE is to first approximation proportional to $1/r^6$, with r being the distance between the protons: the correlation between two protons depends on the distance between them, but normally a signal is only observed if their distance is smaller than 5 Å. The NOESY experiment correlates all protons that are close enough. It also correlates protons that are distant in the amino acid sequence but close in space and this is the most important information for the determination of protein structures.

Tertiary structure of many calcium binding proteins in solution was elucidated by the NMR method. Thus, several crystal and NMR structures of calmodulin in complex with fragments derived from calmodulin-regulated proteins, which have been reported recently, reveal novel ways for calmodulin to interact with its targets (reviewed in [65]). Now it is possible to discuss and compare features of the interaction between calmodulin and its target domains derived from the plasma membrane Ca^{2+}-pump, the Ca^{2+}-activated K^+-channel, the Ca^{2+}/calmodulin-dependent kinase kinase, and the anthrax exotoxin. Unexpected aspects of calmodulin/target interaction observed in these complexes include: (a) binding of the Ca^{2+}-pump domain to only the C-terminal part of calmodulin, (b) dimer formation with fragments of the K^+-channel, (c) insertion of calmodulin between two domains of the anthrax exotoxin, (d) binding of Ca^{2+} ions to only one EF-hand pair, and (e) binding of calmodulin in an extended conformation to some of its targets. The mode of interaction between calmodulin and these targets differs from binding conformations previously observed between calmodulin and peptides derived from myosin light chain kinase and calmodulin-dependent kinase IIα. In the latter complexes, calmodulin engulfs the calmodulin-binding domain peptide with its two Ca^{2+}-binding lobes and forms a compact, ellipsoid-like complex.

NMR data allowed determination of details of calcium-induced structural changes in calmodulin (reviewed in [66]). NMR spectroscopy provided the three-dimensional structures of the S100 family members S100A6 and S100B in both the apo- and calcium-bound forms (reviewed in [67]).

Two-dimensional ^1H-^{13}C shift correlation NMR spectra of recoverin containing a ^{13}C-labeled myristoyl group were obtained to selectively probe the effect of Ca^{2+} on the environment of the attached myristoyl group [68]. These data demonstrate that the binding of Ca^{2+} to recoverin induces the extrusion of its myristoyl group into the solvent, which would enable it to interact with a lipid bilayer or a hydrophobic site of a target protein. On the basis of a ^2H NMR order parameter analysis performed on recoverin containing a fully deuterated myristoyl group, the N-terminal myristoyl group of calcium-loaded recoverin appears to be located within the lipid bilayer [69]. Two-dimensional solid-state NMR (^1H-^{15}N PISEMA) spectra of uniformly and selectively ^{15}N-labeled recoverin show that the Ca^{2+}-bound protein is positioned on the membrane surface such that its long molecular axis is oriented approximately 45 degrees with respect to the membrane normal. The N-terminal region of recoverin points toward the membrane surface, with close contacts formed by basic residues K5, K11, K22, K37, R43, and K84.

Proteins contain a number of hydrogens that can exchange with hydrogen in the surrounding solvent. If the normal H_2O solvent is changed to D_2O, the protein gradually becomes deuterated. Since deuterium and hydrogen differ in mass by 1 dalton, the incorporation of deuterium into a protein can be monitored. Traditionally, hydrogen exchange was used in conjunction with NMR, but it is also possible to use mass spectrometry to measure hydrogen exchange (reviewed in [70-72]). The rate of hydrogen exchange depends on hydrogen bonding and solvent accessibility. Folded proteins can have amino acids with hydrogen exchange rates as much as 10^9 times slower than the same amino acid that is not in a folded protein. Protein folding and unfolding and also smaller structural changes essential for protein function can also be probed with hydrogen exchange mass spectroscopy. Besides the dependence of the exchange rates on pH, hydrogen exchange rates also depend on the accessibility of the hydrogens to solvent, the stabilization by hydrogen-bonded secondary and tertiary structure and on the local fluctuations of the protein. Amide hydrogens can be exchanged with deuterons as a result of local unfolding (local fluctuations) of the protein; whereas others are exchanged only upon global unfolding of the protein. The corresponding local and global thermodynamic stability parameters can be derived from the hydrogen exchange rates. The hydrogen/deuterium exchange is initiated by dissolving the protonated protein of interest into D_2O. The subsequent decrease in NMR amide signal intensity with time enables the determination of the exchange rates. Calcium induced changes of protein structure can be detected by the deuterium exchange method.

Various protons on the side chains and main chains of a fully denatured protein have their own ^2H exchange rates, characteristic of the pH of the solution. The exchange rates differ in a non-denatured protein and reflect the extent to which that site is protected from the solvent.

Zhu et al. [73] reported a new method for quantification of protein-ligand interaction by mass spectrometry, titration and H/D exchange (PLIMSTEX) for determination of the binding stoichiometry and affinity of a wide range of protein-ligand interactions. They described the method for analyzing the PLIMSTEX titration curves and evaluate the effect of various models on the precision and accuracy for determining binding constants using H/D exchange and a titration. The titration data were fitted using a 1:n protein:ligand sequential binding model, where n is the number of binding sites for the same ligand. The interaction of intestinal fatty-acid-binding protein (IFABP) with a fatty-acid carboxylate and that of calmodulin with Ca^{2+} were used as two tests.

16. Conclusion

Experimental physical methods provide valuable quantitative information about properties of calcium binding proteins. This information gives a solid basis for understanding the operation of these proteins under physiological conditions. Only combined use of various physical methods can give reliable data on structure and properties of the proteins. Especially interesting results can be obtained upon the use of physical methods in combination with the modern genetic engineering methods.

Acknowledgements

This work was supported in part by the Programs of the Russian Academy of Sciences "Molecular and cellular biology" and "Fundamental science for medicine", and by grant 04-04-97322 from the Russian Foundation for Basic Research.

References

[1] Jackson, K.W., Mahmood, T.M. 1994, *Anal. Chem., 66*, 252R.
[2] Harris, D.C. 1999, *Quantitative Chemical Analysis*, W.H. Freeman and Company, NY.
[3] Villarroya, M., Lopez, M.G., Cano-Abad, M.F., and Garcia, A.G. 1999, *Methods Mol. Biol., 114*, 137.
[4] Skoog, D.A., Holler, E.J., and Nieman, T.A. 1998, *Principles of Instrumental Analysis*. Brooks/Cole Thomson Learning.
[5] Bers, D.M. 1996, *New Horiz. 4*, 36.
[6] Baudet, S., Hove-Madsen, L., and Bers, D.M. 1994, *Methods Cell Biol., 40*, 93.
[7] Bers, D.M., Patton, C.W., and Nuccitelli, R. 1994, *Methods Cell Biol., 40*, 3.
[8] Permyakov, E.A., Murakami, K., and Berliner, L.J. 1987, *J. Biol. Chem., 268*, 3196.
[9] Roe, M.W., Lemasters, J.J., and Herman, B. 1990, *Cell Calcium, 11*, 63.
[10] Demaurex, N., and Frieden, M. 2003, *Cell Calcium, 34*, 109.
[11] Vysotski, E.S., and Lee, J. 2004, *Acc. Chem. Res., 37*, 405.
[12] Alvarez, J., and Montero, M. 2002, *Cell Calcium, 32*, 251.
[13] Jelesarov, I., and Bosshard, H.R. 1999, *J. Mol. Recognit., 12*, 3.
[14] Leavitt, S., and Freire, E. 2001, *Curr. Opin. Struct. Biol., 11*, 560.
[15] Cliff, M.J., and Ladbury, J.E. 2003, *J. Mol. Recognit., 16*, 383.
[16] Nielsen, A.D., Fuglsang, C.C., and Westh, P. 2003, *Anal. Biochem., 314*, 227.
[17] Plum, G.E, and Breslauer, K.J. 1995, *Curr. Opin. Struct. Biol., 5*, 682.
[18] Sanchez-Ruiz, J.M. 1995, *Subcell. Biochem., 24*, 133.
[19] Freire, E. 1995, *Methods Mol. Biol., 40*, 191.
[20] Shnyrov, V.L., Sanchez-Ruiz, J.M., Boiko, B.N., Zhadan, G.G., and Permyakov, E.A. 1997, *Thermochim. Acta, 302*, 165.
[21] Mach, H., Volkin, D.B., Burke, C.J., and Middaugh, C.R. 1995, *Methods Mol. Biol., 40*, 91.
[22] Permyakov, E.A., Yarmolenko, V.V., Emelyanenko, V.I., Burstein, E.A., Gerday, C., and Closset, J. 1980, *Eur. J. Biochem., 109*, 307.
[23] Barth, A., and Zscherp, C. 2002, *Q. Rev. Biophys. 35*, 369.
[24] Barth, A., and Zscherp, C. 2000, *FEBS Lett., 477*, 151.
[25] Jung, C. 2000, *J. Mol. Recognit. 13*, 325.
[26] Pelton, J.T., and McLean, L.R. 2000, *Anal Biochem., 277*, 167.
[27] Tamm, L.K., and Tatulian, S.A. 1997, *Q. Rev. Biophys. 30*, 365.
[28] Laberge, M., Wright, W.W., Sudhakar, K., Liebman, P.A., and Vanderkooi, J.M. 1997, *Biochemistry, 36*, 5363.
[29] Cooper, E.A., and Knutson, K. 1995, *Pharm. Biotechnol., 7*, 101.

[30] Jackson, M., and Mantsch, H.H. 1995, *Crit. Rev. Biochem. Mol. Biol., 30*, 95.
[31] Fabian, H., and Vogel, H.J. 2002, *Methods Mol. Biol., 173*, 57.
[32] Ozawa, T., Fukuda, M., Nara, M., Nakamura, A., Komine, Y., Kohama, K., and Umezawa, Y. 2000, *Biochemistry, 39*, 14495.
[33] Permyakov, E.A. 1993, *Luminescent Spectroscopy of Proteins*. CRC Press, Boca Raton, Ann Arbor, London, Tokyo.
[34] Permyakov, E.A. 2003, *Method of Intrinsic Protein Luminescence*. Moscow, Nauka.
[35] Sreerama, N., and Woody, R.W. 2004, *Methods Enzymol., 383*, 318.
[36] Kelly, S.M., and Price, N.C. 2000, *Curr. Protein Pept. Sci. 1*, 349.
[37] Greenfield, N.J. 2004, *Methods Enzymol., 383*, 282.
[38] Permyakov, S.E., Cherskaya, A.M., Senin, I.I., Zargarov, A.A., Shulga-Morskoy, S.V., Alekseev, A.M., Zinchenko, D.V., Lipkin, V.M., Philippov, P.P., Uversky, V.N., and Permyakov, E.A. 2000, *Protein Eng., 13*, 783.
[39] Heck, A.J., and Van Den Heuvel, R.H. 2004, *Mass Spectrom. Rev., 23*, 368.
[40] Yates, J.R. 3rd. 2004, *Annu. Rev. Biophys. Biomol. Struct., 33*, 297.
[41] Page, J.S., Masselon, C.D., and Smith, R.D. 2004, *Curr. Opin. Biotechnol., 15*, 3.
[42] Yan, X., Watson, J., Ho, P.S., and Deinzer, M.L. 2004, *Mol. Cell. Proteomics, 3*, 10.
[43] Svergun, D.I., and Koch, M.H. 2002, *Curr. Opin. Struct. Biol., 12*, 654.
[44] Doniach, S. 2001, *Chem. Rev., 101*, 1763.
[45] Perkins, S.J., Ashton, A.W., Boehm, M.K., and Chamberlain, D. 1998, *Int. J. Biol. Macromol. 22*, 1.
[46] Lecomte, C., Guillot, B., Muzet, N., Pichon-Pesme, V., and Jelsch, C. 2004, *Cell. Mol. Life Sci., 61*, 774.
[47] Burley, S.K., and Bonanno, J.B. 2002, *Annu. Rev. Genomics Hum. Genet., 3*, 243.
[48] Esposito, L., Vitagliano, L., and Mazzarella, L. 2002, *Protein Pept. Lett. 9*, 95.
[49] Heinemann, U., Illing, G., Oschkinat, H. 2001, *Curr. Opin. Biotechnol., 12*, 348.
[50] Wiencek, J.M. 1999, *Annu. Rev. Biomed. Eng., 1*, 505.
[51] Kretsinger, R.H. 1976, *Annu. Rev. Biochem., 45*, 239.
[52] Heizmann, C.W., and Hunziker, W. 1991, *Trends Biochem. Sci., 16*, 98.
[53] Ikura, M. 1996, *Trends Biochem. Sci., 21*, 14.
[54] Fallon, J.L., and Quiocho, F.A. 2003, *Structure (Camb)., 11*, 1303.
[55] Wilson, M.A., and Brunger, A.T. 2003, *Acta Crystallogr. D Biol. Crystallogr., 59*, 1782.
[56] Yamauchi, E., Nakatsu, T., Matsubara, M., Kato, H., and Taniguchi, H. 2003, *Nature Struct. Biol., 10*, 226.
[57] Aoyagi, M., Arvai, A.S., Tainer, J.A., and Getzoff, E.D. 2003, *EMBO J., 22*, 766.
[58] Kumar, T.K., and Yu, C. 2004, *Acc. Chem. Res., 37*, 929.
[59] Dyson, H.J., Wright, P.E. 2004, *Chem. Rev., 104*, 3607.
[60] Wuthrich, K. 1995, *Acta Crystallogr. D Biol. Crystallogr., 51*, 249.
[61] Opella, S.J., and Marassi, F.M. 2004, *Chem. Rev., 104*, 3587.
[62] Pickford, A.R., and Campbell, I.D. 2004, *Chem. Rev., 104*, 3557.
[63] Tugarinov, V., Hwang, P.M., and Kay, L.E. 2004, *Annu. Rev. Biochem., 73*, 107.
[64] Clarkson, J., and Campbell, I.D. 2003, *Biochem. Soc. Trans., 31*, 1006.
[65] Vetter, S.W., and Leclerc, E. 2003, *Eur. J. Biochem., 270*, 404.

[66] Zhang, M., and Yuan, T. 1998, *Biochem. Cell. Biol., 76*, 313.
[67] Smith, S.P., and Shaw, G.S. 1998, *Biochem. Cell. Biol., 76*, 324.
[68] Ames, J.B., Tanaka, T., Ikura, M., and Stryer, L. 1995, *J. Biol. Chem., 270*, 30909.
[69] Valentine, K.G., Mesleh, M.F., Opella, S.J., Ikura, M., and Ames, J.B. 2003, *Biochemistry, 42*, 6333.
[70] Hoofnagle, A.N., Resing, K.A., and Ahn, N.G. 2003, *Annu. Rev. Biophys. Biomol. Struct., 32*, 1.
[71] Lanman, J., and Prevelige, P.E. Jr. 2004, *Curr. Opin. Struct. Biol., 14*, 181.
[72] Garcia, R.A., Pantazatos, D., and Villarreal, F.J. 2004, *Assay Drug. Dev. Technol., 2*, 81.
[73] Zhu, M.M., Rempel, D.L., and Gross, M.L. 2004, *J. Am. Soc. Mass. Spectrom., 15*, 388.

In: Neuronal Calcium Sensor Proteins
Editors: Pavel Philippov and Karl Koch, pp. 351-370
ISBN 1-59454-978-8

Chapter XX

IN VITRO ACYLATION OF NEUROCALCIN AND EXPLORATION OF ITS MEMBRANE BINDING PROPERTIES WITH SURFACE PLASMON RESONANCE SPECTROSCOPY

***Daniel Ladant*[1], *Claire Rossi*[2], *Laure Beven*[2] *and Joël Chopineau*[3,*]**
[1]URA 2185 CNRS, Département de Biologie Structurale et Chimie, Institut Pasteur, 28 rue du Docteur Roux, 75724 Paris cedex, France;
[2]U MR 6022 CNRS, Université de Technologie de Compiègne, BP 20529, 60205, Compiègne cedex, France ;
[3]Centre de Biochimie Structurale, UMR 5048 CNRS, Centre Universitaire de Nîmes, rue Docteur Georges Salan 30021 Nîmes cedex, France.

Abstract

Neurocalcin is a member of a novel family of neuronal calcium sensors (NCS) that belongs to the superfamily of EF-hand Ca^{2+}-binding proteins. Neurocalcin, like several other NCS, is myristoylated on its *N*-terminus *in vivo* and can associate with biological membranes in a calcium and myristoyl-dependent manner, a process known as "Ca^{2+}-myristoyl switch". Here, we describe an *in vitro* approach to chemically acylate non-myristoylated neurocalcin at its *N*-terminus with fatty acids of different lengths (from C12 to C16). Several neurocalcin derivatives with a single fatty acid were thus prepared and their membrane binding properties were then characterized by cosedimentation with phospholipid vesicles and direct binding to lipid biomimetic membranes by surface plasmon resonance spectroscopy (Biacore). Our results show that neurocalcins monoacylated with lauric, myristic or palmitic acid were able to associate with membrane

* Correspondence concerning this article should be addressed to Joël Chopineau Centre de Biochimie Structurale, UMR 5048 CNRS, Centre Universitaire de Nîmes, rue Docteur Georges Salan 30021 Nîmes cedex, France. Tel: 33 4 66 27 95 63, fax: 33 3 66 27 95 54. E-mail: joel.chopineau@unimes.fr.

in a calcium-dependent manner. This indicates that the Ca^{2+}-myristoyl switch can function with different lipid moieties and is not strictly restricted to myristate. The ability to modify at will the fatty acid linked to the *N*-terminal glycine should be useful to analyze the contribution of the fatty acid moiety to the biological function of this family of neuronal calcium sensors.

Keywords: Neurocalcin, chemical acylation, SPR spectroscopy, membrane binding

Abbreviations

NCS	neuronal calcium sensors
ADMS	aminopropyl-dimethylethoxysilane
AOT	sulfosuccinic acid bis [2-ethylhexyl] ester
DSPE-PEG-NHS	distearoylphosphatidyl ethanolamine-.-poly-(ethyleneglycol) -N-hydroxysuccinimide
DPPE-NBD	1,2-dipalmitoyl-*sn*-glycero-3-phosphoethanolamine-N-7-nitro-2,1,3-benzoxadiazol-4-yl
DTT	dithiothreitol
EGTA	ethylene glycol-bis(b-aminoethyl ether)-*N,N,N',N'*-tetraacetic acid
ESI-MS	electrospray ionisation mass spectrometry
FPR	fluorescence photobleaching recovery
HBM	hybrid bilayer membrane
HEPES	(*N*-{2-hydroxyethyl}piperazine-*N'*-{2-ethanesulfonic acid})
MADLI-TOF MS	matrix-assisted laser desorption time-of-flight mass spectrometry
myr-neurocalcin	*in vivo* myristoylated neurocalcin
NCS	neuronal calcium sensors
NHS	N-hydroxysuccinimide
PC	phosphatidyl choline
PEG	-poly-(ethyleneglycol)
Q TOF MS	quadrupole time-of-flight mass spectrometry
RP-HPLC	reverse phase-high performance liquid chromatography
SDS-PAGE	sodium dodecyl sulfate-polyacrylamide gel electrophoresis
SPR	surface plasmon resonance
TLB	tethered lipid bilayer.

1. Introduction

Neuronal calcium sensors (NCS) constitute a subgroup of EF-hand Ca^{2+}-binding proteins that are expressed predominantly in neurons and neuroendocrine cells. The NCS family contains more than 40 (homologues and orthologues) members including recoverin, S-modulin, visinin, frequenin, GCAPs (guanylyl cyclase-activating proteins), GCIP (guanylyl cyclase inhibitory protein) and multiple isoforms of visinin-like proteins, including VILIP, hippocalcin, and neurocalcin. These calcium sensors participate in a wide range of biological functions such as phototransduction, pain perception, and learning by regulating various signal-transduction pathways. Their molecular mechanisms of action include the modulation of neurotransmitter release, the regulation of cyclic nucleotides or phosphoinositides metabolism, modulation of calmodulin-dependent signaling pathways, regulation of G-protein coupled receptor kinase activity and/or the regulation of Ca^{2+} and K^{+} channels [1].

The NCS proteins typically contain between 180 and 210 amino acids, with more than 40 % sequence identity [2]. They exhibit a consensus signal for *N*-terminal myristoylation and 4 EF-hand Ca^{2+}-binding motifs, although in all cases the first EF hand is unable to bind calcium. Several of these proteins have been shown to be myristoylated either *in vivo* and/or *in vitro*. In several instances, it was also demonstrated that the covalently attached myristoyl group at the *N*-terminus allows the NCS proteins to interact with biological membranes in a Ca^{2+} dependent manner in a process called "Ca^{2+}-myristoyl switch". The molecular mechanisms of the Ca^{2+}-myristoyl switch have been mostly described for recoverin, a photoreceptor specific protein [3-7]. 3D structures (obtained by X-ray analysis for the non-myristoylated and NMR for both forms) of both Ca^{2+}-bound and Ca^{2+}-free myristoylated recoverin have revealed that in the Ca^{2+}-free state, the myristoyl group is sequestered in a deep hydrophobic pocket where it is clamped by residues from the three EF-hands. Upon Ca^{2+} binding, the myristoyl group is extruded and therefore free to bind to membranes and/or protein targets. Ca^{2+} binding is accompanied by a 45° rotation of the *N*-terminal domain, which exposes hydrophobic residues that might contribute for additional free energy to interact with lipids. Crystal structure of recombinant bovine neurocalcin [8] and the NMR structure of GCAP-2 [9] strongly suggests that all NCS proteins share a similar three-dimensional structure. Recently, Burgoyne and colleagues [10] used fluorescently-tagged proteins to image *in vivo* the translocation of hippocalcin and neurocalcin δ from the cytosol to membranes upon intracellular elevation of Ca^{2+}. Interestingly, another NCS isoform NCS-1=was found to be constitutively membrane-associated even at low Ca^{2+}, as a result of selective mutations within the *N*-terminus that locks the myristoyl group in an open conformation [11]. These studies indicate that closely related Ca^{2+} sensor proteins use their myristoyl groups in distinct ways *in vivo* to ensure various spatial and temporal distributions in response to calcium mobilization.

In this work, we describe two different approaches that have been used to explore the biochemical mechanisms of the Ca^{2+}/myristoyl switch of neurocalcin. Firstly, we have established an *in vitro* procedure that allows to selectively acylate the *N*-terminus of the protein with different acyl chains. Secondly, we describe the use of two different types of biomimetic supported membranes assembled at the surface of an optical biosensor system, to characterize the membrane binding properties of the different acylated proteins. The association of biomolecules with such artificial membranes can be detected by surface plasmon resonance (SPR) spectroscopy. These techniques may find applications in the study of the interactions of acylated proteins with biological membranes.

2. Chemical Acylation of Neurocalcin

N-myristoylation represents one of the most common fatty acid modifications of proteins. This modification, that consists in the covalent attachment of a myristic acid to the *N*-terminus of a polypeptide chain, is catalyzed by the myristoyl-CoA:protein *N*-myristoyltransferase and occurs during or soon after translation of the proteins [12,13]. *N*-terminal myristoylation is important for the biological functions of the corresponding proteins: their structure, thermodynamic stability, enzymatic activity, or ability to interact with specific partners can be critically modulated by *N*-myristoylation. In addition, in many cases, *N*-myristoylation facilitates the binding of the corresponding proteins to biological membranes. Numerous myristoylated proteins are indeed involved in signal transduction, such as protein kinases, heterotrimeric G proteins, ADP-ribosylation factors, and the NCS calcium binding proteins [10,14-18].

Some of NCS proteins display a characteristic Ca^{2+}-myristoyl switch that confers a calcium-dependent control on the subcellular localization of the proteins: the *N*-terminally attached myristic group, sequestered in an hydrophobic pocket in the calcium-free form of the protein, is expelled upon calcium binding and can insert into lipid bilayers to stabilize the association of the protein with biological membranes [1,7,19]. One intriguing observation is that different members of the NCS family exhibit distinct sub-cellular distribution and calcium responsiveness although they have an identical *N*-terminal myristoylation modification and similar amino acid sequence [10]. How subtle differences in the structure of the NCS proteins can determine their selective targeting to different types of cellular membranes remains to be clarified.

To explore the contribution of the fatty acid group in the membrane binding properties of NCS proteins, we developed a generic procedure that allows to chemically acylate neurocalcin in a controlled manner with any desired acyl chain. The acylation reaction took place in reversed micelles that can be viewed as nanoreactors. In this system, the apolar solvent allows the dissolution of fatty acid chlorides of different chain length whereas, the protein dispersed in the microaqueous phase, is protected by a molecular surfactant barrier, resulting in a good accessibility of the protein to the lipophilic reagent. This procedure allows a fine-tuning of the protein containers by controlling the hydration degree [water]/[AOT]. Using optimal conditions, neurocalcin can be monoacylated in a highly specific manner on its *N*-terminus by different fatty acid chains, as described below.

2.1. Acylation of Neurocalcin in Reversed Micelles

Unmyristoylated and myristoylated forms of bovine neurocalcin δ were obtained as recombinant proteins from *E. coli*. Expression and purification of both proteins were performed as previously described [20]. About 10-20 mg of recombinant myristoylated or unmyristoylated neurocalcin δ were obtained from one liter of culture. The neurocalcin concentration was routinely determined from the absorbance at 280 nm using a molar extinction coefficient of 20 mM^{-1} cm^{-1}. Protein purity was visualized by sodium dodecyl sulfate-polyacrylamide gel electrophoresis (SDS-PAGE). Unmyristoylated neurocalcin was the starting material for protein acylation while *in vivo* myristoylated neurocalcin was used as a positive control.

Acylation was carried out using AOT reversed micelles as reactors for protein modification. The procedure previously established for ribonuclease A [21,22] was adapted and optimized for neurocalcin. The acylation procedure using reversed micelles comprised three major steps:

(1) Chemical modification of the protein encapsulated into reversed micelles. The protein was first incorporated into reversed micelles, then an organic solution of the acyl chloride was added to the mixture. After diffusion into the reversed micelles the acyl chloride could reach the encapsulated protein and react with accessible NH_2-groups of the polypeptide. In particular conditions of the reaction, that is at an intramicellar basic pH, the *N*-terminus is, by far, the most reactive amino group as compared to the amino group of lysine side chain.

(2) Recovery of the modified protein as a water soluble fraction from the reaction mixture was obtained by cold acetone precipitation.

(3) Purification of neurocalcin acylated derivatives was performed using a Triton X-114 temperature-induced phase separation at 30°C. The overall procedure was optimized with myristic acid chloride, and the optimal conditions for neurocalcin acylation were the following: a water content corresponding to [water]/[AOT] = 12, a molar ratio [acyl chloride]/[neurocalcin δ] = 2, 50 mM borate buffer, pH 10.1. Neurocalcin acylation was performed with lauric, myristic and palmitic acid chlorides.

2.2. Characterization of the Chemically Modified Neurocalcins by Reversed Phase HPLC

Protein analysis was performed at 30°C on a Si-C18 Nucleosil column. The protein mixture obtained after acetone precipitation contained two major fractions in a 50/50 ratio. The first one corresponds to the non-modified neurocalcin (identified by its retention time) while the second one may be identified as the putative acylated derivative.

After chromatography through a reverse C18 column, the recovered myristoylated neurocalcin was no more able to bind membranes in a calcium-dependent manner (data not shown), probably as result of its denaturation in these HPLC conditions. Therefore, in order to preserve the native structure of the chemically acylated proteins, they were purified by using a Triton X-114 temperature-induced phase separation at 30°C. In brief, the protein mixture in presence of Triton X-114 was layered onto a sucrose cushion, successive centrifugations were performed in order to separate the non-acylated protein from the acylated one [23]. The acylated protein was recovered in the detergent solution containing about 1.5% Triton X-114 that was then removed by using Bio-Beads (less than 0.04% of Triton X-114, as determined by absorption at 275 nm, was present in the recovered protein solution). RP-HPLC analysis was used to control purity of the acylated proteins. On average, 20% of the starting material was recovered as pure mono-acylated neurocalcin.

Myristoyl and palmitoyl derivatives were recovered as nearly pure proteins: the initial non-modified neurocalcin represents less than 10 % of the total protein in these preparations. However, in the case of the lauryl derivative, the Triton X-114 purification procedure was unable to efficiently separate the acylated neurocalcin from the unmodified protein. Thus, the mixture containing non-modified neurocalcin and its lauryl derivative (50/50) was retained for subsequent membrane binding experiments. As shown in Fig. 1, the retention time of the derivatives increased with the length of the acyl chain of the fatty acid. As expected, the chemically modified C14-neurocalcin retention time was identical to that of the *in vivo* myristoylated neurocalcin (myr-neurocalcin).

All the above-mentioned compounds showed the same electrophoretic mobility on SDS-PAGE.

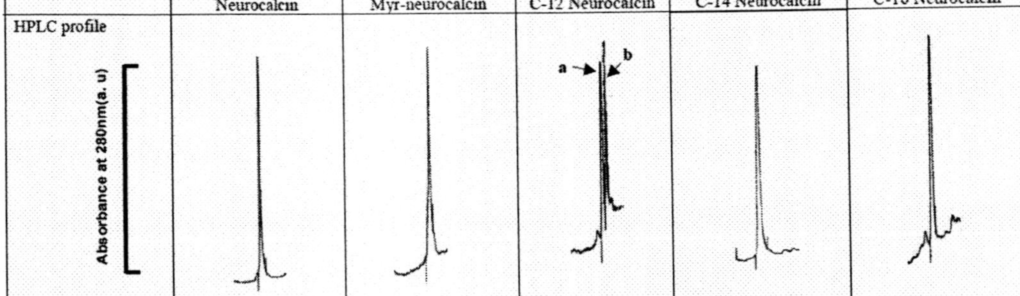

	Neurocalcin	Myr-neurocalcin	C-12 Neurocalcin	C-14 Neurocalcin	C-16 Neurocalcin
Retention time (min)	18.93	19.51	a: 18.93 b: 19.27	19.53	19.83
Molar mass ($g.mol^{-1}$ ± SD)	22 110 ± 2	22 325± 2	a: 22 110 ± 2 b: 22 300 ± 2	22 326 ± 2	22 354 ± 2
Expected mass variation ($g.mol^{-1}$)	0	210	182	210	238
Number of fatty acid	0	1	1	1	1

Figure 1. HPLC and mass spectrometry characterization of neurocalcin, myr-neurocalcin, C12-neurocalcin , C14-neurocalcin and C16-neurocalcin. Line 1: RP-HPLC profiles of the indicated protein separated on a C-18 grafted silica column; line 2: retention time of the corresponding peaks; line 3: molecular masses of the isolated protein found by ESI-MS; line 4: mass variation expected for one acyl moiety linked per polypeptide chain; line 5: number of acyl moieties per protein molecule as deduced from the difference of molecular mass of the sample as compared to non-modified neurocalcin.

2.3. Characterization of the Chemically Modified Neurocalcins by Mass Spectrometry

The extent and location of the protein chemical modification were sequentially investigated by mass spectrometry. Our strategy was first to determine the number of chemically modified amino acid residues per protein in the neurocalcin derivatives obtained. To check the extent of protein modification, positive ion electrospray ionization mass spectrometry (ESI-MS) was performed using a Finnigan SSQ710 (San José, CA). The MS analysis of the modified and non-modified proteins showed a mass increase of 182, 210 or 238 for the C12-, C14- and C16- derivatives, respectively (see Fig. 1). These values are in agreement with the expected masses for neurocalcins bearing a single acyl chain per molecule.

The second step of the analysis was to characterize the location of the grafted acyl group on the polypeptide chain. Matrix-assisted laser desorption/ionization mass spectrometry (MALDI-MS) of fragments resulting from the enzymatic cleavage of proteins was used to identify the peptides bearing the chemical modification. Protein hydrolysis was achieved by treatment with Asp-*N* endoproteinase produced by a *Pseudomonas fragi* mutant (Boerhinger Mannheim, Germany) with an enzyme:protein weight ratio of 1:200. Hydrophobic pipette tips were used to obtain desalted peptide solutions. To perform the MALDI-MS analysis, the matrix chosen for co-crystallisation with peptides dissolved in methanol/water/acetic acid (50/50/0.5) was a dihydroxybenzoic acid saturated solution in 0.1% trifluoroacetic acid. The peptide mass analysis was carried out using positive and reflector modes, on a MALDI Time Of Flight (TOF) spectrometer Voyager-STR biospectrometry workstation (Perspectives Biosystems, Framingham, MA). We compared the theoretical masses of peptides produced after the proteolytic cleavage of non-modified neurocalcin with the masses which were experimentally obtained with the hydrolysis products of the modified neurocalcin preparations obtained. The resulting spectra for non-modified, myr- and C14-neurocalcin are shown in Fig. 2.

The comparison of experimental and theoretical masses of modified and non-modified fragments allowed us the identification of the chemically modified peptides. Finally, the more precise location of the grafted acyl group was determined using a hybrid quadrupole time-of-flight electrospray ionization (Q-TOF 2, Micromass, Manchester, UK) mass spectrometer. The desalted peptide mixture, deposited with a gel loader tip, was analyzed using a nanospray golden glass needle (Protana, Odense, DK). Among the many different fragments which could be identified, the 14-residues *N*-terminal peptide GKQNSQKLRPEVMQ could be detected for the non-modified protein. For Myr- and C14- neurocalcin, the mass spectra data indicated that the relative intensity of the peak corresponding to the *N*-terminal peptide is either null or consistently reduced. Also, a peak corresponding to an experimental mass ($M+H^+$) of 1725 occurs for the myristoyled proteins and is absent in the neurocalcin spectrum. Thus, this latter peak is due to the occurrence of the *N*-terminal peptide bearing the myristoyl chain. It must be underlined that no additional acyl-group could be detected for all the other identified peptides, confirming the previous ESI-MS analysis which showed the monoacylation of the modified neurocalcins. Furthermore, these results indicate that our acylation procedure is selective for the *N*-terminal peptide. An identical strategy was adopted for the characterization of the peptide bearing the fatty acid chain in the cases of C12- and C16-neurocalcin derivatives (data not shown), for which similar results were obtained as only the *N*-terminal peptide proved to be acylated (with a single acyl chain). To further identify the modified amino acid residue in the *N*-terminal peptide, we performed a MS/MS electrospray analysis: the ions corresponding to the *N*-terminal peptide were selected and submitted to fragmentation in order to analyze the produced b (*N*-terminal end) and Y'' (*C*-terminal end) ions; b, Y'' and internal fragments were then identified by comparison with the theoretical masses. The data obtained with the biologically modified myr-neurocalcin (data not shown) indicated that the *N*-terminal residues GK are linked to the acyl chain, as confirmed by the occurrence of the ions corresponding to fragments QNSKLRPE and

NSKLRPEVM. The identification of Y''12 ions in both C12- and C14 spectra and of the fragment KQNSKLRPEVM in the C16- spectrum demonstrates the linkage of the acyl chain via an amide bond to the *N*-terminal glycine residue. This study also confirmed that the various neurocalcin derivatives were monoacylated as no other ions corresponding to different overlapping internal fragments could be identified.

Altogether these results strongly support a selective *N*-terminal chemical monoacylation of neurocalcin. The possibility of modifying at will the length of the *N*-terminal fatty acid allowed us to study its role in the Ca^{2+}- and acyl-dependent binding of neurocalcin to membranes.

A

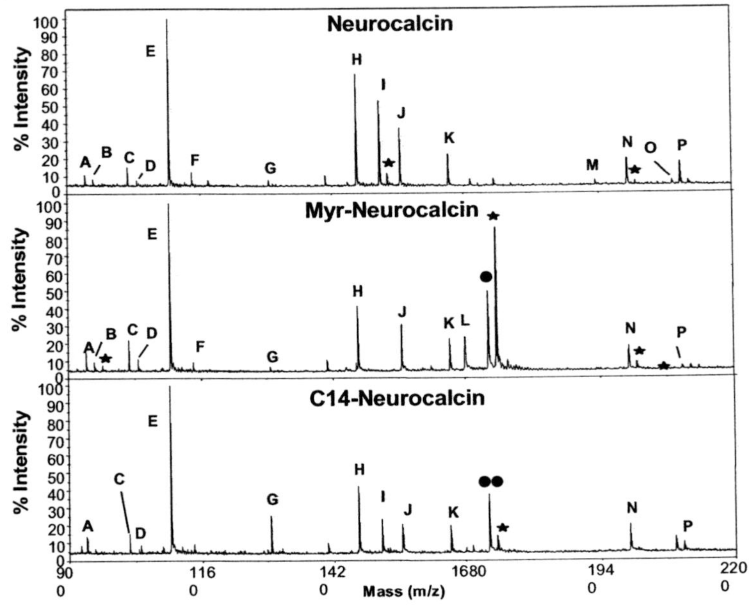

B

Peak	Theorical position	Experimental	Peptide Sequence	Fragment	mass	mass
A	935.47	935.37 ± 0.13	EHVFRTF	65-71		
B	951.51	951.51 ± 0.06	EKIFRQM	149-155		
C	1018.45	1018.39 ± 0.02	DFTEHEIQ	20-27		
D	1036.54	1036.46 ± 0.07	EEFIRGAKS	166-174		
E	1098.57	1098.54 ± 0.03	EWYKGFLR	28-35		
F	1143.62	1143.49 ± 0.10	DPSIVRLLQC	175-184		
G	1297.61	1297.54 ± 0.03	DGNGYISKAEML	110-121		
H	1468.76	1468.74 ± 0.03	EIQEWYKGFLR	25-35		
I	1514.81	1514.79 ± 0.02	GKQNSKLRPEVMQ	1-13		
J	1554.77	1555.75 ± 1.02	DASKFAEHVFRTF	59-71		
K	1649.89	1649.63 ± 0.32	DGKLSLEEFIRGAKS	160-174		
L	1678.78	1680.58 ± 1.17	EHVFRTFDANGDGTI	65-79		
M	1932.95	1934.22 ± 0.55	DPSIVRLLQCDPSSAGQF	175-192		
N	1995.00	1995.95 ± 0.02	DESTPEKRTEKIFRQM	140-155		
O	2083.06	2083.81 ± 0.45	EIVQAIYKMVSSVMKMPE	122-139		
P	2098.50	2099.78 ± 1.84	DFTEHEIQEWYKGFLR	20-35		

Figure 2. MALDI-TOF analysis of peptide mixtures released from the digestion of neurocalcins by Asp*N* endoproteinase (*Pseudomonas fragi*). (A) Mass profiles of Asp*N* protease digest neurocalcin, Myr-neurocalcin and C14-neurocalcin. Peaks noted ● and ●● correspond to the (1-13) peptides bearing the myristoyl group resulting from the digestion of Myr-neurocalcin and C14-neurocalcin, respectively (see text for details). (B) Theoretical and experimental masses of excepted peptide fragments. The MALDI-TOF identity of peptides was established by comparing experimental and theoretical masses of peptides resulting from Asp*N* protease digestion of non-modified neurocalcin.

2.4. Vesicle Binding Properties of the Chemically Modified Neurocalcins

The membrane binding properties of neurocalcin derivatives were examined by sedimentation experiments. The proteins were incubated with sucrose-loaded vesicles (bovine brain lipid) in the presence or in the absence of Ca^{2+} and pelleted by high speed centrifugation (100,000 × *g*). An airfuge (Beckman) was used for sucrose loaded vesicles preparation and for protein-lipid binding studies.

Vesicles were obtained by hydration of dried lipidic films with10 mM HEPES-Na, pH 7.4, containing 20 mM KCl and 180 mM sucrose. After ultrasonic disruption, the suspension was centrifuged twenty minutes at 5000 × *g* in order to eliminate titanium particles coming from the sonication tip. The supernatant was subjected to four cycles of freeze-thawing and then centrifuged for 30 min at 100,000 × *g*. The liposome pellet was then resuspended in HEPES buffer. For sedimentation binding assays, 15 μg of neurocalcin were incubated with liposomes (about 500 μg of brain lipids) in 300 μl of 20 mM HEPES-Na, pH 7.5, containing 100 mM KCl and either 2 mM EGTA[1] or 1 mM $CaCl_2$. After 30 min incubation at 22 °C, the mixtures were centrifuged for one hour at 100,000 × *g* and 22 °C. Both supernatant and pellet fractions were denaturated in 100 μl of SDS sample buffer, and identical volumes were electrophoresed on a 12.5% SDS gel, which was stained with Coomassie Blue.

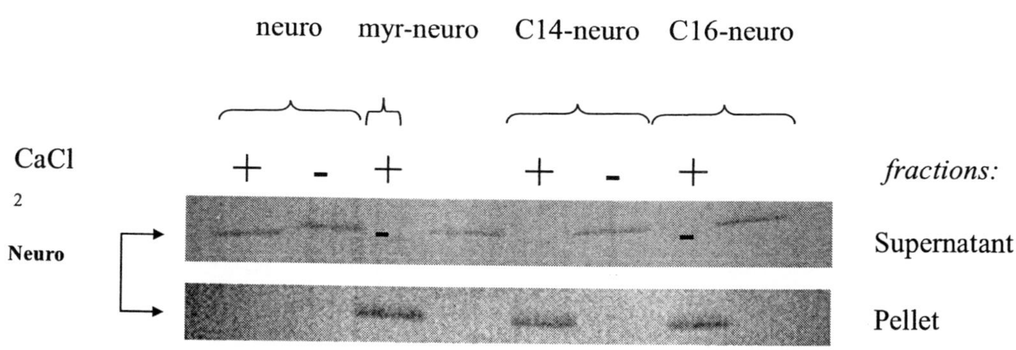

Figure 3. Binding properties of neurocalcins toward liposomes. Neurocalcins and derivatives were incubated with bovine brain liposomes either in the absence of Ca^{2+} (+ EGTA) or in the presence of Ca^{2+} (+CaCl2). The suspension was sedimented by high speed centrifugation. Pellet and supernatant fractions (equal volume) were electrophoresed on a 12,5 % SDS gel which was stained with Coomassie Blue.

As shown in Fig. 3, the non-acylated neurocalcin (*i.e.* the starting material for acylation), did not bind to liposomes and remained in the supernatant both in the presence and in the absence of Ca^{2+}. Myr-neurocalcin was recovered in the lipid pellet, when incubated in the presence of Ca^{2+} but remained in the supernatant in the absence of Ca^{2+} (*i.e.* in the presence of EGTA). C14 and C16-neurocalcin derivatives were associated with the vesicles in the presence of Ca^{2+} while they were recovered in the supernatant in the presence of EGTA. The C12-neurocalcin reaction mixture containing unmodified neurocalcin and acylated neurocalcin was evidenced both in the pellet and in the supernatant in the presence of Ca^{2+}, while in the presence of EGTA, no protein was found in the pellet (data not shown). This suggests that the C12-neurocalcin derivative is also able to bind to membrane in a Ca^{2+}

dependent manner. Control experiments showed that chemically modified neurocalcins centrifuged at high speed in the presence Ca^{2+} but without lipids remained soluble (data not shown).

3. SPR Spectroscopy Analysis of Neurocalcin Binding to Artifical Membranes

To study the membrane binding properties of the acylated neurocalcin derivatives, we have designed two different types of supported biomimetic membranes that were assembled on a gold surface mounted in a SPR apparatus. Protein binding to such biomimetic membranes can then be characterized by SPR spectroscopy, a well-known technique for measurements of thin films [24]. The method is based on excitation of the surface plasmons by p-polarized light at the noble metal-dielectric interface. Surface plasmons are described as electromagnetic evanescent waves propagating along the surface with a typical penetration depth of 200 nm. The SPR signal is highly sensitive to the refraction indexes of the materials at the film surface, and therefore to the amount of material (protein and/or lipids) deposited on this surface [24]. Hence both the biomimetic membrane construction and the binding of proteins (or other types of molecules) to the artificial membranes can be monitored by SPR spectroscopy.

Two different types of biomimetic membranes on gold surfaces have been examined: a hybrid bilayer membrane (HBM) and a supported tethered lipid bilayer (TLB) (Fig. 4). HBM is obtained by deposition of a single layer of phospholipids onto a hydrophobic surface of octadecanethiol molecules coated onto the bare gold surface. This type of structure only partly mimic an authentic biological membrane as it contains only a single phospholipid leaflet; however, this construction is very robust and can be easily assembled onto a classical Biacore chip. The second type of artificial membrane is a polymer tethered membrane, obtained by covalent coupling of lipid vesicles doped with amino-reactive PEGylated-phospholipids onto a monolayer of cysteamine precoated onto the gold surface. Once covalently coupled to the surface, the vesicles spontaneously disrupt to form a continuous phospholipid bilayer tethered to the supporting surface by a PEG cushion. TLB is a much better model for biological membranes as compared to the hybrid bilayer. In particular, we have shown that it is much more appropriate to study the association properties of proteins that insert into the membrane bilayer and it should be useful to reconstitute integral membrane protein in a functional state [25]. Moreover, because this type of covalently tethered bilayer membrane permits to define an aqueous compartment between the membrane and the supporting surface, it can be used to study transport of molecules across a biological membrane.

3.1. SPR Spectroscopy

Membranes were assembled on bare glass or on gold substrate evaporated onto glass. A chromium adhesion layer of 2 ± 0.5 nm was deposited before the gold layer of 47 ± 1 nm. For

SPR measurements, the gold-coated glass slide was vertically assembled with a 0.6 ml polytetrafluoroethylene sample cell and a 90° LaSFN9 prism (n = 1.85) was mounted with the glass slides using terphenyl monobromonaphtalene (Cargille Laboratories Inc, NJ) as a matching fluid (n = 1.656). In this work, the Kretschmann configuration was used to create an evanescent field on the gold surface. Monochromatic p-polarized light from a He-Ne laser beam (λ = 633 nm) was reflected from the backside of the gold-coated glass slide. The SPR curve is a function of the refraction indexes of the materials at the film surface. A minimum reflectance corresponds to the excitation of electron plasmons, and the minimum angle shifts if the refraction indexes of the layer at the surface are modified by a value of $\Delta n = n_{layer}-n_{buffer}$. The geometrical thickness (Δd) of the adsorbed material can be determined from the angle shift ($\Delta\theta$) and the refractive indexes of each layer must be known for calculations. We used previously described values of n = 1.5 for long chain alkyl-thiols on gold [26] for phospholipids layers [27,28] and proteins [29,30]. Reflectivity was recorded as a function of incident angle, and optical thicknesses were determined according to Fresnel equations using the Winspall program (Max-Planck Institute for Polymer Research, Mainz, Germany). Kinetics were measured at a fixed angle of 1° below the minimum angle. To determine the amount of protein bound to the biomimetic membrane, the optical thickness were converted to protein surface concentration using 1 ng/mm^2 for an optical thickness of 10 Å [24,31]. All measurements were performed at 25°C.

3.2. Assembly of Hybrid Bilayer Membrane

HBM was composed of an inner monolayer obtained by adsorption of 1-octadecane thiol on top of which vesicles, obtained by extrusion, were fused leading to the formation of the outer leaflet. The SPR curve fits with Winspall program gave a 1-octadecane thiol inner monolayer thickness of 29 ± 2.5 Å and a refractive index of 1.5, which is in agreement with layer thicknesses of 29 ± 2.5 Å determined from experiments of 1-octadecane thiol adsorption onto gold surfaces [32-34]. After 1-octadecane thiol monolayer formation and careful drying, the coated-substrate was mounted with the Teflon SPR cell and a spectrum was recorded. A lipid (brain lipid or egg PC) vesicle suspension (1 mg/ ml) was injected onto the hydrophobic 1-octadecanethiol monolayer and the fusion was recorded during one hour. After rinsing the surface with buffer (20 mM HEPES-Na, pH 7.4, containing 150 mM NaCl) Winspall calculations gave an average thickness of 22 ± 2.5 Å for the egg PC layer and 22 ± 4 Å for the brain lipid one. These thicknesses are in agreement with the formation of a lipid monolayer [33,35]. In both cases (brain lipid or egg PC), the overall thickness of the biomimetic membrane was 50 ± 5 Å.

3.3. Assembly of Polymer-Tethered Membrane Bilayer

The polymer-tethered phospholipid bilayer was obtained in two steps. A monolayer of 1-aminoethane thiol (cysteamine) was first deposited on the gold surface, then vesicles containing 1,2-distearoyl-*sn*-glycero-3-phosphoethanolamine-poly-ethyleneglycol-N-hydroxy

-succinimide (DSPE-PEG_{3400}-NHS) as anchoring molecule were added for formation of the supported bilayer (see Fig. 4).

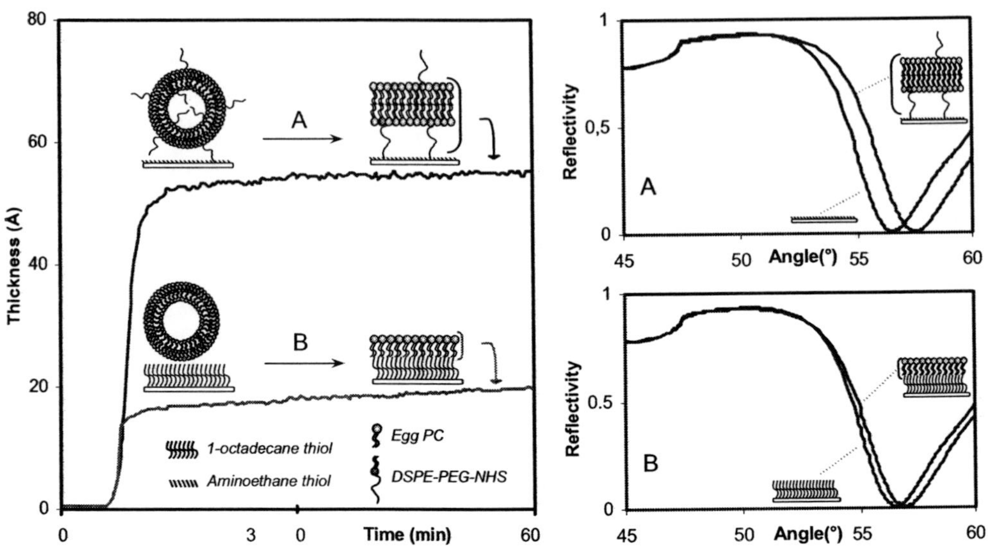

Figure 4. Construction of biomimetic membranes. The assembly of biomimetic membranes was followed by SPR spectroscopy. Left side: kinetic of membrane formation as deduced from SPR signal during lipid deposition. (A) Tethered membrane. The mixed EggPC/DSPE-PEG_{3400}-NHS vesicles were chemically linked onto the cysteamine gold coated surface, osmotic stress led to disruption and bilayer formation. (B) Hybrid bilayer. The lipid monolayer was obtained by adsorption and hydrophobic fusion of the lipid vesicles onto the preformed 1-octadecane thiol monolayer. Right side: raw SPR spectra before and after TLB (A) or HBM (B) which were used to determined membrane thickness.

To obtain an amine covered surface, the gold-coated glass slides were immersed overnight in a 2-mercaptoethylamine degassed solution (5 mM in pure ethanol). The coated slides were thoroughly rinsed with ethanol and dried under a nitrogen stream. They were stored under nitrogen before being used. The SPR data fitting gave a value of $n = 1.45$ for the cysteamine monolayer refractive index when the thickness of the layer adsorbed on gold was considered to be 5 Å [32]. These two parameters (n = 1.45, e = 5 Å) were used for all subsequent SPR calculations. The coated slide was assembled with the SPR cell, then 4 ml of vesicle suspension of egg PC/DSPE-PEG_{3400}-NHS (egg PC final concentration of 1.3×10^{-3} M with DSPE-PEG_{3400}-NHS molar ratios from 0 to 10 mol%) in 20 mM HEPES-Na, pH 8, containing 150 mM NaCl, were injected over the surface at a flow rate of 3 ml/min. A one hour minimum reaction time was necessary for vesicles anchoring and disruption. The cell was rinsed with the same buffer (10 min at a flow rate of 1 ml/min), then with 4 ml of 4 mM NaOH (5 min) and finally flushed during one hour with 20 mM HEPES-Na, pH 7.4, containing 150 mM NaCl, at a flow rate of 1 ml/min. DSPE-PEG-NHS containing vesicles were prepared from a mixed dried film obtained from lipids containing the anchoring molecule in chloroform. Covalent coupling of the reactive NHS moiety with the amino group on the surface was achieved after about 2 hours of reaction at room temperature. Then, 20 mM HEPES buffer, pH 7.4, was flushed over the surface in order to obtain a continuous phospholipid bilayer as a result of osmotic stress. Analysis and fitting of the SPR reflectivity curves gave a thickness between 50 Å to 55 Å for the adsorbed bilayer (see Fig. 4 A). For molar ratios of anchoring molecules between 0.4 and 6 mol%, the vesicle adsorption kinetics were highly reproducible which was not the case when the molar ratios were higher than 6 mol%. The formation of vesicles, with such high density of anchoring molecules, might be more difficult to achieve because of the limited miscibility of DSPE within egg PC bilayers. Altogether, HBM and tethered bilayer membranes thickness calculated from SPR experiments were similar. An overall thickness of 48 or 51 Å (29 + 22 Å) was found for the HBM whereas that of the tethered bilayer was determined to be 50 to 55 Å. These two membrane-like constructions share a similar outer layer of egg PC while the inner layer and the coupling to the substrate are different. Yet, the outer layer of polymer-tethered bilayer may contain up to half of the initial vesicular molar ratio of DSPE-PEG (*i.e.* 3 mol%), which is different from the hybrid bilayers (that are free of DSPE-PEG). Formation of hybrid bilayers incorporating DSPE-PEG is difficult to obtain because the presence of PEG chain impairs hydrophobic vesicular fusion.

3.4. Characterization of the Artificial Membranes Fluidity by Fluorescence Microscopy

The long-range fluidity of the lipid layer of both HBM and TLB constructions was assessed from fluorescence microscopy observation of the diffusion of fluorescent lipid tracers incorporated in the lipid layer (FRAP, fluorescence recovery after photobleaching). Fluorescence microscopy experiments were performed using a confocal scanning light microscope (LSM 410, Micro Systems Zeiss, Germany). A transparent surface was required to perform the fluorescence experiments. The organic lipid layers were assembled on a glass support, transparent for the laser beam; therefore the alkylthiols, cysteamine and 1-octadecane thiol, used to coat the gold surface for SPR experiments, were replaced by silanes of corresponding length.

For HBM formation, the cleaned glass slides were coated by a monolayer of self-assembled 1-octadecyltrichlorosilane: the slides were immersed 15 min in a freshly prepared solution of octadecyltrichlorosilane (2% (w/w) in anhydrous hexadecane). The coated glass slides were extensively rinsed with toluene and ethanol and finally dried under a nitrogen flow. An open cell (top of a spectrophotometric cuvette) was glued on the slide. 300 µl of the vesicle suspension (egg PC/DPPE-NBD) were added onto the dried surface to form the hemimembrane. For TLB formation, the amino-covered surface was obtained from self-assembling aminopropyl-dimethylethoxysilane (ADMS): the slides were immersed into an ADMS solution (2%, v/v) in toluene for 4 h, thoroughly rinsed with toluene, chloroform, ethanol and water before being dried in an oven at 110°C. Similarly to the hybrid membrane, an open cuvette was glued to the surface. The cell was washed with 20 mM HEPES-Na, pH 7.4, containing 150 mM NaCl, and 300 µl of a vesicle suspension (egg PC/ DSPE-PEG$_{3400}$-NHS/ DPPE-NBD) were then added onto the surface. After an 8 hours of incubation, the surface was rinsed with 20 mM HEPES-Na, pH 7.4, containing 150 mM NaCl.

After this first silanization step of the glass surface, the procedures for HBM or TLB constructions were identical as those used for the gold surfaces. The mobile fluorofore DPPE-NBD (1-2 mol%) was added in the liposomes preparation. The lateral diffusion coefficients and mobile fractions of the lipid probe were estimated from FPR (fluorescence photobleaching recovery) experiments carried out on a confocal microscope according to a method described previously [36]. A small area (45×45 μm^2) of the lipid surface (several mm^2) of the biomimetic membrane was bleached and FPR images (320 × 320 μm^2) were recorded at different time intervals in order to determine fluorescence intensity recovery.

In both cases (HBM and TLB), the fluorescent lipid probes were found to be mobile and fluorescence recovery was observed. For HBM, more than 95% of mobility was observed and a lateral diffusion coefficient of 2.5×10^{-8} $cm^2.s^{-1}$ was determined. For the supported tethered membrane (prepared with 2 mol% of anchoring molecules), we found that 95 to 98% of the fluorescent lipid probes were mobile with a lateral diffusion coefficient of 2.4×10^{-8} $cm^2.s^{-1}$.

An egg PC solid supported bilayer directly fused onto a clean surface according to Mc Connell protocol [37] was also included in a control experiment. The mobile fraction (100 ± 5 %) and the lateral diffusion coefficient ($D = 2 \times 10^{-8}$ $cm^2.s^{-1}$) found for the NBD-DPPE lipid probe were indeed consistent with the formation of a continuous lipid bilayer.

In summary, the fluorescence results obtained for both the hybrid bilayer and the tethered membranes indicate the formation of a homogeneous and most likely continuous biomimetic membrane, in good agreement with the data obtained in SPR experiments.

3.5. Interactions of Acylated Neurocalcin Derivatives with Biomimetic Membranes

Binding of myr-neurocalcin to artificial membranes was monitored by SPR spectroscopy. The proteins were diluted in 20 mM HEPES-Na (pH 7.4), containing 150 mM NaCl, and added to the hybrid bilayer or tethered membrane either in the absence or in the presence of calcium. The amount of protein bound per membrane area was determined from optical thickness, using Winspall calculations.

Figs. 5 A and 5 B show a typical SPR record of myr-neurocalcin binding to either tethered or HBM membrane, respectively. Myr-neurocalcin bound efficiently to both types of artificial membranes in the presence of Ca^{2+} while only a residual binding was observed in the absence of calcium. Interestingly, the myr-neurocalcin bound to the artificial membranes could be almost completely removed upon washing by carbonate buffer (data not shown), which suggests that the protein was associated with the outer phospholipid layer without insertion. Similar binding assays were carried out on both types of membranes at different protein concentrations, as shown in Figs. 5 A and 5 B. At the highest protein concentration tested, the maximal amount of myr-neurocalcin per unit of surface, as determined from the SPR data, were similar on both type of membranes and correspond to about 150-160 ng/cm^2 (*i.e.* 7 – 8 pmol protein/cm^2). In the same conditions, unmyristoylated-neurocalcin did not bind to the lipid coverage neither in the presence nor in the absence of Ca^{2+}.

Kinetic measurements.

The kinetics of binding of acylated neurocalcins to a brain lipid hybrid layer were investigated by using a BIAcore X apparatus (Biacore AB, Uppsala, Sweden). The self assembled monolayer of 1-octadecanethiol on a gold-coated glass surface was glued onto the support chip. The surface of the sensor chip was cleaned by a 2 min injection of 40 mM octyl-β-D-glucoside at a flow rate of 20 μl/min. A vesicle suspension, brain lipids or egg PC, (80 μl at a flow rate of 2 μl/min) was injected over the octadecane thiol surface to generate the hybrid membrane structure. This coating procedure might result in a surface that had multiple lipid layers and carried partially fused liposomes. Such structures could be easily washed out by injection of 40 mM NaOH (20 μl/min) during 2 min [34,35,38]. Coating of the surface typically took two hours and led to a signal of 1700 ± 100 RU (resonance unit). This result is in accordance with the formation of a lipid monolayer [33,35].

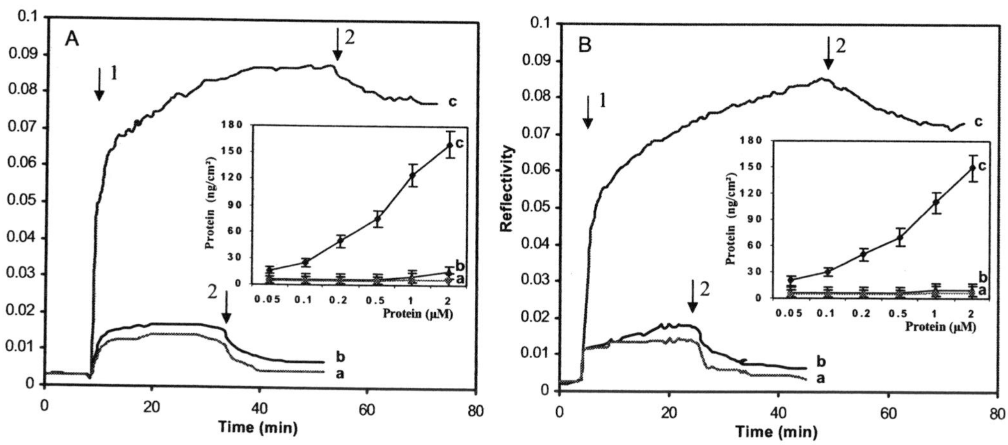

Figure 5. Interactions of non-acylated neurocalcin and myr-neurocalcin with tethered membrane (A) and hybrid bilayer (B). Curves correspond to SPR signal recorded after injection of the indicated protein (concentration 2 μM), arrow (1), or after injection of the corresponding buffer (20 mM HEPES, pH 7.4, plus or minus 2 mM Ca^{2+}), arrow (2). Curve a, non-acylated neurocalcin in the presence of 2 mM Ca^{2+}; curve b, myr-neurocalcin in the absence of Ca^{2+} (2 mM EGTA); curve c, myr-neurocalcin in the presence of 2 mM Ca^{2+}. Insets show protein bound quantities as a function of concentration of injected proteins (0.05 - 2 μM).

An analytical cycle of protein binding to artificial membranes consisted in injection of protein (0.02-5 μM), dissociation of protein in buffer flow and regeneration of the surface with 40 mM NaOH. Proteins samples in HEPES buffer were centrifuged 15 min at 100,000 × *g* and injected at flow rates ranging from 5 to 20 μl/min. Reversion to buffer flow was done 6 min after protein injection. A regeneration time of two minutes (40 mM NaOH, 20 μl/min) was necessary for complete regeneration of the phospholipid monolayer between two successive protein injections and equilibration with HEPES buffer. All solutions in HEPES-Na buffer were filtered and thoroughly degassed. Temperature was maintained at 25°C. Analyses of SPR curves were carried out by using BIAeval 3.1 global analysis software. Both association and dissociation processes were analyzed: association phase was taken from 20 s to 320 s after injection and dissociation phase was considered in the time interval 80 to 240 s after reversion to buffer flow. Equilibrium constants (K_D) were determined from Scatchard plots as done for recoverin [39] and artificially acylated RNases [38]. In our setup, the complex formation between a flowing analyte (the acylated protein) on top of the biomimetic membrane is observed. Concentration of the free analyte, under continuous flow, is equal to the bulk protein concentration (C), concentration of the complex (protein bound) can be measured directly as the steady state response (R_{eq}). The steady state level (in RU) is related to the free protein concentration by the equation: $R_{eq} = R_{max} - K_D R_{eq} /C$; where R_{max} is the maximum protein binding capacity.

Myr-neurocalcin

Under flow conditions, myr-neurocalcin bound efficiently to the lipid monolayer in the presence of Ca^{2+} whereas only a residual binding was observed in the absence of Ca^{2+}. As a control, we also checked that myr-neurocalcin did not bind to the bare alkane-thiol layer (i.e. without biomimetic membrane). Fig. 6A displays sensorgrams at increasing myr-neurocalcin concentrations (from 0.02 to 2 μM) in the presence of 2 mM Ca^{2+}. The steady-state values of sensorgrams (corrected for bulk effects) were analyzed with the 'separated association and dissociation phase' method in BIA evalution 3.0 software. The values of resonance units at equilibrium (R_{eq}) obtained were used for determination of equilibrium constants, Scatchard plots derived from three independent sets of experiments give a K_D value of 0.35 ± 0.05 μM.

C14 and C16-neurocalcins.

The binding properties of the chemically acylated C14 and C16-neurocalcins were similarly examined (see Fig. 6B). In the absence of Ca^{2+}, the chemically modified C14- and C16-neurocalcins did not bind significantly to the lipid monolayer: at the highest protein concentration maximum tested, less than 250 RU could be detected. In the presence of Ca^{2+}, C14- and C16-neurocalcins bound efficiently to the phospholipid coated chips. Importantly, at a given protein concentration (e.g. 0.5 μM), the SPR signal obtained with C14-neurocalcin (900 RU) was similar to that measured with myr-neurocalcin (850 RU) while the signal obtained for C16-neurocalcin (2000 RU) was about two-fold higher. The apparent equilibrium constant deduced from Scatchard analysis of the binding curves gave a K_D value of 0.39 ± 0.05 μM for C14-neurocalcins and 0.28 ± 0.05 μM for C16-neurocalcins. These results indicate that the binding of C16-neurocalcin is facilitated compared to the C14/myr-neurocalcins although the increase in affinity of 16-neurocalcin for lipid monolayer appeared less than expected for the presence of two additional methylene groups.

4. Concluding Remarks

We have developed a generic procedure that allows to chemically acylate neurocalcin in a controlled manner with any desired acyl chain. The acylation reaction was performed in reversed micelles in which the protein is dispersed into the microaqueous phase and surrounded by an apolar solvent phase containing reactive fatty acid chlorides of different chain length thus resulting in a good accessibility of the protein to the lipophilic reagent. Under optimal conditions, neurocalcin was monoacylated by lauric, myristic and palmitic acid as demonstrated by RP-HPLC and ESI-MS. Mass spectrometry analysis of peptides obtained after proteolytic cleavage indicates that the single fatty acid is linked to the *N*-terminal glycine of the polypeptide chain through an amide bond. The high selectivity of acylation on the *N*-terminus of the neurocalcin polypeptide chain may be due to the particular reactivity of the *N*-terminal amino group as compared to ε-amino groups of lysine as previously observed with ribonuclease A [22]. Besides, the *N*-terminal chain of unmyristoylated neurocalcin is flexible [20] and could interact with the surfactant shell. The neurocalcin derivatives (C12, C14 and C16-neurocalcins) were purified using a Triton X-114 temperature-induced phase separation method, and shown to bind to membranes in a calcium-dependent manner, as demonstrated by cosedimentation with phospholipid vesicles and/or direct binding to supported lipid membranes. Our results indicate that the calcium-acyl switch can function with C12 or C16 fatty acids in addition to the standard C14. This approach may be used to probe the role of different fatty acid in the biological functions of acylated proteins. Indeed, although in most tissues, proteins are acylated with only the 14:0 myristic acid, in some cells like the photoreceptor cells, proteins are heterogenously acylated with 12:0, 14:1n-9, and 14:2n-6 in addition to 14:0 [40]. The biological significance of such heterogeneity is not clearly established. Diversity in the fatty acid chain attached to the NCS proteins may also contribute to their specific sub-cellular distribution and/or interactions with targets. The approach described above should permit to resolve these issues.

The calcium-dependent membrane binding properties of the acylated neurocalcins were then investigated by SPR spectroscopy. For this purpose, we designed two different types of supported biomimetic membranes: HBM obtained by deposition of a single layer of phospholipids onto a hydrophobic surface of octadecanethiol molecules coated onto the bare gold surface, and TLB obtained by covalent coupling to the gold surface of lipid vesicles that spontaneously disrupt into a continuous phospholipid bilayer that is tethered to the supporting surface by a PEG cushion [25]. The fluidity and continuity of the phospholipid leaflet of both TLB and HBM were established by fluorescence recovery experiments that clearly demonstrated the mobility of a fluorescent lipid probe. The overall thickness of both type of biomimetic structures determined from SPR measurements were in accordance with a bilayer thickness (about 5 nm). The calcium and acyl-dependent binding of neurocalcin to both TLB and HBM biomimetic membranes was characterized by SPR spectroscopy. Myristoylated neurocalcin was found to bind similarly on TLB and HBM in the presence of calcium while almost no binding was observed in the absence of calcium. The membrane-bound protein could be removed upon washing with carbonate buffer, suggesting that the acylated protein associated to the outer phospholipids leaflet without insertion, as expected for an extrinsic membrane protein. Kinetics of membrane binding for the differently acylated neurocalcins were examined on HBM mounted onto a Biacore chip. A kinetic analysis of the association and dissociation of the proteins with the artificial membranes could be easily determined by SPR spectroscopy, and apparent affinity constants of the protein/membrane association could be derived [23].

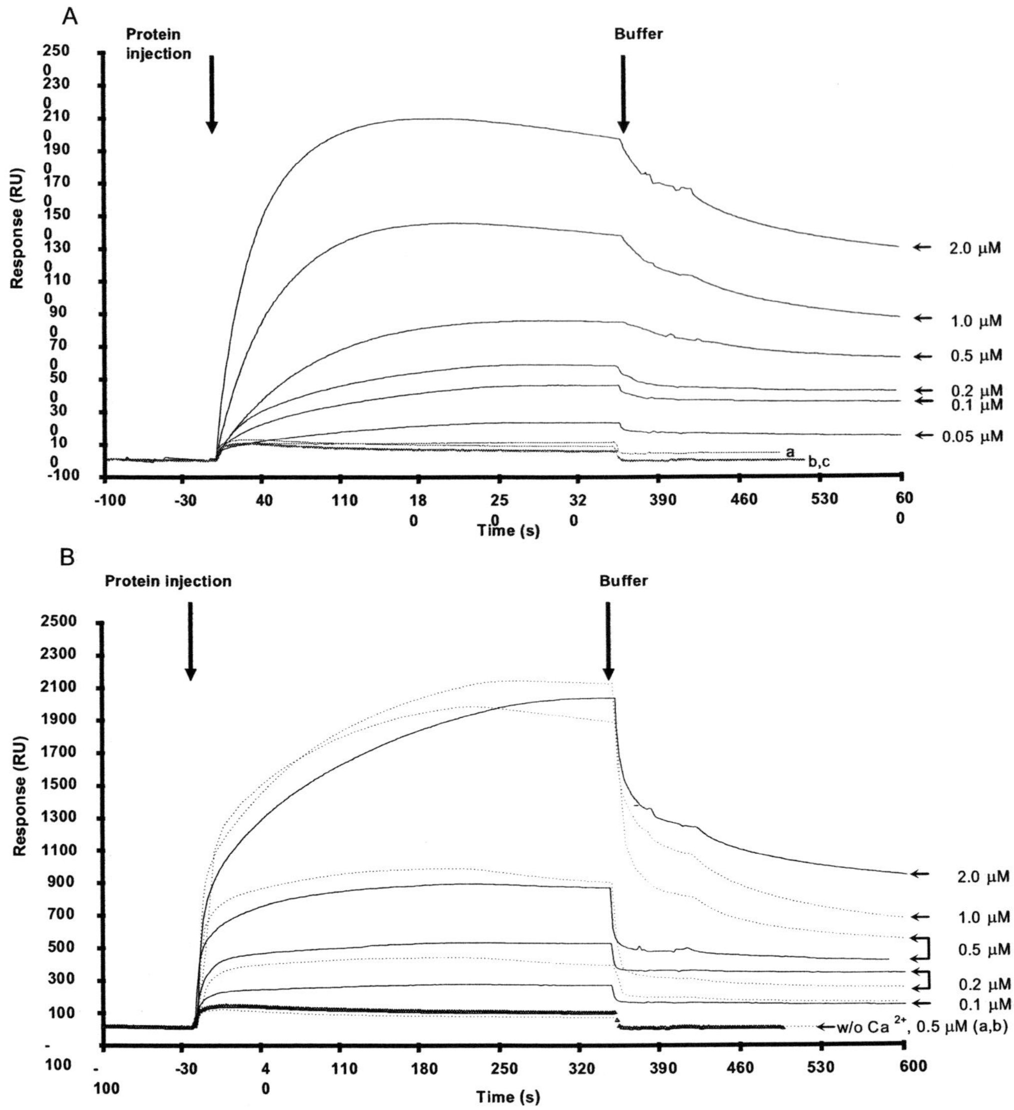

Figure 6. Binding of acylated neurocalcins to a bovine brain lipid monolayer studied by SPR spectroscopy (Biacore). (A) Sensorgram plots obtained with Myr-neurocalcin with concentrations ranging from 0.05 µM to 2 µM in the presence of 2 mM Ca^{2+}. Control sensorgrams were recorded for 0.5 µM Myr-neurocalcin in the absence of Ca^{2+} (a), for 0.5 µM non-modified neurocalcin in the absence of Ca^{2+} (b), and in the presence of 2 mM Ca^{2+} (c). (B) Sensorgram plots of the C14-neurocalcin (solid lines) and C16-neurocalcin (dashed lines) binding to the phospholipid monolayer in the presence of 2 mM Ca^{2+}, with protein concentrations ranging from 0.1µM to 2 µM. Control sensorgrams were obtained in the absence of Ca^{2+} with 0.5 µM C14-neurocalcin (a) or C16-neurocalcin (b). Proteins were injected at a flow rate of 5 µl/min.

Altogether these studies showed that biomimetic membranes of two types, TLB and HBM, could be useful to characterize protein/membrane associations. Our earlier studies suggested that comparison of binding properties of proteins with these two different types of membranes should permit to characterize the mode of association of protein with membrane and in particular to discriminate extrinsic membrane protein (that binds similarly to both

HBM and TLB, e.g. calcium-bound myr-neurocalcin) from intrinsic membrane protein (that binds only to TLB, e.g. bacterial toxin) [25].

The family of NCS calcium sensors offers an intriguing example of how a single post-translational modification, *N*-terminal myristoylation, has been tuned during evolution to mediate distinct spatial and temporal patterns of Ca^{2+} signal transduction. The approaches described here should help to understand the molecular basis of these important biological processes. Besides, they should be also applicable for the study of other myristoylated proteins and/or membrane associated polypeptides. Indeed, fatty acid modifications play a key role in the subcellular localization of proteins. Many acylated proteins are involved in signaling networks that are spatially organized by specific localization of their protein constituents to distinct membrane systems or microdomains. Deciphering the relationships between particular fatty acid modifications and the sub-cellular targeting of proteins represents an important step to understand cellular physiology.

Acknowledgment

This work was supported by Centre National de la Recherche Scientifique and by Institut Pasteur.

References

[1] Burgoyne, R. D., O'Callaghan D. W., Hasdemir B., Haynes L. P. and Tepikin A. V. 2004, *Trends Neurosci, 27*, 203.

[2] Braunewell, K. H. and Gundelfinger E. D. 1999, *Cell. Tissue Res., 295*, 1.

[3] Zozulya, S. and Stryer L. 1992, *Proc Natl Acad Sci U S A, 89*, 11569.

[4] Tanaka, T., Ames J. B., Harvey T. S., Stryer L. and Ikura M. 1995, *Nature, 376*, 444.

[5] Hughes, R. E., Brzovic P. S., Klevit R. E. and Hurley J. B. 1995, *Biochemistry, 34*, 11410.

[6] Ames, J. B., Tanaka T., Stryer L. and Ikura M. 1996, *Curr Opin Struct Biol, 6*, 432.

[7] Ames, J. B., Ishima R., Tanaka T., Gordon J. I., Stryer L. and Ikura M. 1997, *Nature, 389*, 198.

[8] Vijay-Kumar, S. and Kumar V. D. 1999, *Nat Struct Biol, 6*, 80.

[9] Ames, J. B., Dizhoor A. M., Ikura M., Palczewski K. and Stryer L. 1999, *J Biol Chem, 274*, 19329.

[10] O'Callaghan, D. W., Ivings L., Weiss J. L., Ashby M. C., Tepikin A. V. and Burgoyne R. D. 2002, *J Biol Chem, 277*, 14227.

[11] O'Callaghan, D. W. and Burgoyne R. D. 2004, *J Biol Chem, 279*, 14347.

[12] Bhatnagar, R. S., Futterer K., Waksman G. and Gordon J. I. 1999, *Biochim. Biophys. Acta., 1441*, 162.

[13] Rajala, R. V., Datla R. S., Moyana T. N., Kakkar R., Carlsen S. A. and Sharma R. K. 2000, *Mol. Cell. Biochem., 204*, 135.

[14] Johnson, D. R., Bhatnagar R. S., Knoll L. J. and Gordon J. I. 1994, *Annu. Rev. Biochem., 63*, 869.

[15] Casey, P. J. 1995, *Science, 268*, 221.

[16] Boutin, J. A. 1997, *Cell. Signal., 9*, 15.

[17] Resh, M. D. 1999, *Biochim. Biophys. Acta., 1451*, 1.

[18] O'Callaghan, D. W. and Burgoyne R. D. 2003, *Biochem Soc Trans, 31*, 963.

[19] Valentine, K. G., Mesleh M. F., Opella S. J., Ikura M. and Ames J. B. 2003, *Biochemistry, 42*, 6333.

[20] Ladant, D. 1995, *J. Biol. Chem., 270*, 3179.

[21] Robert, S., Domurado D., Thomas D. and Chopineau J. 1993, *Biochem Biophys Res Commun, 196*, 447.

[22] Roy, M.-O., Uppenberg J., Robert S., Boyer M., Chopineau J. and Jullien M. 1997, *Eur. Biophys. J., 26*, 155.

[23] Beven, L., Adenier H., Kichenama R., Homand J., Redeker V., Le Caer J. P., Ladant D. and Chopineau J. 2001, *Biochemistry, 40*, 8152.

[24] Knoll, W. 1998, *Annu. Rev. Phys. Chem., 49*, 569.

[25] Rossi, C., Homand J., Bauche C., Hamdi H., Ladant D. and Chopineau J. 2003, *Biochemistry, 42*, 15273.

[26] Spinke, J., Liley M., Schmitt F. J., Guder H. J., Angermaier L. and Knoll W. 1993, *J. Chem. Phys., 99*, 7012.

[27] Lang, H., Duschl C. and Vogel H. 1994, *Langmuir, 10*, 197.

[28] Reiter, R., Motschmann H. and Knoll W. 1993, *Langmuir, 9*, 7012.

[29] Cuypers, P. A., Hermens W. T. and Hemker H. C. 1978, *Anal. Biochem., 84*, 56.

[30] Spinke, J., Liley M., Guder H. J., Angermaier L. and Knoll W. 1993, *J. Chem. Phys., 1821.*

[31] Stenberg, E., Persson B., Roos H. and Urbaniczky C. 1991, *J. Colloid Interface Sci., 143*, 513.

[32] Porter, M. D., Bright T. B., Allara D. L. and Chidsey C. E. D. 1987, *J. Am. Chem. Soc., 109*, 3559.

[33] Lingler, S., Rubinstein I., Knoll W. and Offenhäusser A. 1997, *Langmuir, 13*, 7085.

[34] Pierrat, O., Lechat N., Bourdillon C. and Laval J. M. 1997, *Langmuir, 13*, 4112.

[35] Cooper, M. A., Try A. C., Carroll J., Ellar D. J. and Williams D. H. 1998, *Biochim. Biophys. Acta, 1373*, 101.

[36] Berquand, A., Mazeran P. E., Pantigny J., Laval J. M. and Bourdillon C. 2003, *Langmuir, 19*, 1700.

[37] Brian, A. A. and McConnell H. M. 1984, *Proc. Natl. Acad. Sci. USA, 81*, 6159.

[38] Roy, M. O., Pugniere M., Jullien M., Chopineau J. and Mani J. C. 2001, *J. Mol. Recognit., 14*, 72.

[39] Lange, C. and Koch K. W. 1997, *Biochemistry, 36*, 12019.

[40] DeMar Jr, J. C., Rundle D. R., Wensel T. G. and Anderson R. E. 1999, *Prog. Lipid Res., 38*, 49.

INDEX

A

B

C

D

I

N

O

P

Q

R

S

T

U

V

W

X

Y

Z